HORTICULTURE:

PRINCIPLES
AND
PRACTICAL APPLICATIONS

HORTICULTURE:
PRINCIPLES
AND
PRACTICAL APPLICATIONS

RAYMOND P. POINCELOT

Department of Biology
Fairfield University
Fairfield, Connecticut

PRENTICE-HALL, INC., *Englewood Cliffs, New Jersey 07632*

Library of Congress Cataloging in Publication Data

Poincelot, Raymond P   date
   Horticulture.

   Bibliography: p.
   Includes index.
   1. Horticulture. I. Title.
SB91.P65   635   79-28307
ISBN 0-13-394809-9

Editorial/production supervision and
interior design by *Virginia Huebner*
Page layout by *Rita Schwartz*
Art direction: *Lorraine Mullaney*
Cover photo: *Webbphotos*
© *1979 Ted McDonough*
Manufacturing buyer: *Ed Leone*

Printed in the United States of America

10 9 8 7 6 5 4 3 2 1

PRENTICE-HALL INTERNATIONAL, INC., *London*
PRENTICE-HALL OF AUSTRALIA PTY. LIMITED, *Sydney*
PRENTICE-HALL OF CANADA, LTD., *Toronto*
PRENTICE-HALL OF INDIA PRIVATE LIMITED, *New Delhi*
PRENTICE-HALL OF JAPAN, INC., *Tokyo*
PRENTICE-HALL OF SOUTHEAST ASIA PTE. LTD., *Singapore*
WHITEHALL BOOKS LIMITED, *Wellington, New Zealand*

Contents

Preface

This book is a comprehensive introduction to horticulture. The material is covered in sufficient detail for students who have little or no background in the plant sciences. Depending on the approach used by the instructor, it can be used as the text for a one or two semester course.

I have attempted a logical progression, one that is scientifically correct but pragmatic, and yet readable throughout. In the first section, the student is introduced to the basic science that forms the foundation of horticulture. These concepts, once mastered, enable the student to perceive the factual basis for applied horticultural science, *i.e.,* the horticultural practices in the second section. In logical sequence this leads the student to the third section, horticultural production, with special emphasis on horticultural crops. The sections on horticultural practices and horticultural production are sufficiently detailed, to provide a stimulating challenge to those students who already possess some background in horticultural science.

My belief is that horticulture can only be understood when it is seen as the sum of the foregoing three parts. However, an appreciation of horticulture depends upon the weaving of one more element in the horticultural fabric: artistry. It is my hope that the student will achieve both an understanding and appreciation of horticulture through the use of this text.

The following text features will help to achieve the foregoing purpose and set it apart from other horticultural texts.

I. General Features

1. Contemporary nature. The latest published research and research in progress are both included. For example, the involvement of mycoplasmas and rickettsia in horticulture crop diseases, environmental problems and restrictions on pesticides, natural control of plant pests, and energy conservation are just some recent events cited that make this text an ongoing examination of horticulture.

2. Comprehensive but detailed. There is much subject matter on applied horticulture overall, and the detail is great, but it is detail with a practical base. The text's potential is large, simply because the detail is there when wanted.

3. Scientifically correct but pragmatic. Plants are cited by common and scientific names. Plant nomenclature is based on the authority established by *Hortus Third: A Concise Dictionary of Plants Cultivated in the United States and Canada* (The staff of the Liberty Hyde Bailey Hortorium, Macmillan Publishing Co., Inc., 1976). In certain cases beyond the scope of this reference, specialized monographs were utilized. Scientific names were used for insects and diseases, as well as for plants, whenever possible. Pesticides and growth regulators are cited by chemical and trade or common names. Although chapters one through five contain basic science, chapters six through sixteen contain pragmatic applications that follow from, and are directly related to, the science, leading to a better understanding of these applied practices.

4. Balanced coverage. Many texts have somewhat regionalized coverages, but this book provides a national approach to practices and horticultural crops.

Coverage of both home gardening and commercial horticulture is presented, to satisfy the needs of both types of horticulturists.

Some division exists between horticulturists favoring the "organic" or natural approach versus those favoring the "chemical" approach. Both sides are presented here. The author personally believes both sides have good and bad points, and the best path is the blending of the good from both into an integrated horticulture.

Crops are covered uniformly. There is no excessive emphasis on any crop: fruits, ornamentals, and vegetables. All are important to horticulture.

Both city, suburban and country horticulture are discussed. All exist and the well trained horticulturist must have a knowledge of all of them.

5. Designed for learning. An extensive number of figures and photographs are included to illustrate subjects that some other texts leave out or skim over on the assumption that the new student already knows them. Coverage is especially heavy for practices, equipment, and horticultural crops. The crop photographs are useful for identification purposes.

The book contains extensive listing in table form of horticultural crops by ornamentals, vegetables, and fruits. Ornamental tables are broken down into 13 categories to reflect complete coverage of ornamental crops. All available data on the horticultural crops, listed by scientific and common names, is presented in concise form. Normally several sources would be necessary to locate such data. The extensive listing of most horticultural crops with collective data will make this text a continuing valuable reference to the student. Other tables cover flowering and effect of photoperiod, propagation media, fertilizers, biological controls, fungal diseases, pesticides, and crop preservation.

Difficult concepts are presented in a concise, simplified manner which is thoroughly readable, without sacrificing accuracy. The flow is often initially general followed by specific detail and cited examples. Wherever needed the relevancy of what appears at first glance to be irrelevant is shown to be relevant. My aim is to accelerate learning and to improve retention by special emphasis on the above.

An extensive bibliography is given at the conclusion of each chapter. The listings are complete and recent, except for certain older books which have become "horticultural classics." The range is such as to include both simple and highly technical references to satisfy the needs of all students. These books are worth purchasing or consulting when the student is interested enough in a topic to desire more information.

II. Specific Features.

1. Part one contains basic plant science. Five chapters are covered. First is plant classification with special emphasis on scientific classification and taxonomy, as it applies to horticulture. Horticultural classification, a valuable communications tool, is also examined. Plant structures are next with special attention being placed on macrostructures, rather than microstructures. Plant metabolic processes, particularly those associated with good development of plants, come next, followed by plant development. Development emphasis is placed on macroaspects: vegetative and reproductive cycles. The section concludes with the relationships between environmental factors and how the plant responds to them. Completion of this section enables the student to understand why certain practices are used and how they produce good results.

2. Part two contains the applied aspects which are divided into seven chapters. This section begins with extensive coverage of asexual and sexual propagation and interim care, a basic and important horticultural practice. After propagation plants are placed in a more or less permanent environment, either outdoors or indoors. This forms the basis of the next two areas of coverage, an in depth analaysis of environmental management outdoors and indoors. Energy conservation, a critical factor now in greenhouse management is given special attention. Next follows control of plant development, with stress on pruning practices and growth regulators. Plant protection comes next with examination of vectors causing plant problems and how to treat them. Unlike some books which tread lightly in this area, biological and chemical control are examined in detail. Several tables of biological/chemical controls are presented. The section concludes with plant improvement presented clearly and concisely, followed by harvesting and post-harvest techniques.

3. Section three deals largely with horticultural crops which are grouped by ornamentals, vegetables, fruits, and other crops. The latter includes nuts and herbs, which are sometimes ignored or minimized in other texts. Career opportunities in the horticultural industries associated with these crops are pointed out. The economic significance of these fields and their basic thrust is examined. A surprisingly comprehensive amount of information is concisely expressed in this last section.

I would welcome constructive criticism and any comments. If possible I will incorporate such response in future editions of this text. My closing thought is the reason why I wrote this book. The fields of horticulture are a joy to the eye and mind, but a deeper appreciation is yours only through questioning, study, and involvement.

Acknowledgments

My special thanks go to my wife, Marian, and our children, Raymond, Daniel, and Wendy. My family understood and gave their love through the long, busy preparation of this text.

I thank all those professionals, friends, and students who discussed horticulture with me. You are too numerous to be cited here, but your thoughts may be found here.

I thank the following professors who critically reviewed the entire manuscript: Charles Jenkins at the University of Michigan, Lyle Littlefield at the University of Maine, and Emil Pierson at the University of Nebraska. I also thank those professors who reviewed part of the manuscript, Harold Tukey at Cornell University and A. Carl Leopold.

Many thanks to those individuals, organizations, institutions, and industries that supplied photographs or permission to quote from their publications. I wish to cite two people who worked especially hard behind the scenes: Eugene Memmler, Chairman of the American Society for Horticultural Science Slide Collection, and Marcia Suber in the Photography Division, Office of Communication, U.S. Department of Agriculture.

Thank you to the people who typed my manuscript, Marianne Esposito, Patricia O'Brien, and Marian Poincelot.

Finally, my thanks to all those people at Prentice-Hall, Inc. who helped make this book possible; especially Paul J. Feyen, Biology/Agriculture editor, Virginia Huebner, College Book Editorial Production, and Robin Bartlett, Marketing Manager.

RAYMOND P. POINCELOT

Fairfield, Connecticut

part one

INTRODUCTION TO HORTICULTURAL SCIENCE

Horticulture has its etymological origins in the Latin words hortus *and* cultura, *which mean garden and culture, respectively. Today its meaning goes far beyond the culture of gardens. Horticulture is that branch of plant agriculture concerned with the intense cultivation of garden crops produced for food and medicine and for enjoyment, recreation, and general environmental improvement. This is in contrast to those branches of plant agriculture called agronomy and forestry. Agronomy pertains to field crops, such as cereals and fodders, and forestry is associated with tree crops and their products.*

Garden crops grown in the horticultural sense include vegetables, fruits, and plants raised for ornamental use or for medicinals and spices. These garden crops and their purposes are the basis for the three divisions of horticulture: Olericulture or vegetable culture; pomology or fruit culture; and ornamental horticulture, which includes three general areas: (1) floriculture, or the culture of cut flowers, potted flowering and foliage plants, and greenery; (2) landscape horticulture, or the design and construction of sites along with the planting and maintenance of woody and herbaceous ornamentals, bulbs, and related crops; and (3) production nursery, or the production of landscape horticultural crops. Some consider propagation to be a fourth division, and, indeed, it can be considered alone or as an integral part of the preceding three divisions. The area of propagation is associated with the reproduction of horticultural crops by seed and vegetative techniques, as well as the improvement of plant material through plant breeding.

It is obvious that horticulture and its three divisions touch upon many disciplines. These include the physiology, biochemistry, genetics, and pathology of plants, as well as botany, soil science, ecology, climatology, physics, taxonomy, entomology, economics, food technology, art, architecture, landscape design, and others. In a sense, horticulture is a blend of science, art, technology, and esthetics that has stimulated interest over the centuries.

Nevertheless, it is the scientific part, the mixture of sciences providing information on the plant and its environment, that forms the foundation of horticulture. Horticultural science provides the means to understand the art of horticulture, to improve its technology, to increase its production, and to deepen our appreciation of its esthetic value. It is with this part of horticulture that Part One is concerned.

Plant Classification

Early humans were undoubtedly bewildered by the array of plants in the environment. By trial and error, some of these plants were probably divided into those that were good and those that were harmful. With time, these groups were subdivided into plants that were edible or sources of poison, ornament, flavoring, medicine, hallucinogens, and dyes. These plants became necessary for survival and presumably developed into articles of barter. Thus, communication with others in regard to these plants was necessary, so they needed names. The naming and classification of plants started before recorded history.

Our present system for classifying plants has its origin in ancient Greece. The most important contribution came from Theophrastus of Eresus (370 to 285 B.C.), who was a pupil of Plato and Aristotle and later an assistant to Aristotle. Much of his time was spent in the botanical gardens established by Aristotle at Athens. Although knowledgeable in many fields, he is remembered for his work *Historia Plantarum* or *Enquiry into Plants.*

The *Historia Plantarum* is considered to mark the start of scientific botany. It covers observations by Theophrastus on germination of seeds, seedling development, internal plant structures, leaf classifications, roots, stem and root modifications, plant distributions, and the relations of plants to the environment. However, it is his contributions to classification that concern us here. Theophrastus classified plants into four groups: herbs, half-shrubs, shrubs, and trees. He noticed and recorded families among the flowering plants, such as the parsley family, which is a designated family today, Umbelliferae. Other relationships among the birches, poplars, and alders, as well as among conifers, cereals, and thistles, were noted by him. Finally, he established groupings of species, not unlike the genera in use today, and applied Greek names to them. One of these, *Asparagus,* is the scientific name of a genus today.

Little progress was noted after Theophrastus until the work of the German, English, and Italian herbalists (approximately 1470 to 1670). They dealt mainly with the practical use of plants as medicinals. Their printed works, or herbals, described native and foreign plants. Some attempts at classification based on natural relationships are seen in the later works of this period.

It was not until the eighteenth century that our present methods of plant nomenclature were developed. Credit goes to the Swedish naturalist and amateur horticulturist, Carl Linnaeus (ennobled, von Linné). He lived from 1707 to 1778 and is remembered today as the father of taxonomy. As a physician and later a professor of botany and medicine at the University of Uppsala, he established a two-word naming system, the binomial system of nomenclature, for each plant species.

Previous to the binomial system, plants were named with a single name followed by descriptive nouns and adjectives. The ever-increasing number of recognized plants became ever more difficult to name by the long descriptive method. Linnaeus set forth the much needed reform of the nomenclature in 1753 in his publication, *Species Plantarum* (Fig. 1-1).

Linnaeus also established a plant classification called the "sexual" system. It was based on the number of stamens and their relation to one another and to other floral parts. With this "sexual" system, Linnaeus grouped the flowering plants of his time into 23 classes. Those without flowers were put into a twenty-fourth group, which contained the algae, mosses, ferns, and fungi.

Other classifications followed in time. These classifications, like that employed by Linnaeus, are called artificial classifications, since they are based on a few obvious morphological characteristics (morphology is the study of form and its development). Some of these artificial classifications are still with us today. Often wildflowers are grouped according to flower color or time of blooming.

Eventually, the artificial approach was supplanted by the natural system, which is based upon overall resemblances in external morphology. The natural system differs from the artificial, with its few morphological characters, in that as many morphological characters as possible are utilized. With the natural system the classification is dependent upon degrees of similarity and not evolutionary relationships. The fundamental units

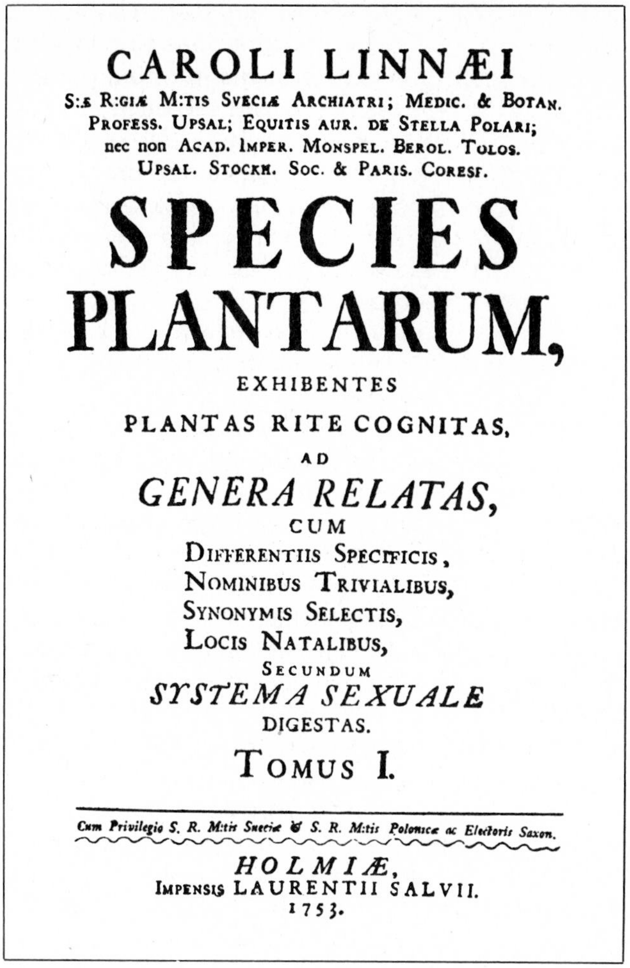

CAROLI LINNÆI

S:ᴇ Rːɢɪᴀ Mːᴛɪs Sᴠᴇᴄɪᴀ Aʀᴄʜɪᴀᴛʀɪ; Mᴇᴅɪᴄ. & Bᴏᴛᴀɴ.
Pʀᴏғᴇss. Uᴘsᴀʟ; Eǫᴜɪᴛɪs ᴀᴜʀ. ᴅᴇ Sᴛᴇʟʟᴀ Pᴏʟᴀʀɪ;
nec non Aᴄᴀᴅ. Iᴍᴘᴇʀ. Mᴏɴsᴘᴇʟ. Bᴇʀᴏʟ. Tᴏʟᴏs.
Uᴘsᴀʟ. Sᴛᴏᴄᴋʜ. Sᴏᴄ. & Pᴀʀɪs. Cᴏʀᴇsғ.

SPECIES PLANTARUM,

Exʜɪʙᴇɴᴛᴇs

Pʟᴀɴᴛᴀs ʀɪᴛᴇ ᴄᴏɢɴɪᴛᴀs,

ᴀᴅ

GENERA RELATAS,

ᴄᴜᴍ

Dɪғғᴇʀᴇɴᴛɪɪs Sᴘᴇᴄɪғɪᴄɪs,
Nᴏᴍɪɴɪʙᴜs Tʀɪᴠɪᴀʟɪʙᴜs,
Sʏɴᴏɴʏᴍɪs Sᴇʟᴇᴄᴛɪs,
Lᴏᴄɪs Nᴀᴛᴀʟɪʙᴜs,
Sᴇᴄᴜɴᴅᴜᴍ

SYSTEMA SEXUALE

Dɪɢᴇsᴛᴀs.

Tᴏᴍᴜs I.

Cum Privilegio S. R. M:tis Sueciæ & S. R. M:tis Polonica ac Electoris Saxon.

HOLMIÆ,
Iᴍᴘᴇɴsɪs Lᴀᴜʀᴇɴᴛɪɪ Sᴀʟᴠɪɪ.
1753.

Fig. 1–1. This is a reproduction of the title page from *Species Plantarum* in 1753 by Linneaus in which he set forth the binomial system of nomenclature and a "sexual" classification system for plants. From Samuel R. Rushforth, *The Plant Kingdom: Evolution and Form,* © 1976, p. 7. Adapted by permission of Prentice-Hall, Inc., Englewood Cliffs, New Jersey.

are the species. Similar species are grouped into genera, similar genera in turn are arranged in families, and so on.

After Darwin's *Origin of Species* in 1859, the theory of evolution had far-reaching effects. It became apparent that the discontinuities between species, genera, and families with the natural system of classification resulted from processes such as polyploidy (production of more than two sets of complete chromosomes, that is, 3, 4, 5, or more sets), extinction of intermediate forms, and reproductive isolation. This gave rise to the phylogenetic system of classification.

The *phylogenetic system* is based upon relationship by descent; that is, similarities in morphology are indicative of evolutionary relationships. Under this form of classification, species within a genus or family would be related by descent from a common ancestry. Groups with the most morphological characters in common would be closely related, as opposed to those with few common characters, which would be distantly related.

It is apparent that such a system has inherent difficulties. As of now a classification based on relationship by descent has not been perfected. Problems arise on two levels, the collection and the interpretation of data. The taxonomist is overwhelmed by the sheer numbers of living plants, which is further complicated by the appearance of new species and the extinction of older ones. Another troublesome aspect is the genetic instability and variation encountered with some species. To determine the lineage, a good fossil record is required. At best the available fossil record can be described as fragmentary. It is assumed that the trend in evolution has been progressive, from simple to complex structures. However, for some groups and characters, the trend may be a retrogression toward reduction or greater simplicity. Is the character in question simple by virtue of being primitive or is it because of reduction? Convergent evolution, or the independent development of similar structures, by unrelated organisms can result in problems. Is one dealing with a direct or convergent evolutionary relationship? Finally, some disagreements exist on the relative importance of various characters and groups of characters. In effect, the phylogenetic classification is a dynamic one, subject to change as our knowledge increases.

Today taxonomists must rely heavily on evidence collected by biochemical, genetic, cytological, ecological, or physiological approaches. Many scientific techniques, such as gel electrophoresis for the examination of proteins, scanning electron microscopy for observation of plant structures, or gas chromatography for the study of natural products, have become important aids to the taxonomist who wishes to extend a phylogenetic classification. Because of the complex and extensive array of morphological characteristics and supplementary scientific data, the taxonomist must call upon the computer for interpretation.

Where does this leave us? Classification is still dependent upon the natural system, but phylogenetic information is slowly integrated as it becomes available. Much of the natural classification in use is based upon the efforts of August Eichler at the close of the nineteenth century and, later, Adolph Engler and Karl Prantl. The latter published their plant classification in a twenty-volume series. Although many parts are phylogenetically incorrect, it is still widely used. As the volume of phylogenetic data accumulates, we can expect a conversion from the natural classification to one that is predominately phylogenetic.

Scientific Classification of the Plant Kingdom

The functions of plant classification are the arrangement of plants into decreasingly smaller groups with increasingly similar characteristics for identification, and to suggest possible relationships among plants. The format of the scientific classification is shown next. Each category, regardless of size or rank is referred to as a *taxon* (plural, *taxa*).

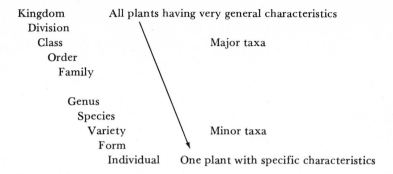

Kingdom	All plants having very general characteristics
Division	
Class	Major taxa
Order	
Family	
Genus	
Species	
Variety	Minor taxa
Form	
Individual	One plant with specific characteristics

Each taxon is successively separated into parts by the taxon following it. Sometimes intermediate separations of a taxon are made and denoted by the prefix "sub," for example, subclass.

Taxonomists have made great progress within the last 200 years. However, disagreements do exist in the scheme of plant classification. On the kingdom level, the older two-kingdom system has been challenged with a three-, four-, and five-kingdom concept. This causes no problem for horticulturists, since the plants of interest to them remain in the Kingdom Plantae, regardless of which approach is used.

Another problem centers around the concept of divisions (equivalent of phyla used by zoologists) and classes. Some taxonomists feel there should be a few divisions with many classes, as opposed to those who favor more divisions with fewer classes. The horticulturist can always place the plants of interest correctly by class and division in whichever approach is used, if the following points are kept in mind.

Most of the plants of interest to the horticulturist are vascular plants, or those with specialized supporting and conducting connective tissue. The vascular plants of primary interest are some seedless plants, specifically the ferns, and most of the seed-bearing plants. The seed-bearing plants can be divided into (1) *gymnosperms,* or those whose seeds are not enclosed in an ovary, specifically the cone-bearing plants or the conifers, and (2) *angiosperms,* flowering plants, whose seeds are enclosed in a fruit or matured ovary. The lycophytes (seedless) and cycads (gymnosperms) are of minor interest, since a few of these plants are seen in the houseplant trade. *Selaginella* and *Zamia* species are respective examples of the lycophytes and cycads.

Some examples of the placement of horticultural plants in the various approaches to classification are as follows. If all the vascular plants are placed in one division, it is designated as Division Tracheophyta. If they are spread over several divisions, those of interest to the horticulturist are Lycophyta (lycophytes) and Filicophyta (ferns), the seedless vascular plants; Cycadophyta (cycads) and Coniferophyta (conifers), vascular plants with exposed seeds or gymnosperms; and Anthophyta, flowering plants with protected seeds or angiosperms. It should be apparent that "phyta" is the ending indicating division. If only one division exists, the same plants can be described as subdivisions, that is, Lycophytina, Filicophytina, and Spermatophytina (seed plants, both angiosperms and gymnosperms). Obviously, "phytina" is indicative of subdivision. Spermatophytina could then be divided into classes: Coniferinae (conifers) and Cycadinae (cycads), the gymnosperms; and Angiospermae, the angiosperms. If the horticulturist recognizes the basic Latin root words for the plants of interest and learns the endings indicative of division, subdivision, and class, it should not be difficult to pick out the groups of interest in most classification schemes.

The phylogenetic lines of the ferns are relatively well known because of good fossil records. Both extinct and living ferns are classified. More than 10,000 species of living ferns are known. Many species are used as garden or house plants and by the floral industry.

The phylogenetic lines of the conifers are also well characterized from their fossil records. The conifers are important landscaping plants. Their use also in the Christmas tree business, lumber industry, and paper industry makes them an important horticultural crop plant.

The fossil record of the flowering plants is incomplete, so the phylogenetic lines are not as well characterized. The flowering plants are commonly known as *angiosperms,* which means enclosed seed. Their seeds are enclosed in a matured ovary or fruit, unlike *gymnosperms* (meaning naked seed), such as the conifers, which have exposed seeds on cone scales. Angiosperms are the most highly evolved plants on earth. Their success is evident, since there are at least 250,000 species of angiosperms.

Angiosperms are broken down into monocots and dicots. *Monocots* have one cotyledon (seed leaf); *dicots* have two. Their primary vascular system arrangement in the stem also differs (see Chapter 2). Other differences occur in the numbers of flower parts, type of venation, and secondary growth ability. There are about four times as many dicots as monocots (Figs. 1-2, and 1-3). These may be grouped as classes or subclasses of the angiosperms, depending on the classification system.

Angiosperms are very important plants to the horticulturists and to humanity in general. They are significant food plants for humans and their livestock. Some are utilized as lumber and paper pulp; others are used in landscaping and in the floral and drug industries.

Classes are broken down into orders, which consist of groups with some shared characteristics called families. Families are groups familiar to many horticulturists. For example, the class Dicotyledonae contains orders, of which one is Ranales. In this order are families, such as Magnoliaceae (magnolia family) and Ranunculaceae (crowfoot family). Family names, both common and scientific, are often derived from the most important genus of those genera which constitute the family.

Some families are well known to horticulturists. Dicot families of note are the two families previously mentioned and, among many others, the Cactaceae (cactus family), Caryophyllaceae (pink family), Compositae (composite family), Cruciferae (mustard family), Ericaceae (heath family),

Fig. 1-2. The lily (*Lilium*) is a well-known example of a monocot. Two distinguishing features of the monocot are clearly visible: parallel leaf venation and flower parts in threes or multiples thereof. These are mixed Imperial hybrids, courtesy of Oregon Bulb Farms.

Fig. 1-3. The dahlia (*Dahlia*) is one of many dicots. Netlike leaf venation and flower parts in four(s) or five(s) are characteristic of dicots. This is a smaller dahlia useful for bedding plants. U.S. Department of Agriculture photo.

Labiatae (mint family), Leguminosae (pea family), Malvaceae (mallow family), Papaveraceae (poppy family), Rosaceae (rose family), Scrophulariaceae (figwort family), Solanaceae (nightshade family), and Umbelliferae (parsley family). A few well-known monocot families are the Bromeliaceae (pineapple family), Gramineae (grass family), Liliaceae (lily family), Orchidaceae (orchid family), and Palmae (palm family).

However, it is the taxonomy below the family level that largely concerns the horticulturist: genus (plural, genera), species (abbreviated sp., singular, or spp., plural), variety (abbreviated var.), form (abbreviated f.), and individual. These are concepts encountered often by those who work with or read about plants.

The subdivision of a family is the *genus*. It contains plants that have more morphological or phylogenetic characteristics in common with each other than they do with plants of other genera in the same family. The number of plants within a genus is small enough to allow studies of their ecological, cytological, genetic, and biochemical relationships. Sometimes these studies result in a shifting of plants from one genus to another, or even the elimination of an entire genus, as the modern taxonomist attempts to base the genus on relationships other than the older morphological criteria.

The modern approach to taxonomy has produced some new terminology with which the horticulturist should have some familiarity. For example, a group of plants with varying ability to breed among each other, but completely unable to breed with plants of another group, is called a *comparium*. This is frequently the same as the traditional genus. Successive subdivisions of the comparium are (1) *cenospecies,* a group whose uniqueness is sustained under natural conditions by genetic incompatibility backed by ecological barriers (sometimes the same as the traditional genus or species); (2) *ecospecies,* a group whose uniqueness is supported by differing ecological requirements backed by partial genetic restrictions (sometimes the same as the traditional species); and (3) *ecotype,* a group whose uniqueness is sustained only by differing ecological requirements (sometimes equivalent to a traditional subspecies, variety, or form). Although these terms may not be encountered frequently, their usage will undoubtedly increase in the future, as the modern taxonomist attempts to base the classification of the minor taxa on a genetic and ecological basis.

The subdivision of the traditional genus is the *species,* whose members share greater morphological and phylogenetic similarities among themselves than with other members of the genus. The species is written always in the context of the genus (see the following discussion on nomenclature). The species (Fig. 1–4) may be assumed to represent a continuing succession of like plants from one generation to the next.

Certain degrees of morphological or phylogenetic variance from the essential identifying features of the species occur. If a group within the species shows minor but consistent differences in conjunction with specific geographical or ecological distribution, it may be classed as a subspecies. If the differences are not associated with distribution, it may only warrant the status of *variety* (Latin, *varietas*). The term *form* (Latin, *forma*) is for very minor variances, such as flower or fruit color. These differences are inheritable and should show predominately in the offspring. These fine distinctions may be lost in practice, but they do exist and serve to designate variances from the species norm. In fact, many of our interesting horticultural plants are a variety or form of an otherwise common species (Fig. 1–5).

Sometimes an individual plant, or *individuum,* appears with characteristics that attract the eye of the horticulturist. This plant, an unusual result of chance mutation or genetic recombination, is known as a *sport.* The sport is propagated vegetatively, since it might not reproduce true from seed. The group of plants propagated by vegetative means is called a *clone* (Fig. 1–6). The clonal progenitor is the *ortet,* and the individual offspring is designated as a *ramet.*

Introduction to Horticultural Science / Part I

Fig. 1-4. *Vanilla planifolia* (top left). This species is the chief commercial vanilla source. It is a vine orchid grown in tropical America. U.S. Department of Agriculture photo.

Fig. 1-5 (top right). Below the species level is the variety. Shown here is *Crassula lycopodioides variegata*, which differs from the all green species. *C. lycopodioides* in that it has some silver to white variegation. Author photo.

Fig. 1-6 (right). Named roses are usually vegetatively propagated and hence are clones. Rose clones are usually the result of deliberate plant breeding rather than chance mutation. Each clone is usually given a cultivar name (see text and 1-7), such as this award winning clonal cultivar, 'Paradise'. Courtesy of All-America Rose Selections.

Certain plants of the large number grown have a practical significance to humans and especially to the horticulturist. A special horticultural taxonomic category, the cultivar, is used to designate these cultivated plants. The cultivar, a contraction of cultivated variety, should be differentiated from the naturally occurring botanical variety discussed previously. *Cultivars* may be thought of as plants that have originated and persisted under cultivation; they are clearly distinguished by characters of a morphological, biochemical, cytological, or physiological nature (Fig. 1-7). These characteristics are retained through asexual and often sexual reproduction. Different cultivars of a particular plant are often collectively termed a *group,* another horticultural category.

Fig. 1-7. This award winning cultivar is sexually reproduced rather than asexually as in Fig. 1-6. Specifically, this winter squash is a hybrid cultivar designated 'Sweet Mama.' Courtesy of All-America Selections.

The cultivar can be the progeny of a minor variant within a species or of a clone, or a hybrid between species. Asexually reproduced cultivars come from clones (*clonal cultivar*) or apomicts (*apomictic cultivar*). Apomicts are unique plants produced from seed that did not arise through the usual sexual means, but from an asexual process. Sexually produced cultivars arise from self-fertilizing plants that breed true naturally (*line cultivars*), from cross-fertilizing plants maintained true through selection and isolation (*inbred line cultivars*), or from selectively cross bred or hybridized plants (*hybrid cultivars*).

Introduction to Horticultural Science / Part I

Scientific and Common Names of Plants

Many plants have common or vernacular names, especially if they are commonplace, useful, or interesting plants. Common names are helpful, but they often lead to confusion. This confusion arises since the choice of a common name is not necessarily logically conceived or consistent. Often the same common name refers to several different plants, or one plant has two or more common names. For example, *Agastache cana, Azolla caroliniana,* and *Cynanchum ascyrifolium* are all commonly called mosquito plant, whereas Moses-in-a-boat, Moses-in-the-bulrushes, Moses-in-the-cradle, and Moses-on-a-raft all refer to *Rhoeo spathacea.* Common names sometimes vary by geographical region or differ from one language to another.

These problems were eliminated by the adoption in the eighteenth century of the scientific or binomial system of nomenclature. The concept is not difficult to learn; in fact, it is very similar to the way we have named ourselves for centuries. The first word, always a noun, refers to the *genus,* and the second, usually a descriptive adjective and sometimes a noun, refers to the *specific epithet.* The generic name is capitalized and the specific epithet is not. Together, the two parts are known as the name of the *species.* The *species* is usually indicated by italics or underlining. The use of only Latin (sometimes Latinized Greek) for these words provides for understanding by scientists of all nationalities, and since it is no longer a spoken language, Latin does not change. Scientific names offer uniqueness and stability. They cannot be changed except for valid, recognized reasons and only then by established international rules of nomenclature.

However, even with the scientific name some minor discrepancies can arise. Sanctioned name changes are sometimes accepted slowly, which leads to the use of two scientific names during the period of transition. The problems of naming some plants can be so complex as to cause disagreement among the experts. Nevertheless, professionals and amateurs now use scientific names as a matter of course. The need for proper identification far outweighs any resistance to their use. In cases where there is no ambiguity, the scientific name is complex (especially with cultivars), or to avoid repetition, I have used the common name. However, the scientific name for all such cited common names can be found in Part Three.

Those of us who have studied Latin and/or have good memories may find the scientific name easy to learn. Many do not find it easy beyond the few cases where the Latin word suggests the English equivalent or where the scientific name is already used in the common sense. For example, the scientific name for the red maple is *Acer rubrum. Rubrum* is Latin for red, which is the color of the spring flowers and the autumn foliage. Even the fruit and twigs have a reddish cast. Those scientific names used in the common sense include such genera as *Chrysanthemum, Delphinium, Geranium, Calceolaria, Begonia, Philodendron, Caladium, Coleus, Cyclamen,* and *Dracaena.* For those who find it difficult, the only help is the association of the scientific name with exposure to plant material over a long period of time, reinforced by reference to the literature.

At times scientific names are written as *Agave parviflora* Torr. or *Sedum frutescens* Rose. The nonitalicized part refers to the authority who has established the scientific name. Most of the names of the authorities, unless short, are abbreviated. The authority is not cited during popular writing or common speech, but only when a sense of botanical or historical accuracy is desired.

As we have seen in the previous section, plants of horticultural significance are often forms, varieties, subspecies, or cultivars. As such it is often necessary to include them in the scientific name. Forms may be designated as *Astrophytum myriostigma* f. *nudum*, varieties as *Astrophytum myriostigma* var. *coahuilense* (or *Astrophytum myriostigma coahuilense*), subspecies as *Pachypodium lealii* ssp. *saundersii*, and cultivars as *Sedum guatemalense* cv. Aurora (or *Sedum guatemalense* 'Aurora'). The cultivar name is usually in a common language (not Latin) and is not italicized. The names of hybrids are designated by a preceding X, such as *Clematis* X *jackmanii*, a hybrid between C. *lanuginosa* and C. *viticella*.

Horticultural Classification of Plants

Horticulturists have devised other schemes for categorizing the many plants that constitute the horticultural realm. These schemes collectively form the basis of a horticultural plant classification. This system of classifying plants is useful, since it is based on practical criteria such as usage, growth habits, and ability to tolerate environmental conditions. As such, it enables the horticulturist to select the proper plant for any given use, condition, or site and to predict the final appearance of the mature plant.

Plants can be conveniently classified on the basis of growth habit or other significant physiological characteristics. *Nonwoody plants* with comparatively soft tissues, of which the aerial portion is relatively short lived, are called *herbaceous* or *succulent* plants. *Woody plants* have long-lived, dense, strong aerial tissues, which increase in length and diameter each year. Herbaceous plants can be divided further on the basis of their stem growth habit. Seed-producing herbaceous plants with self-supporting stems are *herbs* (Fig. 1–8). Herbaceous plants with a trailing or climbing habit are *vines* (Fig. 1–9). Woody plants with self-supporting stems are either *trees* or *shrubs;* those that climb or trail are correctly designated as *lianas* (Fig. 1–10). Often the term liana is overlooked and the term vine is substituted. *Shrubs* (Fig. 1–11) are usually low with several shoots or trunks produced at the base, as opposed to the single trunk and distinct, elevated head of the *trees* (Fig. 1–12). The distinction between trees and shrubs can be blurred by horticultural practices like pruning or training or by the effects of environmental conditions. Plants that grow on trees or other plants, but do not take moisture or nutrients from the host, are known as *epiphytes* (Fig. 1–13). They obtain their essential nutrients from decaying organic matter that collects around their aerial roots.

Fig. 1-8. Bamboo is a good example of an herb. It is a monocot, hence it never has true secondary woody growth with age as do many herbaceous dicots. U.S. Department of Agriculture photo.

Fig. 1-9. The morning glory (cultivars of *Ipomoea purpurea*) is a typical vine. It has a twining vine habit. Courtesy of National Garden Bureau, Inc.

Fig. 1-10. All roses are not shrubs. Some are climbers, such as this climbing rose. Since it becomes woody with age, it can technically be considered a liana. Other well-known lianas are *Wisteria* and certain species of *Clematis* and *Allamanda*. U.S. Department of Agriculture photo.

Fig. 1-11. Forsythia is an attractive spring flowering shrub of deciduous habit. This particular one, *Forsythia suspensa* forma *fortunei*, is popular because of its gracefully arching branches and brilliant yellow flowers. This one was pruned the season prior to blooming. William P. Raffone Jr. photo.

Fig. 1-12. Deciduous trees enhance this home in early spring. Their deciduous nature is apparent by comparison of the right large tree which has not yet leafed out versus that on the left which has. U.S. Department of Agriculture photo now in National Archives (No. 16-N-31743).

Foliage-retention habits form the basis of a major distinction between plants. Those that are always in leaf are called *evergreens* (Fig. 1-14); those that lose all their leaves during a portion of the year (usually winter) are known as *deciduous* (Fig. 1-12). Evergreens do shed their leaves, but at a gradual rate over one or more years, and therefore are never without leaves. The persistence of leaves can vary with climate, as some plants that are evergreen in the tropics become deciduous in cooler regions.

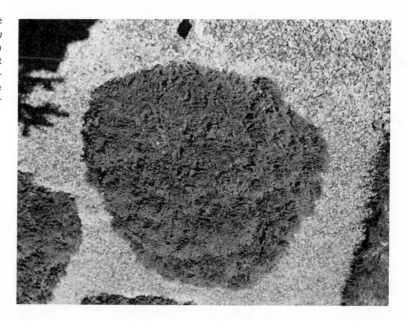

Size is frequently an important factor in horticultural classification. Terms like *dwarf, small, medium,* and *tall* may be applied to shrubs or trees. These categories play an important part when a plant is chosen for landscape purposes. Another factor of importance to landscapers is the growth habit of the plants. Classifications for trees include *globular* (roundheaded), *fastigiate* (columnar), *pyramidal, picturesque* (irregular), *weeping* (drooping), and *horizontally branching.* Shrubs can be classed as *weeping, upright,* or *horizontally branching.* Vines can be described as *trailing, clinging,* or *twining.*

Plants are often classified on the basis of their tolerance toward environmental conditions such as shade, sun, drought, wet soils, hot or cold temperatures, infertile soils, air pollution, road or sea salts, acidity, and alkalinity. This information is helpful when we select plants.

Some of these categories are especially detailed, such as the cold-temperature or acid-to-alkaline tolerances of plants. In the cold-temperature category, we find a division of large regions, such as the United States and Canada, into hardiness zones based on the limits of the average annual minimal temperatures (see Part Three for a map). Plants would be called *hardy* or *tender,* depending on their ability to survive the minimal temperature of a zone. Additional refinements of this category are the distinctions made for woody plants of *wood hardiness* and *flower-bud hardiness.* A woody plant might be capable of survival at certain low temperatures, but flower buds might not. *Root hardiness* is also important, especially in container-grown plants wintered over. Temperatures in the root zone could be lower in a container than would be encountered in the ground; failure to recognize this point could result in extensive winterkill. Such factors can restrict plants to certain areas, which explains the centralization of the apricot industry in California. Horticultural crops are even classified further on the basis of temperature preference during the growing season. Thus, corn, tomatoes, peppers, eggplants, and melons are referred to as *warm-season* crops, whereas lettuce, spinach, and cole crops are known as *cool-season* crops.

Horticulturists are often concerned with classification of the life spans of plant material. Plants that finish their growth cycle from seed through maturity and death in one growing season are called *annuals* (Fig. 1–15). Plants that require two seasons to complete their growth cycle are known as *biennials* (Fig. 1–16). Vegetative growth occurs during the first growing season, and blooming, seed formation, and death occur in the second growing season. Plants that continue to grow for several or more growing seasons and usually do not die after seed formation are designated as *perennials* (Fig. 1–17). These life-span classifications are not absolute. Subtropical perennials grown in cooler areas cannot live through the winter; thus they become annuals in those areas. Certain root crops are biennials, but they are harvested after one season, so they are treated as annuals. Annuals and biennials are usually herbaceous, whereas perennials can be herbaceous or woody.

Fig. 1–15. The Spiderflower (*Cleome spinosa*) (right) is just one of many annuals. Courtesy of National Garden Bureau, Inc.

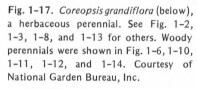

Fig. 1–17. *Coreopsis grandiflora* (below), a herbaceous perennial. See Fig. 1–2, 1–3, 1–8, and 1–13 for others. Woody perennials were shown in Fig. 1–6, 1–10, 1–11, 1–12, and 1–14. Courtesy of National Garden Bureau, Inc.

Fig. 1–16. The pansy (above) is a popular biennial. This one is an award winning hybrid cultivar, 'Orange Prince.' Courtesy of All-America Selections.

21

The classification of plants by use is probably the most important to the horticulturist. Essentially, three use categories exist: (1) edible plants, (2) ornamental plants, and (3) plants that are sources of natural products used as medicines, drugs, condiments, beverages, oils, insecticides, and gums or resins.

Each of these groups can be divided further. The *edible* plants consist of *fruits* and *vegetables*. Generally, *fruits* are considered to be the mature, ripened, seed-containing ovary and sometimes accessory parts of a flowering plant (botanical fruit), which when edible tastes sweet and is served as a dessert (Fig. 1-18). *Vegetables* are any edible part of a herbaceous plant (including actual botanical fruit), which is served raw or cooked with the main course of the meal (Fig. 1-19). These definitions are not botanical and are not absolute, since certain exceptions arise. For example, freshly picked sweet corn fits the definition of a botanical fruit, but is eaten as a vegetable. Rhubarb fits more closely the definition of a botanical vegetable (plant parts other than botanical fruit), but its use as a dessert makes it a fruit to most of us. The concepts of fruits and vegetables can also vary with nationalities and geographical regions.

Fruit plants are frequently woody perennials, with those of the temperate zone usually being deciduous and those of the tropics evergreen. Fruit subdivisions include *tree fruits, small fruits,* and *nuts.* Deciduous tree fruits include the *pome fruits* (apple and pear) and the *drupe fruits* (stone fruits such as peach, plum, apricot, and cherry). Nuts harvested from deciduous trees include the walnut, pecan, almond, and filbert. Deciduous small fruits encompass grapes, red and black raspberries, blackberries, loganberries, cranberries, strawberries, currants, gooseberries, and blueberries. Evergreen tree fruits consist of citrus fruits (lime, lemon, orange, and grapefruit) and miscellaneous fruits (fig, persimmon, date, avocado, mango, guava, and papaya). Tropical evergreen nut trees include the coconut, brazil nut, cashew, and macadamia. Other evergreen fruits of the warmer areas are the banana and pineapple.

Vegetables can be divided into two categories, those harvested for (1) aerial parts or (2) underground portions. The category of aerial parts is divided further into stems and/or leaves plus fruits and/or seeds. Underground portions may be roots, tubers and rootstocks, or bulbs. Vegetables grown for their stems and/or leaves include the *cole* or *cabbage crops* (cabbage, Chinese cabbage, Brussels sprouts, kohlrabi, kale, cauliflower, collards, mustard, and broccoli), *greens* or *pot herbs* (New Zealand spinach, Swiss chard, spinach, and turnip greens), *salad crops* (lettuce, endive, celery, cress, watercress, and parsley), and *miscellaneous crops* (asparagus, rhubarb, and artichoke). Vegetables grown for their fruits and/or seeds are the *legumes* or *pulse crops* (snap beans, lima beans, soybeans, and peas), the *solanaceous fruit crops* (tomato, pepper, and eggplant), the *vine crops* or *cucurbits* (squash, pumpkin, watermelon, cucumber, and melon), and *miscellaneous crops* (sweet corn and okra). Vegetables grown as *root crops* include the radish, carrot, turnip, rutabaga, salsify, sweet

Fig. 1-18. The grape (cultivars of *Vitis*) is an example of a fruit, both in the common and horticultural sense. It is also a woody deciduous perennial of vining habit, a liana. Woodiness is minimized by pruning as old wood is not productive. U.S. Department of Agriculture photo.

Fig. 1-19. Lettuces (cultivars of *Lactuca sativa*) are typical vegetables, both in the common and horticultural sense. They are herbaceous and annuals. Courtesy of National Garden Bureau, Inc.

potato, cassava, and parsnip. Those harvested for their *tubers* and *root-stocks* are the potato, Jerusalem artichoke, and yam. Vegetables used for their *bulbs* or *corms* include the onion, garlic, shallot, leek, and taro.

Ornamental plants are usually separated into two groups: (1) *nursery plants* and (2) *flowering* plus *foliage plants.* Flowers and foliage plants are separated according to their life span: annuals, biennials, and perennials. Popular annuals include marigolds, petunias, impatiens, zinnias, and snapdragons. Some biennials of note are sweet william, Canterbury bells, and foxglove. A few perennials would be chrysanthemum, delphinium, hyacinth, columbine, and iris.

Nursery plants are an extensive division of the ornamentals. Subdivisions of this group are *lawn* or *turf plants,* such as bluegrass, bentgrass, bermuda grass, fescues, and zoysia; *ground covers* like pachysandra, *Hosta,* periwinkle (*Vinca minor*), and sedum; *vines* such as English ivy (*Hedera helix*), clematis, glorybower (*Clerodendrum*), and wisteria; *shrubs* (usually deciduous) like viburnum, flowering quince, lilac, and hydrangea; *evergreens* (trees and shrubs) such as blue spruce, rhododendron, andromeda, and Japanese holly (*Hex crenata*); and *trees* (usually deciduous) like maple, oak, mountain ash, and dogwood.

Plants are frequently sources of natural products. This use category has several divisions. Medicines and drugs such as witch hazel, digitalis, quinine, and cocaine are derived from *medicinal plants.* Condiments such as pepper, basil, oregano, thyme, ginger, and cinnamon come from *herb* and *spice plants.* Coffee, tea, and cocoa are derived from *beverage plants.* Cooking oils can come from *oil-bearing plants.* Insecticides such as pyrethrin, rotenone, ryania, sabadilla, hellebore, and nicotine were first isolated from plants, although most of them have been replaced by synthesized analogues. Plants are also sources of gums, resins, and turpentine.

Plant
Identification
Keys

At times the horticulturist encounters plants that are difficult to identify. A first recourse is to consult some of the excellent books of plant descriptions. If a careful search fails, the plant may not be in these plant manuals because the taxonomist or horticulturist felt it was of little value or too rare to be included. The possibility that it has not yet been given a name exists too. Under these circumstances the plant can be sent to a professional taxonomist at a well-known university, herbarium, or botanical garden. Once identified, the horticulturist can verify the identity by comparison with dried, pressed specimens found in herbarium collections. If indeed the plant has not been named previously, a specimen of it too will enter the herbarium collection.

Plants can be identified through some manuals by an analytical key arranged in dichotomous form. The *key* confronts the reader with series of contrasting, paired choices. As one choice of the pair is eliminated, the correct one determines which set of choices is consulted next. Again one choice is eliminated, and the next set of choices is chosen, and so on, until the plant is identified or keyed out. At times the key leads to confusion on the species level, since the diagnostic features of the key are morphological characteristics (leaf, twig, seed, flower, fruit, or others), unlike the sexual, phylogenetic, or other features used for plant classifications. On questions of identity with taxa below species, such as variety or form, it often becomes necessary to consult specialized monographs dealing with the plant group under study. Successful use of keys requires an understanding of the morphological terms (often complex), patience, practice, and a knowledgeable person to verify your answers on your initial attempts.

FURTHER READING

American Horticultural Society, *Plant Sciences Data Center Master Inventory.* Mount Vernon, Va. An annual, updated, computerized listing of plant records from twenty-nine North American botanical gardens and arboreta.

Backberg, C., *Cactus Lexicon* (English ed.). London: Blandford Press, 1977.

Bailey, L. H., *Manual of Cultivated Plants Most Commonly Grown in the Continental United States and Canada* (rev. ed.). New York: Macmillan, Inc., 1949. Plant identification manual.

Bailey Hortorium at Cornell University Staff, *Hortus Third.* New York: Macmillan, Inc., 1976. The last word on correct nomenclature and an excellent source book for the serious horticulturist.

Beck, Charles B., ed., *Origin and Early Evolution of Angiosperms.* New York: Columbia University Press, 1976.

Benson, Lyman, *Plant Classification* (2nd ed.). Lexington, Mass.: D.C. Heath and Company, 1979.

Blake, S. F., and A. C. Atwood, *Geographical Guide to Floras of the World; Annotated List with Special Reference to Useful Plants and Common Plant Names. I. Africa, Australia, North America, South America, and Islands of the Atlantic, Pacific, and Indian Oceans. II. Western Europe: Finland, Sweden, Norway, Denmark, Iceland, Great Britain with Ireland, Netherlands, Belgium, Luxembourg, France, Spain, Portugal, Andorra, Monaco, Italy, San Marino, and Switzerland.* Washington, D.C.: U.S. Department of Agriculture Miscellaneous Publications 401 and 797, 1942, 1961. Very extensive plant bibliography.

Bold, H. C., *The Plant Kingdom* (4th ed.). Englewood Cliffs, N.J.: Prentice-Hall, Inc., 1977.

Cronquist, A., *The Evolution and Classification of Flowering Plants.* Boston: Houghton Mifflin Company, 1968.

Gilmour, J. S. L., and others, *International Code of Nomenclature for Cultivated Plants.* Utrecht, Netherlands: Regnum Vegetabile, Vol. 64, 1969.

Graf, A. B., *Exotica: Pictorial Cyclopedia of Exotic Plants* (8th ed.). E. Rutherford, N.J.: Roehrs Co., 1975. Comprehensive coverage of worldwide ornamental plants used indoors, in greenhouses, and on patios.

Jacobsen, H., *Lexicon of Succulent Plants* (English ed.). London: Blandford Press, 1974. Extensive worldwide coverage of succulent plants other than *Cactaceae.*

Ouden, P. den and B. K. Boom, *Manual of Cultivated Conifers* (English ed.). Boston: Martinus Nijhoff, 1978.

Plowden, C. C., *A Manual of Plant Names* (2nd ed.). New York: Philosophical Library, Inc., 1970.

Rehder, A., *Manual of Cultivated Trees and Shrubs Hardy in North America Exclusive of the Subtropical and Warmer Temperate Regions* (2nd ed.). New York: Macmillan, Inc., 1940. Good coverage of trees, shrubs, lianas, and semiwoody plants of the indicated areas.

Rushforth, S. R., *The Plant Kingdom: Evolution and Form.* Englewood Cliffs, N.J.: Prentice-Hall, Inc., 1976.

Stafleu, F. A., and others, *International Code of Botanical Nomenclature.* Utrecht, Netherlands: Regnum Vegetabile, Vol. 82, 1972.

Stearn, W. T., *Botanical Latin: History, Grammar, Syntax, Terminology, and Vocabulary.* New York: Hafner Press, 1966.

2

Plant Structures

At first glance there would appear to be few essential similarities among the vascular plants of the horticultural realm. Variety and diversity would seem to be the obvious descriptive factor. However diverse they may look, all these plants carry out the same basic physiological processes and possess great similarity in the construction of their seemingly diverse structures.

Whole Plant

Basically, the plant body (Fig. 2-1) is comprised of an integrated axis, called the *root* when below ground and the *shoot* when above ground. *Roots* anchor the plant, absorb mineral salts and water from the soil, and translocate them to the stem base. The shoot has functions of support, food manufacture, and conduction. Stems and leaves form the shoot. The joint where a leaf may be borne or is borne on the stem is called a *node*. The space between the nodes is known as the *internode*. *Lateral buds* are usually located at the leaf base between the leaf and stem angle. These buds are capable of growing into branches that duplicate the existing shoot structures. A *terminal bud* is present at the stem apex where new stem and leaf tissues are produced. *Flowers* are essentially specialized stems whose leaves have been modified through evolution for reproduction.

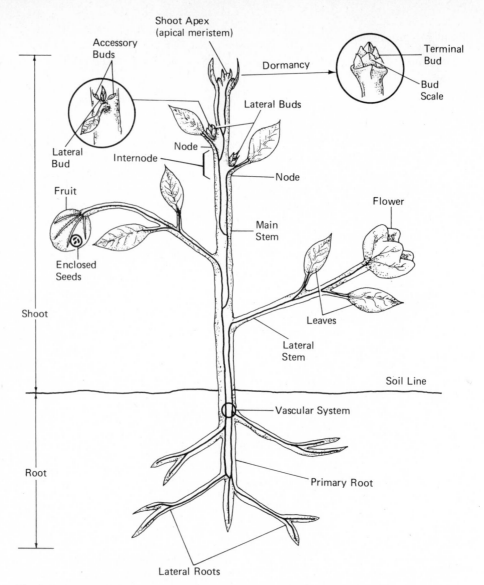

Fig. 2-1. A generalized diagram of the vascular plant body found with most horticultural plants. See text for explanation. The dormant terminal bud and accessory ("contingency") buds would be more characteristic of woody rather than herbaceous plants and fruit-seeds would not be found with the few seedless vascular plants of horticultural interest, such as ferns. Cone-seeds rather than fruit-seeds would be found with conifers.

The whole plant can be broken down into smaller and smaller units, as we go from the human eye to the light microscope and finally to the electron microscope. In order of decreasing size, these anatomical features are morphological structures, tissue systems, tissues, and the cell with its cellular components. Each of these will be examined in turn, starting with the macrostructures of most interest to the horticulturist.

Plant Morphology

The cells, tissues, and tissue systems are organized into plant organs with distinctive form and structure: roots, shoots, buds, leaves, flowers, fruits, seeds, and modifications thereof.

ROOTS

The *root* arises from the lower end or radicle of the embryo during germination of the seed. This early or primary root system grows downward into the soil and forms the root system of the plant. If the system is formed mostly by lateral branching of the primary root, it is a *fibrous root system*. When the system consists predominately of the primary root growing downward with minimal branching, it is a *taproot* system.

Taproots are found on plants such as the carrot, dandelion, or hickory nut, as opposed to fibrous-rooted plants like lawngrass. Plants with fibrous root systems respond quicker to fluctuations in the water and nutrient supply than do plants with taproots, since the fibrous root system tends to be shallow. Some plants have both a lower taproot and upper fibrous roots; the taproot increases access to the water supply, and the fibrous upper roots provide better nutrient uptake in the more fertile upper layers of soil.

Another form of roots, *adventitious roots,* is produced by many plants at locations other than the site of the primary root system. These roots develop from aerial portions of the stem, such as the prop roots of corn, or from rhizomes, cuttings, stolons, and corms. This type of root is important to plant propagation (see Chapter 6). Adventitious roots originate within or near the vascular tissues of the stem. The latter are involved in food and water conduction.

Root systems can account for over half the dry weight of the whole plant. The primary function of the roots is absorption of water and nutrients and transport of absorbed materials to the base of the stem. Secondary functions are anchorage and support of the shoot. In addition, roots serve as food-accumulating organs in some plants.

Root systems are ideally suited to absorption, since their irregular arrangement of complex branching presents a very large surface area in contact with the soil. Much of this surface area derives from the numerous root hairs associated with areas of new growth. Root hairs slough away as they get older and farther back on the maturing root. Much of the absorption is through the root hairs, although some occurs in young roots through their epidermal cells (the outermost single layer of cells). Fungi in contact with roots, or *mycorrhizae,* increase absorption of water and nutrients by the roots of many plants of horticultural importance. Mycorrhizae can be especially important in those mature root areas that no longer have root hairs.

Some roots become rich in accumulated foods in the form of sugars and starch. These roots become enlarged and fleshy, such as the taproots

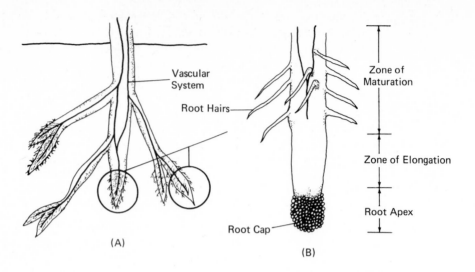

Fig. 2-2. A generalized diagram of vascular plant roots. See text for explanation. The zone of maturation contains the root hairs where much of the water and mineral absorption occurs prior to translocation upward in the xylem (part of vascular system). The majority of plants may also contain fungi in a symbiotic relationship with the roots (mychorrhizae) which enhances water/nutrient absorption. Food can reach the roots via phloem in the vascular system.

of some plants. Plants with edible food-storage roots formed from taproots are the carrot, beet, turnip, parsnip, and radish. In other plants the branch roots can become swollen and function as storage organs. Roots of this type are known as *tuberous roots* and occur on the dahlia and sweet potato. Some storage roots can also be useful for propagation, if they have the capability of producing adventitious shoot buds.

The general regions of a young root are shown in Figure 2-2. The processes of cell division and elongation, initiated by the *root apex* (an apical meristem, or physiologically young cells that form new tissues and maintain themselves by cell division), occur primarily in the *zone of elongation*. The *root cap* protects the root apex from mechanical injury and acts as a lubricant as the root pushes through the soil particles. In the *zone of maturation* the cells have or almost have attained physiological maturity. Root hairs are produced from this zone. The tissues of the zone of maturation are highly specialized for functions of absorption and translocation of water, minerals, and food (see the later discussion on xylem, phloem, and vascular systems, and Figs. 2-3 and 2-4).

SHOOTS

Shoots are composed of stems and leaves. The stem supports the leaves; it also conducts the nutrients and water absorbed by the roots to the leaves. Some stems have limited food-production abilities when young and green. Stems and stem modifications may also act as food-storage organs. Plant forms are also determined by the structure and growth of the stems.

Stems have many specialized tissues (Figs. 2-3 and 2-4), because of stem functions such as growth, support, and conduction. Stems grow in length and diameter when woody. Growth in length is initiated by the apical meristem (Fig. 2-5), and growth in diameter by lateral meristems (*vascular cambium* and *cork cambium*). Upright growth of plants resulting from a rigid stem with one actively growing point is the norm. Multiple stems and several growing points produce shrubby or bushy growth.

Fig. 2-3. Cross-sections through the mature monocot stem and root (A, C) and the herbaceous dicot stem and root (B, C). Monocots do not develop a vascular cambium, so they never become woody in the true sense. Dicots can be either since they develop a vascular cambium (see woody dicot—Fig. 2-4). The monocot can be distinguished from the herbaceous (or woody) dicot on the basis of unsymmetrical versus symmetrical arrangement of the vascular system in the stem cross-section. The tissues shown here were derived from the apical meristem (shoot apex) which increases stem length. Tissues are discussed in detail under the section, Cell Organization and Tissues, in this chapter.

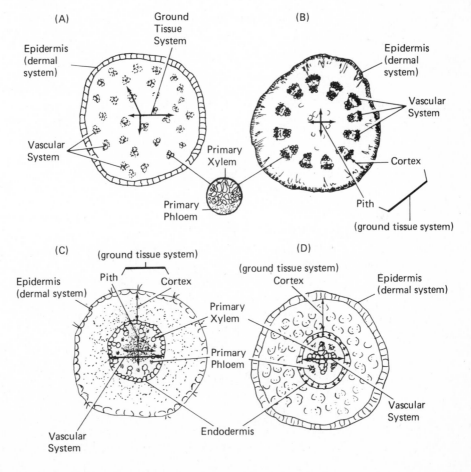

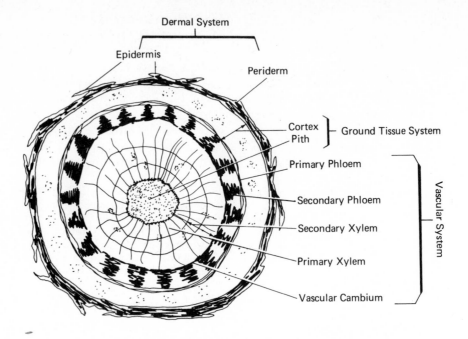

Dermal System

Epidermis

Periderm

Cortex
Pith } Ground Tissue System

Primary Phloem

Secondary Phloem

Secondary Xylem

Primary Xylem

Vascular Cambium

Vascular System

Fig. 2–4. Cross-section through a woody dicot stem (upper) and root (lower). The woodiness (and increase in stem diameter) are the result of activities by the laterial meristems (vascular and cork cambiums). In the stem increasing woodiness results in the destruction of the epidermis and usually the cortex and primary phloem. Likewise the epidermis, cortex (including endodermis) and primary phloem can be destroyed in mature woody roots.

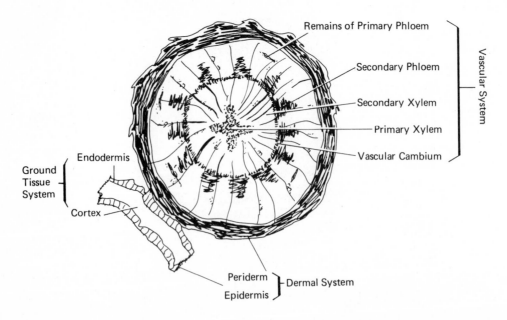

Remains of Primary Phloem

Secondary Phloem

Secondary Xylem

Primary Xylem

Vascular Cambium

Vascular System

Endodermis

Ground
Tissue
System

Cortex

Periderm } Dermal System

Epidermis

32

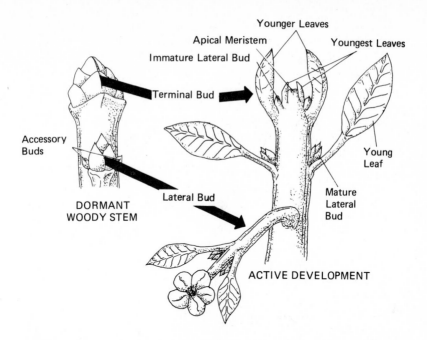

Younger Leaves

Apical Meristem

Immature Lateral Bud

Youngest Leaves

Terminal Bud

Accessory Buds

Young Leaf

Lateral Bud

Mature Lateral Bud

DORMANT WOODY STEM

ACTIVE DEVELOPMENT

Fig. 2-5. Buds and shoot development. Terminal buds contain an apical meristem and are found on tips of growing shoots. These are involved in increase of stem length and production of leaves and lateral buds (leafy stems). They may also give rise to flowers in some cases. Lateral buds give rise to leafy stems and/or flowers. Accessory buds can produce (when present) similar structures in event of lateral bud destruction. While buds are found both on herbaceous and woody plants, they are more pronounced in the latter case, especially dormant buds.

BUDS

Buds (Fig. 2-5) are embryonic stems, which can grow actively or maintain the potential of growth while in a dormant stage. *Terminal buds* occur at the tips of stems and *lateral (axillary) buds* are found in the leaf axils (angle where leaf is attached to the stem). Some species have *accessory buds,* which are located above or to the side of the axillary bud. Many buds develop into leafy stems; these are termed *leaf buds. Flower buds,* such as found on elm or cherry, produce flowers, and *mixed buds,* such as found on the apple, can give rise to leafy stems and flowers.

Most woody plants have their terminal and lateral buds covered by overlapping *bud scales.* These probably reduce water loss from the bud and protect it from mechanical injury. The buds of some woody plants have no bud scales, and such buds are called *naked buds.* In these buds the outer embryonic leaves are well developed and covered with hairs or scales; such leaves protect the younger, enclosed leaves. Nonwoody or herbaceous plants have less conspicuous terminal buds than woody plants.

Adventitious shoot buds arise from areas other than expected, such as from a root, or at developmental stages other than expected, such as from mature cells no longer in the meristematic condition. Many cells, even mature ones, can return to the meristematic state and produce adventitious buds (or roots). This makes vegetative propagation possible. In root cuttings, adventitious buds originate from the pericycle (outermost tissue of root vascular system). In leaf cuttings, secondary meristems arise from mature cells at the base of the leaf blade or petiole. More information on propagation can be found in Chapter 6.

STEM MODIFICATIONS

Several modifications of the normal, upright stem exist. These include both above- and belowground modifications. Stem alterations are of interest to the horticulturist, since some are important as food sources or for propagation purposes.

Aboveground modifications of major proportions include climbing and succulent stems. *Climbing stems* are of several types: those that lean or clamber over supports (*Allamanda*), those that twine around supports (*Bougainvillea, Clematis, Wisteria*), and those that grasp supports with tendrils (grape) or holdfasts (English ivy). *Succulent stems* are enlarged, with much reduced or absent leaves; this modification permits maximal food manufacture and water storage in the stem, and minimal water loss through leaves. Examples are cacti and *Euphorbia*.

Lesser aboveground modifications of the stem include offsets, runners, and spurs (Fig. 2-6). The first two modifications arise from the *crown,* which is that portion of the stem near the ground. It may simply be a point of transition between the roots and stem, such as with trees and shrubs, or short, compact stems that give rise to new shoots, such as with herbaceous perennials. Crowns of many plants (chrysanthemum, rhubarb, forsythia) can be divided to produce new plants. *Offsets* (sometimes called *offshoots*) are shortened, thickened stems of rosettelike appearance produced from the crown of some plants, such as many bromeliads and succulents. These are particularly useful for propagation. *Runners* are specialized stems that arise from leaf axils at the crown of a plant (strawberry, *Ajuga*); they grow horizontally above the ground, and new plants arise at nodes. *Spurs,* found on woody branches, are woody plant stems of lateral habit characterized by restricted growth; flowering and fruit production occurs mostly on the spurs of some fruit trees, such as the apple.

Belowground modifications include bulbs, corms, rhizomes, stolons, suckers, and tubers (Fig. 2-7). A *bulb* consists of a short, fleshy stem axis (*basal plate*) that has a flower rudiment or growing point at its apex, plus an extensive overlayment of fleshy scales. Bulbs are produced by some monocots for food storage and reproduction; they are valued for purposes of propagation. Scales of bulbs may be present in continuous, concentric

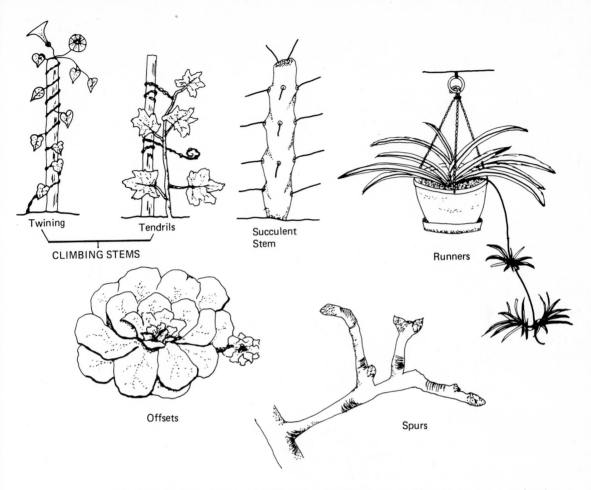

Fig. 2–6. Above-ground stem modifications. Climbing stems are represented here by morning glory (left) and grape (right). Grape (*Vitis* cultivars) tendrils are also modified stems, but most climbing stem tendrils are modified leaves. Others are succulent stems (*Euphorbia*), runners (Spider plant, *Chlorophytum*), offsets (*Echeveria*) and spurs (apple). See text for details.

layers (*tunicate* or *laminate bulbs*), which are dry and papery outside and fleshy inside. Examples are the Amaryllis, onion, and tulip. Bulbs without an outer dry covering and a less structured arrangement of the scales are known as *scaly* or *nontunicate bulbs*. An example is the lily (*Lilium* sp.). Miniature bulbs developed below ground from the parent bulb are referred to as *bulblets* and *offsets* when mature, or as *bulbils* when formed above ground on the stem.

A *corm* is the enlarged base of a stem axis enclosed by a few dry, scalelike leaves. Bulbs are primarily leaf scales, in contrast to the corm, which is mostly stem structure with obvious nodes and internodes. The corm, like the bulb, is valued for its use in propagation. Examples are the *Crocus, Colocasia,* and *Tritonia.*

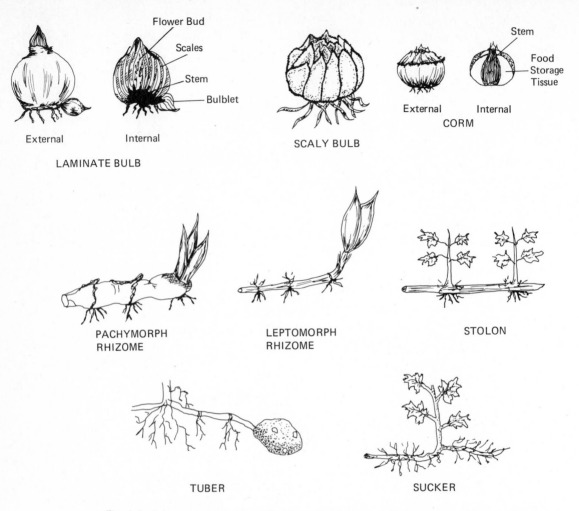

Fig. 2–7. Below-ground stem modifications. The laminate bulb is represented here by hyacinth (*Hyacinthus*), scaly bulb by lily (*Lilium*), corm by *Crocus*, pachymorph and leptomorph rhizomes by *Iris* and lily-of-the-valley respectively, stolon by mint (*Mentha* sp.), sucker by red raspberry (cultivars of *Rubus*), and tuber by white potato (cultivars of *Solanum tuberosum*).

A *rhizome* is a specialized primary stem structure, which grows horizontally at the ground surface or underground. It is distinguished from roots by the presence of nodes and internodes. Buds and scalelike leaves are sometimes found at the nodes. Most rhizomes are produced by monocots and the lower plant groups. Two types of rhizomes are known: *pachymorph*, which is thick, fleshy, short, and found as a many-branched clump, and *leptomorph*, which is slender with long internodes. Examples are the *Iris* and lily-of-the-valley (*Convallaria*), respectively. Intermediate forms are called *mesomorphs*. The production of roots, shoots, and flowering stems from rhizomes makes them useful for propagation.

Stolons are modified stems similar to that of the runner. In fact, the two terms are often confused. It has been suggested that runners should refer to the aboveground specialized stems and stolons to the belowground form. This usage will be followed here. Stolons are useful for propagation, such as with mint (*Mentha* sp.).

A *sucker* is a shoot that arises below ground from an adventitious root bud. However, in practice, shoots arising from the crown or stem below ground are also referred to as suckers. Suckers can be used for propagation purposes, such as with the red raspberry.

A *tuber* is a belowground modified stem structure, essentially a swelling of a stolon or rhizome. It too can be useful for propagation. A true tuber should be distinguished from a *tuberous root.* A true tuber has all the parts of a stem, that is, eyes that represent nodes and contain buds, whereas a tuberous root has the external and internal structure of a root. The white potato and *Caladium* have tubers; the sweet potato and dahlia have tuberous roots.

LEAVES

Leaves may be thought of as a flattened or expanded portion of the stem. Many of the kinds of cells and tissues found in the stem are present in leaves. However, the growth of the leaf, unlike the stem, is limited. The design of the leaf serves to present a large surface area for the absorption of light energy needed for photosynthesis (see Chapter 3). The main two leaf functions are food manufacture (photosynthesis) and water loss by evaporation (transpiration).

The typical leaf has two parts: the *blade,* or thin, flattened expanded portion, and the *petiole,* or stalk. Leaves without petioles are said to be *sessile.* The base of the petiole sometimes has leaflike or scalelike structures called *stipules.*

On the lower and often the upper surface of the leaf are lines or ridges called *veins* (Fig. 2–8). These are a continuation of the vascular system of the stem. The vein arrangement (venation) is usually parallel in monocots and netlike in dicots. Netlike veins may assume either the form of a feather, a strong main vein with lateral smaller veins (*pinnately veined*), or the form of the fingers spreading from the palm of the hand (*palmately veined*).

Leaves have either a simple or compound form (Fig. 2–8). A *simple leaf* has an individual blade, which may be indented. The *compound leaf* consists of a blade divided into several parts called *leaflets.*

Compound leaves can be either *palmately* or *pinnately* compound (Fig. 2–8). In the former, all the leaflets are attached at the petiole tip, in contrast to the latter, where the leaflets are attached to the sides of a central stalk like the barbs of a feather.

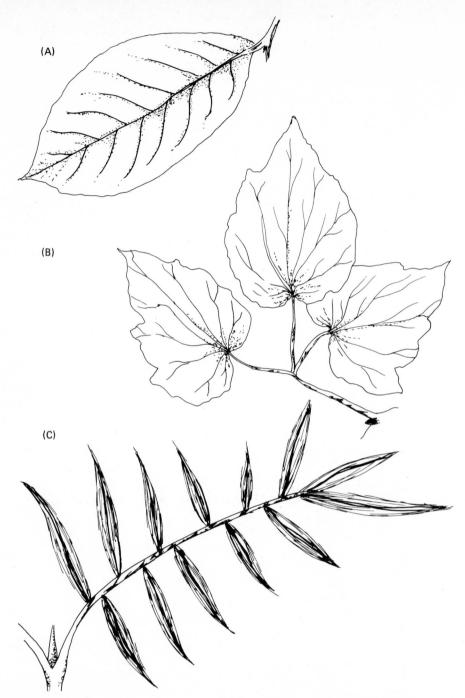

Fig. 2–8. Leaf form and venation. A. Form: simple, Venation: net-like pinnate, Examples: Elm, Birch (*Ulmus, Betula*). B. Form: compound palmate, Venation: net-like palmate, Example: *Clematis*. C. Form: compound pinnate, Venation: parallel, Example: Parlor Palm (*Chamaedorea elegans*).
The position of the lateral bud establishes A as a leaf and B, C, as leaflets comprising a compound leaf since the bud is found in the angle between the stem and leaf petiole.

38

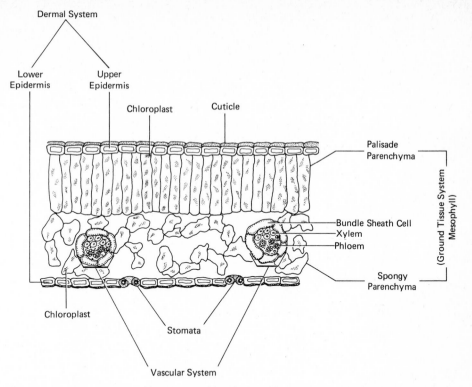

Fig. 2-9. Leaf cross-section. Most leaves of horticultural interest have this appearance, except for certain desert plants which may have an upper and lower palisade parenchyma layer with spongy parenchyma between, or some plants such as grasses and sweet corn where no clear cut distinction between palisade and spongy parenchyma cells is apparent.

The leaf (Fig. 2-9) is surrounded by an upper and lower single cell layer (*epidermis*), which is covered by a waxy coat (*cuticle*). The lower epidermis is rich in pores (*stomata*) through which gases and water vapor move. The upper layer has fewer or no stomata. Between the epidermal layers is the mesophyll tissue, where photosynthesis takes place.

LEAF MODIFICATIONS

Leaf modifications are of special interest to the horticulturist. These include thickened petioles, basal leaf portions that become bulb scales, thickened fleshy leaves, tendrils, and spines. These may be edible, such as the *thickened petioles* of rhubarb or celery, the *bulb scales* of onions, or the *fleshy leaves* of cabbage. Others may be ornamental, such as the thick, fleshy leaves of succulents, or functional, like the *tendrils* of vines or *spines* of cacti. Spines, which are found just below an axillary bud or shoot, should be differentiated from thorns and prickles. *Thorns* are modified branches, which arise from or just above the leaf axil; *prickles* are random outgrowths of the surface stem tissues (epidermis periderm).

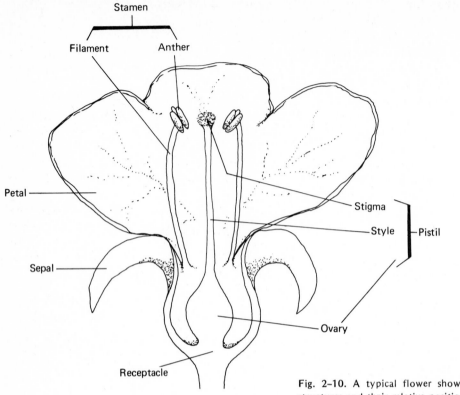

Stamen

Filament   Anther

Petal

Sepal

Stigma

Style — Pistil

Ovary

Receptacle

Fig. 2–10. A typical flower showing all possible structures and their relative positions. See text for details.

FLOWERS

The reproductive structures or *flowers* of plants (Fig. 2-10) vary greatly in appearance. However, they share fundamental similarities in their basic structural plan. Flowers are carried at the end of a specialized stalk or *pedicel.* The *receptacle* is a somewhat enlarged area at the apex of the pedicel; the floral parts are attached to the receptacle.

Four kinds of floral organs, which may be thought of as modified leaves, arise from the receptacle. These are the sepals, petals, stamens, and pistils (Fig. 2-10).

Sepals are found at the base of the flower, since they enclosed the blossom when it was in bud. Sepals are usually small, green, and leaflike (sometimes colored); collectively, the sepals are called the *calyx.* Above the calyx and toward the center are the *petals,* which are collectively known as the *corolla.* Petals are usually brightly colored, and sometimes fragrant from perfume and sticky with a sugary substance (nectar) produced from glands. Together the calyx and corolla are called the *perianth.* In some flowers the sepals and petals are similar in color and size. In that case the individual parts of the perianth are not called sepals and petals, but are referred to as *tepals* (example, tulip).

The *stamens* (male reproductive organ) are located above the petals. Typically, they consist of an elongated stalk, the *filament,* which supports an enlarged, pollen-containing structure called the *anther.* In the center of the flower is the female reproductive organ(s) or the *pistil*(s). This structure is somewhat flask shaped. The swollen, basal part (the *ovary*) is connected by a stalklike portion (the *style*) to an enlarged, terminal portion (the *stigma*). The mature pollen grains are released from the disintegrating anther and carried to the stigma by insects, wind, and rain (pollination). If the stigma is receptive, fertilization follows. The ovary, fertilized contents, and sometimes accessory parts then give rise to the fruit and seeds contained therein.

Pistils may be either simple or compound, depending upon the numbers of carpels. *Carpels* are leaflike structures that bear the ovaries (or ovules). A *simple pistil* has one carpel, and a *compound pistil* bears two or more fused carpels. The arrangement of the ovules on the carpels is a useful characteristic from the viewpoint of classification.

The floral organs are usually arranged in whorls. In dicots (magnolia, columbine, carnation, poppy, cacti, maple, apple, cabbage) the number of floral organs (sepals, petals, and so on) in each whorl is four or five or multiples thereof. In monocots (lily, palm, corn, orchid) the factor is three or multiples thereof. Some families of both may have more than ten stamens or pistils arranged in a spiral, rather than a whorl. This arrangement is considered to be more primitive in an evolutionary sense.

Modifications of the floral organs exist (Fig. 2-11). In some flowers the petals are fused to form a tube called a *corolla tube,* which is lobed. Examples are the petunia and morning glory. These lobes correspond to the number of petals. Fusion of the other floral organs is possible. The position of the ovary in relationship to the fused or unfused floral organs is also significant in plant classification.

Flowers that contain all four floral organs are called *complete flowers* (lily, cherry); *incomplete flowers* (grasses, willow, maple) lack some or have only vestigial floral organs (Fig. 2-12). Flowers that have both functional stamens and pistils, such as the tomato flower, are said to be *bisexual* (also *perfect* or *hermaphroditic*); flowers lacking either the stamens or pistils are termed *unisexual* (Fig. 2-12). Those without pistils are called *staminate* (tassel of corn); those without stamens are called *pistillate* (silk of corn). The perianth may be present or absent in unisexual and bisexual flowers. Plants with both pistillate and staminate flowers are termed *monoecious* (corn, begonia, cucumber, squash, oak, walnut); *dioecious* plants (*Ilex* species, spinach) have pistillate flowers and staminate flowers on separate plants. The latter conditions are important to the horticulturist in regard to ease of pollination, productivity of fruits and vegetables, and production of seeds (especially hybrid seed). These will be touched upon in later chapters.

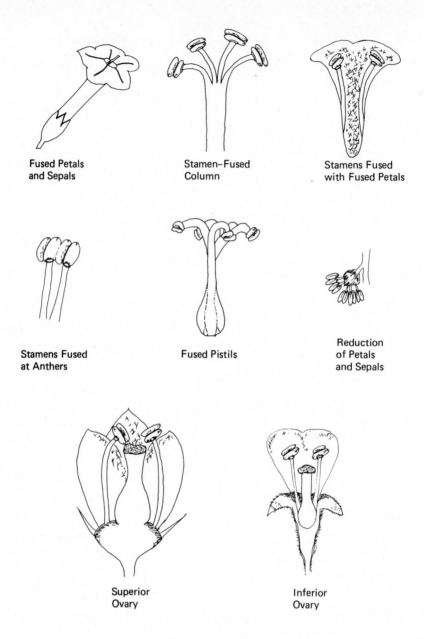

Fused Petals and Sepals

Stamen–Fused Column

Stamens Fused with Fused Petals

Stamens Fused at Anthers

Fused Pistils

Reduction of Petals and Sepals

Superior Ovary

Inferior Ovary

Fig. 2–11. Floral organ modifications. Fusion and reduction of flower parts are considered more advanced evolutionary trends. The inferior ovary (sepals, petals, stamens appear to grow from the top of the ovary) is also considered more advanced than the superior ovary.

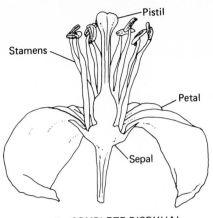

Pistil

Stamens

Petal

Sepal

A. COMPLETE BISEXUAL
FLOWER

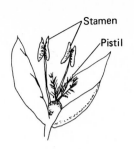

Stamen

Pistil

B. BISEXUAL
INCOMPLETE
FLOWER

Fig. 2–12. Flower organs and sexuality. See text for details. Examples shown: A. grapefruit (*Citrus* x *paradisi*). B. Grasses. C. Squash (cultivars of *Cucurbita*). D. Asparagus (cultivars of *Asparagus officinalis*). Cross- and self-pollination are both possible for types shown by A, B, and C. Only cross-pollination is possible in D. Self-pollination is not always assured, however, in A, B, and C, since some plants have preventative mechanisms against it. These include dichogamy, stamens and pistil mature at different times such that a receptive pistil occurs on another plant; heterostyly, differences in style length within a species which favor pollen removal by insects in one flower and stigma contact in another; and incompatibility, pollen tube fails to reach ovules on same plant but does on separate plant within the species.

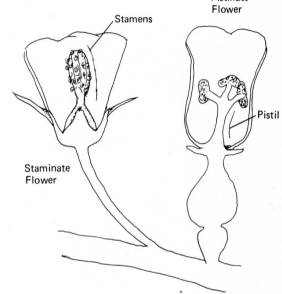

Pistillate
Flower

Stamens

Pistil

Staminate
Flower

C. UNISEXUAL INCOMPLETE
FLOWERS, MONOECIOUS

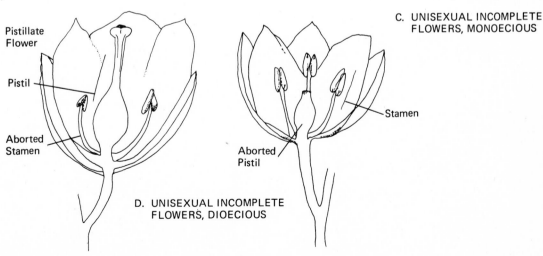

Pistillate
Flower

Pistil

Aborted
Stamen

Aborted
Pistil

Stamen

D. UNISEXUAL INCOMPLETE
FLOWERS, DIOECIOUS

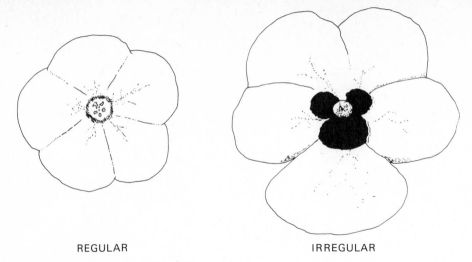

REGULAR   IRREGULAR

Fig. 2–13. Regular (Petunia) and irregular flowers (Pansy) have radial and bilateral symmetry, respectively. The latter is considered to be more advanced evolutionarily.

Fig. 2–14. Various forms for inflorescences. The main stalk of an inflorescence or of a single flower (not in a group) is termed a peduncle, while the stalks of individual flowers in an inflorescence is termed a pedicel. Most are easily distinguished, but some confusion arises with the corymb vs. the cyme. The outer flowers open first in the former, whereas the central flower opens first in the cyme.

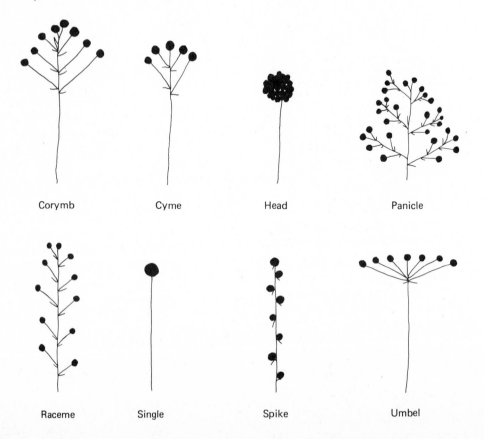

Corymb   Cyme   Head   Panicle

Raceme   Single   Spike   Umbel

Flower arrangement, both in terms of its individual structure and placement on the stem, varies (Figs. 2-13 and 2-14). Flowers that can be divided into two similar parts by more than one longitudinal plane (radically symmetrical) are called *regular* or *actinomorphic* (petunia). Flowers that can be divided into two similar parts by only one longitudinal plane (bilateral symmetry) are termed *irregular* or *zygomorphic* (pansy). Flowers may be singular or borne in groups or clusters termed *inflorescences*. The arrangement of the flowers in the inflorescence varies, which provides the basis for several recognized forms: *catkin, corymb, dichasium, head, panicle, raceme, spike,* and *umbel.* Some of the more common forms with examples are shown in Figure 2-14. Since these may have a degree of constancy for a genus or even family, the type of inflorescence is often useful in plant identification.

FRUITS

A *fruit* is the ripened ovary or group of ovaries and its contents. It can also include any parts adjacent to the ovary that are fused to it. Seeds may or may not be present in the ovary. Fruits arise only from flowering plants, since floral organs are needed for fruit production. This is the definition of the botanical fruit and should be distinguished from the popular usage of fruits and vegetables. In the popular sense some botanical fruits are called vegetables (tomato, cucumber, corn, eggplant, and squash).

Fruits may be classified as simple, aggregate, or multiple fruits. Simple fruits (Fig. 2-15) in turn may be fleshy or dry. Dry fruits are either dehiscent (open at maturity) or indehiscent (not open at maturity).

A *simple fruit* develops from the ovary of one simple or compound pistil. When these simple fruits have other floral parts fused to the ovary, they are termed *simple accessory fruits.* The ripened, enlarged ovary wall is termed the *pericarp;* it may contain up to three distinct layers from the outside and inward: the *exocarp* (skin), *mesocarp,* and *endocarp.* The soft or hard pericarp can be either dry or fleshy. Dry tissue consists of dead sclerenchyma cells with suberized or lignified walls. Fleshy tissue is composed of living parenchyma cells.

Simple dry fruits are either *dehiscent* or *indehiscent.* Dehiscent forms include the *legume* (found in the pea), which has one carpel and two dehiscing sutures (open at maturity); *follicle* (found in columbine), one carpel and one dehiscing suture; *silique* (found in mustard), two carpels and two dehiscing halves with membranous portion remaining with attached seeds; and *capsule* (found in Brazil nut), two or more carpels and dehiscing pores, slits, or top. Dehiscent forms usually release several to many seeds when opening upon maturity.

Indehiscent forms include the *achene* (found in sunflower), one carpel and one loose seed; *samara* (found in maple), a winged achene with sometimes two carpels; *nut* (found in walnut), a large achene with a thick, hard

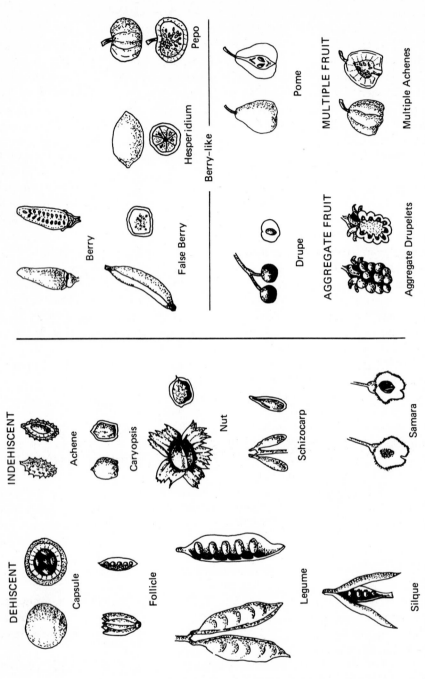

Fig. 2–15. Various types of botanical fruits. Seed portions are darkened. See text for explanation of each. Examples illustrated here are as follows: Brazil nut or *Bertholletia excelsa* (capsule), garden columbine or *Aquilegia vulgaris* (follicle), mimosa tree or *Albizia julibrissin* (legume), cole crops or cultivars of *Brassica* (silique), Transvaal daisy or *Gerbera jamesonii* (achene), sweet corn or cultivars of *Zea mays rugosa* (caryopsis), filbert or *Corylus avellana* (nut), chervil or *Anthriscus cerefolium* (schizocarp), American elm or *Ulmus americana* (samara), tabasco pepper or *Capsicum frutescens* (berry), banana or cultivars of *Musa* (false berry) lime or other cultivars of *Citrus* (hesperidium), pumpkin or other cultivars of *Cucurbita* (pepo), cherry or other stone fruit cultivars of *Prunus* (drupe), quince or cultivars of *Cydonia oblonga* (pome), blackberry or cultivars of *Rubus* (aggregate), and fig or cultivars of *Ficus carica* (multiple).

pericarp; *caryopsis* or *grain* (found in corn), one carpel and one seed, with the seed and pericarp joined at all points; and *schizocarp* (found in parsley), two or more carpels with each containing usually one seed, the carpels separate at maturity, but each retaining its seed.

Simple fleshy fruits include the berry, drupe, and pome. The *berry* (found in tomato, grape, and eggplant) has a fleshy pericarp with a thin skin. The *pepo* is a berry fruit with a hard rind (pumpkin, squash, and cucumber), and the *hesperidium* is a berry fruit with a leathery skin and radial partition (citrus fruits). It should be noted that the botanical berry is not the same as the edible fruits that are termed berries in the horticultural sense.

Drupes or *stone fruits* have thin-skinned exocarp, a fleshy mesocarp, and a stony endocarp. Examples are the coconut, olive, peach, plum, cherry, apricot, and almond. *Pomes* have an endocarp that forms a dry, paperlike core, such as the apple and pear. Since pomes have other floral parts fused to the ovary, they are simple *accessory* fruits. Other simple accessory fruits that are completely fleshy are termed *false berries* (banana, cranberry), since in the true berry the ovary is the complete and only fleshy part.

Aggregate fruits are clusters of individual fruits that develop from the several pistils of a single flower (Fig. 2-15). These pistils share a common receptacle. These fruits may consist of individual fruits, which are drupes or achenes. Examples are the raspberry, blackberry, and strawberry.

Multiple fruits (Fig. 2-15) arise from the many flowers found on a compact inflorescence. The fleshy parts of these fruits are often fused floral organs (*accessory structures*). The pineapple is an example.

Parthenocarpic fruits are seedless fruits. These result from the phenomenon called *parthenocarpy,* or the production of fruit even though fertilization does not happen. Some seedless fruits are not parthenocarpic, since they have resulted from the abortion of the young, fertilized ovaries, such as by spraying with growth regulators. Certain varieties of seedless grapes are nonparthenocarpic. Other varieties of seedless grapes and cultivated varieties of the pineapple, banana, the Washington navel orange, and some kinds of fig are parthenocarpic fruit.

SEEDS

A *seed* is a structure developed by flowering plants after fertilization, which has the potential to develop into a plant. Seeds differ so much in shape, size, seed coats, and other characteristics that it is difficult to form any meaningful classification.

Seeds (Fig. 2-16) are bounded by a *seed coat,* which has developed from the integuments of the ovule. Seed coats vary from thin and papery, such as found on a peanut, to hard and thick, such as with the coconut.

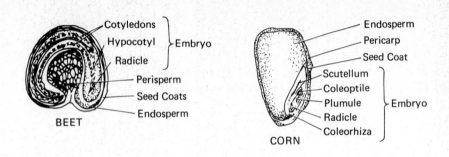

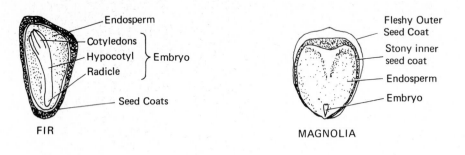

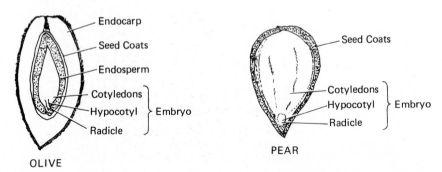

Fig. 2–16. Structures of various seeds of horticultural crops. See text for details. Perisperm (beet) is stored food like the endosperm in the embryo sac, but it is found outside the embryo sac. From Hudson T. Hartmann and Dale E. Kester, *Plant Propagation: Principles and Practices*, 3rd edition, ©1975, p. 59. Adapted by permission of Prentice-Hall, Inc., Englewood Cliffs, New Jersey.

On some seed coats the *micropyle*, the opening in the integuments of the ovule, is still seen as a small pore. The *hilum*, or scar left by the stalk that attached the seed to the *placenta*, is usually visible.

The *seed embryo* is usually fully developed when the seed is mature and ready for dispersal. A few plants, such as the orchid, have immature embryos that develop after dispersal. The structural features of the embryo are its cotyledons, epicotyl or plumule, hypocotyl, and radicle.

Cotyledons are the specialized seed leaves; one is found in seeds of monocots and two in seeds of dicots. The *plumule* or *epicotyl* is a potential
shoot apex, the *hypocotyl* a transition zone between the rudimentary
shoot and root, and the radicle a potential root apex.

A nutrient tissue, the *endosperm,* is present in many seeds; it is used
by the embryo and seedling during germination. In other seeds it has been
utilized during the development of the seed. Endosperm is rich in oil or
starch and protein. Seeds that have depleted the endosperm utilize the
cotyledons as food-storage organs.

Microstructures

The morphological structures are composed of cells, which in
turn are organized into tissues, and these in turn into tissue
systems. Each of these will be examined, starting with the basic
structure, the cell.

The Cellular Level

Cells are the structural units of plants and animals. This is a
concept that is universally accepted. Some plants are composed
of one cell, but the more complex plants contain many billions
of cells. These cells are fundamentally alike, but as they grow and differentiate, changes in size and form become commonplace. Sizes vary from
0.5 micrometers for single-cell plants to 0.5 meters for the largest bark
fiber cells. Cell forms are spherical, irregular, cylindrical, spindle shaped,
and brick shaped. Cells vary considerably in their functions. Some of these
functions include food manufacture, transportation of food and water,
food storage, reproduction, prevention of water loss, and support. There
are even unspecialized cells that give rise to other cells.

With this diversity it is apparent that there is no such thing as a typical
plant cell. However, two groups can be defined: (1) cells involved in the
metabolic activities of the plant, and (2) cells responsible for mechanical
support and the conduction of fluids. Cells involved in the latter functions
developed from the first group through growth and differentiation. Therefore, a metabolically active plant cell may be considered as a typical plant
cell for purposes of this section. A diagrammatic illustration of such a
plant cell is shown in Figure 2–17.

Plant cells, exclusive of their cell walls, are referred to as protoplasts.
These may be produced easily by enzymic digestion of the cell wall.
Protoplasts consist of two parts, the nucleus and cytoplasm. Although less
used now, the term *protoplasm* denotes the nucleus and the living cell
material in which it is situated.

The *cytoplasm,* which is 75 to 90 percent water, contains proteins,
lipids (fatty materials), carbohydrates, and lesser amounts of other organic
and inorganic compounds. The chemical compounds are present in dis-

50

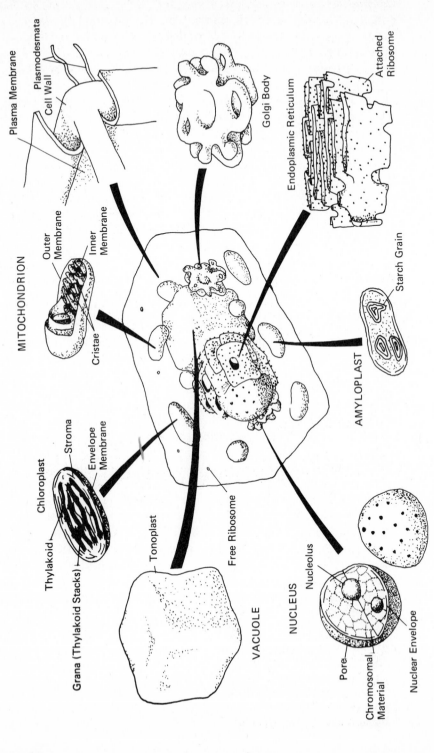

Plasma Membrane

Plasmodesmata

Cell Wall

Golgi Body

Endoplasmic Reticulum

Attached Ribosome

MITOCHONDRION

Outer Membrane

Inner Membrane

Cristae

Starch Grain

AMYLOPLAST

Chloroplast

Stroma

Thylakoid

Envelope Membrane

Grana (Thylakoid Stacks)

Tonoplast

Free Ribosome

VACUOLE

NUCLEUS

Nucleolus

Pore

Chromosomal Material

Nuclear Envelope

Fig. 2-17. A typical metabolically active plant cell showing various structures and organelles. See text for details.

solved, colloidal, and particulate states. This viscous fluid phase (sometimes called the *cytosol*) contains various specialized cytoplasmic bodies called *organelles*. Much emphasis has been placed on the cytoplasmic protein present as enzymes, since these have very important functions in the vital processes of the cell. The organelles have been studied extensively for the same reasons.

The *cell wall* surrounds the plant cell and is generally permeable to most solvents and solutes. It is a relatively rigid structure, which gives support and protection to the cell. Cell walls are produced by the cytoplasm of the cell.

Cells of immature and soft tissues have only a *primary cell wall* (Fig. 2-18). It is composed of pectin, cellulose, and hemicellulose. In fibrous and woody tissues a *secondary wall* is formed inside the primary wall (Fig. 2-18). The secondary wall contains cellulose, hemicellulose, and lignin. It has large numbers of roughly circular holes called *pits*. When a secondary wall is present, the primary wall will contain lignin, as will the intercellular space (*middle lamella*), a pectinaceous layer found between adjacent cell walls (Fig. 2-18). Lignin increases cell-wall rigidity, which accounts for the greater stiffness of fibers such as wood, hemp, and jute.

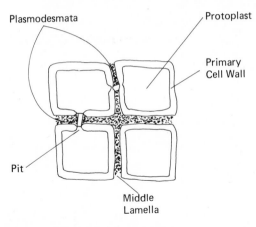

Fig. 2-18. Primary and secondary cell walls. See text for details. Proptoplasts are connected through cell wall pits by membrane-cytoplasm strands termed plasmodesmata. The middle lamella is rich in pectic materials and helps to "glue" cells together. In time the middle lamella may partially disappear, leading to intercellular spaces through which gases and water vapor may flow.

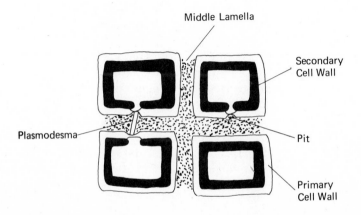

The chemical constituents of the cell walls are polymers. *Cellulose* is a homopolymer that consists of linked simple sugar units (glucose) from which one oxygen and two hydrogen atoms have been removed. Hundreds or thousands of these linked sugars are present in a single cellulose molecule. *Hemicelluloses* are heteropolymers of galactose, mannose, xylose, and arabinose (simple sugars); *pectins* are heteropolymers of modified forms of glucuronic and galacturonic acids. *Lignin* is a very complex polyphenolic polymer and is not a carbohydrate.

These cell-wall materials are useful horticultural products. Paper and its products are derived from cellulose. Pectins are used in jellies and medicines. Lignin has many uses in the synthetic rubber, food, drug, and ceramic industries. The manufacture of synthetic resin, adhesives, vanilla, pigments, and tanning agents depends upon lignin.

The *plasma membrane* (*plasmalemma*) is found immediately adjacent to the cell wall (Fig. 2-17). It is easily exposed by enzymic digestion of the cell wall. The plasma membrane contains lipids, proteins, and smaller amounts of carbohydrates. Through this limiting membrane the cell maintains selective permeability with the external environment, as well as the organization and maintenance of its internal environment.

Another membrane system of the cell is known as the *endoplasmic reticulum* (Fig. 2-17). It appears as a network of channels and vesicles bounded by membranes. The endoplasmic reticulum varies in amount, size, and shape and appears to extend from the plasma membrane to the nucleus, mitochondria, and Golgi bodies. It actually appears to be continuous with the Golgi bodies, nuclear membrane, and the plasma membrane, but only surrounds the mitochondria. Rough and smooth forms of endoplasmic reticulum are known. Ribosomes are attached to the former, but not the latter. Ribosomes, when present in aggregates, are called *polysomes*. *Ribosomes* can also be found in the cytoplasm, nuclei, mitochondria, and chloroplasts; they are involved in protein synthesis.

The function of the endoplasmic reticulum is not clear; enzymic processes may occur in them, or they may be viewed as an intracellular circulatory system. The endoplasmic reticulum is also thought to form a network between cells. The cell-wall pits are penetrated by cytoplasmic strands called *plasmodesmata.* These appear to link neighboring cells, and it is thought that the endoplasmic reticulum extends through these plasmodesmata.

Another membranous system is denoted as the *Golgi bodies (dictyosomes).* They appear as a complex organization of netlike tubules and smaller vesicles (Fig. 2-17). Evidence is strong that they are involved in cell-wall biosynthesis and intracellular transport.

The *nucleus* is a spherical or elliptical organelle bounded by the nuclear membrane (Fig. 2-17). Numerous pores are present on the nuclear membrane, which permit passage of products from nuclear biosynthesis into the surrounding cytoplasm. The nucleus contains the *chromosomes,* which are composed of *deoxyribonucleic acid* (DNA) in association with protein. These contain the genetic information for the cell. This information is transmitted to the cytoplasm by *ribonucleic acids* (RNA) for subsequent protein biosynthesis at the ribosomes. Bodies composed of RNA nucleoproteins are also found in the nucleus. They are called *nucleoli* and are involved in the formation of ribosomes. The role of DNA in determining the physiological functions of the cell through its control of morphology and metabolism in present and future generations will be discussed further in Chapters 4 and 11.

Plastids are cytoplasmic organelles present in large numbers. These disc-shaped structures can be colorless (*leucoplasts*) or pigmented (*chromoplasts*). The *amyloplast* is a leucoplast involved in starch formation and storage. The potato tuber contains numerous amyloplasts. Some yellow flowers and the red tomato owe their color to the presence of pigmented chromoplasts.

The most important plastids are the *chloroplasts* (Fig. 2–17), which contain protein, lipids, and pigments such as carotenoids and chlorophyll. Chlorophyll gives the chloroplasts and the leaves their green color. Chloroplasts are surrounded by a limiting membrane, the *envelope membrane,* which is selectively permeable. Inside they contain series of disclike membranes, which contain chlorophyll. These are the *thylakoids*, which are embedded in a protein matrix called the *stroma*. The importance of the chloroplast lies in its production of food and energy during the photosynthetic process, which is discussed in Chapter 3. Starch grains that are formed from the photosynthetic process can accumulate in the chloroplasts.

Two kinds of chloroplasts exist. Some plants such as spinach, have one type only, the *mesophyll chloroplast,* which contains thylakoids packed in stacks called *grana*. In addition to this type, other plants, such as lilies and sweet corn, have a *bundle sheath chloroplast* in which the lamellae are not stacked, but rather extend the length of the chloroplasts. These two plant groups show a difference in their photosynthetic pathways, which is discussed in Chapter 3.

Chloroplasts contain DNA, RNA, and ribosomes. Some proteins are synthesized within the chloroplast, and other proteins of the chloroplast are synthesized in the cytoplasm. Chloroplasts develop from proplastids, which are triggered into chloroplast formation by light.

Just as chloroplasts are the centers of food production, the *mitochondria* (Fig. 2–17) are the centers of energy production. Mitochondria contain lipids and protein and are bounded by an outer membrane. An inner membrane extends into the mitochondrial matrix through invagination to form folds called *cristae*. Enzymic processes related to oxidative metabolism occur in the mitochondria, which result in the formation of an energy carrier, adenosine triphosphate (ATP). DNA, RNA, and ribosomes are present in mitochondria. Evidence suggests the synthesis of some mitochondrial protein in the mitochondria, but most of the protein comes from the cytoplasm.

Vacuoles (Fig. 2-17) are liquid inclusions of the cytoplasm bound by a membrane, the *tonoplast*. The membrane is relatively impermeable, so solutes must be "pumped" into it at an energy cost (active transport). Vacuoles have possible functions such as waste repository or food storage. Vacuoles containing pigments in their cell sap also contribute to the color of various plant tissues, such as the red petals of the tulip. In a mature cell the vacuole occupies a large portion of the cell. The water content of the vacuole also helps maintain the turgidity of the cell.

Other minor inclusions are present in the cytoplasm. These are *aleurone grains* or stores of reserve protein, *oil droplets,* and enzyme-containing particles. The latter are called *spherosomes, glyoxysomes,* and *peroxisomes.*

Cell Organization and Tissues Masses of cells arise through the process of cell division. These cells enlarge and undergo differentiation (change to a more specialized cell). The end result is an organism consisting of cells that differ in their structure and physiological functions. Despite these cellular differences, groups of similar cells exist in ordered aggregates. These cellular aggregations are known as *tissues.*

Tissues can consist of *undifferentiated cells,* which are undergoing active growth and division, or *differentiated cells* that are no longer carrying out division. The first are denoted by botanists as *meristematic tissues* and the second as *permanent tissues.* Permanent tissues may be *simple,* composed of only one kind of cell, or *complex,* composed of more than one type of cell.

MERISTEMATIC TISSUES

The cells of *meristematic tissue* are physiologically young and capable of continual division. Generally, these cells have thin primary cell walls, rather large nuclei, and abundant cytoplasm; they initiate the formation of other tissues and also maintain themselves. Under some conditions, permanent tissues achieve meristematic activity, so the distinction between the two tissues is not absolute.

Meristematic tissues (Fig. 2-5), or meristems, are classified into two groups: apical meristems and lateral meristems. *Apical meristems* are found at the tips of shoots (growing points) and roots. Tissues initiated from these meristems are *primary tissues. Lateral meristems* lie along sides of roots and stems. In addition, *intercalary meristems* are zones found in elongating areas below the apex and not restricted to the sides of roots and stems. These have prolonged, but not permanent, meristematic activity and may be found in plants such as grasses. Tissues arising from the activities of lateral meristems are *secondary tissues.*

PERMANENT TISSUES: SIMPLE
AND COMPLEX

Permanent tissues result from meristematic activity. Those of one-cell type, or simple tissues, include the collenchyma, parenchyma, and sclerenchyma (Fig. 2–19).

Collenchyma is a primary living tissue composed of elongated cells with tapering ends. These cells start to develop thick primary walls of cellulose and pectin at an early age, mainly in the corners of the walls.

Fig. 2–19. Various cells involved in the formation of simple and complex tissues. The meristematic cell is found in areas of active cell division (meristems) and through differentiation matures into the other cells which compose the various simple and complex tissues. From Arthur W. Galston, *The Life of the Green Plant,* 2nd edition, © 1964, p. 32. Adapted by permission of Prentice-Hall, Inc., Englewood Cliffs, New Jersey.

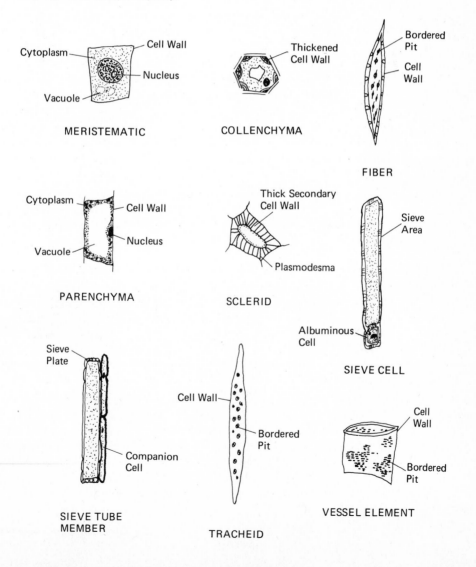

Their function of mechanical support is most characteristic of flexible, actively growing herbaceous dicots, such as sunflower (*Helianthus annuus*). Collenchyma tissue is the primary component of the ridges visible on the surface of celery petioles.

Parenchyma tissue is the least highly specialized and most abundant tissue. This abundance and the frequent presence of chloroplasts make parenchyma tissue the site of the largest part of the plant's metabolic activity. Parenchyma cells with chloroplasts are referred to as *chlorenchyma cells.* Parenchyma cells often have similar widths and lengths (isodiametric). Their walls are thicker than those of meristematic cells, and they have large vacuoles. They are living cells at maturity. The edible tissues of most fruits consist mainly of parenchyma tissue.

Sclerenchyma cells are not alive at maturity when functioning as rigid supporting tissue, such as in the nongrowing stem parts of corn (*Zea mays*). They are essentially elongated cells with tapering ends and thick secondary walls. Their size and shape differ, but their tissue form is either *fibers* or *sclerids.* The latter tend to be more isodiametrical than the fibers. Fibers are important in the manufacture of rope and cloth from plants such as hemp, jute, and cotton. Hard tissues of plants, except for wood, are composed partly or entirely of sclerids. Examples are nut shells, hard seed coats, and fruit pits. The hard grains of pears are compact clusters of sclerids known as stone cells.

Complex tissues are another form of permanent tissues. These complex or combined simple and specialized tissues are phloem, xylem, epidermis, and periderm.

Phloem may be either primary or secondary tissue; the former arises from the apical meristem and the latter from a lateral meristem, the *vascular cambium.* The function of phloem is translocation of solutes, or food conduction. The functional life of phloem is not long-lasting in many plants; it can collapse from pressures of surrounding cells.

The essential cells of the phloem are the *sieve elements,* which are regarded as the most specialized plant cells. In angiosperms, sieve elements called *sieve tube members* exist in an end-to-end arrangement in the form of a long, multicellular tube called the *sieve tube* (Fig. 2–19). *Sieve cells* are found in gymnosperms and lower vascular plants, but they do not form a definite sieve tube. Numerous pores exist at the cell ends of sieve tube members in what is known as a *sieve plate.* As sieve elements mature, the nucleus disappears. However, the cells may remain alive and active for up to a year afterward.

Sieve tube members are associated with a form of parenchyma cells, the *companion cells* (Fig. 2–19). Their role is unclear, but they may keep the sieve cells alive after the disappearance of the nucleus. Sieve cells are associated with an *albuminous cell.* Other parenchyma cells and sclerids and fibers may be part of the phloem.

Xylem, like phloem, may be either primary or secondary tissue; the origin of the primary form is the apical meristem, and the secondary form arises from the vascular cambium. Its function is water conduction. Xylem is a long-lasting tissue composed of living and nonliving cells; it is present at higher levels in woody plants than in herbaceous ones. The wood of woody plants is mostly xylem.

Tracheids and *vessel elements* (Fig. 2-19) are the cells of xylem through which water flows. Gymnosperms and the lower vascular plants do not have vessel elements in their xylem. The predominate water-conducting cells of angiosperms are vessel elements, and tracheids may only be present in some angiosperms. Fibers and parenchyma cells may also be present in xylem.

Tracheid cells are elongated, with pointed ends and not especially thick walls. *Pits* are found in the cell walls, which aid the flow of water (as cell sap) through the *lumens,* or the space formerly filled by the protoplasts. Vessel elements are shorter and wider than tracheids and have truncated ends. Their end-to-end arrangement and the eventual loss of their protoplasts and end walls lead to tube formations (*vessels*).

Disease organisms and their slimy excretions can clog xylem and phloem; the subsequent disruption of water and food movement leads to wilting and eventual death. Functional phloem usually exists in the outer portions of woody plants. Sufficient girdling disrupts the phloem, halts food movements, and results in eventual death. If only minor phloem disruption occurs, such as by ringing or scoring (see Chapter 9), the phloem regenerates. The temporary stoppage of food movement may produce dwarfing, earlier flowering and fruiting, higher yields, or higher sugar contents of fruits.

Epidermis consists of epidermal and accessory cells, trichomes (hairs), guard cells, and sometimes sclerenchyma; periderm contains cork and cork cambial cells. These are both outer, protective tissues.

Tissue Systems

Levels of organization in the multicellular plant obviously exist beyond the cell and ordered aggregates of cells discussed in the previous section. If not, the coordination of cells into a structurally organized, functional organism would not result. This section will be concerned with plant structures and functions at the supracellular level of organization. Tissue systems at this level include the ground, dermal, secretory, and vascular tissue systems.

GROUND TISSUE SYSTEM

In leaves the ground tissue system consists of only mesophyll tissue, which is found between the epidermal layers. The upper part of the mesophyll tissue (Fig. 2-9) can have elogated cells at right angles to the surface. These cells, called *palisade parenchyma,* are rich in chloroplasts.

Below the palisade parenchyma, going toward the lower epidermis, are irregularly shaped cells and large intercellular spaces. These cells constitute the *spongy parenchyma,* and the intercellular spaces there provide for gaseous exchange and transpiration. These cells contain chloroplasts also, but in fewer numbers. Collectively, the palisade and spongy parenchyma cells or indistinct parenchyma in other leaves are the *mesophyll tissue,* and their chloroplasts are *mesophyll chlorplasts.* Vascular bundles run through the mesophyll cells. A sleeve of parenchyma cells surrounds the bundle sheath. These *bundle sheath cells* may contain chloroplasts, which are designated as *bundle sheath chloroplasts.* These chloroplasts are found in plants that possess greater photosynthetic efficiency (see Chapter 3).

The ground tissue system in stems of conifers and woody and herbaceous dicots is divided into an inner part (*pith*) and outer part (*cortex*) by the symmetrical arrangement of the vascular system (Figs. 2–3 and 2–4). In the monocots the unsymmetrical arrangement of the vascular system (Figs. 2–3 and 2–4) produces no clear-cut division, so the tissue is simply called *ground tissue.* In the root of horticultural plants the ground tissue system consists only of cortex, except for some monocots that have a pith in the center of their vascular system.

The pith is entirely parenchyma cells, and the cortex is mostly parenchyma cells. Some chlorenchyma cells may be present on the outermost part of the cortex of woody twigs and herbaceous stems. This may give a greenish color to the cortex. In addition, collenchyma and/or sclerenchyma cells may occur as strands in the outer regions of stem and petiole cortex.

DERMAL TISSUE SYSTEM

The outermost layers of the cortex give rise to a lateral meristem called the *cork cambium (phellogen),* which to a lesser extent can arise in the epidermis, as discussed shortly. The cork cambium initiates the formation of *cork (phellem)* on the outside of stems on most woody and some herbaceous dicot plants. Cork is also produced by woody roots. Most of the cork cells are produced outward, although some may be produced inward (*phelloderm*). Collectively, the cork cambium and cork cells are known as the *periderm* (Fig. 2–4), which when present is part of the dermal system.

In some woody plants the formation of cork leads to the death and disappearance of the cortex and epidermis. As the outer cork cells mature, their cell walls are impregnated with *suberin,* a mixture of waves and other lipids. This prevents excessive water loss from woody stems, but also cuts off the epidermis from its food and water supply, causing it to die and scale away (see Fig. 2–4). Cork cells laid inward with time also disrupt the cortex. Outermost layers of cork and included cells die as the area of cork cambial activity shifts. Since they are unable to increase with increases of diameter, these outer corky layers crack from internal pressure to form a rough bark. *Bark* includes all tissues outside the vascular cambium. In other trees with smooth bark, the cortex is not disrupted, and the area of cork cambial activity remains stationary.

Rough, raised areas seen on the bark are *lenticels,* or loosely arranged cells produced by the cork cambium, which provide for gaseous diffusion. Lenticels are also found on stems, roots, and other plant parts.

The *epidermis* is a tissue that is usually only one cell thick (see Figs. 2-3, 2-4, and 2-9). It comprises the surface area (dermal tissue system) of young or herbaceous plant parts, except for woody stems and roots, where periderm formation results in a gradual disappearance of epidermal tissue. In woody parts of plants, the dermal system may eventually consist of only periderm. Several types of cells are found in the epidermis. Cell size and shape vary considerably among species.

Epidermis cells, except for the *guard cells,* have no chloroplasts. A pair of guard cells surrounds a pore; these structures are *stomata.* Stomata are bounded by accessory cells and then epidermal cells. Diffusion of water vapor, carbon dioxide, and oxygen into and out of the plant occurs through the stomata.

Water loss is restricted to the stomata (see Fig. 2-9), which are found mostly on leaves and in lesser numbers on stems and flower parts. Loss of water through other parts of the epidermis is prevented by the *cuticle,* a layer of waxes and other lipids secreted by the epidermal cells. This secreted material is called *cutin.*

The cuticle is absent on the epidermal cells of roots. Long projections of the root epidermis are denoted as *root hairs;* they are involved in absorption of water and nutrients. Some epidermal cells of stems and leaves may become hairs called *trichomes.*

SECRETORY TISSUE SYSTEM

Some trichomes found on the epidermis are glandular. They secrete materials characteristic of a species. In addition, other glands arise from the epidermis or tissues below it. These structures secrete complex metabolic products: mucilages, gums, rubber, resins, and essential oils.

Many of these secretions are used commercially. *Resin canals* of conifers produce a resinous substance that yields turpentine and resin. *Lactifers* are found in some vascular plants; they secrete latex, which is a milky to clear liquid. Latex from the rubber tree yields rubber; poppy latex contains well-known alkaloids, and pawpaw latex contains papain, a proteolytic enzyme used in meat tenderizers. *Essential oils,* produced by various glands, are valued for their odors and flavors.

VASCULAR TISSUE SYSTEM

Xylem and phloem are the major tissues of the *vascular system,* whose function is that of water and food conduction and, to a lesser extent, support. The structural arrangement of the stem vascular system differs for the herbaceous monocots, herbaceous dicots, woody dicots, and conifers.

Vascular systems appear as scattered or circularly arranged bundles in cross sections of stems from herbaceous monocots and dicots, respectively (Fig. 2-3). Vascular bundles of herbaceous monocotyledons have no secondary phloem or xylem, since there is no vascular cambium. In herbaceous dicots a cross section would show the vascular bundles arranged in a ring, or as a cylinder in longitudinal sections (Fig. 2-3). The ring of vascular bundles becomes a continuous cylinder of vascular tissues in some mature herbaceous dicotyledons that possess vascular cambial activity. The woody dicotyledons and conifers, because of vascular cambial activity, have a cylindrical arrangement of vascular tissues (Fig. 2-3), which appears as a ring in cross section. Ribbonlike sheets of parenchymous tissue can be observed to extend radially through the vascular cylinder in woody plants. These are *vascular rays,* which are important in food storage and lateral conduction of water and food. The vascular cylinder is arranged around a parenchymous tissue called the *pith* (Fig. 2-3). The pith is not present in herbaceous monocotyledons or roots.

The vascular systems of roots are more clearly separated from the cortex than are those of stems (Figs. 2-3 and 2-4). In roots the *pericycle* and *endodermis* (innermost limit of cortex) surround the vascular cylinders. The pericycle, a parenchymous tissue, can become meristematic and give rise to branch roots, or to lateral meristems that can increase girth and cork production in roots with secondary growth. The endodermis cells have a strip of lignin and suberin in their walls called a *Casparian strip.* Its function is unclear, but some scientists believe it plays a role in the movement of water and solutes between the cortex and vascular tissues.

The vascular system is not restricted to roots and stems, but is continuous into the petiole (leaf stalk) through the *leaf trace,* or vascular bundle that connects the vascular tissue of the stem and leaf. The vascular bundles continue into the leaf blade as veins. The vascular system is also present in flowers and fruits; the arrangement in these plant parts is quite complex.

FURTHER READING

Aronoff, S., J. Dainty, P. R. Gorham, L. M. Srivastava, and C. A. Swanson, eds., *Phloem Transport.* New York: Plenum Publishing Corp., 1975. An advanced text.

Bierhorst, D. W., *Morphology of Vascular Plants.* New York: Macmillan, Inc., 1971.

Cutter, E. G., *Plant Anatomy: Experiment and Interpretation, Part I: Cells and Tissues.* Reading, Mass.: Addison-Wesley Publishing Co., Inc., 1969.

——, *Plant Anatomy: Experiment and Interpretation, Part II: Organs.* Reading, Mass.: Addison-Wesley Publishing Co., Inc., 1971.

Esau, K., *Anatomy of Seed Plants* (2nd ed.). New York: John Wiley & Sons, Inc., 1977.

Greulach, V. A., *Plant Function and Structure.* New York: Macmillan, Inc., 1973.

Heyward, H. E., *The Structure of Economic Plants.* New York: Macmillan Inc., 1938. Good treatment of many horticultural crop plants.

Raven, P. H., R. F. Evert, and H. Curtis, *Biology of Plants* (2nd ed.). New York: Worth Publishers, Inc., 1976.

Smith, H., ed., *The Molecular Biology of Plant Cells.* Botanical Monographs Vol. 14. Berkeley: University of California Press, 1977.

Torrey, J. G., and D. T. Clarkson, eds., *The Development and Function of Roots.* New York: Academic Press, Inc., 1976. An advanced text.

Weir, T. E., C. R. Stocking, and M. G. Barbour, *Botany: An Introduction to Plant Biology* (5th ed.). New York: John Wiley & Sons, Inc., 1974.

Wilson, C. L., W. E. Loomis, and T. A. Steeves, *Botany* (5th ed.). New York: Holt, Rinehart, & Winston, 1971.

Plant Physiology and Biochemistry

The life of a plant from sexual or asexual propagation through maturity is a complex process. Growth occurs by cell division and enlargement, which serves to increase plant size. At the same time the plant develops its mature form through cellular differentiation to produce organs such as stems, leaves, roots, flowers, fruits, and seeds. Both of these steps are dependent upon a series of complex and well-integrated chemical changes. These metabolic and other processes are reasonably understood through the continuing efforts of plant physiologists and biochemists. A basic understanding of essential plant processes—photosynthesis, respiration, photorespiration, water absorption, nutrient absorption, translocation, transpiration, and other metabolic processes—will heighten the horticulturist's appreciation of the plants used and provide greater understanding of the need for specific horticultural practices.

Photosynthesis

Few chemical events in our biosphere equal photosynthesis in importance and scope. Without photosynthesis the evolution and maintenance of terrestrial life as we know it would not be possible. Land plants alone are estimated to assimilate between 20 and 100 billion metric tons of carbon dioxide per year during the photosynthetic process.

62

The complex series of integrated processes called photosynthesis, which takes place in the chloroplast, is not easily defined, since it begins with the mathematics of radiation physics and terminates in the science of ecology. In its simplest sense, *photosynthesis* is a series of processes whereby light energy is converted to chemical free energy, which is utilized for carbohydrate biosynthesis. This definition is also frequently described by the equation

$$CO_2 + 2H_2O \xrightarrow[\text{chlorophyll}]{\text{light}} (CH_2O) + H_2O + O_2$$

carbon dioxide   carbohydrate   water   oxygen
and water

Both explanations are oversimplifications, since they are only indicative of the overall process and give no idea of the photochemical or carbon-fixation steps involved during the process. In addition, only some parts of photosynthesis require light (*light phase*); the remaining reactions do not require it (*dark phase*). The dark phase is sometimes called the *Blackman reaction*. A diagrammatic representation of photosynthesis is shown in Figure 3–1.

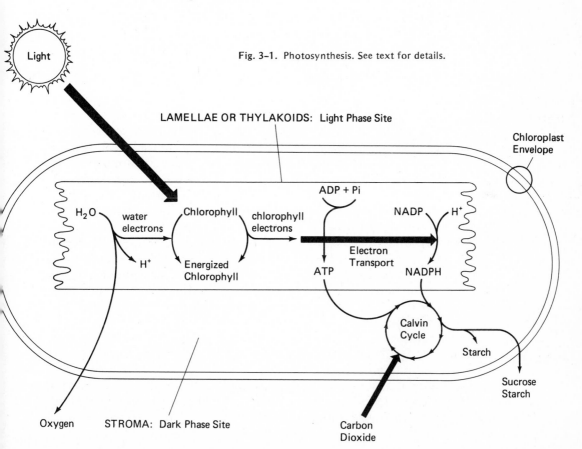

Fig. 3–1. Photosynthesis. See text for details.

Photosynthesis is initiated in the chloroplast (see discussion of chloroplast in Chapter 2) with the absorption of light by the chloroplast pigments, chlorophyll a and chlorophyll b. This process is independent of temperature. Some of this trapped radiant or photochemical energy is used to remove electrons from chlorophyll and from water; the water then splits into gaseous oxygen (O_2) and hydrogen (H^+), a weak reductant. The electrons are transferred through a series of compounds and energy is released. Some energy is utilized in the reduction of nicotinamide adenine dinucleotide phosphate (NADP) to form a source of reducing power, NADPH. These events are the well-known *Hill reaction,* where NADP is the natural Hill reagent. The remaining photochemical energy is involved in the conversion of adenosine diphosphate (ADP) to adenosine triphosphate (ATP) in a process called *photophosphorylation.* The Hill reaction and photophosphorylation are collectively known as the *light phase* of photosynthesis. This phase occurs in or on the lamellar membrane system of the chloroplast.

ATP may be viewed as an energy carrier containing "trapped" photochemical energy. This energy can be released and utilized in cellular processes, when ATP is converted back to ADP. NADPH can transfer electrons and hydrogen atoms to other compounds. Together these crucial products of the light phase are utilized during the *dark phase* of photosynthesis to supply the energy and reducing power required for the conversion of CO_2 to carbohydrate. The dark phase is temperature dependent, light independent, and takes place in the stromal part of the chloroplast.

While the light phase is the same for all horticultural crops, the dark phase is not. There are three different pathways by which CO_2 is reduced to carbohydrate (photosynthetic carbon reduction cycle); each is favored by different environmental conditions, which are reflected in the culture best suited for optimal photosynthesis in each group. The biochemistry of each pathway is illustrated in Figures 3-2 through 3-4.

The photosynthetic carbon reduction pathway common to the majority of horticultural crops is the Calvin cycle (Fig. 3-2). Plants with this pathway have less efficient photosynthesis, one type of chloroplast (mesophyll), and high rates of photorespiration (wasteful of photosynthetic intermediates; see the discussion later in this chapter). Plants with only an operative Calvin cycle are frequently referred to as C_3 *plants,* because the first product formed after the initial carboxylation contains three carbons.

A few horticultural crops, of which sweet corn is the most familiar, and many obnoxious weeds have a second pathway, the Hatch and Slack cycle (Fig. 3-3). This is in addition to the Calvin cycle. Plants of this group have both mesophyll and bundle sheath chloroplasts (see Chapter 2), more efficient photosynthesis, and low to zero rates of photorespiration. These are often called C_4 *plants* because the first product produced from the initial carboxylation has four carbons.

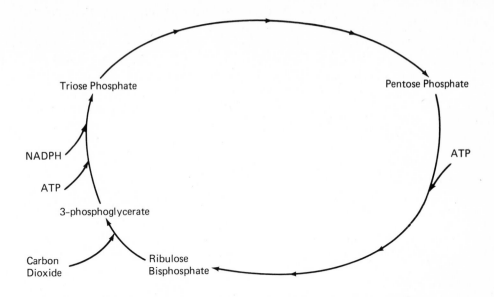

Triose Phosphate

Pentose Phosphate

NADPH

ATP

ATP

3-phosphoglycerate

Carbon
Dioxide

Ribulose
Bisphosphate

Fig. 3-2. Calvin cycle shown in simplied form; *i.e.*, a number of intermediates are left out. The cycle in C_3 plants is initiated in mesophyll chloroplasts by the addition of CO_2 (carboxylation) to a five carbon phosphorylated sugar (Ribulose bisphosphate). This splits into two 3 carbon compounds (3-phosphoglycerate), which undergo a number of reactions, eventually regenerating the starting compound, ribulose bisphosphate. Some triose phosphate is drawn off and metabolized into sucrose and starch, but enough remains to keep the pathway cyclic.

Fig. 3-3. Hatch and Slack cycle shown in simplified form. This pathway in C_4 plants starts in mesophyll cells by the carboxylation of phsophoenol pyruvate to produce a 4 carbon compound, oxaloacetate, which is converted to malate or aspartate (dependent on plant species). This compound is shuttled to the bundle sheath chloroplasts where it is decarboxylated with the released CO_2 now going through the Calvin cycle (Fig. 3-2). The pyruvate is returned and converted to phosphoenol pyruvate to restart the cycle.

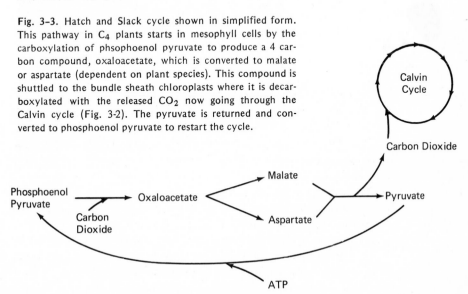

Calvin
Cycle

Carbon Dioxide

Malate

Phosphoenol
Pyruvate

Oxaloacetate

Pyruvate

Carbon
Dioxide

Aspartate

ATP

The C_4 plants with their more efficient photosynthesis are better adapted than C_3 plants to conditions of high temperature, bright light, and dryness. This explains why many weeds outcompete horticultural crops during July and August. C_4 and C_3 plants are not genus specific, as one genus may contain both. Examples of horticultural genera with both C_3 and C_4 species are *Atriplex, Euphorbia,* and *Kochia.*

Another pathway is known as CAM or crassulacean acid metabolism (Fig. 3-4). Horticultural crops with this pathway are *xerophytes* like the cacti, many succulents of arid regions, and the pineapple. These plants

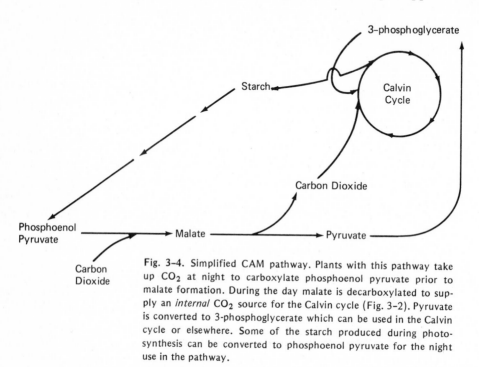

Fig. 3-4. Simplified CAM pathway. Plants with this pathway take up CO_2 at night to carboxylate phosphoenol pyruvate prior to malate formation. During the day malate is decarboxylated to supply an *internal* CO_2 source for the Calvin cycle (Fig. 3-2). Pyruvate is converted to 3-phosphoglycerate which can be used in the Calvin cycle or elsewhere. Some of the starch produced during photosynthesis can be converted to phosphoenol pyruvate for the night use in the pathway.

exist under very hot, bright, dry conditions. Their stomata are closed during the day to conserve water. Carbon dioxide enters the plant at night when the stomata are open, not during the day as in the preceding two groups. The CO_2 is fixed at night into an organic acid, which is oxidized internally during the day to produce CO_2, which is utilized to produce carbohydrate via the Calvin cycle. Photosynthesis essentially occurs then with internally supplied CO_2.

Carbohydrates produced during the photosynthetic carbon reduction cycle are usually converted enzymatically to more complex forms for initial storage. In leaves the main two storage forms are starch and sucrose. Further metabolic processes convert carbohydrates into lipids, proteins, nucleic acids, and other organic compounds. These organic molecules are assimilated into the tissues, organs, and ultimately all parts of the plant.

Introduction to Horticultural Science / Part I

Respiration

The maintenance of life requires the continuous use of energy. In plants, most of the needed energy has its origin in the photosynthetic process, whereby light energy is converted into a chemical form. This stored energy can be released by the biological oxidation of the organic compounds resulting from photosynthesis, or from compounds synthesized later from the photosynthetic products. The release of energy through the biological oxidation of organic compounds is termed *respiration.*

Respiration in plants provides for a slow, controlled release of energy with a minimal amount of heat as a by-product. In a sense it may be viewed as a slow, controlled form of combustion. Control is only possible through a number of biochemical steps catalyzed by many enzymes. About 40 percent of the energy released during the respiration of glucose is trapped in the form of an energy carrier, ATP. This energy can be utilized later or elsewhere in cellular processes, when it is released upon the conversion of ATP to ADP.

Carbohydrates are the primary substrates of respiration in higher plants. The most important ones are glucose, fructose, sucrose, and starch. Other substances to a much lesser extent may serve as respiratory substrates in certain organs or circumstances. These include the fat reserves stored in the endosperm of some seeds, organic acids, and proteins.

The most common substrate is glucose, which is oxidized in the presence of oxygen (*aerobic respiration*) to produce energy in the form of thirty-eight molecules of ATP per molecule of glucose.

$$\underset{\text{Glucose}}{C_6H_{12}O_6} + \underset{\text{Oxygen}}{6O_2} \longrightarrow \underset{\substack{\text{Carbon} \\ \text{dioxide}}}{6CO_2} + \underset{\text{Water}}{6H_2O} + \text{energy}$$

This reaction, for all practical purposes, is the reverse of photosynthesis. It is an oversimplification, since it is only indicative of the overall process and gives no idea of the sequence of reactions or intermediates that occur. Many of the intermediate compounds formed during respiration may be used as precursors for the biosynthesis of numerous cellular constituents. These will be considered in a later discussion in this chapter.

The sequence of reactions during respiratory metabolism includes phosphorylations, oxidations, hydrations, decarboxylations, group transfers, isomerizations, and cleavages. Respiratory metabolism is summarized in very brief form in Figure 3–5. Pyruvic acid is produced from glucose in a series of enzyme-catalyzed reactions collectively called *glycolysis.* This occurs in the cytoplasm of the cell.

Pyruvic acid is oxidized in the mitochondria through a series of enzymic reactions termed the *tricarboxylic acid cycle* (also known as the *Krebs* or *citric acid cycle*). The net result of this cycle is the production of carbon dioxide and hydrogen atoms; the hydrogen atoms are used in the reduction of the hydrogen acceptor, nicotinamide adenine dinucleotide

(NAD⁺), to form NADH. Some H^+ is also produced. Electrons are carried by NADH and transferred (some protons transferred at some steps) through a series of cytochromes and other electron acceptors at progressively lower energy levels. The final electron acceptor is oxygen, which in its reduced form combines with protons to form water. The energy lost during the electron transfer is trapped by the formation of ATP. This process is termed *oxidative phosphorylation* and takes place also in the mitochondria.

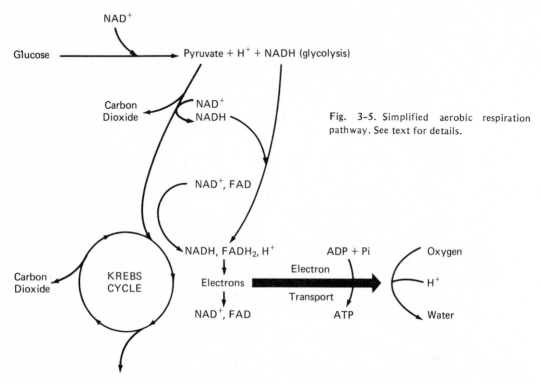

Fig. 3–5. Simplified aerobic respiration pathway. See text for details.

The respiratory metabolism described occurs when oxygen is present and is termed *aerobic respiration*. When oxygen is lacking, the respiration process becomes *anaerobic*. In that process, pyruvic acid is converted to alcohol or lactic acid. Anaerobic respiration will take place in roots deprived of oxygen because of waterlogged soil or in plants submerged by flood waters. Vascular plants generally cannot survive on anaerobic respiration and usually die. Those plants that thrive even though they have submerged roots (rice, cattails, willows) probably do not survive on anaerobic respiration, but on their ability to carry out aerobic respiration with the levels of oxygen dissolved in the water. Anaerobic respiration may occur in the center of bulky tissues or in germinating seeds until oxygen becomes available.

Gaseous exchanges in respiration and photosynthesis are in opposite directions. In respiration, oxygen is used and carbon dioxide is a by-product, as opposed to the use of carbon dioxide and the production of oxygen in photosynthesis. During darkness, no photosynthesis occurs, but in the light both processes are taking place. The question arises as to which gaseous exchange predominates when both processes are operating.

Evidence indicates that oxygen produced during photosynthesis may be used in respiration, and carbon dioxide from respiration may be used in photosynthesis. Under low light intensities a point can be reached where the two processes are balanced; that is, oxygen and carbon dioxide neither enter nor leave the leaf. The light intensity at which this happens is termed the *light compensation point.* This light intensity is less than full sunlight, so the rate of photosynthesis is faster than respiration during the day. Even including the night, when only respiration occurs, the net effect is that plants liberate more oxygen than they use and release less carbon dioxide to the air than they utilize. If this were not so, the plant would be unable to have food available for immediate use or for storage and ultimately would die. From the preceding and other studies of oxygen usage and carbon dioxide production, it is clear that plant respiration is unlikely to have an adverse effect upon the air of those rooms that are used for sleeping.

Several external and internal factors affect the rate of respiration in plants. Temperature, water, and levels of O_2 and CO_2 are the most important environmental factors. Internal factors include injury, the levels of available food, the age or kind of tissues concerned, and levels of intermediates utilized in the respiratory process. Increasing levels of O_2 and CO_2 increase and decrease, respectively, the rates of respiration. An increase in temperature causes an increase in the respiratory rate; the rate of respiration is approximately doubled for each $10°C$ rise in temperature between $5°$ and $36°C$. These factors and the effect of others present a matter of practical concern in connection with the transportation and storage of fruits, vegetables, and cereal grains, which continue to respire after harvest. This problem will be discussed further in Chapter 12.

Photorespiration

The accumulation of carbohydrate during photosynthesis is partially offset by losses of CO_2 during respiration:

$$C_6H_{12}O_6 + 6O_2 \longrightarrow 6CO_2 + 6H_2O$$

Therefore, net CO_2 fixation or "apparent" photosynthesis is equal to the gross or "true" photosynthesis minus the loss resulting from respiration. Since 90 to 95 percent of the dry weight of plants is derived from photosynthetically fixed CO_2, it is logical that plant productivity correlates closely with rates of net CO_2 fixation.

Much of the respiratory loss of CO_2 stems from the respiratory process discussed in the previous section. This form of respiration, sometimes called *dark respiration,* takes place mostly in the mitochondria and is necessary for providing energy for biochemical reactions concerned with growth and ATP synthesis. However, some respiration may be wasteful, as under some conditions, for example, high night temperatures, an excess of energy beyond the needs of the plant is produced. In addition, some plant species also lose CO_2 rapidly through another respiratory process (usually considered as CO_2 evolution) called *photorespiration,* which accounts for the higher rates of respiration observed with many plants in the light as opposed to the dark. Net CO_2 uptake is greatly diminished by photorespiration (which appears to serve no useful function) in these species, since recently fixed photosynthetic products are oxidized, often at rates three to five times greater than in dark respiration. Photorespiration, unlike dark respiration, requires light and occurs better at high levels of O_2 in the atmosphere. Like dark respiration it is very temperature dependent and is inhibited by high concentrations of CO_2.

Plants with high rates of photorespiration are C_3 species (see discussion on photosynthesis) and generally have photosynthetic rates two to three times slower than C_4 species at 25° to 35°C (77 to 95°F) and high light intensities. The C_4 plants also have low rates of photorespiration. This explains why weeds, usually C_4 plants, have a competitive edge. A few species, such as sunflower (*Helianthus annuus*) and cattail (*Typha latifolia*), are exceptions in that they are C_3 species with high rates of photorespiration, yet they also have high rates of net CO_2 fixation.

It is well established that the primary substrate of photorespiration is glycolic acid ($CH_2OH-COOH$). A number of reactions are known that produce glycolate during photosynthesis, some of which occur simultaneously. Whether one or multiples of these pathways account for the production of glycolate, and hence photorespiration, is unclear presently.

The only direct biochemical reaction of glycolate known is its enzymatic oxidation to glyoxylate ($CHO-COOH$). Glyoxylate in turn may be further oxidized enzymatically or by hydrogen peroxide into formate and CO_2. Additional CO_2 may also be formed during a step on the glycolate pathway of carbohydrate synthesis, when glycine (produced by transamination of glyoxylate) is converted to serine. The question of which one is or whether both are the main source of photorespiratory CO_2 is uncertain at present.

Since plant productivity correlates closely with rates of net CO_2 fixation, processes such as photorespiration, which diminish the assimilation of CO_2, are considered wasteful. Measurements indicate that plants with high rates of photorespiration are usually less productive than those with lower rates. Many scientists share the belief that photorespiratory processes can be regulated through biochemical means. Recent advances in somatic cell genetics (tissue culture) suggest the usefulness of the technique to selectively screen plant cells so that only those with low rates of photorespiration will survive. Control of photorespiration would appear to be a promising possibility for the future horticulturist.

Water Absorption

In vascular plants, most water absorption is from the soil through the roots (Fig. 3-6). However, epiphytes can absorb water through aerial roots, leaves, or other organs. Most water is absorbed in the root-hair zone in young roots with primary tissues, which is a short distance back from the root tip. Here the water-transport system, the xylem, is well differentiated, water permeability has not yet been reduced by suberization, and the root hairs present a large surface area for water absorption. Water absorption also occurs through root-associated fungi, the mycorrhizae.

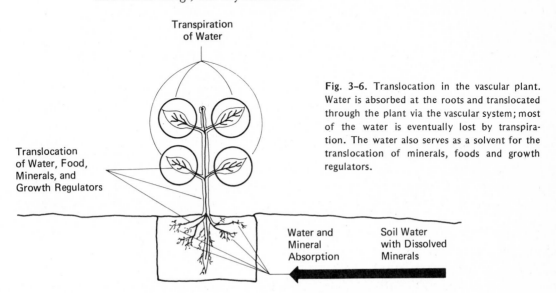

Transpiration of Water

Translocation of Water, Food, Minerals, and Growth Regulators

Fig. 3–6. Translocation in the vascular plant. Water is absorbed at the roots and translocated through the plant via the vascular system; most of the water is eventually lost by transpiration. The water also serves as a solvent for the translocation of minerals, foods and growth regulators.

Water and Mineral Absorption

Soil Water with Dissolved Minerals

Older roots with secondary tissues impregnated with suberin can also absorb water through breaks and cracks in the suberized secondary tissues. This is particularly true for large trees, in which the suberized tissues are a large part of the total root system, and in winter when unsuberized root surfaces are minimal.

Water may be absorbed by two mechanisms: *active* and *passive absorption. Active absorption* of water implies an essential participation of the roots in water absorption. This process occurs when the soil moisture is high and the rate of transpiration or loss of water from the plant tissues as vapor is low (see discussion on transpiration rates in this chapter). Active absorption of water constitutes a smaller part of the total water absorption of the plant than does passive absorption.

Active absorption generates root pressure, and a number of theories have been proposed to explain the active absorption–root pressure complex. These are the *secretion theories, electroosmotic theories,* and *osmotic theories.* Evidence favors the osmotic viewpoint.

The inward movement of water across the differentially permeable membranes of the living cells of the root by osmosis implies that a higher solute concentration is present in the xylem vessels than in the soil solution. Undoubtedly, solutes leak or are secreted from the living root cells into the adjacent xylem tissues. The resultant diffusion of water into the xylem cells of the root is responsible for root pressure. This root pressure is responsible for the loss of liquid water drops from the leaf (*guttation*) and the bleeding of xylem sap from cut or broken stems. If the solute level in the soil solution becomes higher than that in the xylem vessels, which can result from deicing salts or too much soluble fertilizer, the direction of water movement by osmosis will reverse. Unless checked, this will result in wilting, desiccation, and death of the plant.

Passive absorption of water occurs when plants are transpiring rapidly. It does not require participation of the root cells, but comes about from forces arising at the top of the plant. In fact, passive absorption of water will take place with a dead root system. Active absorption of water does not happen during rapid transpiration, since the rapid movement of water from passive absorption dilutes and washes away the solutes in the xylem upon which active absorption depends.

The flow of water in the passive mode is dependent upon two important properties of water molecules: adhesion and cohesion. Water molecules hold firmly to molecules of other substances (*adhesion*) that contain large numbers of oxygen or nitrogen atoms. Water molecules also hold very firmly to each other (*cohesion*), which accounts for the surface tension of water.

Upward movement of water from the root xylem to the leaves, or the transpiration stream, is initiated by the loss of water vapor through the stomata. This causes local tissue dehydration, and the movement of water into these tissues results because of the cohesive properties of water. This in turn causes partial dehydration deeper into the plant tissues, which brings about water diffusion from the leaf xylem. Again the cohesive property of water in turn helps to "pull" the water through the xylem of the stem and eventually the root. This generates a presssure potential between the roots and the soil solution. This hydrostatic pressure differential is the driving force for the movement of water into the roots. The adhesive property of water prevents the too rapid removal of water from the xylem, and along with cohesion maintains continuity of the water column.

Nutrient Absorption

Sixteen elements are necessary for plant growth. Three of these, carbon, hydrogen, and oxygen, are drived from water, oxygen, and carbon dioxide. The remaining thirteen are absorbed by roots as inorganic salts from the soil solution. They are nitrogen, potassium, calcium, magnesium, phosphorus, sulfur, iron, copper, manganese, zinc, molybdenum, boron, and chlorine (Fig. 3–6).

All these mineral elements are present in the soil solution as ions or absorbed onto soil particles. To reach the interior of the root, these ions have to cross several layers of different substances. The mechanisms by which this passage occurs are complex and not as well understood as many other plant processes.

The ions first cross the cell walls of the root hairs and other root tissues. Plant cell walls offer little hindrance to the movement of gaseous molecules, dissolved nutrients, or ions, except for having relatively weak attractive forces for *cations* or positively charged ions (K^+, Mg^{2+}, Ca^{2+}, and others). These cations move across the cell wall in a leapfrog manner. Cations absorbed at a negatively charged site can be displaced by another cation with a higher absorptive affinity for that site. The released cation will move inward to another cation absorptive site. Negatively charged ions (*anions*), some cations, gases, and dissolved nutrients will diffuse through the cell wall dissolved in water. The area comprised by the cell walls is termed *free space* because of unopposed ion movement.

Once the ions reach the plasma membrane, they encounter an effective barrier or nonfree space. This probably does not occur until the ions reach the endodermis around the root's vascular system. Ion movement across this and subsequent membranes occurs primarily by active transport. Gases, small polar molecules, and fat-soluble molecules can cross the plasma membrane or others by passive transport. These two processes are differentiated as follows. Membranes may be visualized as gateways with selective or discriminatory abilities, which allow passage of ions, gases, neutral molecules, and the like, at rates of passage that vary from very low to rapid.

Passive transport refers to transport resulting from physical driving forces. The movement or diffusion of solutes from an area of higher concentration to an area of lower concentration is a form of passive transport. Some molecules may cross the plasma membrane or other membranes by passive transport. These molecules include gases such as oxygen or carbon dioxide, small polar molecules like ammonia, and fat-soluble molecules. It is speculated that the first two types may move through the membrane, if it acts as a *molecular sieve;* the fat-soluble molecules are soluble in the lipid portions of membranes, so it is suggested that diffusion occurs with the membrane acting as a *selective solvent.*

Active transport is dependent upon metabolic driving forces. Most of the ions probably cross the plasma membrane or other membranes by active transport. Evidence indicates the involvement of carriers, which bind ions, transport them across the membrane, and discharge them on the other side. These carriers have properties which suggest that they are large protein molecules analogous to enzymes. Evidence strongly suggests that Ca^{2+} is transported across plasma membranes in roots by a carrier that is thought to be adenosine triphosphatase (ATPase), which is a well-characterized enzyme.

Once into the cytoplasm, diffusion brings about further movement of the ions. Here the ions may undergo several fates, which result in their removal. This is necessary if ion absorption by the roots is to continue. Ions may be transformed during the course of a metabolic reaction into another substance. They may be absorbed by protein molecules or transported both in the active and passive mode through other cells until they reach the xylem. They may also be accumulated in vacuoles or cellular organelles.

Translocation

The movement of water, minerals, and food from one part of a plant to another is called *translocation* (Fig. 3-6). Minerals absorbed by the roots are translocated to the stems, leaves, and reproductive organs of the shoots. Once there, minerals may be further translocated up or down the vascular system, such as from older leaves to younger leaves. Fertilizers applied as foliar sprays are translocated to metabolically active tissues. Most minerals are translocated in the form of ions; however, nitrates absorbed in the roots are converted there and translocated as amino acids.

Most food is translocated in the form of sugars. Much of the sugar is translocated as sucrose, except for a few species in which sorbitol, raffinose, stachyose, and verbascose are also present in large quantities. The predominant flow is from leaves to metabolically active tissues, such as meristems or areas of growth, where sugar demand exceeds the synthesized supply. Translocation of sugars from storage organs after hydrolysis of starch also occurs when the leaf supply is insufficient.

Other substances can be translocated, but to a much lesser extent. Amino acids are translocated; the peak flow is probably from senescent leaves prior to abscission. Auxins, vitamins, and other plant-growth substances are translocated through the vascular system. Although their concentration is low, their translocation has a profound effect upon the development of the plant.

Xylem is clearly the part of the vascular system through which water flows. Minerals absorbed at the roots are translocated upward primarily through the xylem. Some lateral movement of minerals occurs from the xylem to the phloem. Minerals can also be translocated from older leaves through the phloem in an upward or downward direction. Some translocation of minerals through the phloem may be important in deciduous trees in the winter, when water flow through the xylem is slow.

Small amounts of sugars may move upward in the xylem during certain seasons, but the bulk of the sugar translocation, both upward and downward, takes place in the phloem. Sugars may also move laterally through the vascular rays. Amino acids, organic acids, soluble proteins, auxins, and vitamins are translocated through the phloem.

Translocation through the xylem is primarily upward and dependent upon passive transport. Minerals and small amounts of other solutes are simply carried along with the upward flow of water. In the section on water absorption, it was indicated that water movement was dependent upon the loss of water through the stomata and the resulting transpiration stream.

Phloem translocation has been explained by several hypotheses. The most widely accepted is the pressure-flow hypothesis. Sugar movement from the site of availability, the *source,* to the site of utilization, the *sink,* is dependent upon a concentration gradient. First, the sugar leaves the parenchyma cells by active transport adjacent to the phloem and enters the sieve tubes. This increases the solute concentration in the sieve tubes, causing osmotic movement of water from the xylem into the sieve tubes. Water movement carries the sugar along the sieve tubes to the sink. Adjacent parenchyma cells remove the sugar by active transport, and the reduced solute level causes water movement by osmosis out of the sieve tube.

Transpiration

Only about 1 percent or less of the water absorbed by plants is utilized in biochemical processes such as photosynthesis and hydrolysis, in the hydration of cell walls, or in the maintenance of the swollen conditions of cells (*turgor*) through internal water pressure (*turgor pressure*). Most of the water is lost from the aerial parts of the plant by evaporation, followed by diffusion of water vapor into the air. This process is called *transpiration* (Fig. 3-6).

Much water is lost through the stomata and, to a much lesser extent, the cuticle and lenticels. The major part of stomatal transpiration is from the leaves. Under conditions of high soil moisture and humidity, a small fraction of water may be lost by *guttation,* that is, as exuded drops from terminal ends of veins in the leaf margin.

When the loss of water by transpiration exceeds that of replacement by absorption, a water deficit occurs within the plant. Water deficits reduce the turgor pressure of the plant cells, and eventually the leaves and herbaceous stems will droop. A decrease in turgor pressure of the guard cells does cause the stomatal aperture to decrease and slow wilting, but environmental factors may be such that this stomatal response is too slow to prevent wilting. Wilting is not observed in plants with extensive mechanical support, such as in magnolia. Wilting can be either temporary or permanent. When adequate soil moisture is present, a water deficit can result in *temporary wilting,* which disappears at night when the rate of transpiration slows. Inadequate soil moisture produces *permanent wilting,* which can cause death by desiccation if prolonged. Recovery from permanent wilting is possible if water is added to the soil. Transpiration rates are usually higher in the day than at night. An exception is many succulents with crassulacean acid metabolism (see discussion of photosynthesis), since their stomata are open at night, rather than during the day. Several environmental factors affect the rate of transpiration: internal plant factors, light, humidity, temperature, wind, and soil water content. These will be discussed in Chapter 5.

Other Metabolic Processes

Metabolism in the plant is the sum total of all the biochemical reactions that take place in the plant body (Fig. 3–7). Most of these reactions are catalyzed by enzymes. Synthetic metabolism is called *anabolism;* degradative metabolism is termed *catabolism.* Photosynthesis and respiration are examples of anabolic and catabolic processes, respectively. Growth occurs only when anabolism exceeds catabolism. Some of the more well-known metabolic processes have been discussed already in this chapter. However, other equally important forms of plant metabolism exist, and brief discussions of these will follow. No more than a brief, simplistic view of these many metabolic pathways is possible, since a complete consideration would require a book in itself. Many of the metabolites involved are important enough to man to be or to have been isolated from plants.

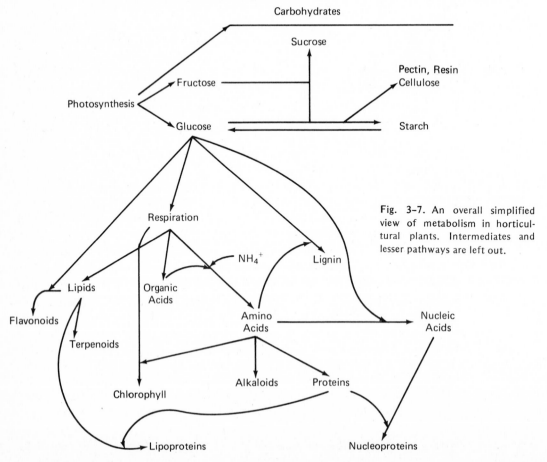

Fig. 3–7. An overall simplified view of metabolism in horticultural plants. Intermediates and lesser pathways are left out.

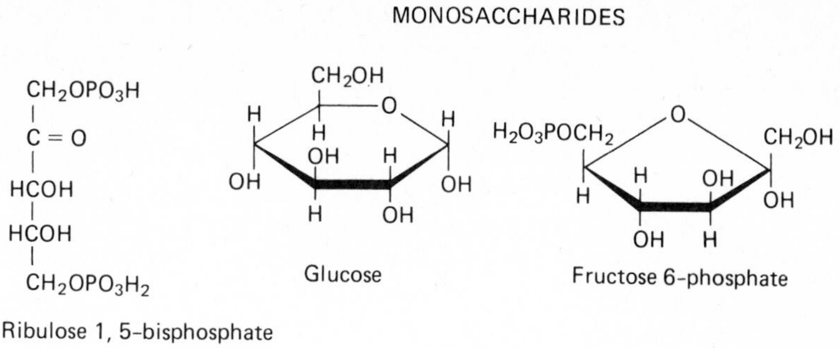

$$CH_2OPO_3H$$
$$|$$
$$C = O$$
$$|$$
$$HCOH$$
$$|$$
$$HCOH$$
$$|$$
$$CH_2OPO_3H_2$$

Ribulose 1, 5–bisphosphate

Glucose

Fructose 6–phosphate

Fig. 3–8. Some examples of plant carbohydrates.

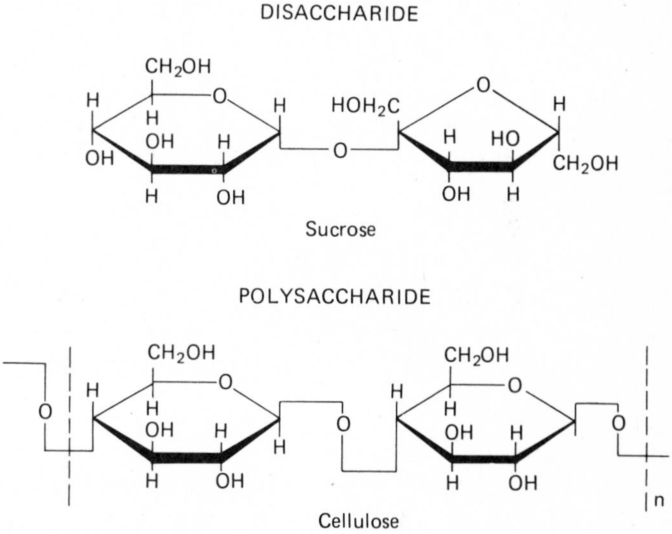

DISACCHARIDE

Sucrose

POLYSACCHARIDE

Cellulose

CARBOHYDRATES

The ultimate source of the carbon in carbohydrates is atmospheric carbon dioxide. Simple sugars arise from the fixation of carbon dioxide during photosynthesis and from pyruvic acid by what is essentially a reversal of glycolysis (see respiration). These simple sugars (Fig. 3–8), or *monosaccharides* (glucose, fructose, ribose, and others), are produced in the form of monosaccharide phosphates during the two processes just mentioned. These can be converted to free sugars by additional enzymic reactions. The monosaccharide phosphates can undergo several interconversions through further enzymic processes.

Disaccharides (Fig. 3–8) are produced when two monosaccharides are linked together. Sucrose, the major product of photosynthesis, is a disaccharide composed of glucose and fructose. Sucrose is the major translocation form of carbohydrate in plants. It is a significant storage form in some plants, such as sugarcane and sugar beet, which are major sources of sucrose.

Polysaccharides (Fig. 3-8), or long chains of carbohydrates, result from the enzymic linkage of many monosaccharides. Starch and cellulose, horticultural products of commercial importance, are the most prevalent polysaccharides. Starch is synthesized in chloroplasts and amyloplasts; it is an important storage carbohydrate. Cellulose is an important cell-wall carbohydrate. Other polysaccharides are hemicellulose and pectin, which are also cell-wall constituents.

Storage forms of carbohydrate can be degraded enzymically in the plant by a process called *digestion*. After digestion the resulting soluble foods can be assimilated, respired, or transported elsewhere. Starch is broken down to glucose when needed, such as during seedling germination or development of vegetative tissue. Inulin, a polysaccharide accumulated instead of or in addition to starch in some plants, is degraded to fructose. Pectin is degraded to galacturonic acid.

PROTEINS

Nitrogen is taken up by plants mostly as the nitrate ion (NO_3^-). In the legumes and a few other vascular plants, nitrogen is fixed by symbiotic bacteria located in root nodules. These bacteria convert free nitrogen to ammonium ion (NH_4^+), which is readily used in subsequent metabolic pathways. Nitrate, once absorbed by plants, is also enzymically converted to NH_4^+.

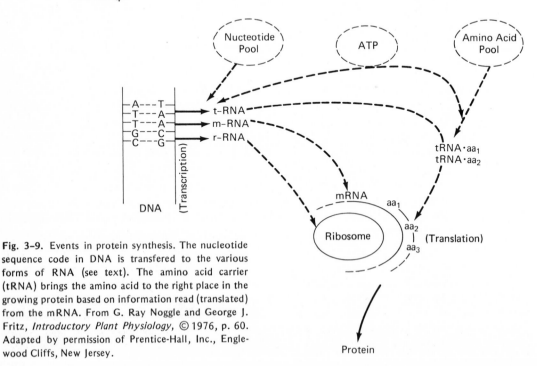

Fig. 3-9. Events in protein synthesis. The nucleotide sequence code in DNA is transfered to the various forms of RNA (see text). The amino acid carrier (tRNA) brings the amino acid to the right place in the growing protein based on information read (translated) from the mRNA. From G. Ray Noggle and George J. Fritz, *Introductory Plant Physiology*, © 1976, p. 60. Adapted by permission of Prentice-Hall, Inc., Englewood Cliffs, New Jersey.

Ammonium ion does not accumulate, since it is used rapidly in the enzymic synthesis of amino acids. The initial synthesis of amino acids is the reaction of α-ketoglutaric acid (carbon framework derived from glucose, a photosynthetic product), which arises from the tricarboxylic acid cycle (see respiration), with NH_4^+ to produce glutamic acid:

$$
\begin{array}{cccc}
\text{COOH} & & \text{COOH} & \\
| & & | & \\
\text{C}=\text{O} & & \text{H}-\text{C}-\text{NH}_2 & \\
| & & | & \\
\text{CH}_2 & + NH_4^+ + NADH \rightleftharpoons & \text{CH}_2 & + NAD^+ + H_2O \\
| & & | & \\
\text{CH}_2 & & \text{CH}_2 & \\
| & & | & \\
\text{COOH} & & \text{COOH} & \\
\text{α-Ketoglutaric acid} & & \text{Glutamic acid} &
\end{array}
$$

Other amino acids are synthesized by transamination, or the transfer of an —NH_2 group:

$$
\begin{array}{ccccccc}
\text{COOH} & & \text{COOH} & & \text{COOH} & & \text{COOH} \\
| & & | & & | & & | \\
\text{H}-\text{C}-\text{NH}_2 & & \text{C}=\text{O} & & \text{C}=\text{O} & & \text{H}-\text{C}-\text{NH}_2 \\
| & + & | & \rightleftharpoons & | & + & | \\
\text{CH}_2 & & \text{CH}_2 & & \text{CH}_2 & & \text{CH}_2 \\
| & & | & & | & & | \\
\text{CH}_2 & & \text{COOH} & & \text{CH}_2 & & \text{COOH} \\
| & & \text{Oxaloacetic} & & | & & \text{Aspartic} \\
\text{COOH} & & \text{acid} & & \text{COOH} & & \text{acid} \\
\text{Glutamic} & & & & \text{α-Ketoglutaric} & & \\
\text{acid} & & & & \text{acid} & &
\end{array}
$$

There are several enzymes, transaminases, that catalyze the various transaminations needed for the different amino acids. Amino acid metabolism is dependent upon nitrogen and photosynthetic carbohydrate, so both carbohydrate and protein metabolism are linked. These compounds are also the most abundant in plants. Any environmental factor that affects one indirectly affects the other, thereby resulting in an extensive effect upon plant development.

About twenty or so amino acids are combined in various ways to synthesize *proteins* (Fig. 3-9). A larger number (over 200) exist only as free amino acids; the metabolic role of the nonprotein amino acids is unclear. The sequence of the amino acids in the proteins is determined by the arrangement of the nucleotides (see other nitrogenous compounds in this chapter) in the *DNA molecule.* Various combinations of three nucleotides code for the different amino acids. A form of RNA, messenger RNA, reads the triplet sequence and carries this information to the cytoplasm. The information is contained in the nucleotide sequence of the mRNA, which becomes attached to the ribosomes. Amino acids are carried and oriented to the ribosomal template by another form of RNA, *transfer*

RNA. This form of RNA matches the correct amino acid to the portion of the ribosomal template that codes for it. The amino acids are polymerized through the formation of peptide bonds.

Proteins can also be enzymically degraded into their constituent amino acids. Both synthesis and degradation probably occur simultaneously. The meat tenderizer, papain, is an enzyme derived from plants that is used by plants for protein degradation. In actively growing tissue, protein synthesis exceeds breakdown. However, protein degradation occurs to a larger extent during dehydration, senescence, and starvation.

LIPIDS

Lipids (Fig. 3-10) are much more diverse than either carbohydrates or proteins, since they are not composed of repetitive, well-defined monomeric units. As a group they are defined more in an operational sense; that is, they are insoluble in water but soluble in "fat" solvents such as chloroform, benzene, and petroleum ether. As such, lipids include the fats, phospholipids, glycolipids, waxes, and sterols.

Triglyceride

Phospholipid

Fig. 3-10. Some examples of plant lipids.

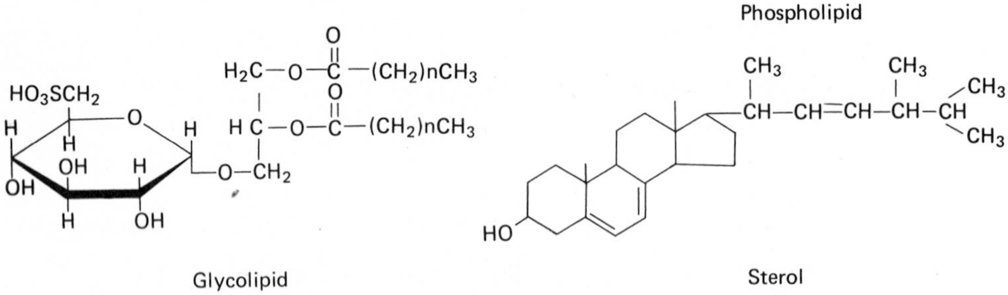

Glycolipid

Sterol

Fats are found in plant cells, but the highest concentrations are present in the endosperms of seeds. If the fat is a liquid at room temperature, it is called an *oil.* The oils derived from corn, soybean, olive, sunflower, cottonseed, peanut, coconut, and safflower are horticultural products of great value.

The synthesis of fats can be summed up in three steps: the production of glycerol from carbohydrates, the production of fatty acids, and the esterification of glycerol with the fatty acids. This fat is correctly called a *triglyceride,* since three fatty acids are esterified to the glycerol.

Phospholipids and *glycolipids* differ from triglycerides in that the third —OH of the glycerol is substituted with a group other than a fatty acid. For phospholipids this group is either phosphoric acid or a substituted phosphoric acid, and for glycolipids it is a sugar or a substituted sugar. Glycerol can be phosphorylated to glycerol phosphate, which in turn can be esterified with two fatty acids to produce phosphatidic acid. Enzymic reactions of glycerol with two fatty acids followed by interactions with nitrogenous compounds or sugars lead to the synthesis of phospholipids and glycolipids.

Plant *waxes* are long-chain esters of fatty acid and alcohols of great complexity. Most of the waxes are found in the cutin and some in suberin. *Sterols* are complex alcohols with a tetracyclic ring structure. The best-known plant sterol is erogosterol, which can be converted to vitamin D by irradiation.

Lipids can be enzymically digested to glycerol and fatty acids. Further metabolism of glycerol can produce sugars for respiration or other uses. Fatty acid may be oxidized to produce ATP. For example, lipid digestion occurs during seed germination.

OTHER NITROGENOUS COMPOUNDS

Much of the nitrogen in the plant is present in the amino acids and proteins. A few phospholipids, glycolipids, and carbohydrates also contain nitrogen. Nitrogen is also found in a wide variety of other minor compounds. These include purines, pyrimidines, nucleic acids, porphyrins, alkaloids, vitamins, coenzymes, and hormones.

Purines and *pyrimidines* (Fig. 3-11). are heterocyclic nitrogenous bases. They are involved in the synthesis of a number of essential compounds, such as plant hormones, nucleic acids, DNA, RNA, ATP, ADP, coenzymes, vitamins, and alkaloids. A considerable number of purines and pyrimidines are synthesized by plants, but the most important are adenine, cytosine, guanine, thymine, and uracil.

Nucleic acids are DNA, the genetic code carrier, and RNA, the genetic code transcriber. Nucleic acids are usually linked to proteins, forming compounds called *nucleoproteins.* The nucleic acid portion is a high-molecular-weight polymer of nucleotides; each *nucleotide* consists of a purine or pyrimidine, a sugar, and phosphoric acid. Both DNA and RNA contain the purines, adenine and guanine, and the pyrimidine, cytosine. However, DNA and RNA differ, too. DNA has the pyrimidine, thymine, and the sugar, deoxyribose, whereas RNA has the pyrimidine, uracil, and the sugar, ribose.

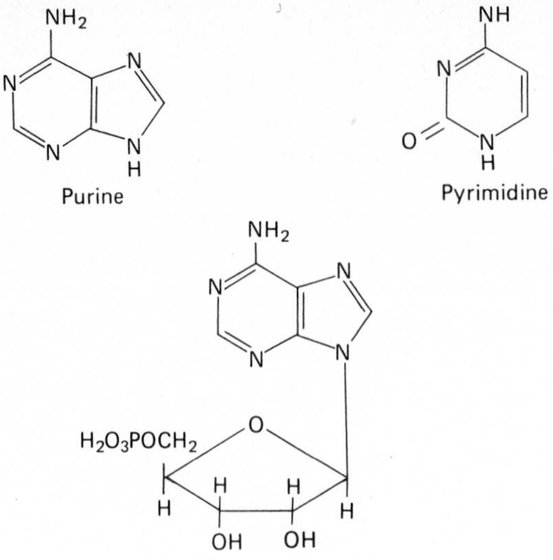

Purine

Pyrimidine

Nucleotide

$H_2O_3POCH_2$

OH   OH

Fig. 3-11. Purines and pyrimidines.

Porphyrins (Fig. 3-12) are composed of pyrrole rings. The best-known plant porphyrins are chlorophyll and phytochrome. These are involved in photosynthesis and photomorphogenetic responses, respectively. Porphyrins are also found in cytochromes, which are involved in electron transport during photosynthesis, and in enzymes such as catalase, which converts hydrogen peroxide to oxygen and water.

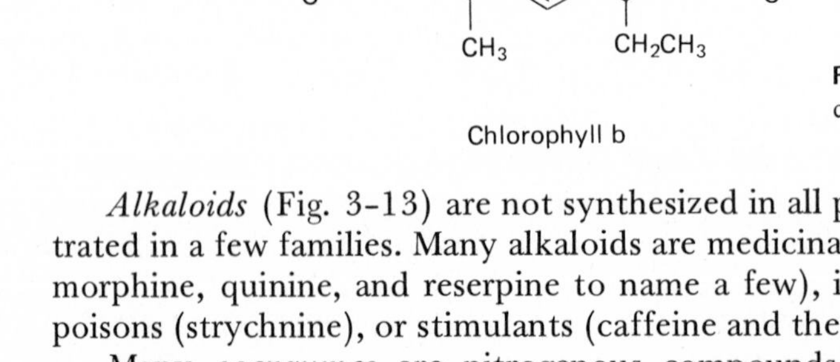

$[C_{20}H_{39}O]$

Chlorophyll b

Fig. 3-12. A plant porphyrin: chlorophyll b.

Alkaloids (Fig. 3-13) are not synthesized in all plants, but are concentrated in a few families. Many alkaloids are medicinals (atrophine, cocaine, morphine, quinine, and reserpine to name a few), insecticides (nicotine), poisons (strychnine), or stimulants (caffeine and theobromine).

Many *coenzymes* are nitrogenous compounds, such as NADP (see photosynthesis). Vitamin B (thiamin) and vitamin B_2 (riboflavin) contain nitrogen (Fig. 3-14), as well as plant hormones (see Chapter 4) like indole-3-acetic acid and kinetin.

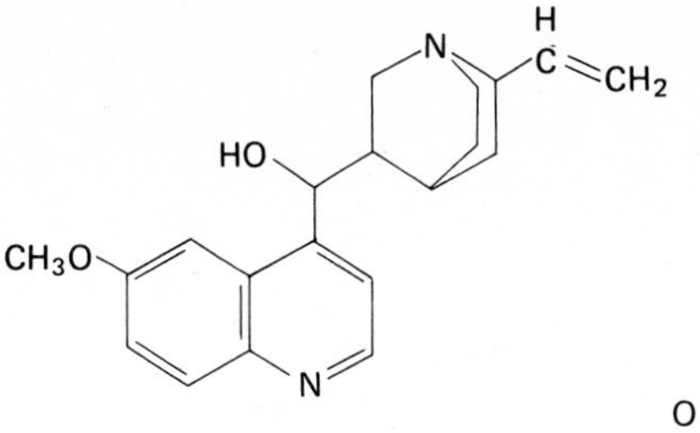

Quinine

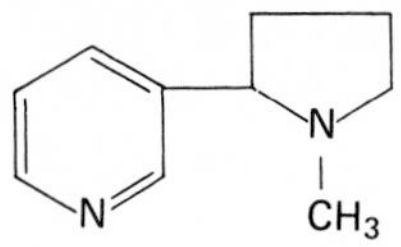

Nicotine

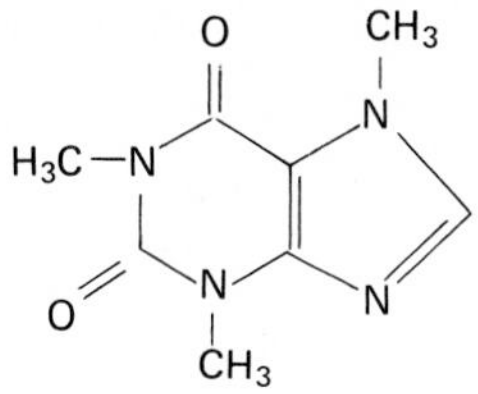

Caffeine

Fig. 3-13. A few well-known plant alkaloids.

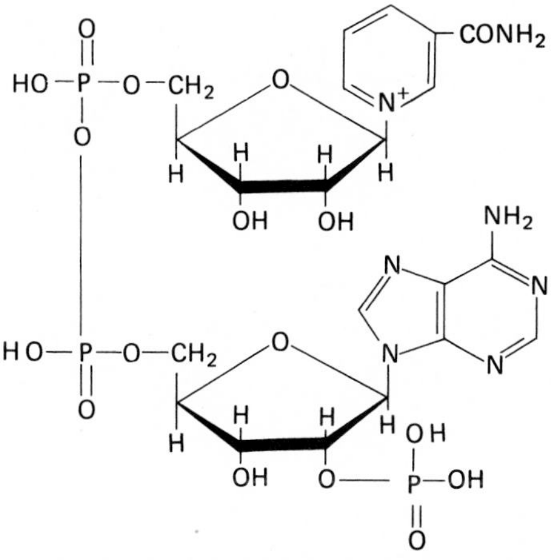

Nicotinamide Adenine Dinucleotide Phosphate

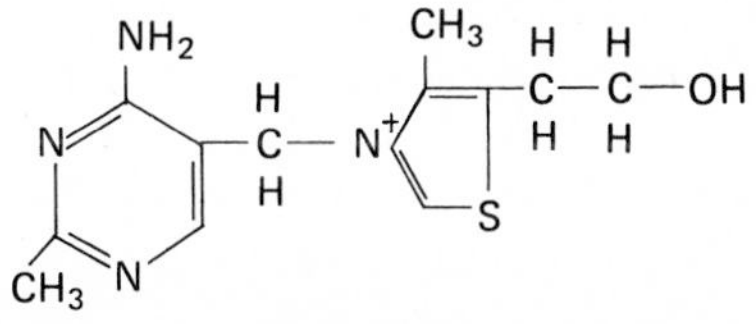

Thiamine

Fig. 3-14. A few plant coenzymes.

OTHER ORGANIC COMPOUNDS

Terpenoids (Fig. 3-15) are an abundant and diverse group of compounds found in plants. Some have important metabolic functions, and others appear to have no known function. As a group, terpenoids include terpenes, essential oils, sterols, pigments, glycosides, carotenoids, and many other related compounds. The isopentane unit is a basic building block for all terpenoids. Added head to tail, long chains can be produced, and cyclization of these chains can produce the ring structure found in some terpenoids.

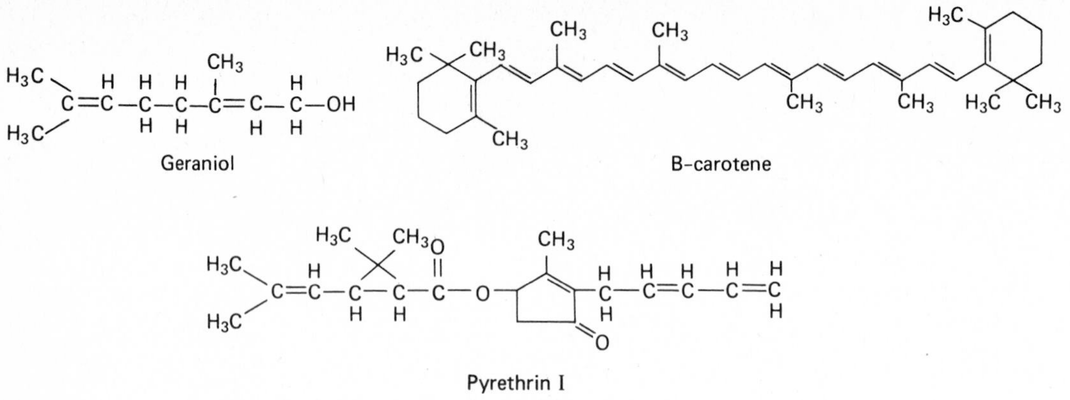

CH3

H3C — C=C — C — C — C=C — C — OH (Geraniol)

Geraniol

H3C, CH3 β-carotene

Pyrethrin I

Fig. 3-15. Terpenoids. Geraniol is an essential oil, β carotene is a carotenoid, and pyrethrin is a mixed terpenoid.

Terpenes of note include the resins, essential oils, rubber, hormones, and carotenoids. Terpenes containing up to twenty carbons are volatile oils. Many are valuable horticultural products. Some of these are *essential oils,* like lemon, rose, peppermint, and lavender oil, and others are *resins,* such as those found in pine trees. Resins and essential oils can be found in combined forms like turpentine. Plant *hormones,* such as gibberellins and abscisic acid, are derivatives of terpenes with less than twenty carbons. Terpenes with forty carbons are termed *carotenoids,* which are yellow to red pigments found in chloroplasts and other plant parts. One plant carotenoid, B-carotene, is a source of vitamin A. Terpenes with many thousands of carbons exist. The best known is rubber, which is derived from the latex of *Hevea brasiliensis.*

Other terpenoids are the sterols (see lipids), the glycosides (some are valuable medicinals such as digitalis), some alkaloids, and the bitter principles (responsible for the bitter taste in cucumbers and citrus fruit). Mixed terpenoids containing a sugar or fatty acid also exist. Pyrethrin I, an important natural insecticide, and cannabidiol, the active ingredient of marihuana, are well-known mixed terpenoids.

Organic acids (Fig. 3-16) arise during some metabolic pathways, such as the tricarboxylic acid cycle (see respiration). Some accumulate in various plant organs, like shikimic and malic acids in apples or citric acid in citrus fruits and grapes. Organic alcohols, ketones, aldehydes, and esters exist; some are part of the characteristic flavor and aroma of fruits.

Another group is denoted as *aromatic compounds* (Fig. 3-17), based on the presence of a benzene ring in their structure. Those in plants include the phenolics, the flavonoids, lignin, and tannin. Some of the aromatics are derived from acetate and others from shikimic acid, a key intermediate in the synthesis of aromatic amino acids. Some of the simpler *phenols* are responsible for the odors and flavors of plant oils, many of which are valuable horticultural products. The *flavonoids* are a group of pigments consisting of *anthocyanidins* and *flavonols*. They contribute color to some plant organs, but their metabolic role is not clear. *Lignin* is a cell-wall constituent (see Chapter 2), and the function of *tannin* is unknown. Tannin is used in the tanning of leather. The taste of tannin is bitter, and its higher level in unripe fruits, such as persimmons and plums, gives rise to their astringent tastes.

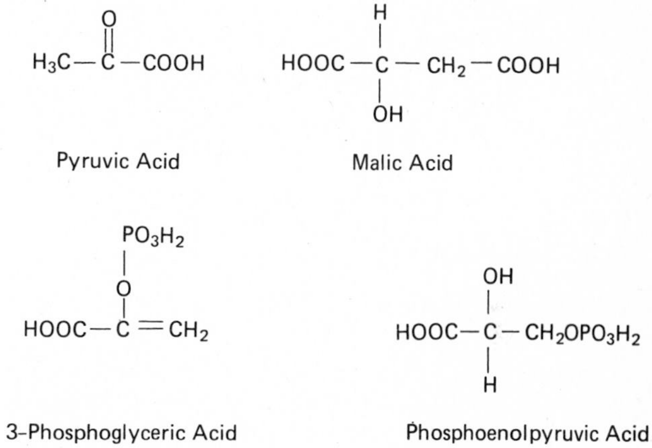

Fig. 3-16. Some organic acids involved in photosynthetic metabolism (see fig. 3-2, 3-3, 3-4).

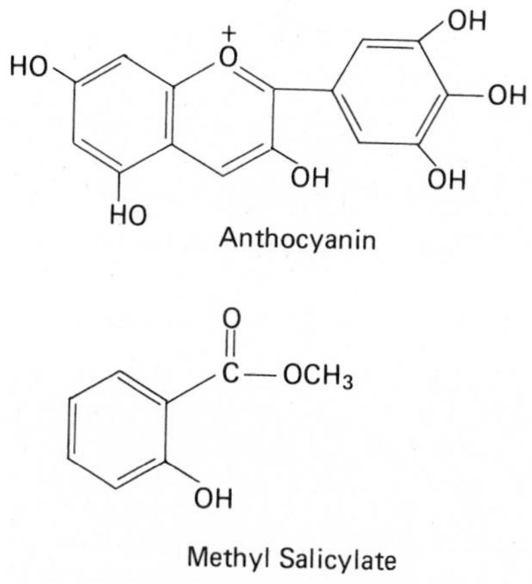

Pyruvic Acid

Malic Acid

3-Phosphoglyceric Acid

Phosphoenolpyruvic Acid

Anthocyanin

Methyl Salicylate

Fig. 3-17. Some aromatic compounds found in plants. Anthocyanin is a pigment involved in part for red, pink, mauve and blue coloration in horticultural plants. Methyl salicylate is responsible for the essence of wintergreen.

FURTHER READING

Bonner, J., and J. E. Varner, *Plant Biochemistry* (3rd ed.). New York: Academic Press, Inc., 1976.

Burris, R. H., and C. C. Black, *CO$_2$ Metabolism and Plant Productivity*. Baltimore: University Park Press, 1976.

Epstein, E., *Mineral Nutrition of Plants: Principles and Perspectives.* New York: John Wiley & Sons, Inc., 1972.

Goodwin, T. W., and E. I. Mercer, *Introduction to Plant Biochemistry.* Elmsford, N.Y.: Pergamon Press, Inc., 1972.

Gregory, R. P. F., *Biochemistry of Photosynthesis.* New York: John Wiley & Sons, Inc., 1977.

Krogmann, D. W., *The Biochemistry of Green Plants.* Englewood Cliffs, N.J.: Prentice-Hall, Inc., 1973.

Noggle, G. R., and G. J. Fritz, *Introductory Plant Physiology.* Englewood Cliffs, N.J.: Prentice-Hall, Inc., 1976.

Salisbury, F. B., and C. W. Ross, *Plant Physiology* (2nd ed.). Belmont, California: Wadsworth Publishing Co., Inc., 1978.

Street, H. E., and W. Cockburn, *Plant Metabolism.* Elmsford, N.Y.: Pergamon Press, Inc., 1972.

Plant Development

The combination of growth and differentiation gives rise to the progressive elaboration of the organized plant body; this process is termed *development*. *Growth* is defined as an irreversible increase in size, which results from a combination of cell division and enlargement. Growth alone cannot lead to the formation of an organized plant body. *Differentiation* denotes the processes involved in the establishment of distinctive differences in the structures and functions of cells and groups of cells. It is the combined integrated activities of both growth and differentiation that lead to plant development.

Development at the Cellular Level

In most organisms, cell division consists of two phases: the division of the nucleus (*mitosis* or *karyokinesis*), followed by the division of the cytoplasm (*cytokinesis*). The details of this process will not be described (Fig. 4-1), except to point out some

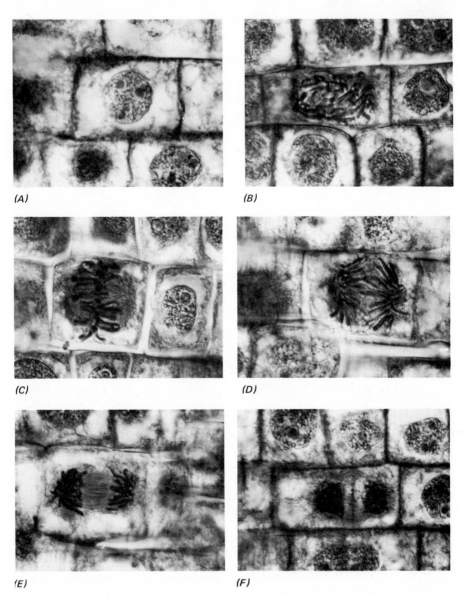

(A)

(B)

(C)

(D)

(E)

(F)

Fig. 4–1. Mitosis in onion root cells: A: interphase, B: prophase, C: metaphase, D and E: early and late anaphase, F: telophase. Cytokinesis is also evident in F where the cell plate is dividing the cytoplasm. From Samuel R. Rushforth, *The Plant Kingdom: Evolution and Form,* ©1976, p. 17. Adapted by permission of Prentice-Hall, Inc., Englewood Cliffs, New Jersey.

events of plant-cell division that are unique to plants. Mitotic plant cells generally do not have the asters and centrioles observed during mitosis of animal cells. The separation of the cytoplasm in animal cells is caused by the invagination of the plasma membrane from both sides toward the center. However, cytokinesis in plant cells starts with the formation of a bisecting pectic cell plate, which becomes the middle lamella after the formation of cell walls and plasma membranes on both sides. Undoubtedly, the biochemical events that trigger cell division in plants and animals differ, since several hormones that have a function in plant-cell division have no effect on animal-cell division.

Enlargement of plant cells can be extensive. For example, the meristematic cells of the shoot apex may be twenty times shorter in length and five times less wide than the palisade parenchyma (in leaf) or cortical parenchyma (in stem) cells that are derived from them. In this case, cell enlargement is essentially cell elongation, but enlargement in other cells comes close to being isodiametric.

A plant hormone, auxin, causes loosening (only one of the many events influenced by auxin) of the cell wall. Turgor pressure produces a subsequent enlargement of the cell. The resulting greater volume produces a decrease in the turgor pressure, which induces the diffusion of more water into the cell. Besides the cell enlargement, a concurrent increase in the vacuole size occurs from the entry of water into the vacuole. Mature cells in some instances also have thicker cell walls than the younger meristematic cells from which they were derived. Upon completion of cell enlargement, cellulose, pectin, and lignin are added to the cell walls to produce this thickening. The volume of the cytoplasm is also greater; hence some assimilation also occurs during cell enlargement.

During and after cell enlargement, biochemical changes under genetic and hormonal control bring about differentiation of the cells at certain times and places. It is much easier to describe these changes than to explain how these complex biochemical changes induce the events of plant development. Some aspects of cell differentiation include development of various cell organelles, alterations in the shape of cells, the production of suberin and cutin in cork and epidermal cells, the deposition of secondary cell walls, the total loss of the nucleus and vacuolar membranes in sieve-tube members of phloem, the disappearance of end walls in vessel elements of xylem, and the loss of the protoplast in the fibers and vessel elements. Most of these changes are in evidence after the differentiating cells go through a parenchymalike state, which follows the enlargement of the meristematic cells. Most cell differentiation is confined to the development of the specialized cells: guard cells, cork cells, collenchyma cells, vessel elements, tracheids, sclerenchyma cells, sieve-tube members, and xylem fibers. Development of the whole plant body on the cellular, tissue, and tissue system level is shown in Figure 4-2.

A

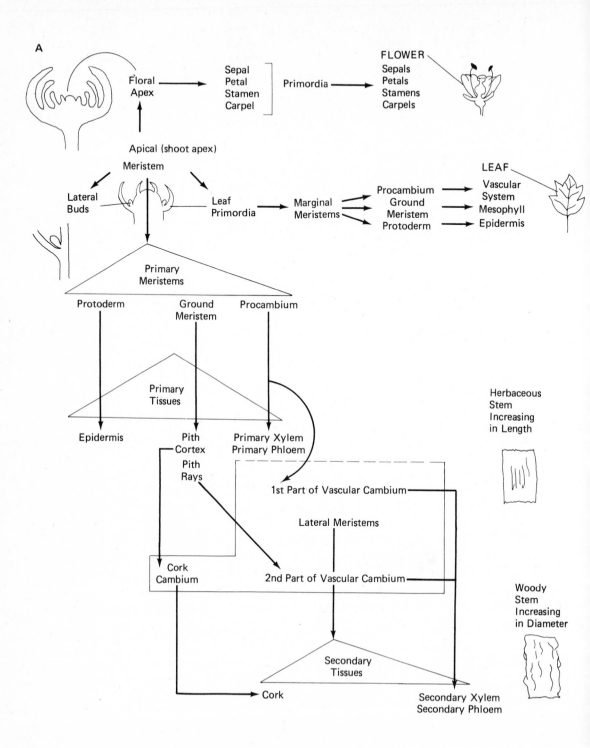

FLOWER
Sepals
Petals
Stamens
Carpels

Floral Apex ⟶ Sepal Petal Stamen Carpel] ⟶ Primordia ⟶ FLOWER Sepals Petals Stamens Carpels

Apical (shoot apex)

Meristem

Lateral Buds

Leaf Primordia ⟶ Marginal Meristems ⟶ Procambium Ground Meristem Protoderm ⟶ Vascular System Mesophyll Epidermis

LEAF

Primary Meristems

Protoderm   Ground Meristem   Procambium

Primary Tissues

Epidermis   Pith Cortex Pith Rays   Primary Xylem Primary Phloem

Herbaceous Stem Increasing in Length

1st Part of Vascular Cambium

Lateral Meristems

Cork Cambium   2nd Part of Vascular Cambium

Woody Stem Increasing in Diameter

Secondary Tissues

Cork   Secondary Xylem Secondary Phloem

90

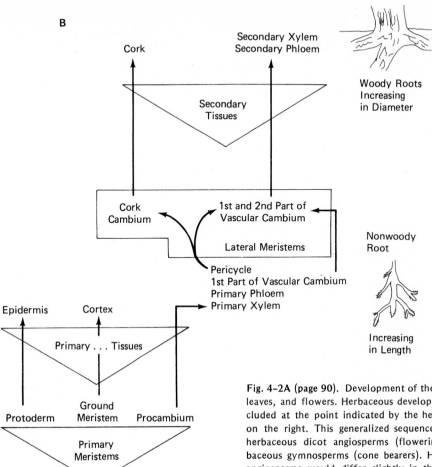

B

Cork

Secondary Xylem
Secondary Phloem

Woody Roots
Increasing
in Diameter

Secondary
Tissues

Cork
Cambium

1st and 2nd Part of
Vascular Cambium

Lateral Meristems

Nonwoody
Root

Pericycle
1st Part of Vascular Cambium
Primary Phloem
Primary Xylem

Increasing
in Length

Epidermis   Cortex

Primary . . . Tissues

Protoderm   Ground
Meristem   Procambium

Primary
Meristems

Apical
Meristem   (root apex)

Root
Cap

Fig. 4-2A (page 90). Development of the vascular plant stem, leaves, and flowers. Herbaceous development would be concluded at the point indicated by the herbaceous stem piece on the right. This generalized sequence would be true for herbaceous dicot angiosperms (flowering plants) and herbaceous gymnosperms (cone bearers). Herbaceous monocot angiosperms would differ slightly in that no vascular cambium would form and the pith, cortex and pith rays would not be clearly distinguished, so they are combined and referred to as ground tissue. Monocots can have additional increases in length from a meristem not found in dicots, the intercalary meristem. Development of woody dicots and woody gymnosperms would go beyond the herbaceous point as indicated. Some thickening of certain monocots, but no true woodiness is possible because of either a primary thickening meristem or a secondary meristem as in palms.

4-2B (this page). Development of the root system of vascular plants. Again the sequence would be true for both herbaceous angiosperms (monocot and dicot) and gymnosperms except for no vascular cambium in monocots. Woody dicots and woody gymnosperms would develop beyond that point as indicated. In some species the primary root tissues may be produced directly from the apical meristem rather than indirectly via primary meristems.

Development of the Plant Body

Development is obviously more noticeable to the horticulturist on the whole-plant level. The developmental changes on the cellular level are expressed through the integrated multicellular plant body in a series of events, which can be divided into two phases: vegetative and reproductive development. Senescence, the deteriorative processes that terminate the functional life of the plant, is a natural consequence of the vegetative and reproductive development. As such it will be described separately. We shall first consider the vegetative phase.

VEGETATIVE DEVELOPMENT IN SEED VASCULAR PLANTS

The vegetative phase can be conveniently divided into the following events: seed dormancy, germination, juvenility, bud dormancy, and maturity.

A period of growth inactivity in seeds is termed *seed dormancy.* Of course, dormancy can also occur in buds, bulbs, or other plant organs. Dormant seed is viable, but will not germinate until the physical or physiological cause of dormancy is negated, even though the proper environmental factors are present. Seed dormancy is a protective mechanism against premature germination. For example, germination shortly after seed maturation prior to winter or a dry season would produce seedlings or young (juvenile) plants that might not be able to withstand a cold or dry season without longer acclimatization. Some seeds, such as many of our cultivated vegetable and flower crops, do not undergo dormancy, suggesting that dormancy can be lost through long-range breeding.

Seed dormancy can be caused by physical and/or physiological reasons. Seed coats that are impermeable to oxygen (basswood, *Tilia americana;* Canadian hemlock, *Tsuga canadensis*) or water (clover, *Meliotus alba*) and/ or are mechanically resistant to embryo enlargement (cherry, peach, raspberry) impose physical restraints upon germination. Impermeability is produced by the impregnation of the seed coat or underlying membranes with waxes or similar substances. Physically induced dormancy is broken naturally through actions against the seed coat, such as weathering (alternate drying and wetting or freezing and thawing), fire, external attack by soil microorganisms, and internal attack by enzymes. Artificial means include moist storage at high temperatures, treatment with acids or organic solvents, and mechanical scarification (abrasion of seed coat). Care must be taken to ensure that the embryo is not damaged by these applied treatments.

Physiological causes of seed dormancy result from partially developed embryos at time of seed dispersal (*Viburnum;* holly, *Ilex;* pine), morphologically mature but physiologically immature embryos (wild ginger, *Asarum; Trillium grandiflorum*), and germination inhibitors (European ash, *Fraxinus excelsa*). Dormancy resulting from partially developed em-

Introduction to Horticultural Science / Part I

bryos is broken naturally simply by the embryo continuing development after seed dispersal. Physiologically immature embryos develop the enzymes needed to catalyze germination and growth after being subjected naturally to moisture and low temperatures for six weeks or more. Physiological changes that take place in dormant seeds after dispersal are termed *after-ripening.* The above two forms of embryo dormancy are broken artificially through refrigeration under moist conditions, a process termed *stratification.*

Germination inhibitors may be produced in the seed coat, the embryo, or the endosperm; they may also arise in the fruit and diffuse into the seed. A number of chemicals have been identified that inhibit germination, and many of them inhibit seedling or plant growth. These include organic acids, alkaloids, phenolics, tannins, coumarin and other unsaturated lactones, essential oils, cyanide- or ammonia-releasing compounds, abscisic acid, and aldehydes. The mode of action for these inhibitors varies. Some are hormonal growth inhibitors, and others provide an unfavorable pH or interfere osmotically. Germination inhibitors are removed naturally by the leaching action of rain or artificially through soaking and washing. In some cases, light exposure or low temperatures may break the hold of certain inhibitors.

In some instances, seeds may be dormant from both physical and physiological causes. This condition is called *double dormancy.* Seeds of this type often show two-year dormancy. One year may be needed for weathering or microbial attack of the seed coat and the second year to leach out an inhibitor. The cherry seed has a seed coat that is mechanically resistant, and the embryo is not fully developed upon seed dispersal.

Germination of the seed starts with the imbibition of water plus favorable environmental conditions. Several morphological and biochemical changes then follow. These include internal hydration, changes in the organization of the embryo and endosperm or cotyledon, light activation of a photoreceptor (phytochrome) and resulting photomorphogentic phenomena, enzyme activation and/or synthesis, respiration and digestion of food reserves, synthesis of organic molecules and subsequent translocation, and cellular development. All these processes underlie the external evidence of germination (Fig. 4-3): the emergence of the radicle, which develops into a root system, and of the plumule, which becomes the shoot system. Germination is considered finished when the plant becomes self-sustaining; that is, photosynthesis commences. Some prefer to view the end of germination as when the radicle appears and to call the remaining period until photosynthesis starts *establishment.*

Juvenility refers to the period of extensive vegetative growth during which time the plant cannot be readily induced into reproductive growth, regardless of environmental effects. During this stage the plant has high rates of metabolic activity and can increase exponentially in size. The juvenile plant differs from the mature plant in terms of leaf size and shape, stem growth patterns, and sometimes the timing of leaf abscission

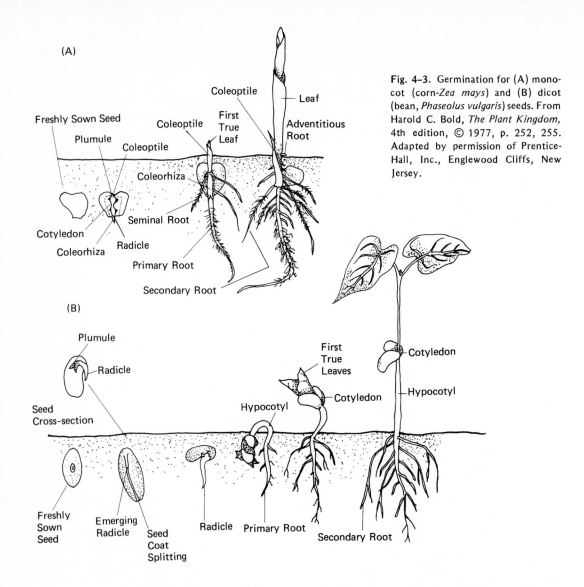

Fig. 4-3. Germination for (A) mono-cot (corn-*Zea mays*) and (B) dicot (bean, *Phaseolus vulgaris*) seeds. From Harold C. Bold, *The Plant Kingdom*, 4th edition, © 1977, p. 252, 255. Adapted by permission of Prentice-Hall, Inc., Englewood Cliffs, New Jersey.

(A)

Freshly Sown Seed
Plumule
Coleoptile
Coleoptile
First True Leaf
Coleoptile
Leaf
Adventitious Root
Coleorhiza
Cotyledon
Coleorhiza
Radicle
Seminal Root
Primary Root
Secondary Root

(B)

Plumule
Radicle
Seed Cross-section
First True Leaves
Cotyledon
Hypocotyl
Cotyledon
Hypocotyl
Freshly Sown Seed
Emerging Radicle
Seed Coat Splitting
Radicle
Primary Root
Secondary Root

(Fig. 4-4). Other differences occur internally, such as with levels of tissue complexity. As a result of juvenility, the plant is able to compete strongly in the plant community.

Much descriptive information exists about differences between juvenile and mature organs, but much less is known about the causes of these differences. The juvenility-to-maturity transition has been attributed to various factors. It has been suggested that progressive aging of the apical meristem results in a loss of the ability to produce juvenile tissue. Others attribute it to the initial presence and subsequent exhaustion of juvenile-inducing substances, such as gibberellin and auxins. Indeed, gibberellin can produce temporary reversion to or prolong the juvenile form, and the morphological characteristics of the juvenile form suggest a higher auxin content than the mature form. That growth substances are involved in the control of juvenility is a most interesting and more likely possibility.

Maturity is preceded by juvenility and succeeded by senescence. It is characterized by morphological differences, a decreased rate of vegetative growth, and the potential for the development of flowers or other reproductive structures. The flowering potential will not be realized unless environmental conditions are favorable. The translocation of food, minerals, and metabolites into the reproductive organs, especially the seeds and fruits, leads in part to the reduced growth of the shoot and root. Removal of flowers can often enhance vegetative growth, suggesting that substances produced by the reproductive organs may possibly inhibit such growth. As stated previously, there is some evidence that the levels of auxins and gibberellin are reduced during the juvenility-to-maturity transition. This reduction may be a factor in causing reproductive development.

In annuals and biennials the period of maturation may last from a few weeks to a few months. It may be measured in terms of several years with herbaceous perennials or up to hundreds of years with trees. These differences are related to different growth patterns: *determinate* and *indeterminate* growth. In the former, growth occurs over a period of time and then stops. The determinate form, largely under genetic control, is characteristic of leaves, fruits, and seeds. The conversion of a bud primordium from the vegetative to reproductive phase on a nonbranching plant, such as the common sunflower (*Helianthus annuus*), results in determinate growth.

Fig. 4–4. These plants show differences in leaf characteristics that can be observed in the lower juvenile and upper mature portions of the shoot. Left: *Acacia melanoxylon* and Right: Eucalyptus. From Hudson T. Hartmann and Dale E. Kester, *Plant Propagation: Principles and Practices*, 3rd edition, © 1975, p. 184. Adapted by permission of Prentice-Hall, Inc., Englewood Cliffs, New Jersey.

Determinate growth can also occur in branched plants, if all buds become flower buds. Indeterminate growth is characterized by continuity and increasing size with age. Growth does slow and cease when the transition from maturity to senescence occurs, but the time period is considerably longer than with determinate growth. Growth of stem and root meristems or the cambium is indeterminate.

VEGETATIVE DEVELOPMENT AND DORMANCY

Periods of growth inactivity, or *dormancy,* can occur with buds, rhizomes, tubers, bulbs, corms, or root systems. Such dormancy is an adaptive mechanism, which allows perennial plants to meet seasonal limitations upon growth. Arrested development can occur throughout the vegetative (and reproductive) phase, but especially during periods of moisture stress and temperature extremes. Some herbaceous perennials may enter a dormant phase after spring flowering, even though conditions are favorable for growth. Woody plants develop vegetative buds at the nodes and shoot tips during active growth and sometimes flower buds that are not developed further as part of the current season's growth. Instead, environmental cues, such as decreasingly colder weather and changes in day length, put these buds into dormant condition, and sequential seasonal cues, such as warmer weather or longer days, break their dormancy. Similar signals can regulate the dormancy of modified stems and root systems.

The causes for starting or breaking dormancy are not completely clear, but it appears that environmental cues induce changes in the levels of plant hormones, which in turn regulate the nucleic acid system. Since the nucleic acid system directs protein synthesis, this would be an effective on-off switch for growth.

REPRODUCTIVE DEVELOPMENT IN SEED VASCULAR PLANTS

Much of the plant development that appeals to the horticulturist is associated with the colorful, spectacular events of reproductive development. The transition from vegetative to reproductive growth signals a major change in the life cycle of plants, especially in annuals. The most obvious events of reproduction are flower formation and fruit and seed development. Reproductive development is also a complex process concerned with physiological, biochemical, anatomical, and morphological events.

A major event in the start of the reproductive phase is the formation of *flower primordia* (Fig. 4-2). This takes place after a period of vegetative growth or some minimal leaf number is attained, but only if certain environmental conditions are satisfied. These environmental factors, light and temperature, will be discussed in Chapter 5, which relates environmental stimuli to plant responses. The relationship of climate to flowering and fruiting (phenology) is important for determining the time of harvest and will be covered in Chapter 12. Phytochrome and growth substances, such as the elusive florigen or the well-known gibberellins, indole-3-acetic acid, and cytokinins play some part in the initiation of flower primordia through their modification of metabolic reactions.

Development and the maturation of floral bud parts (petal, sepal, pistil, stamen, and others) ensue with the shifts in metabolic processes. Environmental effects can influence the sexual expression of unisexual flowers, that is, the constancy or alteration of the ratio of male to female flowers. This will be considered further in Chapter 5. Changes in levels of phytohormones occur, and the synthesis of protein increases. A redistribution of water, nutrients, growth substances, and many other compounds to the flowering portions occurs at the expense of other plant parts. Internal changes prior to the opening of the mature flower (*anthesis*) prepare the flower for pollination and subsequent fertilization. Flower development covers the period from the initiation of floral primordia through anthesis.

Fruit development is considered to start after anthesis, with the initial event being pollination followed by fertilization, growth, maturation, and ripening. However, some aspects of fruit development, even though not obvious to the observer, start soon after flower induction. These earlier cellular-level events are significant in establishing the developmental pattern of the fruit. All stages of fruit development are associated with changes on the cellular and metabolic levels. These changes include cell division and enlargement, activities of growth substances such as gibberellins and cytokinins, and major translocation of materials to the developing fruit.

Pollination in the flowering plants (angiosperms) involves the transfer of the male gametes to the ovary (Fig. 4–5). One important physiological process started by pollination is the prevention of fruit or flower drop. Fusion of the male gametes with the egg nucleus and polar nuclei forms the zygote and endosperm nucleus. This process, denoted as *double fertilization* (Fig. 4–5), is the major impetus to fruit development. The activities of growth substances, such as gibberellins, cytokinins, and indole-3-acetic acid, are involved with subsequent development of the zygote and endosperm. The growth of the surrounding ovary and any accessory tissues, if present, into the mature fruit depends upon activities of the zygote and endosperm, as well as the translocation of nutrients and growth substances from the rest of the plant. Early growth involves cell division; cell enlargement is primarily associated with the latter stages of growth.

There is abundant evidence that seeds may produce regulatory substances that control many aspects of fruit development. Yet many seedless or parthenocarpic fruits exist that grow in a manner similar to seeded

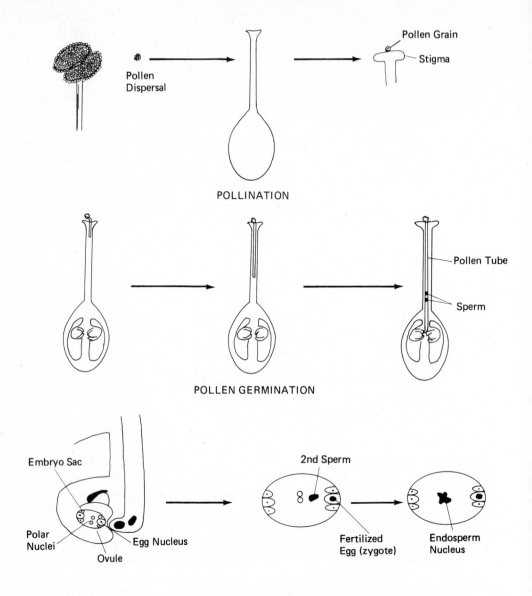

POLLINATION

Pollen Dispersal

Pollen Grain

Stigma

POLLEN GERMINATION

Pollen Tube

Sperm

Embryo Sac

2nd Sperm

Polar Nuclei

Egg Nucleus

Ovule

Fertilized Egg (zygote)

Endosperm Nucleus

DOUBLE FERTILIZATION

Fig. 4–5. During pollination pollen in the anthers is transferred by some vector (usually wind or insects) to the stigma. Here it produces a pollen tube through which sperm make their way downward to the ovule. One sperm fuses with the egg nucleus and the second with the polar nuclei. This double fertilization, unique to angiosperms, is necessary for seed production.

fruits. The effects of growth substances from seeds range from stimulatory to inhibitory, depending upon the stage of fruit growth. In parthenocarpic fruits, other fruit parts may supply the growth signals normally associated with the developing seeds.

Maturation of the fruit is complete when the fruit reaches its full size. Afterward, several subsequent events occur, which include softening of the fruit flesh, chemical changes in pigmentation and flavor, and hydrolytic changes of storage materials into sugars. These changes are collectively called *ripening*. Such changes are dependent upon an increased or *climacteric* rise of respiratory rates. After the climacteric peak, respiration falls, and the fruit enters into the senescent phase. Some fruits, such as citrus fruits, do not appear to undergo a climacteric rise in respiration.

The trigger of ripening appears to be the production of the hormone ethylene. Some nonclimacteric fruits, such as the strawberry, are unaffected by ethylene. The mechanisms of ripening are not clearly established. The modern concept is that it is under the control of DNA. A rise in RNA is observed during the climacteric rise, as well as increased protein synthesis, which is undoubtedly associated with the production of enzymes utilized during the ripening process.

The development of the fruit and seed is concurrent. As would be expected from this, the two processes interact and influence one another. Seed development is dependent upon nutrients (sugar, salts, water) translocated from other plant parts. Other substances, such as amino acids, vitamins, and phytohormones, are synthesized within the seed by the endosperm and later by the embryo. As discussed previously, some aspects of fruit development are dependent upon phytohormones produced by the enclosed seeds. The embryo is important in seed development, as embryo abortion stops seed development. As the seed develops it accumulates stores of starch, hemicellulose, proteins, fats, and, in some species, sucrose. Germination inhibitors are also synthesized in many ripening seeds. Seed coats may become desiccated and hardened and/or impregnated with substances that tend to make them impermeable. Respiration and other processes decrease. These later aspects are involved in seed dormancy and prolonged seed viability.

Seed production in the gymnosperms (Fig. 4-6, conifers) differs somewhat from the angiosperms. Their seeds are not enclosed in an ovary, so there is no development of a fruit. Pollen is produced in a male or staminate cone, and pollination takes place with an ovulate or female cone. Unlike angiosperms, fertilization does not take place shortly after pollination. About 12 to 15 months may elapse between the two events, until the ovules are fully developed. After fertilization the zygote undergoes mitosis. Double fertilization does not occur as in the angiosperms. An endosperm surrounds the developing zygote, but it develops from a reproductive tissue that was not fertilized. The fully developed seed will eventually be dispersed from the ovulate cone.

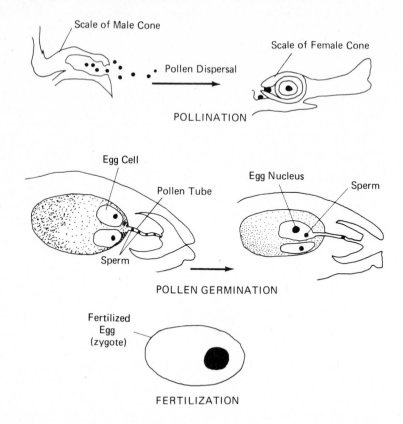

Scale of Male Cone

Scale of Female Cone

Pollen Dispersal

POLLINATION

Egg Cell

Egg Nucleus

Pollen Tube

Sperm

Sperm

POLLEN GERMINATION

Fertilized
Egg
(zygote)

FERTILIZATION

Fig. 4–6. In gymnosperms pollen is transferred from the microsporangia found on the scales of male (pollen-bearing) cones to megasporangia on scales of female (ovulate) cones. Usually male and female cones are found on the lower and upper branches, respectively, of cone-bearing trees. Eventually a pollen tube is produced through which sperm move. Only the fusion of one sperm with the egg nucleus is required for seed production in gymnosperms. In the above diagram one scale of a cone is shown with a microsporangium and megasporangium before pollination. Scales portions are left out thereafter.

VEGETATIVE AND REPRODUCTIVE DEVELOPMENT IN SEEDLESS VASCULAR PLANTS

The preceding discussion was concerned with seed and fruit development in seed vascular plants. But a large group of vascular plants are seedless. Those of interest to the horticulturist are the ferns. Their life cycle is as follows (Fig. 4–7). Ferns reproduce from homospores, which produce a bisexual, free-living gametophyte, as opposed to seed vascular plants in which there are two kinds of spores (heterospores), the microspore and megaspore. These produce male and female gametes, respectively. The gametophyte here is not free-living, but dependent upon the adult plant.

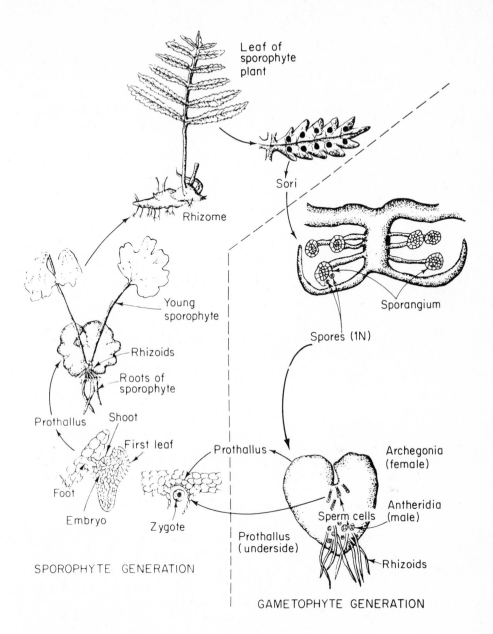

Fig. 4-7. Seedless vascular plants (ferns) differ from seed plants in their life cycle. Most ferns produce homospores which give rise to a bisexual independent gametophyte. In seed plants the gametophyte is much reduced, there is a male and female gametophyte, and these are enclosed in reproductive structures on the adult plant, on which they are dependent. Sperm and egg unite on the fern gametophyte. The fertilized egg develops into the adult fern with which we are most familiar. The gametophyte breaks down as the fern is maturing. From Hudson T. Hartmann and Dale E. Kester, *Plant Propagation: Principles and Practices*, 3rd edition, © 1975, p. 64. Adapted by permission of Prentice-Hall, Inc., Englewood Cliffs, New Jersey.

A very few ferns produce heterospores, which give rise to separate, free-living male and female gametophytes.

The germinating fern homospore produces a green gametophyte, which varies from heart shaped to irregular in form (*prothallus*). It is anchored to the soil by rootlike structures called *rhizoids*. The sperm and egg develop in specialized structures on the gametophyte, and fertilization occurs when the sperm swims through water to contact the egg. The fertilized zygote undergoes mitosis and develops into the vegetative structure we call the fern, during which time the gametophyte disintegrates. The adult fern produces spores on the frond undersides by meiosis in specialized structures, the *sporangia*.

TUBER AND BULB DEVELOPMENT

Although the formation of underground structures is more an aspect of vegetative development, the similarity of several features to reproductive development makes its consideration here more appropriate. First, they share mobilization activities in that the development of tubers and bulbs involves the translocation of nutrients and growth stimulants at the expense of other plant parts. Second, bulbs and tubers are formed as a result of activities in the leaf associated with favorable environmental factors. Finally, tubers and bulbs also undergo a ripening phenomenon. The development processes of these organs are not well understood. It is apparent that several growth hormones are involved.

SENESCENCE

As a plant passes through the various developmental stages, it undergoes various chemical and structural changes, which comprise the overall aging process. At some point these changes become irreversibly degenerative and lead to the death of the plant. The period from that turning point to the death of the plant is known as *senescence*.

Senescence may be overall, such as with the senescence of annuals and biennials after they pass through reproductive development. It can also be partial, as with the loss of top growth on a perennial, the loss of leaves annually by deciduous trees or over several years by evergreens, and the loss of reproductive organs.

During senescence catabolism exceeds anabolism. Declines will occur in photosynthetic and respiratory rates. These factors and subsequent removal of solutes through translocation will cause losses in dry weight. As synthesis slows, losses in protein, lipids, and carbohydrates also occur. Abscission layers will form with leaves and flowers. Most of the changes will be metabolic, rather than morphological. Earlier stages of development, that is, the vegetative and reproductive portions, were extensively characterized by morphological changes.

The causes of senescence are not entirely known. Extensive translocations of foods and minerals to developing seeds, fruits, and vegetative storage organs plays some role, since the removal of young fruits can prevent or slow senescence. However, senescence can also be triggered by young flowers prior to solute mobilization. One possibility is the production of inhibitors of a hormonal nature that induce senescence. However, none has been identified as of now. Some of the known plant hormones can prevent or accelerate the symptoms of senescence, but their role is not clear. External environmental signals may also initiate senescence.

Factors Affecting Development

The development of plants or any organism is controlled by two factors. The first is an internal factor: the hereditary potential. The second is an external factor: the environment. These two factors influence one another in complex and not always obvious ways. Apparent examples are mutations that induce altered hereditary potentialities: these may result from environmental factors such as ionizing radiation. Natural selections and the influence of long-term environmental conditions upon this process are another obvious example. Both heredity and environment together determine the biochemical processes of the plant, and these in turn regulate the pattern of plant development. Simply expressed, heredity determines what the plant can be, and environment determines to what extent these potentials are realized. The effects of heredity on plant development or plant genetics will be covered in this chapter, but the effects of environment will be discussed in the following chapter.

PLANT GENETICS AND DEVELOPMENT

Hereditary potentialities operate in the following manner. Chromosomes in the nucleus contain the information needed to express the hereditary potential. This information is visualized as being used to direct the synthesis of proteins, especially enzymes, by the plant. Each enzyme in turn is associated with the catalysis of a specific biochemical reaction(s). In effect, the control of enzyme synthesis implies biochemical control over development. The *gene* is a unit on the chromosome that directs the synthesis of a given enzyme. The presence or absence of a particular gene in a plant will determine whether or not the plant possesses a specific enzyme along with the associated biochemical reaction(s) and products.

Genes appear to consist of coded information in the form of specific sequences of nucleotides in the DNA molecule. The translation of this coded information into a specific type and sequence of amino acids in protein molecules has been covered in Chapter 3. Of importance to the developmental process is the fact that any particular cell of the plant at a given time does not synthesize all the enzymes for which it has DNA codes. Each vegetative cell contains the information required to produce a whole plant, but what is synthesized depends upon the stage of development and the location of the cell. This implies control or repression of genes. The theory behind mechanisms of genetic repression (and activation) hinges on the binding or removal of histones (a type of protein present in chromosomes) from the genes through the actions of plant hormones (Fig. 4–8). Environmental factors may play a part through their influence on hormone synthesis. Localized substances may be involved when development occurs in only a few cells.

Fig. 4–8. One way in which gene expression might be controlled. 1. The operator gene is an on/off switch which controls the transcription of the structural gene's code by mRNA. The structural gene codes for some enzymic protein. 2. Another gene, the regulator gene, codes for a repressor protein. 3. The repressor, upon combining with the operator gene, shuts off the transcription of the structural gene. 4. The operator gene returns to the on position when an inducer molecule combines with the repressor protein and inactivates it.

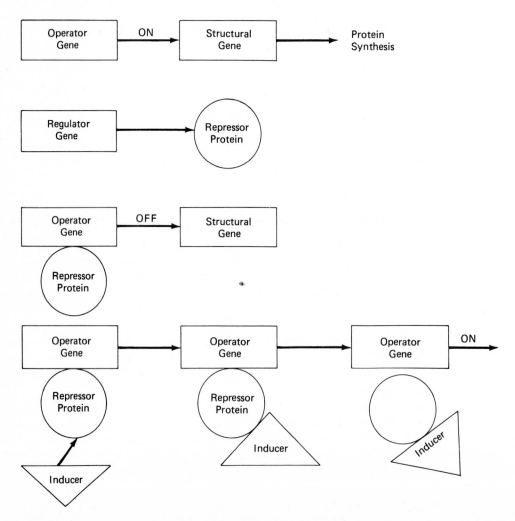

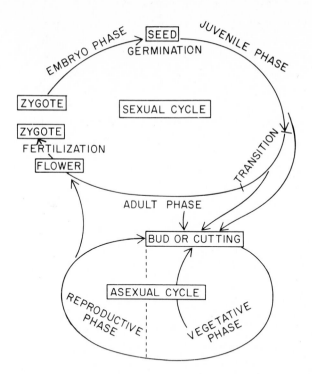

Fig. 4-9. Propagation and sexual/asexual cycles in plant. The sexual cycle of plant propagation is dependent upon seeds and involves some genetic variation among the offspring. During this cycle the plant passes through the juvenile phase, a transition phase, and the maturation (adult) phase. Cuttings or other vegetative propagules may be taken at these points (best point often depends on plant material in question), and carried through the asexual cycle of plant propagation. No genetic variation will be observed for the propagated plants in the asexual cycle. Eventually these plants enter the reproductive phase and subsequent seed formation. From Hudson T. Hartmann and Dale E. Kester, *Plant Propagation: Principles and Practices*, 3rd edition, © 1975, p. 4. Adapted by permission of Prentice-Hall, Inc., Englewood Cliffs, New Jersey.

The influence of heredity on plant development is important in two areas of interest to the horticulturist: propagation and plant improvement. Propagation and plant improvement may be considered as passively and actively dependent upon heredity.

The aim of propagation is to increase the numbers of a plant, while maintaining its essential characteristics. Preservation of a plant's unique characteristics requires the transfer through successive generations of the hereditary potentialities, or genes located on the chromosomes in the cells. The summation of these genes composes the *genotype* and the observable outward characteristics, the *phenotype* of the plant. Propagation is then a passive use of the genes, whereas plant improvement is an active use, since its aim is to produce an improved plant or new phenotype through alteration of the genotype.

Propagation can be achieved either by *sexual* or *asexual* means. Each is possible because of reproductive and vegetative development of the plant (Fig. 4-9). Propagation based on sexual techniques (seeds) produces offspring that are a product of genetic contributions from both parents. Some genotype alteration is expected, but it is minimized through proper handling by the grower. Propagation by the asexual approach makes use of various vegetative methods, all of which preserve the genotype and phenotype of the plant. The applied aspects of propagation will be covered in Chapter 6.

Each plant cell contains all the genes necessary to reproduce the entire plant. Indeed, a whole petunia plant has been produced from petunia plant cells. Asexual propagation is possible because of this fact and cell plus nuclear division, or *mitosis* (Fig. 4–1). During this process the cell divides, the chromosomes duplicate themselves (*DNA* replication), and then split into two identical sets, with each set going to one of the daughter cells produced during the cellular division. The end result is two cells from one, each with the same chromosomes and genes as the original parent cell. The propagated vegetative part undergoes mitosis and produces a whole plant that preserves the genotype and phenotype of the starting plant.

Mitosis is the basic process of vegetative growth, regeneration, and wound healing and takes place in areas of meristematic activity, such as the shoot apex, root apex, the vascular and cork cambium, and intercalary zones (see meristematic tissue in Chapter 2). Mitosis also occurs when callus forms on a wound or when new growth points are started on root or stem cuttings. These new growth points, *adventitious roots* or *shoots,* are the basis of *vegetative* or asexual propagation (see Chapters 3 and 6).

Sexual propagation is possible because of a process termed *meiosis* (Figs. 4–10 through 4–12). To understand meiosis and its implication to the plant breeder, who may wish to minimize genetic variation during propagation or maximize it for plant improvement, we must look at meiosis and further aspects of plant genetics. Meiosis takes place in the reproductive structures of both the seedless and seed vascular plants.

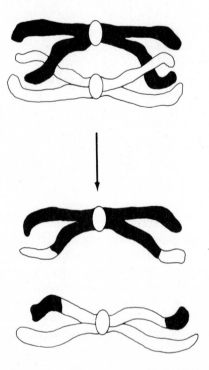

Fig. 4–10. Crossing-over, an exchange of corresponding chromatid segments between homologous pairs. This is a fairly common occurrence during meiosis.

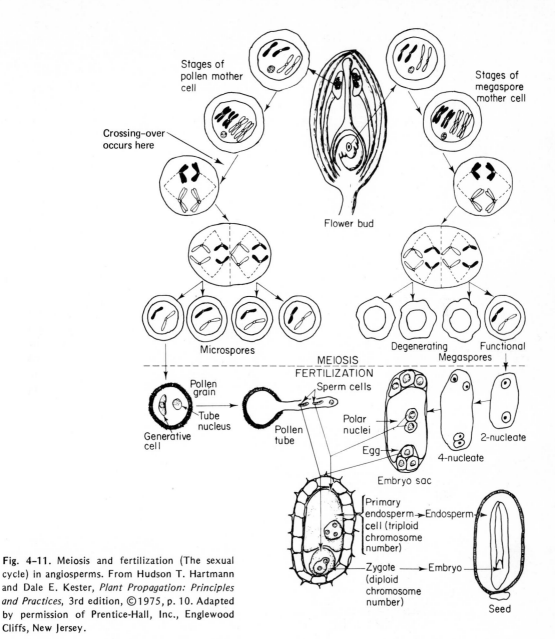

Fig. 4–11. Meiosis and fertilization (The sexual cycle) in angiosperms. From Hudson T. Hartmann and Dale E. Kester, *Plant Propagation: Principles and Practices*, 3rd edition, ©1975, p. 10. Adapted by permission of Prentice-Hall, Inc., Englewood Cliffs, New Jersey.

Labels within figure: Stages of pollen mother cell; Stages of megaspore mother cell; Crossing-over occurs here; Flower bud; Microspores; Degenerating Megaspores; Functional; MEIOSIS; FERTILIZATION; Pollen grain; Sperm cells; Tube nucleus; Polar nuclei; Generative cell; Pollen tube; Egg; 2-nucleate; 4-nucleate; Embryo sac; Primary endosperm cell (triploid chromosome number); Endosperm; Zygote (diploid chromosome number); Embryo; Seed

These are collectively called gametangia. In the fern these structures are the antheridia (male) and archegonia (female). In the flowering plants, the angiosperms, they are the microsporangia (pollen sac) in the anther (male) and the megasporangia (the nucellus) in the ovary (female). In gymnosperms the microsporangia and megasporangia are structures borne on male and female cones, respectively.

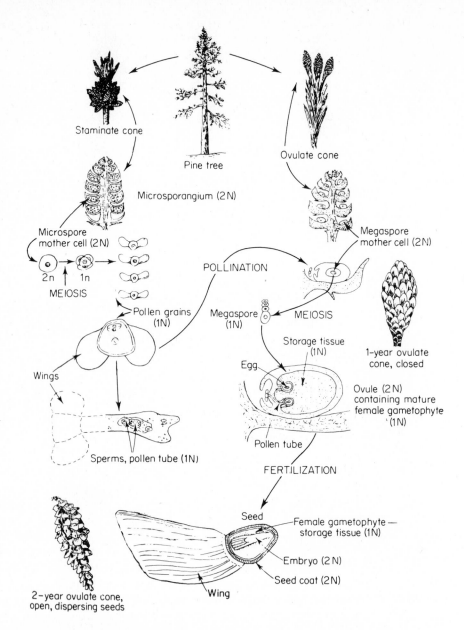

Staminate cone

Pine tree

Ovulate cone

Microsporangium (2N)

Megaspore
mother cell (2N)

Microspore
mother cell (2N)

2n   1n

MEIOSIS

POLLINATION

MEIOSIS

Pollen grains
(1N)

Megaspore
(1N)

1-year ovulate
cone, closed

Storage tissue
(1N)

Egg

Wings

Ovule (2N)
containing mature
female gametophyte
(1N)

Pollen tube

Sperms, pollen tube (1N)

FERTILIZATION

Seed

Female gametophyte —
storage tissue (1N)

Embryo (2N)

Seed coat (2N)

2-year ovulate cone,
open, dispersing seeds

Wing

Fig. 4–12. Meiosis and fertilization (the sexual cycle) in gymnosperms. From Hudson T. Hartmann and Dale E. Kester, *Plant Propagation: Principles and Practices*, 3rd edition, © 1975, p. 11. Adapted by permission of Prentice-Hall, Inc., Englewood Cliffs, New Jersey.

The somatic cells (vegetative cells that compose the plant body) of each plant species have a characteristic number of chromosomes referred to as the 2n or diploid number. For example, the Easter lily (*Lilium grandiflorum*) has a diploid number of 24, and corn (*Zea mays*) is 20. The diploid number of chromosomes in corn consists of 10 pairs (total 20) of homologous chromosomes. Each member of the homologous pair was derived from a different parent. It should be remembered that each chromosome actually consists of two duplicates or *chromatids* that are attached. During meiosis the chromatids of each member of a homologous pair may exchange corresponding segments (termed *crossing-over,* Fig. 4-10), and the homologous pairs are randomly divided in the first division of meiosis; during the second division the chromatids divide into single, unduplicated chromosomes, which eventually duplicate. The end result of meiosis is four cells, each with ten chromosomes, which is the n or haploid number.

Because of the random division and crossing-over, the four haploid cells make it highly unlikely that any would be genetically identical to the preceding haploid generation. Each haploid cell develops by mitosis, such that the adult gamete, sperm or egg, is still haploid. The original diploid number is restored when the chromosomes of the sperm and egg are added together during fertilization. New gene combinations in the offspring (*recombination*) result because of the variance introduced during meiosis. If the fertilization is preceded by cross-pollination, rather than self-pollination, the recombination possibilities are greater.

Genes are located on specific places on the chromosome designated as *loci* (singular, *locus*). Groups of genes (typically two, but possibly more) that affect the same characteristic in different ways (such as flower color, for example, red or white) are called *alleles*. Homologous chromosomes may have at a given locus on each chromosomal member of the pair the identical or different gene of the allele. The first situation is referred to as *homozygous* and the second as *heterozygous*.

In the heterozygous situation, the expression of the two alleles will differ. When one gene expresses itself more strongly than the other, the two alleles are termed dominant and recessive, respectively. A characteristic resulting from the heterozygous condition would be the same as one that resulted from the homozygous condition for the dominant gene. The characteristic controlled by the recessive gene could only be expressed if the homozygous condition for the recessive gene occurred. Alleles may show varying degrees of dominance. When the character is expressed as an intermediate form, with neither allele showing *dominance,* the resulting condition is known as *incomplete dominance.*

If the plant is homozygous for most characteristics, it will breed true if self-pollinated or if pollinated from a genetically similar plant. On the other hand, a plant that is predominately heterozygous will not breed true; the resulting phenotypes may differ from those of the parents as well as each other.

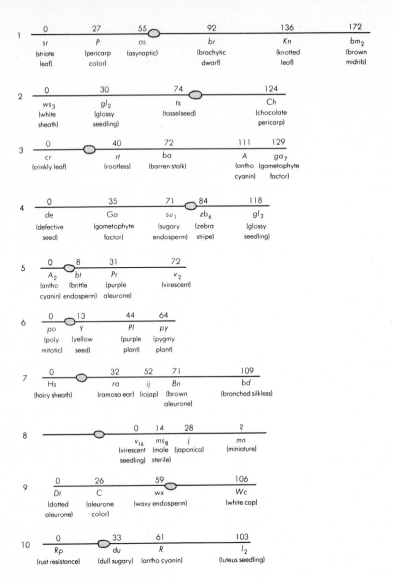

Fig. 4–13. A gene linkage map of corn. The shaded areas on the chromosomes are centromeres, the point where the duplicate chromosomes are attached, *i.e.*, the chromatid. From James L. Brewbaker, *Agricultural Genetics*, © 1964, p. 47, by permission of Prentice-Hall, Inc., Englewood Cliffs, New Jersey.

Genes are arranged on the chromosomes in the manner of a string of beads. Therefore, a tendency exists for genes to be inherited as a group; this tendency is called *linkage*. However, linkage is disrupted somewhat by crossing-over during meiosis. Either condition can help or hinder the plant breeder. The genotype and phenotype will be affected minimally or maximally, depending upon the interplay between linkage and crossing-over. If the linkage occurs between a desirable and undesirable gene, extensive breeding may be necessary such that only the desirable characteristic is expressed. Linkage maps have been developed for some plant species (Fig. 4-13).

A terminology is used to denote the crossing of plants to produce new plants (Fig. 4-14). The starting point for crossing the two parents to obtain the best qualities of each is the P_1 or *parental generation*. The offspring of this cross is the first filial generation and is expressed as F_1. If the F_1 plants are self-pollinated or intercrossed, the next generation is the F_2 generation. Generations produced in the same manner are called the F_3, F_4, and so on, generations. Sometimes the F_1 generation is crossed with either parent, and this is called a *backcross*. Repeated backcrosses help the breeder to accumulate desirable genes more rapidly when only one parent possesses desirable genes.

An example of inheritance is shown in Figure 4-15. This is a simple case with a single gene pair in a monohybrid cross. A *hybrid* is the progeny of a cross between two parents that differ in one or more genes. A somewhat more complicated example of inheritance in a dihybrid cross is shown in Figure 4-16.

Inheritance becomes more complicated beyond the control of a single character by one gene. For example, genes may act together to produce an effect that neither could do alone. Genes of this type are called *complementary* genes. Several genes may influence each other so that anyone can produce the same or similar character; these genes are termed *duplicate* genes. Some genes may mask or act as dominant modifiers of other genes (control their expression) that are nonallelic; this is known as *epistasis*. A simple example of multigenic inheritance is shown in Figure 4-16.

A more complicated situation occurs when numerous genes influence traits of economic importance, such as yield. The influence of many genes upon a character is called *polygenic inheritance*. The resulting phenotypic variation is expressed in a continuous manner. Traits expressed in this manner are *quantitative* traits. Besides the complication of many genes, this condition is made more complex by the influence of environmental conditions upon the phenotypic variation and the occurrence of undesirable genes in some of the linkage groups. The plant breeder must use statistical analysis on large numbers of plants to evaluate quantitative traits.

CROSS-POLLINATION   SELF-POLLINATION

Parental Generation (P₁)

First Filial Generation (F₁)

Second Filial Generation (F₂)

Third Filial Generation (F₃)

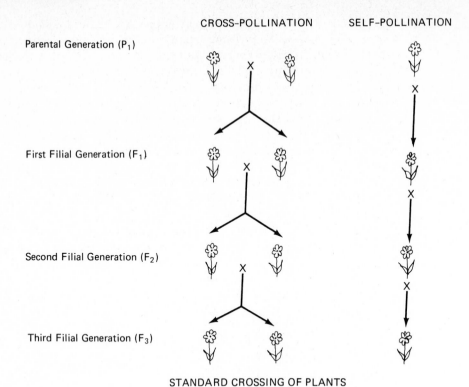

STANDARD CROSSING OF PLANTS

P₁

F₁

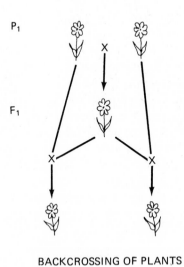

BACKCROSSING OF PLANTS

Fig. 4–14. Terminology used in standard crossing and backcrossing of plants.

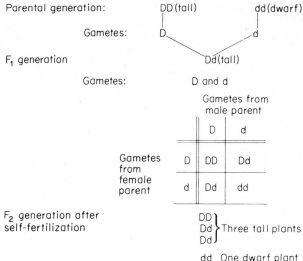

Parental generation: DD (tall)   dd (dwarf)

Gametes: D   d

F₁ generation   Dd (tall)

Gametes: D and d

Gametes from male parent

	D	d
D	DD	Dd
d	Dd	dd

Gametes from female parent

F₂ generation after self-fertilization

DD
Dd } Three tall plants
Dd

dd One dwarf plant

Fig. 4–15. An example of inheritance involving one gene in a monohybrid cross between two garden peas. From Hudson T. Hartmann and Dale E. Kester, *Plant Propagation: Principles and Practices*, 3rd edition, © 1975, p. 12. Adapted by permission of Prentice-Hall, Inc., Englewood Cliffs, New Jersey.

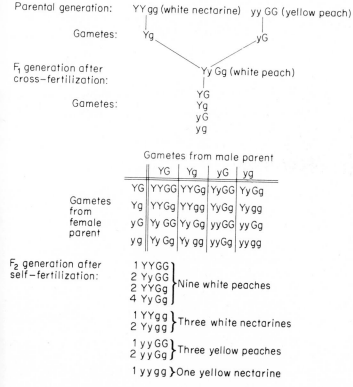

Parental generation: YY gg (white nectarine)   yy GG (yellow peach)

Gametes: Yg   yG

F₁ generation after cross–fertilization: Yy Gg (white peach)

Gametes: YG
          Yg
          yG
          yg

Gametes from male parent

	YG	Yg	yG	yg
YG	YYGG	YYGg	YyGG	YyGg
Yg	YYGg	YYgg	YyGg	Yygg
yG	YyGG	YyGg	yyGG	yyGg
yg	YyGg	Yygg	yyGg	yygg

Gametes from female parent

F₂ generation after self–fertilization:

1 YYGG
2 YyGG
2 YYGg } Nine white peaches
4 YyGg

1 YYgg
2 Yygg } Three white nectarines

1 yyGG
2 yyGg } Three yellow peaches

1 yygg } One yellow nectarine

Fig. 4–16. An example of multigenic inheritance in a dihybrid cross between a peach and nectarine. The dominant genes are G (fuzzy skin) and Y (white flesh), while the recessive genes are g (smooth skin) and y (yellow flesh). This is still relatively simple inheritance, as it does not involve complementary genes, duplicate genes, epistasis, or polygenic inheritance. From Hudson T. Hartmann and Dale E. Kester, *Plant Propagation: Principles and Practices*, 3rd edition, © 1975, p. 12. Adapted by permission of Prentice-Hall, Inc., Englewood Cliffs, New Jersey.

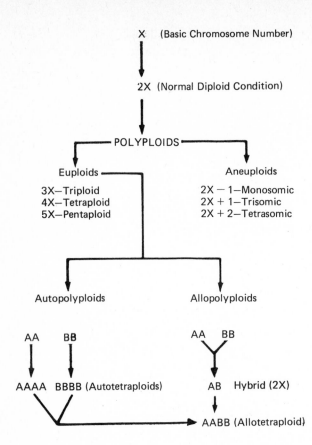

X (Basic Chromosome Number)

2X (Normal Diploid Condition)

POLYPLOIDS

Euploids

3X—Triploid
4X—Tetraploid
5X—Pentaploid

Aneuploids

2X − 1—Monosomic
2X + 1—Trisomic
2X + 2—Tetrasomic

Autopolyploids

Allopolyploids

AA   BB

AA   BB

AAAA BBBB (Autotetraploids)

AB   Hybrid (2X)

AABB (Allotetraploid)

Fig. 4-17. Types of polyploids. The symbol x is used instead of n to avoid confusion, as the 2n diploid condition in many plants has been proven to be actually a natural polyploid such as 4x.

Natural or induced variations in chromosomal number are possible; this is known as *polyploidy* (Fig. 4-17). Many plants of horticultural importance are of polyploid origin. Polyploids that contain multiples beyond the basic two sets of chromosomes are termed *euploids*. Examples are three-chromosome sets or *genomes, triploids;* four *genomes, tetraploids; pentaploids;* and so forth. Differences of one or more chromosomes (fractions of the genome) from the basic diploid number are also possible. These are called *aneuploids;* examples are $2n - 1$, *monosomics;* $2n + 1$, *trisomics,* $2n + 2$, *tetrasomics;* and so on. Euploids can be subdivided on the basis of how the added genomes arose. Direct increases of genomes within a species produce *autopolyploids*, and direct increases of genomes derived from two different species or genera produce *allopolyploids*. The former are usually sterile, and the latter are commonly fertile when the chromosomal number of the hybrids is doubled. As would be expected, the genetics involved in polyploidy are complex; they will not be discussed here.

Often, increased vigor is noted in the progeny of a cross between different inbred lines or unrelated species, varieties, and forms. This is noted as *hybrid vigor* or *heterosis*. It usually appears in the F_1 generation

Introduction to Horticultural Science / Part I

from parents that are nearly homozygous. The consequences of continual crossing of closely related parents, or *inbreeding*, can be the opposite of hybrid vigor. These consequences may range from a reduction of size and vigor to a weak, sterile plant. Fortunately, the loss of vigor during inbreeding can often be recovered upon crossing inbred lines. Hybrid vigor can be viewed as a benefit from cross-pollination, whereas self-pollination leads to a loss in vigor, but a gain in genetic constancy. Selection of the best progeny from two inbred lines and their crossing produces plants with hybrid vigor and a good degree of genetic constancy. Continual inbreeding favors the increase of the homozygous condition and hence increases the chances for expression by recessive genes with deleterious characteristics. Crossing two inbred lines increases the heterozygous condition, and the chances for expression by deleterious recessive genes diminish.

Sudden heritable changes are known as *mutations*. When mutations occur in plants, the resulting altered form is termed a *sport* by horticulturists. These changes may be spontaneous or deliberate through treatment by the plant breeder with reagents known to cause mutation. These reagents, or *mutagens*, include colchicine (a plant alkaloid) and radioactive emissions.

These changes may occur at a specific location on the chromosome (*point mutations*), as multiple duplication of the genome (*polyploidy*), as large structural alterations of the chromosomes (*inversions, duplications, deletions*), and as gains or losses in some of the chromosomes within the genome *(aneuploidy)*. In addition, the chloroplasts and mitochondria contain some DNA and play a role in the determination of plant characteristics *(cytoplasmic inheritance)*. Any changes in these organelles may lead to mutations also. An example of cytoplasmic inheritance is certain forms of variation in leaf color. Mutated chloroplasts, which have no or low levels of chlorophyll, are responsible for this condition. Proplastids in the cytoplasm of the egg carry the determination of leaf color through control of chlorophyll synthesis and can pass any induced changes to the next generation. Mutations can also occur in part of the meristem and produce mutated tissue in parts of what is otherwise a normal plant. Plants of this type are termed *chimeras*. Examples of chimeras are shown in Figure 4–18.

Fig. 4–18. A chimera found in an orange. The skin is partially thickened and yellow instead of orange in the chimera portion. From Hudson T. Hartmann and Dale E. Kester, *Plant Propagation: Principles and Practices*, 3rd edition, © 1975, p. 188. Adapted by permission of Prentice-Hall, Inc., Englewood Cliffs, New Jersey.

Changes in the pattern of development are triggered and regulated by plant *hormones,* or *phytohormones,* which are organic, nonnutrient compounds found in low concentrations. They are usually translocated from the site of synthesis to the site of action, although they can produce responses at the site of synthesis. The broader term, *plant-growth regulators,* includes both natural phytohormones and synthetic substances capable of producing phytohormone responses. Plant-growth regulators are important to the horticulturist in terms of development, plant propagation and improvement, and chemical control. A number of natural or *endogenous* phytohormones and their relationship to development will be discussed. Applications of these and synthetic phytohormones to the areas of propagation, improvement, and control will be covered in later chapters. Phytohormones can promote or hinder development through their action alone or interaction with other hormones. In addition, environmental factors and phytohormones interact and influence development (see Chapter 5). The action of phytohormones is complex and not completely understood. A brief coverage of each phytohormone follows.

Abscisic acid. This phytohormone (Fig. 4–19), once termed dormin, may be considered as a growth inhibitor. Abscisic acid (ABA) is found in most monocots, dicots, conifers, and ferns at levels of 0.01 to 1 part per million (ppm) of fresh tissue. ABA is an acidic sesquiterpene synthesized in the chloroplast from mevalonic acid and has also been synthesized by chemists. The known effects of ABA include promotion of abscission and senescence in leaves, flowers, and fruits; induction of dormancy in buds and seeds; closure of stomates during water stress; and inhibition of growth. Transport is in xylem and phloem. ABA often interacts with other plant-growth substances in an inhibitory manner. The effects of ABA suggest some action at the plasma membrane and inhibition of RNA and protein synthesis.

Auxins. Auxins are the best-understood phytohormones and have been assumed to be universal in vascular plants. The best-known auxin found in most vascular plants is indole-3-acetic acid, or IAA (Fig. 4–20). IAA is a growth promoter at concentrations near 10^{-9} to 10^{-5} *M*. High concentrations can lead to growth inhibition. Tryptophan is considered to be the precursor of IAA. Synthesis of auxins occurs in active meristems, particularly in shoot tips, buds, and developing embryos. Translocation of auxins is by active transport from cell to cell. At the cellular level, auxins affect cell elongation, division, and differentiation. In turn, certain areas of development are influenced. These include cell elongation, tropisms such as phototropism, geotropism, and thigmotropism (see Chapter 5), apical dominance, inhibition of abscission, adventitious and lateral root development, and fruit development. Some of these effects cannot be

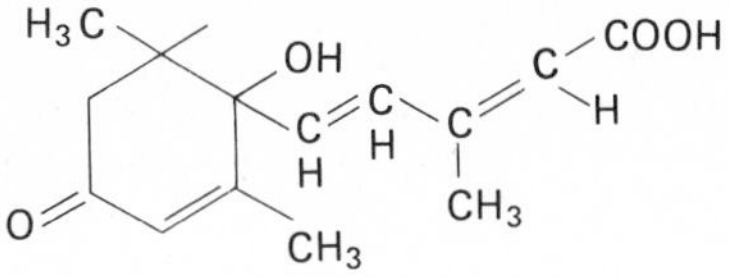

Abscisic Acid

Fig. 4-19. An inhibitory phytohormone which blocks activity
of growth-promoting phytohormones.

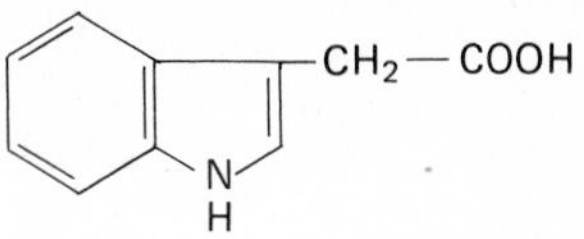

Indole-3-acetic Acid

Fig. 4-20. A growth-promoting
phytohormone widely known as
auxin.

attributed to auxins alone, but are possible through interactions with
other phytohormones. Some of these auxin effects will be described in
more detail later. IAA oxidase degrades IAA, causing it to lose activity.
Monophenols increase and polyphenols inhibit IAA oxidase activity. The
levels of these phenols vary during development and are subject to
environmental influence. As such, they may control auxin levels.

Auxin-induced cell elongation plays a role in stem, coleoptile, and
root elongation. Terminal buds or coleoptiles also have a higher auxin
level on the side that receives less light, which is caused by auxin translocation from the side receiving light. There is a correspondingly greater cell
elongation on the dark side, causing a bending of the plant part toward
the light (*phototropism*). *Geotropism,* or the orientation of plant parts in
response to gravity, also results from gradations in auxin levels. The inhibition of lateral bud growth by the terminal bud (*apical dominance*) can be
traced to the synthesis of IAA at the apical meristem and its subsequent
translocation downward. Auxins can prevent abscission of leaves, fruits,
and flowers, as well as promote the formation of lateral and adventitious
roots. Certain substances appear to work with auxins to promote rooting.
These are termed *auxin synergists*.

Cytokinins. These phytohormones (Fig. 4-21), are probably present
in most vascular and even some nonvascular plants. Their synthesis is
poorly characterized. Cytokinins are effective at low concentrations of
10^{-7} to 10^{-5} *M*. Translocation probably occurs upward in the xylem and
possibly in the phloem. Cytokinins are abundant in root tips and young,
actively dividing tissues (growing fruits, young leaves, seedlings, apical
meristems, and seeds). Synthesis is thought to be in the roots with subsequent transport to the other sites. On the cellular level, cytokinins
primarily enhance cell division and influence cell differentiation in conjunction with other phytohormones such as auxin. Root, bud, and fruit
development are influenced by cytokinins. They also slow the breakdown
of chlorophyll and other degradative reactions, thus slowing senescence.
Evidence indicates that cytokinins affect RNA and protein synthesis, perhaps through gene interactions.

Chapter 4 / Plant Development   **117**

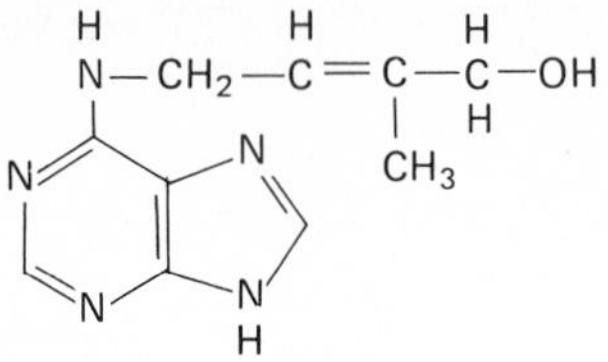

Zeatin

Fig. 4–21. Zeatin, the major cytokinin, is a naturally occuring growth-promoting phytohormone, while the well-known kinetin is a synthetic analogue.

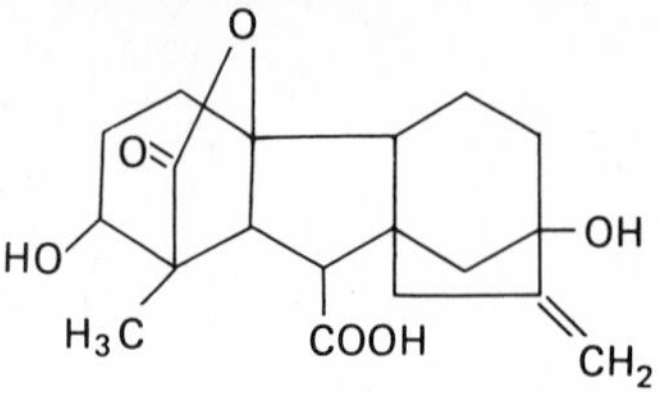

Gibberellic Acid

Fig. 4–22. The gibberellins are growth-promoting phytohormones. Gibberellic acid is probably the best-known one.

Ethylene. Most plant organs produce ethylene ($H_2C = CH_2$), but there is uncertainty in regards to the pathway. Since ethylene is a gas, it moves by diffusive processes. Ethylene appears to be involved closely with auxins. It causes downward bending of leaf petioles (*epinasty*), accelerates abscission of leaves, flowers, stems, and fruits, triggers ripening in many fruits, inhibits root, leaf, and stem elongation, causes stem thickening, inhibits flowering, and influences development in etiolated seedlings. Some evidence suggests that ethylene binds to a metal in an enzyme or membrane protein prior to bringing about responses.

Gibberellins. These phytohormones are present in most vascular plants. Unlike the auxins, for which IAA is the primary auxin, the gibberellins now consist of at least fifty compounds with similar structures and activities (Fig. 4-22). They are related to the terpenoids, and part of their synthetic pathway involves terpenoid biosynthesis. Gibberellins are effective at concentrations as low as 2×10^{-11} M. Their presence is widely distributed in the plant kingdom. Gibberellins are most abundant in developing seeds, growing fruits, apical meristems, and expanding leaves. Young leaves and roots appear to be major sites of synthesis. Translocation is probably up through the xylem and up or down in the phloem. Unlike auxins, they are not inhibitory at concentrations above the effective range.

On the cellular level, gibberellins enhance cell division and, to a lesser effect, cell elongation. Visible effects include stem elongation, increases in leaf, flower, and fruit sizes in some plants, inducement of parthenocarpic fruit, which are commonly seedless, initiation of flowers in certain plants requiring low temperatures followed by long days, and the breaking of dormancy in some seeds and buds. A possibility for the mode of action for gibberellins is that they activate genes, thereby causing increases in enzyme levels or that they influence protein synthesis.

Other phytohormone-like compounds. These include cyclitols, phenols, and vitamins. Their role as possible growth regulators is less well understood than those discussed previously, but some evidence suggests that they function as such. Cyclitols appear to have no effect alone, but stimulate tissue growth in the presence of other phytohormones. Phenolics have been identified in extracts that inhibit cell division and expansion, tissue and organ development, and seed germination. Some of the B vitamins appear necessary for embryo development and root growth in plants.

FURTHER READING

Brewbaker, J. L., *Agricultural Genetics*. Englewood Cliffs, N.J.: Prentice-Hall, Inc., 1964.

Bryant, J. A., *Molecular Aspects of Gene Expression in Plants*. New York: Academic Press, Inc., 1976.

Galston, A. W., and P. J. Davies, *Control Mechanisms in Plant Development*. Englewood Cliffs, N.J.: Prentice-Hall, Inc., 1970.

Hexter, W., and H. T. Yost, Jr., *The Science of Genetics*. Englewood Cliffs, N.J.: Prentice-Hall, Inc., 1976.

Leopold, A. C., and P. E. Kriedemann, *Plant Growth and Development* (2nd ed.). New York: McGraw-Hill Book Co., 1975.

Moore, D. M., *Plant Cytogenetics*. New York: John Wiley & Sons, Inc., 1976.

Steeves, T. A., and I. M. Sussex, *Patterns in Plant Development*. Englewood Cliffs, N.J.: Prentice-Hall, Inc., 1972.

Thimann, K. V., *Hormone Action in the Whole Life of Plants*. Amherst: University of Massachusetts Press, 1977.

Verne, G., *Genetics of Flowering Plants*. New York: Columbia University Press, 1975.

Wareing, P. G., and I. D. J. Phillips, *The Control of Growth and Differentiation in Plants*. Elmsford, N.Y.: Pergamon Press, Inc., 1970.

Plant Environments

Plants have certain environmental needs, which must be satisfied if they are to live and realize their genetic potential. These needs are concerned with four areas: climate, influences resulting from the land profile, soils, and interrelationships with other plants and animals (Fig. 5–1). These factors each play a role in the development of plants. However, it must be remembered that they actually interact in a manner that often approaches great complexity. The application of environmental parameters to plants outdoors and indoors will be covered in Chapters 7 and 8, respectively.

Climate and Microclimate

The role of climate is important, both because of its direct effect on plants and its indirect influence on the other elements of the plant environment. Those aspects of climate that affect plants are precipitation, radiation (heat, light, ionizing), air and soil temperature, carbon dioxide concentration, air pressure, wind, moisture, and cloud cover. Variation of one of these factors can extensively alter the effects of the others upon the rates of photosynthesis and transpiration, two important parameters of plant development.

Climatic variations may be averaged over long periods of time to yield an average annual profile. Plant formations interact so closely with these varied climates that certain natural plant formations and horticultural crops come to be associated with specific types of climates. It must be remembered that these concepts of climate are derived from instrumentation at 2 or more meters above ground. However, many of our horticultural crops grow below this level and are subject to influences arising from the surface profile. These influences result from natural and artificial conditions, such as variations in slope, wind exposure, masses of nearby vegetation, soil characteristics, buildings, and paved surfaces. Climatic conditions tend to be modified by these surface influences. The term *microclimate* is used to distinguish the surface-induced climate from the more unaffected atmospheric macroclimate.

The microclimate is very important to the horticulturist, who must assess its nuances with care. These may be advantageous, such as an area with a favored exposure (land near a large body of water or at the southern side of a building) that has the last killing frost of spring earlier than surrounding areas, or disadvantageous, such as lower areas that become frost pockets earlier in the fall than adjacent sites and large stretches of pavement that alter soil temperature, aeration, and moisture supply. The cause and effect of the various microclimatic factors (that is, temperature, moisture, and the like) will be discussed as they arise in the following coverage of the influence of each climatic factor upon plant development.

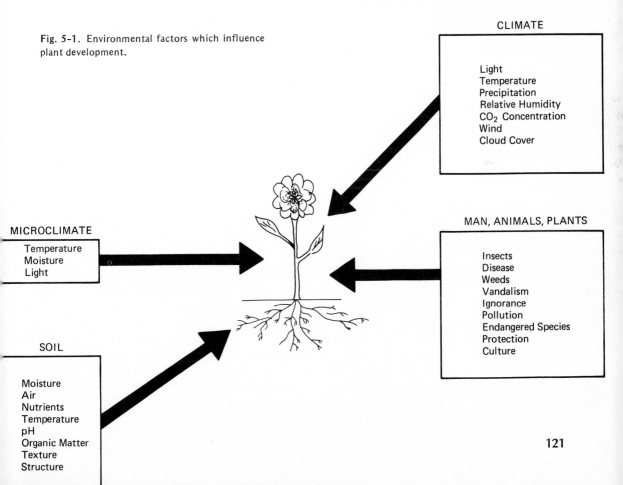

Fig. 5-1. Environmental factors which influence plant development.

CLIMATE

Light
Temperature
Precipitation
Relative Humidity
CO_2 Concentration
Wind
Cloud Cover

MICROCLIMATE

Temperature
Moisture
Light

MAN, ANIMALS, PLANTS

Insects
Disease
Weeds
Vandalism
Ignorance
Pollution
Endangered Species
Protection
Culture

SOIL

Moisture
Air
Nutrients
Temperature
pH
Organic Matter
Texture
Structure

121

WATER AND ITS EFFECTS ON PLANTS

Within wide temperature limits the water factor is probably the most important influence on plant production. In Chapter 3 we saw that water was an essential constituent and biochemical reaction component of plants. Water is absorbed and used as a transport medium for sugars, minerals, and phytohormones. Losses of water occur through transpiration; heat losses during transpiration help to control plant temperatures. The amount of water available to the plant is a function of precipitation (Fig. 5-2), surface drainage and moisture retention of the soil, and losses of water through evaporation from the soil. These soil factors will be covered later.

Thus it is apparent that a water balance exists among the plants, soil, and atmosphere. Certain features of the plant minimize water losses. These include stomata and layers of cutin or suberin. In extremely arid situations, plants have adapted to drought (*xerophytes*) through evolution. These plants include the cacti and many other succulents (such as *Crassula*) with their thickened cell walls, heavy layers of cutin, mucilage-like cellular contents, and extreme reduction of leaf areas. Plants have also adapted to very moist or wet environments. These are known as *hygrophytes.* Bog plants would fall into this category. Examples are the carnivorous plants, such as pitcher plants, sundews, butterworts, and Venus flytrap (*Darlingtonia, Drosera, Pinguicula,* and *Dionaea*).

Certain physiological processes are affected during times of water stress. The most obvious symptom of water stress is wilting (see Chapter 3), which is associated with a water shortage. Wilting can also result from excessive water (see later discussion on excessive water). Water stress also affects photosynthesis, respiration, biosyntheses, and nutrient absorption and translocation. The effects are usually adverse on plant development. These processes are altered because water is the solvent for reactions, and it is an essential reaction component in many cases. It is the solvent for nutrient absorption and translocation and metabolic reactions. It is a chemical reactant in hydrolyses of molecules, such as the reaction of starch with water to produce glucose; it is involved with electron transport during photosynthesis and respiration. A decrease of water causes an increase in reactant (biochemicals) concentration in the cytoplasm and a reduction in turgor pressure (visualized as wilting). There is also evidence that the binding of water molecules to enzymes is altered during water stress, causing configuration changes and hence losses of activity for the enzymes.

Water stress also influences plant development through stimulation of abscisic acid synthesis, which results in growth suppression and hastens

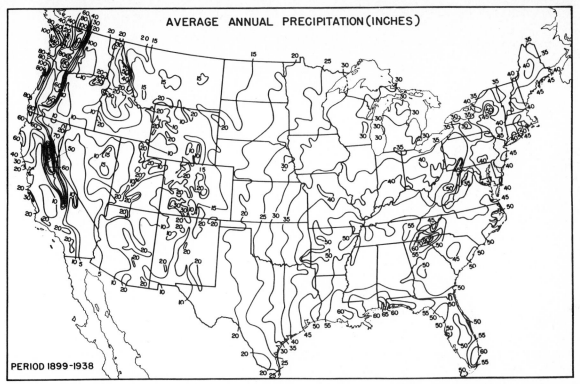

AVERAGE ANNUAL PRECIPITATION (INCHES)

PERIOD 1899-1938

(A)

Fig. 5-2. Average precipitation maps for the United States. A. Annual precipitation. B. Warm season precipitation. Maps of these types indicate the potential water available through precipitation, but not all of this is realized due to climatic variations from the average and soil parameters discussed in the text. Reproduced by permission of the U.S. Department of Agriculture.

(B)

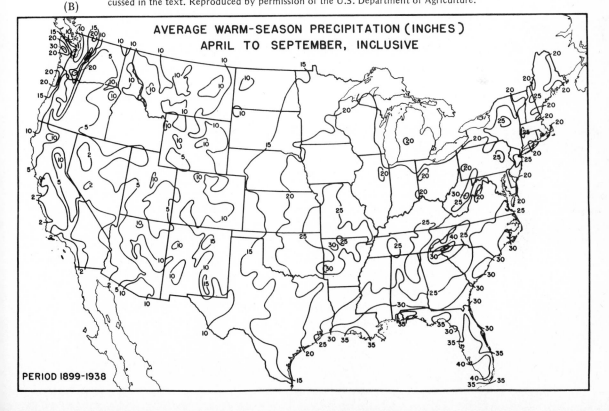

AVERAGE WARM-SEASON PRECIPITATION (INCHES)
APRIL TO SEPTEMBER, INCLUSIVE

PERIOD 1899-1938

leaf senescence. Water can relieve some forms of seed dormancy through the leaching of growth inhibitors.

One would not expect water stress to limit photosynthesis, since less than 1 percent of the absorbed water is used during photosynthesis. However, it does quickly reduce the photosynthetic rate and consequently growth. Unless the water balance is restored rapidly, the growth rate may never return to the original rate. Water stress, as indicated, does affect other processes, but the rapid effect on photosynthesis is probably the most important in terms of plant development. Limitations of photosynthesis probably arise from decreased hydration of the protoplast and chloroplast. In turn, this induces molecular disorientation and decreased enzymic activity, which disrupts light absorption and carbon dioxide fixation. Stomata close in response to water stress, and this limits the supply of carbon dioxide for photosynthesis.

HUMIDITY AND ITS EFFECTS ON PLANTS

Besides the effects of water within the plant, there are certain effects associated with atmospheric moisture around the plant, or *humidity*. The influence of humidity depends on the type of plant, as well as on the availability of soil moisture. For example, xerophytes tolerate reduced humidity, whereas plants of tropical origin often do not. The primary effect of humidity is on the transpiration rate, which tends to be reduced as humidity increases, providing the wind force is not overly large. Increased transpiration and too low humidity result in wilting and desiccation. Humidity becomes of prime concern during propagation with cuttings when no water uptake through roots is possible, yet the leaves carry on transpiration. Water loss through the leaves must be minimized until root formation. This can be done by maintaining a high *relative humidity,* which refers to the ratio of the concentration of existing water vapor in the air to the total amount the air could hold at the given temperature and pressure. High relative humidity is also necessary to prevent desiccation and to promote callus formation with grafts.

SOIL MOISTURE

Precipitation ends up as surface water, soil water, or groundwater (becomes part of the water table). The entry and downward movement of water into the soil are termed *infiltration*. This should not be confused with percolation, which is only the movement of water *through* a column of soil. Excess water over the storage capacity of the soil continues to move downward through capillary forces and gravitation and becomes groundwater. Every soil differs in its *infiltration capacity,* which is the measure of the amount of water infiltrated per unit of time.

Introduction to Horticultural Science / Part I

Many soil parameters are involved in the control of water infiltration. These include soil structure (larger water-stable aggregates shaped like imperfect spheres, the spheroidal type, increase filtration, whereas matted, flattened, platelike aggregates, the platy type, decrease it), the organic matter content (increasing amounts and coarseness increase infiltration in a fine-textured soil, but decrease it in a coarse-textured soil), the amounts of sand, clay, or silt (coarse sands increase infiltration, while fine sand, clay, or silt decrease it), water content (drier soils increase infiltration), temperature (warmer soils increase infiltration), depth (increasing soil depths to hardpan increases infiltration), and degree of compaction (decreasing compaction increases infiltration).

Water is held by the soil in three ways: adhesion, cohesion, and gravity. Films of water that adhere to soil particles are termed *hygroscopic water*. This water is tightly held and unavailable to plants. Other water molecules are held to the hygroscopic water through cohesion or attraction of water molecules for each other. This form, *water of cohesion* or *capillary water*, is less tightly held and supplies much of the plants' water needs (usually found in micropores; see the section on soil). If precipitation or irrigation is heavy, the water may saturate the soil to the extent that it fills all pore space (micropores and macropores). This water will drain at a rate determined by the soil's infiltration capacity. This downward-moving water enters the groundwater and is termed *gravitational* or *free water,* since the soil is unable to retain it against the force of gravity.

Water is retained by the soil with differing degrees of retention energy (indicated above), or *soil-moisture tension.* Moisture retention capacity depends on the texture (function of amounts of sand, clay, and silt; see the section on soil) and/or particle sizes of the soil (sand is bigger than silt; silt is bigger than clay) and also on the organic matter content. The smaller the particles are, the greater the surface area per unit volume; this in turn increases the storage capacity. Increasing organic matter content also increases water-storage capacity. The effect of texture upon water-storage capacity is shown in Figure 5-3.

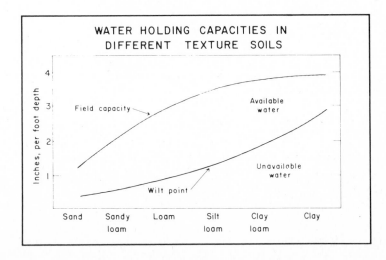

Fig. 5-3. The influence of soil texture upon water storage capacity. From Donald Steila, *The Geography of Soils: Formation, Distribution, and Management,* © 1976, p. 46. Adapted by permission of Prentice-Hall, Inc., Englewood Cliffs, New Jersey.

Together the soil-moisture tension and retention capacity determine the amount of water available to the plant. After sufficient rain or irrigation, the gravitational water will drain into the groundwater, leaving only capillary soil water. Some capillary water will evaporate from the surface of the soil, causing it to dry. The remaining capillary water is the maximal amount that can be retained by the soil. This amount of soil moisture is the *field capacity* of the soil and is capillary water with minimal movement. When the energy required to remove water for the plant becomes too great as the hygroscopic water is approached, the plant starts to wilt. The level of moisture left is called the *wilting point*. The difference between the field capacity and the wilting point is known as *available water* (Fig. 5-4). This is the water available to plants, and it consists of capillary water that has little or no capillary movement.

(A)

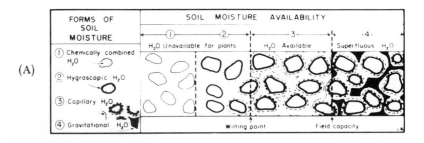

(B)

Fig. 5-4. A. The relationship between forms of soil moisture and water available for use by plants. From Donald Steila, *The Geography of Soils: Formation, Distribution, and Management,* © 1976, p. 45. Adapted by permission of Prentice-Hall, Inc., Englewood Cliffs, New Jersey. B. Plant appearance when wilting point is reached. Unless available water is replenished at this point in sufficient time, the plant will die. Author photo.

DROUGHT AND EXCESSIVE WATER

Too much or not enough soil moisture can cause problems for the horticulturist. Soil moisture is lost through evaporation directly from the soil surface and indirectly through transpiration by the plant. When these combined losses, collectively called *evapotranspiration,* exceed the amount of available soil moisture, we have a water deficit referred to as *drought.* Drought occurs in areas of arid climate with insufficient rainfall, in areas with distinct periods of greatly reduced rainfall, or in areas with unpredictable variations in precipitation. Localized drought can occur in plantings, such as trees, that are surrounded by large expanses of water-impervious pavement. In addition, factors such as high temperatures, winds, and low humidity can speed drought through their increase of evapotranspiration. Drought can be combatted by the use of irrigation, drought-resistant crops, and mulches, and by increasing the soil's field capacity by adding organic matter, which will be discussed in subsequent chapters. Wilting, poor development, and eventual death result from uncorrected drought. Continued drought also reduces soil pH temporarily and increases soluble salts. Fertilizer salts, salts produced by degradation of organic matter, and salts brought to the surface by capillary movement and evaporation are no longer leached downward by water infiltration, so the soluble-salts level is increased in the topsoil. Increased exchange of certain ions (from salts) for hydrogen ions on soil particles frees more H^+ to the soil moisture, which produces a pH decrease.

Excessive soil moisture can result from excessive rainfall, excessive water retention by the soil, overwatering of container plantings, too much field irrigation, poor drainage of soil in the root zone, or high water tables. Increasing soil moisture means decreasing levels of oxygen in the soil. In turn, respiration decreases in the plant roots, causing poorer plant development and eventual death if the excessive water condition is uncorrected. As the plant deteriorates the roots are the first to die, turning from a healthy white to brown in color and rotting from microbial attacks. At this stage, water uptake is greatly reduced and the plant wilts. If mistaken for drought-induced wilting, the remedy is the addition of more water, which accelerates the loss of the plant. In addition, excessive soil moisture increases the leaching of nutrients and may cause nutrient deficiencies.

Although not often considered, excessive soil moisture that may not be great enough to result in death for the plant can affect the plant by the leaching of growth regulators present in the soil. These leachates from the aboveground portions of the local flora are part of the ecosystem, and their removal undoubtedly has some effect (presently not fully researched) upon plant development. If the excessive soil moisture results from excessive rain, an additional effect is to increase the leaching of water-soluble growth regulators from the aboveground portions of the plant. The effect of this is not completely assessed, but one consequence is delayed dormancy due to leaching of abscisic acid (see Chapter 4).

Temperature   In the previous section we saw that water is a medium or essential reaction component for many of the biochemical processes associated with plant development. Heat in turn is necessary for these processes, since any plant has minimal, maximal, and optimal temperature requirements for all aspects of plant development, as well as for individual biochemical processes. The accumulation of heat over time (degree days) can be directly related to number of days needed to harvest. This concept is covered in Chapter 12.

Plants, as well as other biological forms, are generally limited to a temperature range from 32° to 122°F (0° to 50°C). Some plants can survive below this range in a dormant stage, and above it plants die because of protein denaturation. Certain lower plants, such as hot-spring algae, desert lichen, arctic mosses, and arctic lichens can endure temperatures that exceed this range, but higher plants do not.

EFFECTS OF TEMPERATURE ON PHYSIOLOGICAL PROCESSES

Transpiration shows a marked increase as the temperature rises. This increase is maximal if the temperature increase is associated with a decrease in relative humidity and an increase in air movement. Temperature also has an effect on respiration, in that an increase in temperature usually causes higher rates of respiration and photorespiration. As expected, decreasing temperatures slow transpiration and respiration. If temperatures become high enough, about 122° to 131°F (50° to 55°C), disruption of the enzyme systems commences. High temperatures can decrease the photosynthetic rate, but this is probably an indirect effect resulting from direct effects of higher temperatures, such as increases in the photorespiratory rate, damage to the chloroplast membranes, or the opening and closing of the stomata.

EFFECTS OF TEMPERATURE ON PLANT DEVELOPMENT

The influence of temperature extremes on physiological processes in turn causes noticeable effects on plant development. Growth of most plants generally ceases when the soil temperature drops below 41°F (5°C). At these temperatures, water uptake through the roots is limited, and with needled or broad-leaved evergreens the rate of transpiration may exceed the rate of water uptake. Damage may follow from desiccation or from increases in solute concentration caused by reduced levels of water within the plant. If a tender or improperly acclimatized hardy plant freezes, structural disruption of the tissues and even death can occur. Damage is especially severe if the plant is subjected to alternate freezing and thawing. Cold temperatures insufficient to produce freeze damage can still damage or kill plants of subtropical or tropical origin that are sensitive to chilling.

Introduction to Horticultural Science / Part I

The reduction of water uptake at lower temperatures below transpirational needs can cause physiological drought in these plants, which results in the yellowing of leaves. Warm-temperature plants are highly susceptible to chilling damage. Chilling damage can also be worsened or started by the application of cold water to the soil, such as application of 35°F (1.8°C) water directly from the tap in January. Besides the possibility of damage, the chilling of the roots slows development of the plant, and fuel energy is wasted on returning the soil temperature to normal levels. The timing of greenhouse plants to coincide with a holiday could be ruined by cold water.

High temperatures usually speed growth. Damage results mostly from increased evapotranspiration, followed by dehydration, which are secondary effects. However, temperature-moisture interactions must be considered. Increased evapotranspiration and dehydration at high temperatures can be forestalled if the soil and atmospheric moisture are sufficient. At the other end, low temperatures and moisture are associated with damaging frosts (see frost section) and freezes. Winds may speed the damaging effects of high temperatures by increasing rates of evapotranspiration during the summer. Increased winds and/or warmer-than-normal temperatures (a January thaw, for example) can increase transpiration rates to a point where serious stem and foliage desiccation occurs, as the roots cannot take up water from the still-frozen soil to replace that lost by transpiration. This type of damage (*winter kill* or *burn*) is a problem in the North, especially with needled and broad-leaved evergreens.

Microclimates can reduce the effects of temperature extremes on development, such as those to be discussed later under frosts. Less exposed slopes may experience lower temperatures than surrounding areas, thus permitting the growth of plants sensitive to high temperatures. Bodies of water slow the rise of temperatures on their leeward side, which is advantageous for slowing the blossoming of fruit trees until the danger of frost is past. The southern side of a building, because of reflected and radiated heat, may promote more rapid plant development, such as earlier flowering of spring bulbs, than surrounding areas.

Normal variations in temperature (less than the extremes discussed) also affect the various metabolic processes, which in turn affect plant development. This arises from temperature influences that determine the quantity and types of food used by the plant, the amount of ATP produced during respiration, the rates of phytohormone synthesis, rates of translocation, internal levels of hydration, and many metabolic processes.

Optimal temperatures and suitable temperature ranges for plant development are highly variable. These temperature conditions vary among the species, the time of day, and the particular stage in development. Plants do not often proceed to the next stage of plant development until a specific temperature requirement has been satisfied. However, some generalizations may be permitted. Plants of horticultural interest in the temperate zone do not grow below 41°F (5°C) or above 95° to 104°F

(35° to 40°C), and have a growth optimal temperature from 77° to 95°F (25° to 35°C). With tropical plants, these temperatures in respective order are 50°F (10°C), 113°F (45°C), and 86° to 95°F (30° to 35°C). Optimal temperatures for development are generally lower at night than day; this phenomenon is termed *diurnal thermoperiodicity* and will be discussed later. Of course, this difference varies with species, but the day-night difference is often about 9° to 21°F (5° to 12°C). Change in optimal temperature with the various developmental stages is sometimes called *annual* or *seasonal thermoperiodicity*.

More specific effects of temperature upon developmental stages and specific plant parts are of great interest to the horticulturist. Of all the environmental factors that regulate seed germination and seedling growth, temperature is the most important. Seeds have minimal and maximal temperature limits for seed germination, as well as optimal ones. Extensive variation in these temperatures exists among plant species, and they may vary for a particular plant because of diurnal thermoperiodicity, seed age or condition, or interaction with light. Some broad generalizations can be made. Cool-season plants, which include many of the flowers and vegetables that originated in the temperate zones, do not germinate at temperatures above 77°F (25°C). This high-temperature sensitivity is known as *thermodormancy*. Warm-season plants, which include the solanaceous crops, the cucurbits, bean crops, and sweet corn, have a minimal germination temperature of 50°F (10°C) and are highly susceptible to chilling injury. Chilling appears to cause damage and subsequent leakiness of membranes within the chilled cells. Many other seeds of cultivated plants can germinate between 40° and 104°F (5° and 40°C). The optimal temperature for germination and seedling growth is between 68° and 95°F (20° and 35°C) for many plants, with the lower end and higher end being favored for seedling growth and germination, respectively. A diurnal thermoperiodicity temperature differential of about 18°F (10°C) also appears to be most favorable for germination and seedling growth.

Root and shoot development are affected by temperature. The best growth occurs when there is diurnal thermoperiodicity. Both shoot and root growth increase with increasing root temperature and constant shoot temperature up to a point, and then decrease. The shoot-to-root ratio increases at the same time. Small changes in root temperature, therefore, can cause extensive changes in top growth. Situations of interest include container crops and mulched areas. Many plants have mature, and especially young, roots that have less hardiness than the shoot portion. In containers, the soil is at air temperature rather than the warmer field soil. Therefore, damage or death can result if plants are unprotected or insufficiently acclimatized. Applications of cold water can also lower the root-ball temperature and retard or damage growth. Premature application of mulches to insufficiently warmed soils will slow the warming of soil and can retard top growth and even flowering. One possible cause of reduced top growth may be the reduced uptake of water and nutrients associated

130

with decreasing root temperatures. Root temperatures are also important when rooting cuttings, since higher air than soil temperatures favor shoot over root development.

Flower initiation with some plants, such as certain annuals, may occur once the plant is physiologically ready (reached the appropriate stage of development), while in others further stimulus by other environmental factors, temperature and light, may be required. The latter will be discussed later in this chapter. The effect of temperature upon flower initiation is a phenomenon known as *vernalization*. In nature, vernalization begins by the exposure of plants to low temperatures, which induce many biennial and perennial temperate plants to flower during the warm weather following the winter. The exposure of seeds or bulbs (tulip, hyacinth) to low temperatures to induce later flowering is also an example of vernalization. The temperatures and times of exposure vary with species, but a general range is 28° to 50°F (-2° to 10°C) for several weeks. This form of low-temperature conditioning is not a requirement for all plants, such as summer annuals or plants of tropical origin. A few annuals (spinach, China aster) can be vernalized by exposure to high temperatures. In some plants subject to vernalization, a photoperiod with long days must follow after the low temperatures if flowering is to be completed. Short day lengths (see photoperiod in the section on light) can substitute for low temperatures in some plants, but long days afterward are still required. Gibberellin appears to play some role in the vernalization process, but its exact function is unclear. The effects of low-temperature preconditioning can be lost if subsequent exposure to higher temperatures around 86°F (30°C) occurs four or five days after vernalization.

Temperature is also of great importance to the plant propagator. It has an effect upon bulb development and storage, callus production, rooting cuttings, seed dormancy and germination, seedling growth, and seed storage. As would be expected, maximal, minimal, optimal, and harmful temperatures will show great variation for these phenomena and between plant species. These temperatures will be examined more closely in Chapter 6.

FROST

A temperature phenomenon of special interest to the horticulturist is frost. Moisture in the air can drastically affect plant development as the temperature drops and the moisture becomes frost. The last killing frost of the spring and first killing frost of the fall (see frost maps in Part Three) determine the favorable period of plant development.

Frosts can be caused in two ways. *Air-mass frost* results from the appearance of a cold air mass with a temperature less than 32°F (0°C). *Radiation frost* results from rapid radiational cooling of plant material to 32°F or lower. The former is more of a freeze and apt to happen during winter. The main threat to horticultural crops from this frost is to plants limited in winter hardiness. Radiation frosts usually occur in spring or fall and are a spring threat to seedlings and budding trees or shrubs, and a fall threat to plants that have not gone into dormancy and unharvested crops. This frost occurs on clear nights when the sky acts as a black body and absorbs radiant heat from leaves. Cloud covers slow radiant heat loss and thus help prevent frost. Leaves can experience a temperature drop of several degrees below air temperature. Dew formation is encouraged on the cooling leaf when the temperature is reached at which the relative humidity is 100 percent (*dewpoint*). The dew, if present, freezes into ice crystals at 32°F or less; this is termed a *white frost*. If the dewpoint has not been reached and the temperature is 32°F or less, it is called a *black frost*, because no visible signs of frost exist until the plant material is blackened from freeze injuries.

Microclimate effects on frost can be quite significant. Sloping areas experience different temperature regions or thermal belts (Fig. 5–5).

(A)

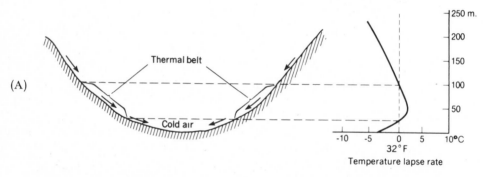

Fig. 5-5. A. A microclimate created by sloping terrain: the thermal belt. From Howard J. Critch-field, *General Climatology*, 3rd edition, © 1974, p. 270. Adapted by permission of Prentice-Hall, Inc., Englewood Cliffs, New Jersey. B. The effect of different day/night temperatures upon the growth of cucumbers is clearly evident. From left to right plants were subjected to day/night temperatures of 75/65, 80/70, 85/75, 90/80 and 95/85 for 20 days. U.S. Department of Agriculture photo.

(B)

Introduction to Horticultural Science / Part I

Cold air, being more dense than warm air, settles into the lower portions and can create frost pockets. The downward movement of cold air displaces the warmer air upward, causing the formation of a *thermal belt* on the slopes. Valleys with north-to-south orientation are shaded quicker than westward ones and have longer times to develop the beneficial thermal belt. Vegetation shades the ground and lowers conduction and convection; consequently, temperatures above bare ground may be up to several degrees higher than above the area with vegetation. An area of clean cultivation could escape a frost, while a nearby vegetated or mulched area would not. Large bodies of water can keep temperatures relatively higher on windward sides. Their heat loss is slower than soil heat loss; thus they can act as a heat sink and slow the advent of killing frosts.

THERMOPERIODICITY

It is known that the optimal development (Fig. 5-5) of most plants occurs when daytime temperature (*phototemperature*) is 9° to 21°F (5° to 12°C) higher than night temperatures (*nyctotemperature*). A possible explanation of diurnal thermoperiodicity may be that carbon dioxide assimilation is better (up to a point) with higher temperatures, and respirational losses of fixed CO_2 decrease with decreasing temperature; thus the plant gains maximal retention of fixed CO_2. The day and night temperatures selected for controlled conditions are a compromise between maximizing photosynthesis and minimizing respirational losses short of depriving the plant of its essential respirational energy. Not all plants respond favorably to this regime. Some plants do best when phototemperature and nyctotemperature are similar; the African violet (*Saintpaulia*) succeeds better when the nyctotemperature is about 10°F (6°C) higher than the phototemperature. Unlike vernalization, the effect of diurnal thermoperiodicity is upon current development. Some effect is exerted on flower initiation, but vegetative development and fruit set and development are influenced more.

Many plants show responses to longer-range fluctuations in temperature, since the various developmental stages have different optimal temperatures. This annual or seasonal thermoperiodicity is most pronounced with temperate-zone plants. Water availability and day length are also involved in determining the responses to seasonal temperature changes.

The most familar response of plants to seasonal thermoperiodicity is *low-temperature preconditioning,* or *hardening,* which enables the plant to enter dormancy and survive the rigors of winter. Continued exposure to low temperatures affects the development of the plant during the following spring and summer. Breaking of seed and bud dormancy and the blooming of biennials are some of the phenomena influenced by low-temperature exposure. The hardening process occurs in steps upon exposure to low temperatures. Development slows, causing simple sugars to

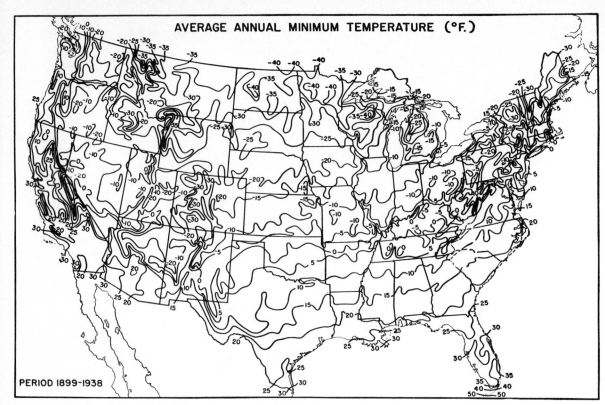

Fig. 5-6. Map showing average annual minimal temperatures in United States. Plants vary in response to these temperatures. Such temperatures are used in part to determine zones of hardiness for different plant material. Reproduced by permission of U.S. Department of Agriculture.

accumulate and increase the osmotic potential. A hardiness-promoting phytohormone is thought to be produced upon exposure to low temperatures and short days. Exposure to frost causes additional changes, such as reduced hydration, increased membrane permeability, and structural alterations of protein. This gives the protoplasm a property that enables it to resist the stresses of further frosts and freezing. Subzero temperatures induce even greater hardiness. The mechanism is not clear, but it may be that the cellular water becomes more tightly bound and hence less available for ice formation.

Seasonal changes in temperature and the ability of plants to meet them tend to determine the distribution of plants. The average annual minimal temperatures vary (Fig. 5-6), and so do the degrees of hardening shown by different species. In addition, the needs of plants vary in terms of chilling requirements. These factors would determine the northern and southern limits of a plant's distribution. On this basis the United States and Canada can be divided in temperature zones, and a knowledge of a plant's response to low temperatures will allow the horticulturist to select areas where the plant will succeed (see hardiness zonal map in Part Three).

Introduction to Horticultural Science / Part I

Light influences several aspects of plant development and physiology. It is not the most important limiting factor in the distribution of plants, but it is an environmental factor of utmost importance in terms of a given plant in a specific environment. Plants respond to a broader light span, 300 to 800 nanometers (nm), than do our eyes. The term light will refer to this broader spectral range. The nature of these responses depends on an interrelationship among the spectral quality, duration, and intensity of light. The actions of light are not always singular, in that effects of temperature and phytohormones are often involved also.

INTENSITY AND ITS EFFECTS ON PHYSIOLOGY AND DEVELOPMENT

The degree or amount of light, the *light intensity,* varies under natural conditions. Cloud cover, leaf canopies, season, pollution, and latitude cause variations in the light intensity reaching plants. Light intensity is often expressed in terms of footcandles. For example, a bright, sunny summer day has an intensity of about 10,000 footcandles as opposed to about 30 for a good reading lamp. However, these units are based upon the spectral sensitivity of the eye. A more useful measure of light energy in terms of plants is the absolute energy of the light wavelengths utilized in the photosynthetic process (the photosynthetically active radiation or PAR, 400 to 700 nm) expressed per unit area. Light sensors are available for these measurements. Units of measurement for PAR are microwatts per square centimeter ($\mu W/cm^2$).

Plants grown from seed under darkness develop until their stored food reserves are exhausted. These plants are yellow, with spindly stems having long internodes. This condition is known as *etiolation.* The yellow color is from carotenoids, which are usually masked by green chlorophyll, except dark-grown plants lack chlorophyll. The absence of light, which promotes some processes and inhibits others, results in a disruption of normal, balanced development.

Plants vary in their response to light intensity. Some plants succeed at low light intensity (shade plants) and others do not (sun plants). A variegated cultivar will require more light than the unvariegated plant, because the absence of chlorophyll in parts of the leaf reduces the surface area capable of photosynthesis per given amount of light; so photosynthetic rates must be increased in the rest of the leaf. Certain morphological and physiological changes can occur within a plant to maximize or minimize utilization of light. Leaves growing in shade or within the interior of a leaf canopy differ in appearance from sun-grown leaves and leaves from the exterior of the leaf canopy (Fig. 5-7). Sun leaves are often thicker, smaller, and have shorter petioles; the thickness usually stems from several extra layers of palisade parenchyma cells. Chloroplasts in shade leaves will often have increased granal development, which increases the lamellar membrane surface to maximize light interception. Alignment of chloroplasts in shade leaves will often be at the upper cell surfaces, with their long axis at right angles to entering light (maximal area for light interception), as opposed to parallel alignment of long axes with the vertical cell walls in sun leaves. Leaf movements also occur, which can expose maximal or minimal surface areas to light, depending on its intensity. Increasing light intensity also activates the development of pigments, such as the anthocyanins found in leaves and stems. Shade-grown leaves may appear less colorful than their sun-grown counterparts.

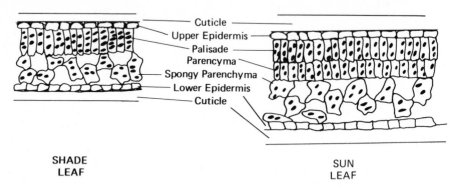

SHADE
LEAF

SUN
LEAF

Fig. 5-7. Cross-sections through shade and sun leaves.

Under some conditions, light intensity can be the limiting factor in photosynthesis. At night, in morning when the light is increasing, and in evening when the light is decreasing are obvious examples. Dense cloud cover or heavy shade are also limiting. Light underneath a forest canopy or large shade tree can be limiting, even though the intensity does not appear greatly diminished, because the wavelengths effective in photosynthesis have been reduced through absorption by the outer leaves. However, the reduced light and lower levels of useful wavelengths provided by large trees or physical obstructions can create a useful microclimate in a sunny area. Plants unable to tolerate high light intensities could succeed under these conditions.

Increasing light intensity will increase the photosynthetic rate until a point of light saturation is reached. The light saturation, and hence photosynthetic rate, varies considerably among species. In shade-adapted species or plants not acclimated to high light intensities, levels of light intensity beyond the light saturation point can damage the photosynthetic apparatus. The less photosynthetically efficient plants are unable to make maximal use of full sunlight, unlike the more photosynthetically efficient species (see discussion on photosynthesis in Chapter 3). Photorespiration also increases with increasing light intensity in the less photosynthetically efficient plants, thus further reducing their competitiveness with the more efficient species.

Introduction to Horticultural Science / Part I

EFFECTS OF SPECTRAL QUALITY AND DURATION ON PHYSIOLOGY AND DEVELOPMENT

The effectiveness of light on plant development and physiology varies with the emitted wavelength or *spectral quality*. A number of pigments in plants absorb wavelengths from specific parts of the light spectrum. These pigments include those involved in photosynthesis, photomorphogenesis (the regulation of plant development by light), and phototropism (growth movement in response to one-sided illumination), all of which will be discussed in more detail later. The photosynthetic pigment, chlorophyll, absorbs light in the red and blue spectral region. Phytochrome, the carotenoids, and flavins are pigments involved in photomorphogenesis; they absorb blue, red, and far-red wavelengths. The pigments responsible for phototropism absorb violet, blue, and green wavelengths.

Spectral quality is important when using artificial light to grow plants. Any light source must have the appropriate spectral quality to initiate and sustain photosynthesis. Chlorophyll absorbs the red (600 to 700 nm) and blue (400 to 500 nm) wavelengths, and lamps designed for plant growth must emit these wavelengths. The application and types of available lamps will be covered in Chapter 8.

Light initiates reactions that control plant development. These types of reactions are termed *photomorphogenetic* reactions; they generally require lower light intensities for initiation than do the photoprocesses, photosynthesis and photorespiration. A very few photomorphogenetic reactions do require high light intensities. Spectral quality is the more important parameter for photomorphogenesis.

Photomorphogenetic processes have a wavelength dependency on blue (450 nm), red (660 nm), and far-red light (730 nm). The pigments involved include the carotenoids, flavins, and phytochrome. The photomorphogenetic functions of the carotenoids include the regulation of phototropism in seedlings. Flavins may also be involved in phototropic curvature and in the transfer of light energy from blue to red wavelengths, which would mean an indirect participation in the phytochrome functions. Phytochrome, activated by red and far-red wavelengths, is involved in numerous photoresponses. Some of these are shown in Table 5-1. Other factors besides phytochrome are probably involved in photomorphogenesis. These include phytohormones, enzymic alterations brought about by repression or derepression of genes, and alterations of membrane properties.

The duration of light-hours in relationship to dark-hours in a 24-hour day is called the *photoperiod*. As the seasons change, so does the photoperiod. Many flowering plants respond to these seasonal changes in the photoperiod; this phenomenon is termed *photoperiodism*. These responses include the change from vegetative to flowering development, leaf drop and autumn coloration, tuber formation, grass tillering, and the onset of winter dormancy. Photoperiodism is one of many phytochrome-mediated events of importance to the horticulturist, because of its relation to flower-

TABLE 5-1. SOME PHYTOCHROME-MEDIATED PHOTORESPONSES[a]

Elongation (leaf, petiole, stem)	Differentiation of primary leaves
Hypocotyl hook unfolding	Formation of tracheary elements
Unfolding of grass leaf	Differentiation of stomata
Sex expression	Changes in rate of cell respiration
Bud dormancy	Synthesis of anthocyanin
Plastid morphology	Increase in protein synthesis
Plastid orientation	Increase in RNA synthesis
Root development	Formation of phenylalanine deaminase (an enzyme)
Rhizome formation	Changes in the rate of fat degradation
Bulb formation	Changes in the rate of degradation of reserve
Leaf abscission	protein
Epinasty	Auxin catabolism
Succulency	Incorporation of sucrose into growing buds
Enlargement of cotyledons	(plumular tissue)
Hair formation along cotyledons	Permeability of cell membranes
Formation of leaf primordia	Photoperiodism
Seed germination	Seed respiration
Flower induction	Lipoxygenase metabolism

[a]From G. Ray Noggle and George J. Fritz, *Introductory Plant Physiology,* © 1976, p. 160. Reprinted by permission of Prentice-Hall, Inc., Englewood Cliffs, New Jersey.

ing. The flowering behavior of plants in response to photoperiod varies. Photoperiodic needs for flower production may be sufficient alone, may not be needed at all, or may be needed in addition to vernalization.

The classification of flowering behavior in response to photoperiod does not depend on day length per se, but upon a factor called the *critical night length* (Fig. 5-8). This is the night duration of the photoperiod required to induce flowering. The actual value has to be determined experimentally for each species. Plants that flower with nights longer than the critical night length are *short-day plants,* and those that flower with nights shorter than the critical night length are *long-day plants.* Those that are unaffected by photoperiod in terms of flowering are called *day-length-neutral* or *indeterminate plants.* Plants that require a night more than the critical night length followed by one of less than the critical night length are *short-long-day plants;* plants requiring the reverse are *long-short-day plants.* Long-day plants usually bloom in spring to summer, short-long-day plants in late spring to early summer, long-short-day plants in late summer to early fall, and short-day plants in the fall. These groups can be categorized further. Those plants with an absolute dependence on day length for flowering are termed *obligate, absolute,* or *qualitative,* whereas those that appear to flower in various photoperiods, but one photoperiod versus another induces flowering either earlier or later, are called *facultative* or *quantitative.* Some examples are shown in Table 5-2. The preceding nomenclature is somewhat incorrect, in that it is the length of the night, and not the day, that is the controlling factor. This was found later, so the original nomenclature (short day, long day) was retained.

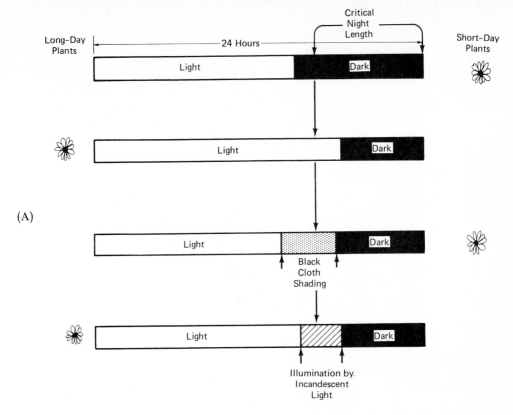

Fig. 5-8. The relationship between critical night length, photoperiod and flowering response. Flowering response can be altered by changes in the photoperiod brought about by natural means (seasonal changes) or artificial means such as darkening plants with black cloth to lengthen the night period or illumination to shorten the night period. B. Response of *Celosia* to variations in the photoperiod. Not only is flowering effected, but so is stem elongation. U.S. Department of Agriculture photo.

TABLE 5-2. PHOTOPERIODIC RESPONSE OF SOME HORTICULTURAL CROPS

Response Unaffected by Temperature

Qualitative Short-Day	Qualitative Long-Day	Quantitative Short-Day
Amaranthus caudatus (Love-lies bleeding)	*Anethum graveolens* (Dill)	*Capsicum frutescens*[a] (Pepper-tabasco)
Cattleya trianaei (Orchid)	*Chrysanthemum maximum*	*Chrysanthemum* × *morifolium*[a]
Chrysanthemum indicum	*Dianthus superbus* (Lilac pink)	*Cosmos bipinnatus*
C. × *morifolium*[a]	*Fuchsia*[a]	*Glycine max*[a] (Soybean)
Coffea arabica	*Hibiscus syriacus* (Shrub althea)	*Senecio cruentus* (Cineraria)
Cosmos sulphureus[a]	*Mentha* × *piperita* (Peppermint)	*Solanum tuberosum*[a] (Potato)
Ipomoea batatus (Sweet potato)	*Nicotiana sylvestris* (Flowering tobacco)	*Zinnia*
Kalanchoe blossfeldiana	*Phlox paniculata*	
K. pinnata	*Rhaphanus sativus* (Radish)	**Quantitative Long-Day**
Phaseolus vulgaris[a] (Kidney bean)	*Rudbeckia hirta* (Black-eyed susan)	*Brassica rapa;* rapifera group (Turnip)
Zea mays[a] (Corn)	*Scabiosa ucranica*	*Camellia japonica*
	Sedum spectabile (Showy stonecrop)	*Dianthus barbatus* (Sweet william)
	S. telephium	*Nigella damascena* (Love-in-a-mist)
	Spinacia oleracea (Spinach)	*Solanum tuberosum*[a] (Potato)
Long-Short-Day		
Cestrum aurantiacum		
C. diurnum		
Kalanchoe tubiflora		
Short-Long-Day		
Echeveria harmsii		
Symphyandra hoffmannii		

[a]Does not include all cultivars.

Photoperiodic response may be altered by temperature; some plants may be short- or long-day plants at some specific temperature range, but day-length neutral plants below that range. The photoperiod effect and temperature relationship are not restricted to flowering. Germination of some seeds can be enhanced or inhibited by long days, but germination of these is also promoted by low temperatures. The formation of dahlia tubers, gladiolus cormels, and potato tubers are promoted by short days, whereas the formation of onion bulbs is a long-day phenomenon. Rooting of cuttings is often influenced by photoperiod: short days favor rooting of certain species of *Abelia, Ilex, Juniperus, Magnolia,* and *Weigela,* whereas other species of *Ilex* and *Juniperus* are favored by long days.

Soil

The roots of a plant receive nutrients, water, oxygen, and mechanical support from their environment—the soil. The capacity of the soil to supply water is a function of its physical properties, and its capacity to supply nutrients is dependent upon chemical

TABLE 5-2. Continued

Response Affected by Temperature[b]		
Day-Neutral	*Qualitative Short-Day*	*Qualitative Long-Day*
Browallia speciosa 'Major'	*Chrysanthemum morifolium*[a]	*Beta vulgaris* (Beet)
Calendula officinalis	*Cosmos sulphureus*[a]	*Brassica rapa;* pekinensis group
Cucumis sativus	*Euphorbia pulcherrima*	(Chinese cabbage)
(Cucumber)	(Poinsettia)	
Fragaria vesca	*Fragaria* × *ananassa*	*Cichorium intybus*
(Woodland strawberry)	(Garden strawberry)	(Common chicory)
Gardenia jasminoides	*Ipomoea purpurea*[a]	*Delphinium* × *cultorum*[a]
Helianthus annuus[a]	(Morning glory)	*Dianthus gratianopolitanus*
Ilex aquifolium	*Salvia splendens*[a]	(Cheddar pink)
(English holly)		
Lunaria annua	*Quantitative Short-Day*	*Quantitative Long-Day*
(Honesty, annual strain)		
Lycopersicon	*Allium cepa*[a] (Onion)	*Antirrhinum majus*
lycopersicum[a] (Tomato)	*Chrysanthemum morifolium*[a]	(Snapdragon)
Pisum sativum[a] (Pea)	*Schlumbergera truncata*	*Begonia semperflorens*
Apium graveolens		*Campanula persicifolia*
(Celery)		(Willow bellflower)
Daucus carota (Carrot)		*Centaurea cyanus* (Cornflower)
Geum bulgaricum		*Dianthus caryophyllus*[a]
Vicia faba (Broad bean)		*Digitalis purpurea* (Foxglove)
		Lactuca sativa[a] (Lettuce)
		Matthiola incana (Stock)
Short-Long-Day		*Petunia* × *hybrida*
Poa pratensis		
(Kentucky bluegrass)		
Campanula medium		
(Canterbury bells)		

[b] For classification of photoperiodic response changes induced by temperatures above and/or below the normal range causing the above listed responses, see D. Vince-Prue, *Photoperiodism in Plants.* New York: McGraw-Hill Book Co., 1975.

properties. These combined properties regulate root extension. These aspects will be examined in more detail.

PHYSICAL PROPERTIES

The solid inorganic portion of soil consists of sand, silt, and clay. Sand ranges in size from 0. 05 to 2.0 millimeters (mm) (very fine to very coarse), silt from 0.002 to 0.5 mm, and clay is less than 0.002 mm. The varying percentages of these particles determines the *soil texture*. A graphic guide for the determination of the textural names of soils based on particulate percentages is shown in Figure 5-9.

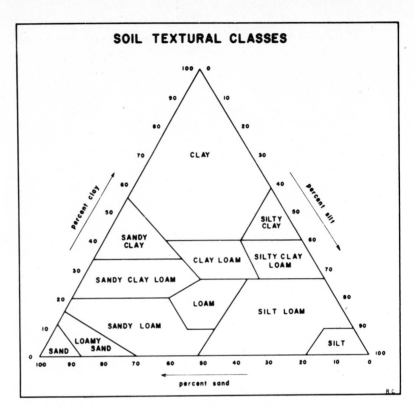

SOIL TEXTURAL CLASSES

Fig. 5-9. Chart showing the percentage of clay (below 0.002mm d.), silt (0.002 to 0.05mm d.) and sand (0.05 to 2.0 mm d.) in the basic soil textural classes.

Sand. Soil material that contains 85 percent or more of sand; the percentage of silt, plus 1 ½ times the percentage of clay, shall not exceed 15 percent. *Loamy sand.* Soil material that contains at the upper limit 85 to 90 percent sand, and the percentage of silt plus 1 ½ times the percentage of clay is not less than 15 percent. At the lower limit it contains not less than 70 to 85 percent sand, and the percentage of silt plus twice the percentage of clay does not exceed 30. *Sandy loam.* Soil material that contains either 20 percent clay or less, and the percentage of silt plus twice the percentage of clay exceeds 30 percent, and 52 percent or more sand; or less than 7 percent clay, less than 50 percent silt, and between 43 and 52 percent sand. *Loam.* Soil material that contains 7 to 27 percent clay, 28 to 50 percent silt, and less than 52 percent sand. *Silt loam.* Soil material that contains 50 percent or more silt and 12 to 27 percent clay, or 50 to 80 percent silt and less than 12 percent clay. *Silt.* Soil material that contains 80 percent or more silt and less than 12 percent clay. *Sandy clay loam.* Soil material that contains 20 to 35 percent clay, less than 28 percent silt, and 45 percent or more sand. *Clay loam.* Soil material that contains 27 to 40 percent clay and 20 to 45 percent sand. *Silty clay loam.* Soil material that contains 27 to 40 percent clay and less than 20 percent sand. *Sandy clay.* Soil material that contains 35 percent or more clay and 45 percent or more sand. *Silty clay.* Soil material that contains 40 percent or more clay and 40 percent or more silt. *Clay.* Soil material that contains 40 percent or more clay, less than 45 percent sand, and less than 40 percent silt. From Donald Steila, *The Geography of Soils: Formation, Distribution, and Management,* © 1976, p. 15-16. Reprinted by permission of Prentice-Hall, Inc., Englewood Cliffs, New Jersey.

Soil particles are grouped into larger, secondary particles termed *aggregates*. The extent of aggregation varies with many factors: organic-matter content, chemical nature and amount of clay, nature of soil micro-organisms, cultivation, temperature, and rainfall. *Soil structure* refers to how soil particles are grouped into aggregates. *Peds* are natural aggregates that are reasonably water stable; *clods* are artificial aggregates that are not water stable. Peds are categorized (Fig. 5-10) according to type (shape and arrangement), class (size), and grade (degree of distinctness). Soil structure influences water infiltration, water-holding capacity, and aeration.

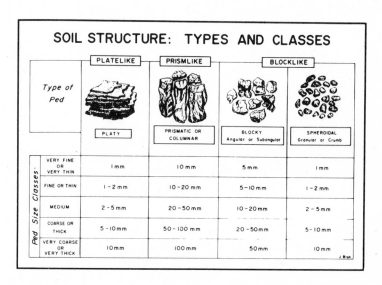

Fig. 5-10. Various types and classes of soil structure. From Donald Steila, *The Geography of Soils: Formation, Distribution, and Management*, © 1976, p. 17. Reprinted by permission of Prentice-Hall, Inc., Englewood Cliffs, New Jersey.

Between the soil particles and aggregates are pore spaces, which comprise 30 to 60 percent of the soil volume. Small pores (micropores) generally are filled with water, and large pores (macropores) contain air alone or air and water together. Sandy soil is dominated by macropores and has good aeration, rapid water absorption, poor water retention, and warms up fast in the spring. Clay soils contain mostly micropores and hence have poor aeration and drainage, but high water retention. They are slower to warm up in the spring. The best soil structure contains both macropores and micropores.

Soil moisture and water movement have been discussed earlier in this chapter. Soil atmosphere is also very important. Roots respire and need oxygen. Their activities along with the animal life of the soil delete the oxygen and raise the concentration of carbon dioxide in the soil. Inadequate pore space will hinder diffusion of oxygen into and carbon dioxide out of the soil. Excess water will reduce the levels of soil oxygen. Inadequate levels of oxygen cause plants to grow poorly or die.

Soils also contain organic matter in addition to the inorganic materials already discussed. *Organic matter* includes all the living (Fig. 5-11) and dead matter associated with the soil. The nondegraded twigs, leaves, or stalks on the surface are referred to as *litter*. Litter reduces the impact of rain, and the resulting reduction of surface runoff and erosion means more water for soil entry. *Humus* includes the stabilized fraction of organic matter left after enzymic decomposition by microorganisms of the plant and animal residues. Humus improves the cation exchange capacity of the soil, and under most conditions its capacity for cation exchange exceeds that of clay.

Organic-matter content is desirable for horticultural purposes. Crop growth depletes the reserves of organic matter, and it becomes necessary to replace it by the techniques discussed in Chapter 7. Organic matter added beyond the naturally present amounts can be used to improve cer-

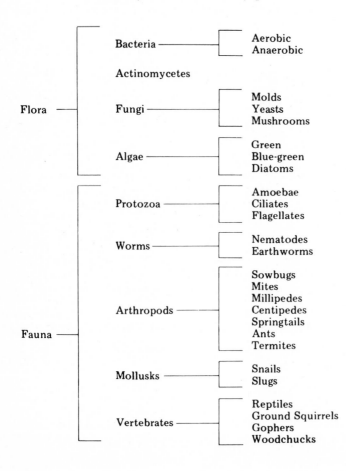

Fig. 5-11. Various types of living organisms found in soils. From Donald Steila, *The Geography of Soils: Formation, Distribution, and Management*, © 1976, p. 23. Reprinted by permission of Prentice-Hall, Inc., Englewood Cliffs, New Jersey.

Introduction to Horticultural Science | Part I

tain soils. Addition of organic matter to a sandy soil (mineral soil with at least 70 percent sand; particle size varies between 0.05 and 2.0 mm, making it coarse textured) improves water retention, mainly because of micropores in the organic matter. An incidental improvement to that of water retention is an improvement of soil structure. The improvement of soil structure in fine-textured soils (particle size of less than 0.05 mm), consisting mostly of clay or silt-clay (40 percent or more), is the main reason for adding organic matter to these soils. The workability or *tilth* of these fine-textured soils is improved by the resulting soil structure improvement, as is aeration and water movement.

In all soils the decomposition of organic matter provides some nutrients and trace elements needed by plants. Organic matter may also act in a buffering capacity against rapid chemical changes when lime or fertilizers are added. The humic fraction is involved in cation exchange (see chemical properties next). Decomposing organic matter releases chemicals that free some of the normally unavailable phosphorus.

CHEMICAL PROPERTIES

The chemically active portion of soil consists mostly of humus and clay particles with diameters of less than 0.002 mm. Particles of this size are called *colloids*. The structure of colloidal clay is crystalline, and humus is amorphous.

Plant roots take up and release ions. Therefore, ions in the liquid and gaseous phase in the soil are in a changing equilibrium with those on the surfaces of the colloidal particles. Colloidal clay or humus carries a negative charge, which absorbs cations. These absorbed cations may be exchanged or replaced by another cation; this process is known as *cation exchange* (Fig. 5–12). This process helps to reduce the loss of nutrients through leaching. Cation exchange may take place between cations adsorbed on colloidal particles and cations in soil solution, cations released by plant roots (usually H^+), and cations on other colloidal particles. Cations are not randomly exchanged, but are displaced by another cation with a higher absorptive affinity for that site. The displacement strengths are in decreasing order: H^+, Ca^{2+}, Mg^{2+}, K^+, and Na^+. Uptake of cations by roots appears to be both by direct exchange of root-released H^+ for another cation bound on an adjacent soil particle (contact exchange) without the cations entering the soil solution, and by H^+ from the root exchanged for a cation that enters the soil solution and is then taken up. The predominant mode depends on several soil variables.

Soils, of course, vary in their cation-exchange capacity, depending upon the amounts and types of clay and amounts of humus present. The cation-exchange capacity of the colloidal particles in increasing order are kaolinite, illite, montmorillonite (all clays), and humus.

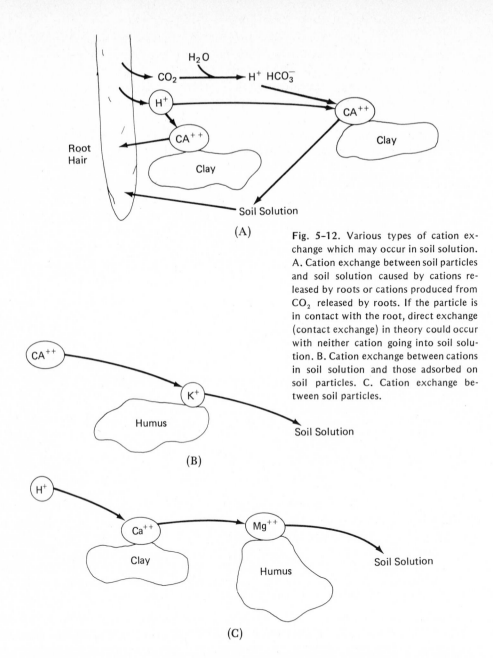

Fig. 5-12. Various types of cation exchange which may occur in soil solution. A. Cation exchange between soil particles and soil solution caused by cations released by roots or cations produced from CO_2 released by roots. If the particle is in contact with the root, direct exchange (contact exchange) in theory could occur with neither cation going into soil solution. B. Cation exchange between cations in soil solution and those adsorbed on soil particles. C. Cation exchange between soil particles.

Under certain conditions small amounts of anions can be held in an exchangeable form. These include NO_3^- and SO_4^-. $H_2PO_4^-$ is an exception, in that it is firmly held with little present in soil solution. Anion exchange can be increased by organic matter, kaolinate, and iron. Anions for the most part exist free in the soil solution and are taken up by roots directly from the soil solution. As such, anions are more susceptible to leaching than cations.

Introduction to Horticultural Science / Part I

The term *soil reaction* or soil pH refers to the degree of acidity or alkalinity. At neutral pH the number of H^+ and OH^- ions are equal, such as in freshly distilled water. If the amount of H^+ ions becomes increasingly larger, the solution increases in acidity. If the OH^- ions increase instead, the solution increases in alkalinity. The actual expression is pH = log $1/(H^+)$, where (H^+) is the active concentration of H^+ ions in grams per liter. The pH scale runs from 0 to 14, and soil pH generally runs from 4 to 8. Extremes include the acid forest humus layers in the Northeast, which can be as low as pH 3.5, and the alkaline desert areas in the West may reach pH 9. In general, soils are acidic in areas of high rainfall in the eastern United States and the Pacific Northwest. Dry areas are usually neutral to alkaline, such as the arid western states. The pH scale is shown in Figure 5-13 with both horticultural and other examples.

Fig. 5-13. The pH scale showing optimal pH range for certain horticultural crops, natural pH of certain soils, and the pH of some common chemicals.

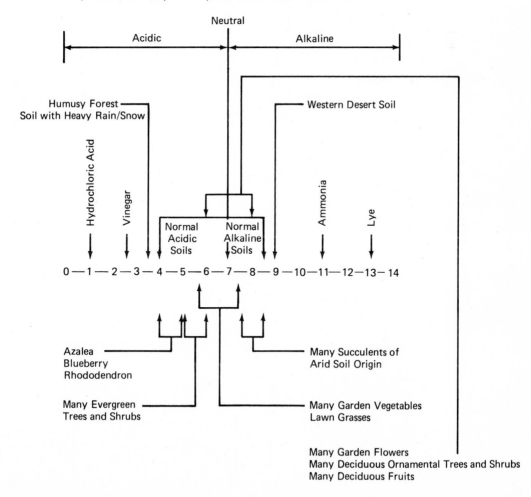

Plant development and pH are important. Plants have an optimal pH or range of pH for development. Higher or lower values result in less than optimal development. The availability of nutrients is also a function of pH. This will be discussed further in the section on nutrients.

Buffers enable solutions to resist changes in pH. Soils have some degree of buffering capacity, which increases with increasing clay and humus contents. Clays vary in their buffering capacity. Alteration of pH by addition of lime or sulfur would require larger amounts in a soil with increased buffering capacity.

PLANTS AND SOIL NUTRIENTS

Essential elements needed by plants and their uptake by roots have been discussed in Chapter 3 under nutrient absorption. Uptake can only occur if the nutrients are present in available forms, and optimal plant development is dependent upon sufficient levels. Nutrient forms and levels are influenced by solubility properties, soil texture, organic matter content, cation-exchange capacity, and pH. Some of these have been discussed previously. The effect of pH on nutrient availability is shown in Figure 5-14.

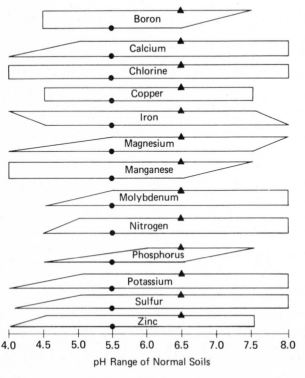

Fig. 5-14. Nutrient availability in the pH range of most normal soils. The top of the bar line is for mineral soils (less than 20% organic matter) and the bottom bar line is for organic soils (greater than 20% organic matter such as peat and muck soils). The optimal nutrient availability in mineral soils occurs around pH 6.5 (▲) and around 5.5 in organic soils (●). Nutrient availability does not stop at the indicated cutoff points, but is substantially below (at least 40% less) that which occurs at the maximum. The optimal pH does not in itself guarantee sufficient nutrient availability, since the actual concentration of nutrient present as well as the plant's requirement for that nutrient must be considered. Some plants develop well at a pH where others fail, simply because their requirement for a nutrient limited by solubility at that pH is greatly reduced compared to the other plant.

Introduction to Horticultural Science / Part I

Nitrogen is a constituent of proteins, amino acids, purines, pyrimidines, chlorophyll, and many coenzymes. As such, the vegetative growth of plants (leaves, stems, and roots) is especially dependent on nitrogen. The atmosphere contains 78 percent nitrogen by volume, yet it is the element that is most often lacking for plant development. This lack happens because plants cannot absorb nitrogen directly as a gas from the atmosphere. It must exist in the soil in a form suitable for nutrient absorption. These forms are NH_4^+ and NO_3^-. Reactions such as these are part of the nitrogen cycle shown in Figure 5-15.

The conversion of atmospheric nitrogen into NO_3^- and NH_4^+ in the soil is called *nitrogen fixation.* Some nitrogen fixation may occur through the incorporation of atmospheric NO_3^- and NH_4^+ into the soil by rainfall. However, most of it results from bacterial action. Nitrogen-fixing bacteria of the genus *Rhizobium* live in nodules on roots of plants in the family Leguminosae. Other nitrogen-fixing microorganisms include unidentified species of actinomycetes in a symbiotic relation with several nonleguminous plants of horticultural interest, such as *Alnus* sp. Through bacterial action, atmospheric nitrogen is converted into NH_4^+, which is utilized by the bacteroids and plants. Nitrogen fixation also occurs in the soil through some free-living bacteria and to a much lesser extent through some actinomycetes and fungi. These microorganisms degrade organic material into amino acids by enzymic means. Upon the death and decay of these microorganisms, other bacteria convert the amino acids into NH_4^+ and NO_3^-.

If NH_4^+ and NO_3^- are limited in the soil, competition for these compounds takes place between plants and microorganisms. The microorganisms win, and symptoms of nitrogen deficiency appear in the plants. This occurs if the soil carbon-to-nitrogen ratio by weight is greater than 10 : 1, such as happens when incompletely degraded organic matter is added to the soil, or nitrogen is low and a heavy organic mulch is used. Addition of external sources of nitrogen will correct the ratio.

Nitrogen is easily lost from the soil by leaching, utilization by plants, and the conversion of nitrates into atmospheric nitrogen. The latter, a bacterial action, is favored by a lack of soil oxygen, such as with a waterlogged soil. Excessive nitrogen promotes vegetative growth at the expense of reproductive growth. Addition of nitrogen is often decreased as the reproductive cycle is nearing.

Phosphorus is found in a number of compounds in the plant, such as nucleic acids, phospholipids, ATP, and NADP. As such, it is important in cell division and growth, especially in areas of rapid development, such as the meristem. Phosphorus is often limiting, even though sufficient amounts are present, because it is present in unavailable forms with iron and aluminum below pH 5 and calcium above pH 7 (Fig. 5-16). Losses of phosphorus by leaching are low. The availability of phosphorus is increased by organic matter, which produces organic acids during decomposition that complex more readily with the iron and aluminum than

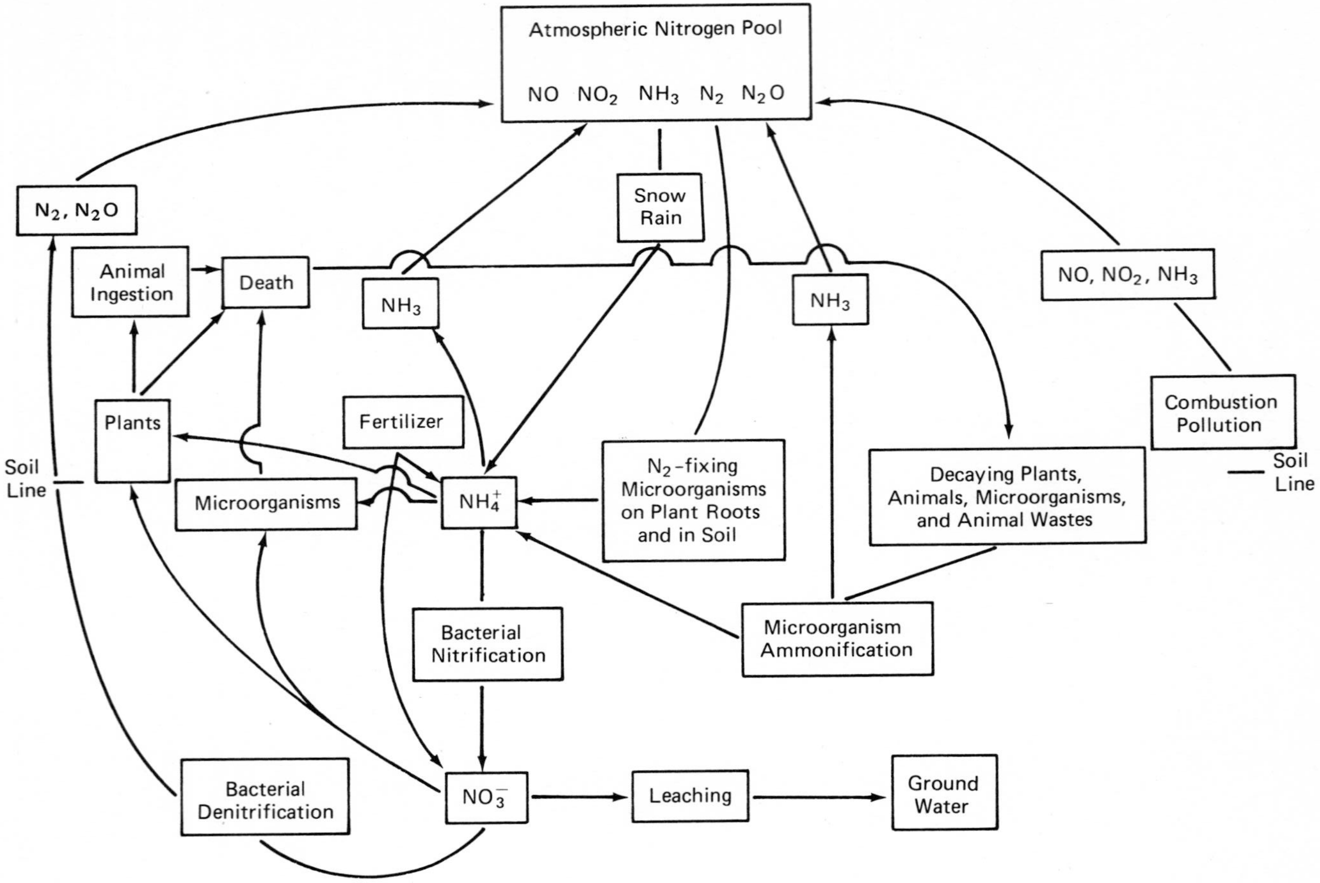
Atmospheric Nitrogen Pool
NO NO2 NH3 N2 N2O
N2, N2O
Snow
Rain
Animal
Ingestion
Death
NH3
NH3
NO, NO2, NH3
Plants
Fertilizer
Combustion
Pollution
Soil
Line
Microorganisms
NH4+
N2-fixing
Microorganisms
on Plant Roots
and in Soil
Decaying Plants,
Animals, Microorganisms,
and Animal Wastes
Soil
Line
Bacterial
Nitrification
Microorganism
Ammonification
Bacterial
Denitrification
NO3−
Leaching
Ground
Water

Fig. 5-15. The nitrogen cycle. Losses are incurred by the leaching of highly soluble NO_3^-, the erosion or blowing away of soil, the removal of plant material prior to its death, fires, and during certain microbial processes. The loss of nitrogen while in the ammonium form (NH_4^+) is considerably less than in the nitrate form (NO_3^-). Soil biochemistry, however, favors the formation of NO_3^- from NH_4^+. NH_3 can be lost from warm soils with an alkaline pH and from decaying organic matter with a carbon/nitrogen ratio below 30. Loss of NO_3^- through denitrification occurs in wet, anaerobic soils.

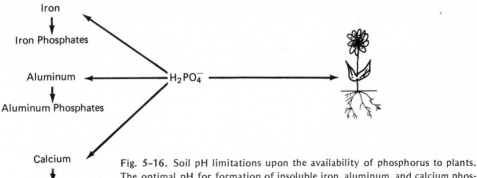

Fig. 5-16. Soil pH limitations upon the availability of phosphorus to plants. The optimal pH for formation of insoluble iron, aluminum, and calcium phosphates is pH 3, 5, and 8, respectively. Not only does acid pH tie up the phosphorus, but it also ties up another essential element: iron. Chelating agents can be used in soils to reduce the loss of iron and other trace elements such as zinc. These form a specific soluble complex with the trace metal and prevent its loss through formation of insoluble compounds. Iron and zinc chelates are widely used in areas where deficiencies are encountered.

phosphorus. Organic matter is also a source of phosphate, which is released by microbial action.

Potassium is very important as an enzyme activator in plants. It is also involved in membrane permeability and translocation. Much soil potassium is present in minerals that dissolve slowly, thereby limiting its availability. The readily available potassium is regulated by cation exchange. Potassium leaching increases as the amounts of clay and humus decrease.

Calcium is needed for the synthesis of calcium pectate in the cell wall. It is also involved in mitosis and is present in membranes as the calcium salt of phosphatidylcholine. Calcium is reasonably soluble and exists as an exchangeable cation on clay and humus. Calcium is derived from organic matter and soil minerals.

Magnesium is a constituent of chlorophyll, ribosomes, and magnesium pectate. The latter is present in cell walls. It is also an enzyme activator. Magnesium is also available as an exchangeable cation, but in amounts less than calcium. It is derived both from organic matter and minerals present in the soil.

Sulfur is present in certain amino acids that are involved in enzymic activities and cross-bonding of proteins. The B vitamins thiamin and biotin contain sulfur. Organic matter and insoluble minerals are the principal sources of sulfur, which is released by microbial action and weathering, respectively.

The remaining essential elements supplied in the soil are iron, copper, manganese, zinc, molybdenum, boron, and chlorine. These are usually classed as micronutrients. Some are present as parts of enzymes, such as

Chapter 5 / Plant Environments

iron, copper, zinc, and molybdenum. Others may be required for the synthesis of certain compounds. Most are derived from weathering of minerals and from organic matter to a lesser extent. Many of these can be toxic to plants if present in more than trace amounts.

Nutrients often become more available to plants because of the activities of soil microorganisms called *mycorrhizae.* These fungi are associated with the roots of many plants, including many deciduous and evergreen trees, shrubs, and herbaceous plants. Ion uptake is greater because of increased absorptive area and a greater efficiency of ion absorption by mycorrhizae. The plant probably supplies carbon compounds such as sugars, amino acids, and probably vitamins to the fungi. In turn, the fungi may supply mineral nutrients and water to the roots. There are two types: ectotrophic mycorrhizae and endotrophic mycorrhizae. The contribution of each of the mutualistic relationship is not well known.

Air Movement and Composition

Wind influences vegetation through physical impact, acceleration of moisture loss, and convective heat transfer. The uptake of carbon dioxide during photosynthesis is also increased, as a lack of air circulation can lead to localized depletion of carbon dioxide in the immediate environment of the leaf. Pollination and seed transfer are also assisted by moderate winds. Severe winds can damage plants, interfere with pollination, cause premature droppage of fruits and nuts, and enhance soil erosion.

Pollutants have adverse effects on plants and their development. Air pollutants include ethylene, lead, sulfur dioxide, ozone, boron, fluorides, hydrocarbons, and photochemical products such as peroxyacetyl nitrate. Plants show different degrees of physiological resistance toward these pollutants. Some appear to suffer no damage, others mild leaf damage (Fig. 5-17), and others die. The relationship between air pollution and plants is important to the horticulturist, especially in the urban environment, where the success of a city tree may rest on its pollution tolerance. In Part Three the pollution tolerance of plants is mentioned where known. It is unfortunate that the pollution susceptibility must be considered when choosing or breeding plants for use in many urban environments.

Plant Tropisms and Nastic Movements

Movement of plant organs takes place in response to environmental stimuli. Movements in which the direction is determined by the direction from which the stimulus originates are termed *tropisms.* Movements that bear no relationship to the direction from which the stimulus arose are termed *nastic* movements. Examples of tropisms include phototropism, geotropism, thigmotropism, and skototropism. Nastic movements include photonasty, thermonasty,

Introduction to Horticultural Science / Part I

Fig. 5-17. Symptoms of air pollution damage on foliage. A. Damage from sulfur dioxide, a major air pollutant. Chlorophyll-containing cells are destroyed, leading to white areas between leaf veins. B. Sun acting on nitric oxide, nitrogen dioxide and certain hydrocarbons produced by combustion of fossil fuels yields photochemical oxidants which damage plants. Flecking in leaves is often indicative of such pollution. U.S. Department of Agriculture photos.

and thigmonasty. The exact causes and mechanisms of these movements are poorly understood. Phytohormones are probably involved in growth movements, which stem from uneven rates of growth on different sides of the organ. Turgor movements caused by losses and gains of cellular water are probably also involved.

These growth movements are defined as follows. *Phototropism* is a growth movement in response to one-sided illumination, such as the bending of the plants toward the sun. *Geotropism* is a growth movement influenced by gravity; that is, the root grows downward and the stem upward even if a bulb is planted upside down. *Thigmotropism* is a growth movement toward an object as a result of contact, such as with tendrils of vine. *Skototropism* is the growth movement of a vine along the ground toward an upright dark object, such as a tree. *Photonasty, thermonasty,* and *thigmonasty* are nastic movements in response to light, temperature, and touch. An example of photonasty is the "sleep" movement of beans: during the day the first pair of leaves above the cotyledon is horizontal, and at night they are folded downward alongside the stems.

FURTHER READING

Bidwell, R. G. S., *Plant Physiology* (2nd ed.). New York: Macmillan, Inc., 1978.

Buckman, H. O., and N. C. Brady, *The Nature and Properties of Soils.* New York: Macmillan, Inc., 1971.

Critchfield, N. J., *General Climatology* (3rd ed.). Englewood Cliffs, N.J.: Prentice-Hall, Inc., 1974.

Daubenmire, R. F., *Plants and Environment: A Textbook of Plant Auto-ecology.* New York: John Wiley & Sons, Inc., 1967.

Etherington, J. R., *Environment and Plant Ecology.* New York: John Wiley & Sons, Inc., 1975.

Hillel, D., *Soil and Water: Physical Principles and Processes.* New York: Academic Press, Inc., 1971.

Larcher, W., *Physiological Plant Ecology.* New York: Springer-Verlag, New York, Inc., 1975.

Rosenberg, N. J., *Microclimate: The Biological Environment.* New York: John Wiley & Sons, Inc., 1974.

Smith, H., *Light and Plant Development.* Woburn, Mass.: Butterworth Inc., 1976.

Steila, D., *The Geography of Soils.* Englewood Cliffs, N.J.: Prentice-Hall, Inc., 1976.

Thompson, L. M., and F. R. Troeh, *Soils and Soil Fertility* (3rd ed.). New York: McGraw-Hill Book Co., 1973.

Vince-Prue, D., *Photoperiodism in Plants.* New York: McGraw-Hill Book Co., 1975.

part two

INTRODUCTION TO APPLIED HORTICULTURAL SCIENCE

Basic horticultural science and applied science may be likened to the artist and the tools of his craft. Science or the artist holds forth the knowledge of what is possible, but unless the science is applied or the artist uses his tools, the potential will not be expressed. And indeed, like the artist, the applied science or technology of horticulture can lead to works that vary from bad to practical to very beautiful.

The competent horticulturist knows that his technology is only as good as his science, and knows that technological improvement arises in large part from advances in the sciences. As horticultural science consists of a blend of many disciplines, so does the area of technology. It is these many diverse aspects that will be covered in the following chapters.

Plant Propagation and Interim Care

The propagation of plants and ultimately their improvement are fundamental to the horticulturist. Success in this area requires a basic knowledge of the science behind plant propagation. This science includes plant structure, development, and metabolism, all of which have been covered previously in Part One. After this, a knowledge of technical skills, or the art of propagation, is essential. However, unless these techniques can be matched with the appropriate plant materials, all is lost. Techniques and accessories germane to both sexual and asexual plant propagation will be considered.

Propagation Media Certain properties have come to be required of materials used in propagation media. In terms of chemical properties, salt contents must be low, exchange capacity must be sufficient to hold and release nutrients, and the composition must remain unaltered during pasteurization or sterilization. The latter becomes important for media that are not naturally sterile or are to be reused. Sufficient nutrients must be present in media used for seed propagation or extended vegetative propagation.

Physical properties must be sufficient to support seeds or vegetative propagules and yet not be restrictive such as to reduce aeration or water drainage. A reasonable degree of water retention is also necessary. Dryness or wetness should not cause any changes in these physical properties. The material should also be free of organisms harmful to plants.

BASIC MATERIALS

Basic materials can be inorganic (Fig. 6-1), such as perlite, pumice, sand, Styrofoam, and vermiculite. Organic forms (Fig. 6-1) include compost, peat, sphagnum, and wood by-products. Costs for all have been increasing, but for perlite and vermiculite even more so, because of energy costs for heating them to prepare a suitable product for horticultural use. The chemical and physical properties of each are summarized in Table 6-1.

The inorganic materials are usually added to mixtures to improve aeration and drainage, whereas the organic materials are utilized to increase water retention and cation-exchange capacity. At present, peat moss, perlite, and vermiculite are the most widely used. As used here the term peat moss refers to moss peat, the moderately degraded remains of hypnum and sphagnum mosses. Sphagnum moss refers to undegraded, fresh, dried remains of *Sphagnum* species.

Fig. 6-1. Basic materials used in propagation. Top row, left to right: perlite, pumice, sand (white quartz), styrofoam and vermiculite. Bottom row, left to right: compost, peat moss, sawdust, sphagnum moss and shredded bark. Author photo.

Introduction to Applied Horticultural Science / Part II

TABLE 6-1. BASIC MATERIALS FOR PROPAGATION

	Source	Wt.[a] (dry)	Sterile	pH	Nutrients[b]	Buffering[c] Capacity	Cation[d] Exchange	Water[e] Retention	Aeration Drainage	Hort.[f] Form (P)	Use[g]	Comments
Inorganics												
Perlite	Heated volcanic silicate	VL	Yes	6.5–7.5	None	None	None	Fair	Very good	1/16–1/8 D_i; 1.6–3.2 D_m	R,M	Nasal irritant
Pumice	Volcanic silicate	M	No	Neutral	Slight (K)	Low	Slight	Fair	Good	1/16–5/16 D_i; 1.6–8 D_m	R,M	Limited availability
Sand	Soil mineral	H	No	Neutral	None	None	None	Poor	Good	0.002–.079 D_i; 0.05–2 D_m	R,M	Best is quartz sand
Styrofoam	Polystyrene	VL	No	Neutral	None	None	None	Poor	Very good	0.16–0.47 D_i; 4–12 D_m	M	Some pasteurization problems
Vermiculite	Heated micaceous mineral	VL	Yes	6.5–7.5	Slight (K,M$_g$)	Moderate	High	Fair	Good	0.003–0.12 D_i; 0.075–3 D_m	S,R,M	Mix with perlite for long use
Organics												
Bark	Softwood tree waste	L	No	3.5–6.5	Slight (N,P,K,Ca,Mg)	Moderate	Low-medium	Good	Good	1/16–5/10 D_i; 1.6–13 D_m	M	Age 30 days with nitrogen
Compost	Plant and animal wastes	L	No	5.5–8.5	Low (N,P,K,Ca,Mg)	Moderate-high	High	Good	Good	1/16–5/10 D_i; 1.6–13 D_m	M	Slow release of nitrogen
Peat moss	Decayed water vegetation	L	No	3.8–4.5	1% N	Moderate	High	Good	Good	1/16–0.4 D_i; 1.6–10 D_m	M	Use wetting agent
Sawdust	Softwood tree waste	L	No	3.5–6.8	Slight (N,P,K,Ca,Mg)	Moderate	Low-medium	Good–very good	Fair	1/16–0.4 D_i; 1.6–10 D_m	M	Age 30 days with nitrogen
Sphagnum	Fresh, dehydrated *Sphagnum* sp.	VL	Yes	3.5	None	Moderate	High	Very good	Fair–good	1/16–3/8 D_i; 1.6–9.5 D_m	S,M	Contains natural fungicides

[a]VL, under 8 lb/ft^3 (128 kg/m^3); L, 8–14 lb/ft^3 (224 kg/m^3); M, 15–30 lb/ft^3 (480 kg/m^3); H, 100 lb/ft^3 (1600 kg/m^3).

[b]None have sufficient nutrients for long-term use.

[c]The higher the buffering capacity, the more lime is needed for pH adjustment.

[d]The higher the cation-exchange capacity, the lesser are problems with salinity and nutrient leaching.

[e]Fair: absorbs 3 to 4 times its weight in water; good: 5 to 10 times; very good: 10 to 20 times.

[f]P, particles; D_i, diameter in inches; D_m, diameter in millimeters.

[g]R, rooting medium alone; M, added to mixtures; S, for seeds alone (smaller particles).

[h]Moderately decayed remains of *Hypnum* and *Sphagnum*, i.e., moss peat.

Styrofoam is an inexpensive substitute being used in place of perlite, but it cannot withstand steam sterilization or chemical sterilization by certain treatments, such as chloropicrin or methyl bromide. Peat moss is difficult to wet, and the use of a wetting agent, such as Aqua-Gro, is advisable. Bark and sawdust are not replacing peat yet, but they are becoming increasingly popular. They must be treated by adding enough nitrogen to supply the needs of microorganisms during decomposition as well as those of the plants, and at least 30 days of aging is needed to minimize phytotoxicity problems. For example, redwood sawdust is amended with 0.5 percent ammonium nitrate on a dry weight basis (3 pounds per cubic yard).

PROPAGATION MIXTURES

Propagation media are often composed of mixtures of the preceding materials for germination of seeds and for vegetative propagation, such as the rooting of cuttings. The materials for these mixtures should be uniform and of correct size; if not, they are screened for uniformity and the elimination of large particles. Some form of fertilizer may be added as an amendment. The ingredients are moistened before mixing, which can be done by hand or mechanized mixers. Mixtures are allowed to stand for at least 24 hours to equilibrate moisture content. Mixtures are usually pasteurized (see later section) unless such treatment will cause harmful alteration.

The oldest propagation mixtures were those based on soil for seed germination and sand for rooting cuttings and other asexual propagation. Their use has declined for a number of reasons. Problems with soil mixtures include increasing scarcity of loam, variations in chemical and physical properties, and heavy weight. Sand when wet has heavy weight, and its water retention is poor.

One soil mixture that persists in use today was developed around 1939 at the John Innes Horticultural Institution in England through the research of W. J. C. Lawrence and J. Newell (see Further Reading). The formulation *by volume* of their mixture for seed propagation is as follows:

John Innes Seed Compost

1 part sand
1 part peat moss
2 parts loam
2 lb superphosphate per cubic yard of mixture
1 lb ground limestone per cubic yard of mixture

The loam originally consisted of nutrient-enriched, composted pasture turf and strawy manure. An effective substitute appears to be soil from bare, fallow land.

Artificial soil mixtures have been developed. These can be prepared on a large-scale basis. They provide a standardized mixture with little variation between batches, and they are more easily produced and maintained in a pathogen-free basis. Some popular mixes are those developed at Cornell University by J. W. Boodley and R. Sheldrake and at the University of California at Los Angeles by K. F. Baker and colleagues (see Further Reading, this chapter). Many commercial variants of these artificial mixtures are also available. The Cornell "Peat-Lite" mixes do not require any decontamination. The Cornell Peat-Lite for seed germination is as follows.

Peat-Lite Mix C (for Seed Germination)

1 bushel horticultural-grade vermiculite (fine, no. 4)
1 bushel shredded peat moss (at least 70% *Sphagnum* sp.)
4 level tablespoons (3 oz) ammonium nitrate
8 level tablespoons (12 oz) superphosphate (20%); best is powdered
10 level tablespoons (7 ½ oz) ground limestone (best form is dolomitic)

Care must be observed in wetting the peat moss during preparation. A nonionic wetting agent, 3 ounces in 5 to 10 gallons of water, is helpful for wetting 1 cubic yard of mix.

The University of California mixture useful for seed germination is as follows.

University of California Mix B

75% sand (0.05 to 0.5 mm in diameter)
25% peat moss

To each cubic yard add the following:

Mixture to Be Stored

6 oz potassium nitrate
4 oz potassium sulfate
2 ½ lb single superphosphate: $CaH_4(PO_4)_2H_2O$
4 ½ lb dolomite lime
1 ¼ lb calcium carbonate lime
1 ¼ lb gypsum

This provides a moderate amount of available nitrogen for about 2 weeks after germination.

Mixture to Be Used within One Week

2 ½ lb hoof and horn meal or blood meal
6 oz potassium nitrate
4 oz potassium sulfate
2 ½ lb single superphosphate

4 ½ lb dolomite lime
1 ¼ lb calcium carbonate lime
1 ¼ lb gypsum

This provides available nitrogen plus a moderate reserve. It cannot be stored, because the organic nitrogen will break down during storage.

Again, care is needed to wet the peat moss. The use of a nonionic wetting agent (see Cornell mixture) is appropriate. The complete University of California mixture can be steam or chemically pasteurized without any harmful effects.

Seeds can also be germinated in straight shredded or milled sphagnum moss, 0.75 to 1.0-mm vermiculite, or sand. Pasteurization for the latter two is needed. Some fertilizer can be added initially to prolong the need for supplemental fertilization. The better results obtained with the mixtures, plus their convenience in that many variations are available commercially and the need for supplemental fertilization is unnecessary until transplanting, account for their widespread use over straight sand, milled sphagnum moss, or vermiculite.

Vegetative propagation, such as the rooting of cuttings, is possible in straight perlite, pumice, vermiculite, or sand. Perlite or pumice are often used to root leafy cuttings under mist because of their good drainage properties. They can be mixed in varying proportions with peat moss or vermiculite. Clean, sharp sand (builder's sand) is widely used to root cuttings, especially those of evergreens (yews, junipers, arborvitaes). Peat moss is often added to sand to improve water retention, such as 2 parts sand and 1 part peat moss. One part sand and 1 part shredded sphagnum moss is also used. Vermiculite is also used alone, but more often on a 1 : 1 basis with perlite or sand, since long-term use of straight vermiculite results in particle collapse and reduced aeration and drainage. Soil is sometimes used for rooting deciduous hardwood cuttings and root cuttings. A well-aerated sandy loam appears to be best. Pasteurization of these mixtures is recommended and necessary if they are to be reused.

The following University of California mixture can be stored indefinitely and used to root cuttings.

University of California Mix C

50% sand (0.05 to 1 mm in diameter)
50% peat moss

To Be Added to Each Cubic Yard

4 oz potassium nitrate
4 oz potassium sulfate
2 ½ lb single superphosphate
7 ½ lb dolomite lime
2 ½ lb calcium carbonate lime

Prepropagation Treatments of Media

Soils, soil mixtures, and fresh or recycled propagating media may be inhabited by some or all of the following: nematodes, fungi, bacteria, and weed seeds. These may prove harmful to plants. For example, the loss of recently germinated seedlings can occur as a result of a disease termed *damping off*. This disease is the result of fungal attack by species such as *Rhizoctonia* and *Pythium*.

Harmful organisms can be destroyed in propagation media by heat or chemical treatment. Heat treatment can be pasteurization or sterilization. Pasteurization primarily destroys harmful organisms, and sterilization destroys both harmful and beneficial organisms. Pasteurization is more often used in actual practice. The moist material is covered in a bin or bench and steamed through perforated pipes 6 to 8 inches below the surface. A temperature of 180°F (82°C) for 30 minutes will destroy nematodes, pathogenic fungi and bacteria, insects, and most weed seeds. Straight steam has a temperature of 212°F (100°C). By mixing air with steam (4.1 : 1), a temperature of 140°F (60°C) can be attained. This temperature for 30 minutes will destroy many pathogenic organisms, but not weed seeds. However, it will not harm many beneficial organisms, which have an antagonistic effect toward pathogens that may appear through recontamination. This lower temperature also reduces the risk of toxicity problems caused by heat-induced breakdown of chemical compounds, such as nutrients.

Chemical treatment may also be utilized to kill organisms in the propagation media. Less physical and chemical disruption occurs, but the toxicity of these fumigation chemicals requires caution in handling them. The mixture should be moist and at 65° to 75°F (18 to 24°C). The prescribed waiting time after fumigation must be observed to allow escape of fumes from the media. The chemicals used include Vapam® (sodium N-methyldithiocarbamate dihydrate), chloropicrin (tear gas), methyl bromide–chloropicrin mixtures, methyl bromide, and formaldehyde. These treatments vary in their effectiveness, depending on the choice of chemical and observaton of proper treatment conditions.

Chemical drenches can be added to the soil to inhibit the growth of many pathogenic fungi, such as *Botrytis, Pythium, Fusarium, Phoma, Phytophthora, Rhizoctonia,* and *Verticillium.* These chemicals include Terraclor® (pentachloronitrobenzene), Dexon® (*p*-dimethylaminobenzine diazo sodium sulfate), Banrot®, Truban® (5-ethoxy-3-trichloromethyl-1, 2, 4-thiadiazole), and Benomyl® (methyl 1- (butylcarbamoyl) 2-benzimidazolecarbamate), just to name a few. The latter is also a systemic fungicide that can also be applied directly to the plant. It will inhibit a number of soil pathogens, but some cases of induced resistance in pathogens have been reported. As with any chemical treatment, the user must be aware of phytotoxicity problems.

Water Supply

Municipal water is usually treated with chlorine or sodium fluoride. The amounts, however, are not high enough to cause phytotoxicity. Levels of soluble salts should not exceed 2 millisiemens per centimeter (1400 ppm), as plant damage can result. High levels of sodium, such as found with water softened with ion-exchange resins like zeolite, are toxic to plants and affect the water absorption rates and physical structures of soil. However, it is desirable to soften hard water, but in another way, since deposit buildup from hard water can harm mist propagation units and evaporate water cooling systems. An alternate way is to use ion-exchange resins that replace calcium, magnesium, and sodium by hydrogen ions and chlorides, sulfates, and carbonates by hydroxyl ions. A hardness over 6 grains (100 ppm) generally necessitates water softening.

No good method exists for removing boron, which should not exceed 1 ppm. Another water supply will be required if it does, as excess boron is phytotoxic. In addition, the water temperature should be watched, as too cold water will retard growth of plants, seed germination, and even cause chilling injury if cold enough. During the winter, unheated tap water may have a temperature of 34° to 40°F (1 to 4.5°C). Propagation mixes watered thusly could take several hours to return to initial temperature.

Propagation Facilities

Successful propagation is dependent upon control of environmental factors. As such the most widely used propagating structure is the greenhouse. Other useful structures include the hotbed and propagating chambers. Light levels are adequate within these structures, and the control of relative humidity and temperature is feasible. Plants propagated in these structures cannot be set directly outdoors without prior conditioning. Structures used for conditioning recently propagated plants include the coldframe and lathhouse. The coldframe is primarily a hardening or conditioning area for plants in transition (see Chapter 8) from the ideal propagating conditions to the more severe conditions outdoors, although it may be used to propagate some of the hardier plants, such as cole crops. The lathhouse, another "halfway house," is used to acclimate propagated plants to reduced light levels prior to setting them outdoors (see Chapter 8).

Greenhouses are either free standing with even-span, gable-roof construction or attached (lean-to) structures surfaced in polyethylene, fiberglass, or glass. Glass is the most permanent and most expensive. Fiberglass, the most widely used, is less expensive and less breakable, but its light transmission deteriorates with age (some better grades will last up to 20 years); thus replacement or refinishing is required. Polyethylene is very inexpensive, but requires replacement often at yearly intervals. However, recent advances appear to be increasing its useful span. Hotbeds and coldframes are much simpler structures of a non walk-in type that are covered with polyethylene, fiberglass, or glass. Propagating chambers are small units where a high humidity level can be maintained. These structures are enclosed in polyethylene, and the light source can be natural, fluorescent, or perhaps high-pressure metal halide lamps. These units are particularly useful for starting seedlings or rooting leafy cuttings. Lathhouses are primarily a shading material, such as wood strips, aluminum strips, or a woven plastic (saran fabric) supported by wood or pipe members. The management of the environment within the propagation structures mentioned and more specific details will be covered in Chapter 8.

Propagation Aids

CONTAINERS

During propagation, some form of container is necessary (Fig. 6–2), unless the plants are being propagated directly in the ground. These containers include pots of clay, plastic, or fiber, flats, propagating blocks of fiber or peat, containers of metal or asphalt-coated felt paper, cups of Styrofoam or paraffined paper, and even polyethlene bags.

Flats or shallow trays are used for rooting cuttings and germinating seeds. Composition is of wood, plastic, or metal. Drainage is provided through bottom holes. An advantage is ease of mobility and storage, especially for the lightweight plastic flats that nest, which are increasingly replacing wood and metal flats.

Pots are usually round or square and come in many sizes and materials. Clay pots are fragile, prone to surface accumulations of salts that can become toxic, expensive, and heavy. However, they are recyclable, easily decontaminated by steam, and their porosity allows good water and oxygen diffusion. Plastic pots offer the advantages of reduced weight, ease of stacking, low cost, reusability, varied colors, and ease of cleaning. Their nonporosity requires different handling, that is, a reduction in watering and fertilization as opposed to clay. Heavy-duty or flexible plastics resist breakage. They cannot be steam sterilized, but can be sterilized by chemical treatment or semipasteurized by hot water for 3 minutes at 158°F (70°C). Chemicals that react with plastic should be avoided. Lightweight plastic square pot packs in attached units of 8 or 12 are convenient and easy to move. Fiber pots are made of pressed peat, wood fiber, and fertilizer, which is suitable for short-term use. They are biodegradable, easy to store, and the joined unit packs are convenient. When wet they tear easily, and dryness of the pot walls discourages good root growth.

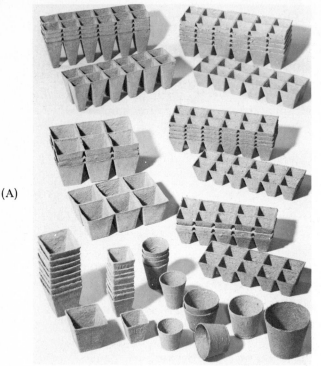

(A)

Fig. 6-2. A. Pots and joined unit pot packs made from a mixture of peat moss, wood fiber, and soluble fertilizer. B. Compressed peat moss/soluble fertilizer wafer encased in plastic net which expands into a combined pot/propagation medium. C. Similar to B, but has no plastic net. Observe well-developed root structure penetrating peat moss. A,B,C are courtesy of Jiffy Products of America. D. Attached square plastic pot packs showing Coleus 'Fiji' ready for planting. Courtesy of Pan-American Seed Co. E. Plastic pots and plastic flats. The latter may be conveniently handled in units of four in ribbed plastic carriers. Courtesy of Lockwood Products Inc. F. Plastic bag used as propagation container for lettuce plant. G. Wooden berry basket used to propagate squash. H. Wooden flat used to hold peat pots. These may be used directly as propagation containers also. Increasingly they are being replaced by plastic units. F,G,H are U.S. Department of Agriculture photos. I. Clay pots, while still used, are being replaced by plastic or peat pots, especially in commercial operations. Author photo.

(B)

(C)

168

(D)

(E)

(F)

(G)

(H)

(I)

Propagating blocks composed of peat or wood fiber mixed with fertilizer are a combined "pot and propagation media." Like fiber pots, the whole container can be set directly into the ground.

A number of containers serve well for temporary use. Asphalt-coated felt paper containers are lightweight, inexpensive, tough, and storable. Styrofoam or paraffined paper cups and even polyethylene bags, when provided with drainage holes, are inexpensive containers for propagating plants.

BOTTOM HEAT AND RELATIVE HUMIDITY

Often the optimal temperature for germinating seeds or rooting cuttings exceeds the optimal temperature for subsequent development. It may be uneconomical to raise the temperature of the entire propagating area, or areas with mixed plants at various developmental stages may preclude such a temperature increase. In such cases lead- or plastic-coated electric soil-heating cables with thermostatic controls are useful for providing bottom heat in the propagating media (Fig. 6–3). Low-voltage systems are available to reduce the hazard of electric shock.

Cuttings can experience severe water loss through transpiration until roots are formed and water uptake commences. Under these conditions, cuttings can die, especially if they are slowly rooting species. One way to reduce water losses by transpiration is to keep the relative humidity of the air around the cuttings at a high level. A simple way to provide these conditions is to place a glass jar over the cuttings or to use a propagating chamber covered with polyethylene. More sophisticated control is provided with intermittent misting equipment. Dips of antitranspirant chemicals have also been used.

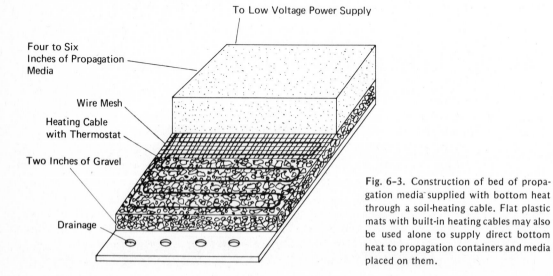

To Low Voltage Power Supply

Four to Six Inches of Propagation Media

Wire Mesh

Heating Cable with Thermostat

Two Inches of Gravel

Drainage

Fig. 6–3. Construction of bed of propagation media supplied with bottom heat through a soil-heating cable. Flat plastic mats with built-in heating cables may also be used alone to supply direct bottom heat to propagation containers and media placed on them.

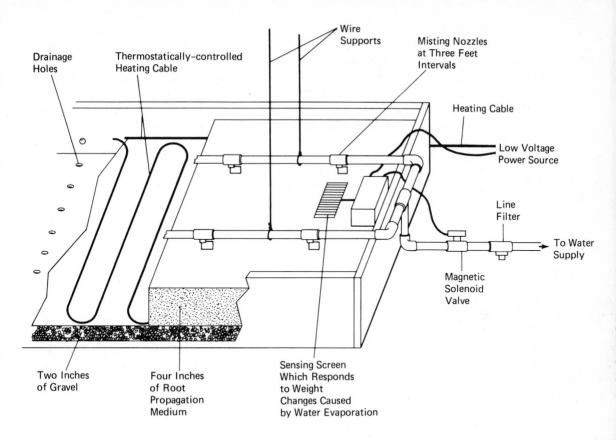

Fig. 6-4. Construction of a misting propagation unit supplied with bottom heat. The cable may be overlaid with heavy plastic mesh to prevent possible cable damage or shock hazard if a tool accidently comes in contact with it. An additional safety feature would be to use a low voltage system. The extent of leaf coverage and not height is the determining factor for the misting water lines. Each line covers about 1 ½ feet on each side, so a bench 4-6 feet wide would need two lines.

Misting increases the relative humidity, as does a glass jar or plastic frame, but the film of water supplied by the mist has the additional advantage of reducing air and leaf temperature, which in turn reduces the rate of transpiration. The propagating media can be maintained warmer than the air with a heating cable. Temperature control with misting allows direct placement of cuttings in full sun, unlike those in an enclosed jar or plastic frame. The resulting increased photosynthesis speeds development of the cutting, especially root formation. In addition, substances important to rooting (carbohydrates, auxins, and flavonoid substances) seem to be increased in cuttings that are under mist compared to those that are not. Intermittent misting provides this kind of environment and is less wasteful of water. Disease problems under mist appear to be minimal. An example of an effective propagating setup is shown in Figure 6-4. Misting can be used indoors and outdoors, both for sexual and asexual propagation.

Water flow through the misting system is regulated by a solenoid (magnetic) valve. These valves should be hooked up so that in the event of a power failure they will be in open or "misting" mode to prevent damage to the cuttings. The solenoid valve in turn is controlled by various mechanisms. These include the very reliable dual variable timers that turn the entire system on and off at preset times, for example, morning and night, and a second timer that determines the length of the misting period in minutes. Another device consists of a thermostat that turns on the system when a preset temperature is reached. It has proved especially effective when used in conjunction with the timer control described previously, since sudden rises in temperature might dry cuttings prior to the timer-activated mist. Another type of control, the "electronic leaf," sits in the cutting bed, and wetting or drying of it breaks or makes electrical contact. A weight-sensitive device can turn the mist on or off when losses or gains in water weight occur from evaporation or misting. This device is especially useful in areas for indoor or outdoor use that experience extensive, rapid weather changes, thus causing rapid variations in the evaporative power of the air (Fig. 6–4). Another control with a photoelectric cell turns mist on in proportion to light intensity. Controls and solenoid valves can be run on low voltage to reduce shock hazards.

Filters in the water line may be necessary to prevent clogging of the mist nozzles. Nutrients can be dispensed through the misting system to counteract nutrient leaching and to improve root formation and subsquent plant growth. Algae growth can be a problem; it is unattractive and slippery. It can be reduced by overnight drying, Bordeaux mixtures, or chemical treatment. Water quality must be watched (see early discussion in this chapter) as deposits can cause clogging or salts can hinder rooting.

Sexual Propagation

Sexual propagation consists of propagation by seed. Annuals and biennials are propagated in this manner, and some herbaceous and woody perennials may be sexually propagated. Sexual propagation would also include the propagation of ferns by spores. Generally, sexual propagation is the method of choice, unless seedling variability cannot be controlled or alternate asexual (vegetative) methods are more rapid, practical, and economical. It is also the best means for producing new cultivars for plant improvement, which along with the production of genetically pure seed will be covered in Chapter 11.

SEED PRODUCTION AND HANDLING

The production of seeds for horticultural use is an important industry upon which plant propagators are dependent, unless they produce their own seed. Generally, commercial seed is raised in areas where the climate is most favorable for the production and harvesting of the desired seed crop. Much seed production centers in the western coastal states, such as California. Foreign areas are also becoming important areas of seed production for the United States, such as Puerto Rico where seed production is favored both for climatic and economic reasons. In addition, seed may be collected from naturally grown plants, such as trees or shrubs, and from fruit-processing industries.

Harvesting of seeds is best done when the seed is mature or can be removed without reduction of its germination potential. Harvesting too early will yield an immature seed with undesirable qualities, and harvesting too late may mean the loss of seed as a result of dropping to the ground, blowing by the wind, or being removed by animals and birds. Harvesting may be by hand or mechanical equipment. Cleaning is by hand or mechnical threshing, followed by air blasts, screening, or gravity separators if needed. These techniques are dependent upon physical differences between the seeds and associated debris. Some drying may be required after harvesting and prior to cleaning. If a fleshy fruit encloses the seed, it must be removed. Mechanical macerators can be used for fruit disruption and the seeds separated by fermentation and floating, screening, or other mechanical means. Many seeds will require drying after harvest to a moisture content of usually 8 to 15 percent.

Usually, some form of seed storage is required after harvest, except for seeds whose viability is very short unless allowed to germinate quickly. Seeds should be of the highest quality when stored, that is, mature and undamaged. Some physiological decline will occur during storage, which affects viability. The amount of decline will depend upon the kind of seed, the length of storage, and environmental storage conditions such as temperature and humidity. Storage conditions are selected to minimize respiration and other metabolic processes within the seed. These conditions will vary, depending upon the plant species of origin. Freeze-drying of seeds prior to storage at temperatures below $50°F$ ($10°C$) with moisture control appears to offer the best storage conditions.

SEED PROPAGATION TECHNIQUES

Many seeds are in a state of dormancy that must be broken prior to germination. The various forms of seed dormancy and examples were discussed in Chapter 4. Some forms of dormancy are broken during storage, but others require physical or chemical preconditioning to stimulate germination. Hard or impermeable seed coats may be abraded, but the embryo must not be injured, to improve permeability to gases and water. This abrasion or mechanical *scarification* can be done with a file, sandpaper, or mechanical tumblers lined with sandpaper or filled with sand or gravel. Hard or impermeable seed coats are also modified by acid treatment, such as a soak in two times their volume of concentrated sulfuric

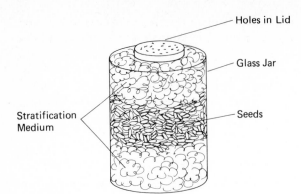

Holes in Lid

Glass Jar

Seeds

Stratification Medium

Fig. 6-5. Seeds prepared for stratification in a glass jar. Seeds and moist stratification media layers are equally thick, *i.e.*, 0.5 to 3.0 inches. The lid must have holes as oxygen is required by the seeds. Polyethylene plastic bags (highly permeable toward oxygen), also make good stratification containers.

acid. The length of time is determined experimentally. A water soak may be helpful in softening the seed coat or removing inhibitors; it also stimulates germination in some cases. The seeds are placed in about five times their volume of water at 170° to 212°F (77° to 100°C). The water is allowed to cool and the seeds removed after 12 to 24 hours. Seeds should be planted soon after the water soak.

Certain seeds must be subjected to cold temperatures prior to germination. This can be done under controlled conditions of moist-chilling (*stratification*). Dry seeds are first soaked in water for up to 24 hours and then placed in a plastic (polyethylene) bag containing a moisture-retaining medium (peat moss, sphagnum moss, vermiculite, or weathered sawdust). These bags are held at 35° to 45°F (2° to 7°C) for periods of 1 to 4 months (Fig. 6-5).

In some cases, two treatments may be necessary to overcome double dormancy of many woody shrub and tree seeds, such as a hard coat and chemically inhibited or dormant embryo. The former can be broken by mechanical or acid scarification followed by water soaking, and the latter by either form of scarification followed by stratification. An effective treatment for many seeds having double dormancy consists of moist-warm conditions to enhance microbial degradation of hard seed coats and stratification afterward to break embryo dormancy.

Phytohormones have been used to stimulate seed germination (and seedling growth) for various seeds. The phytohormones and treatments were (1) gibberellins at 100 to 10,000 ppm for 24 hours, (2) cytokinins at 100 ppm for 3 minutes, and (3) ethylene for various times. Other chemicals used to stimulate germination in some species are potassium nitrate, thiourea, and sodium hypochlorite. Preliminary trials of chemicals and germination tests on small batches of seeds prior to large-scale use are suggested.

Seeds selected for propagation should be of good quality, which is often expressed in terms of viability. The number of seedlings that can be produced from a given number of seeds, the *germination percentage*, is a useful indicator of viability. Additional features of viability are prompt germination, vigorous seedling growth, and normal appearance.

Tests are useful for determining the germination percentage and the number of days required to reach a given germination percentage, the *germination rate*. In addition, seeds should be examined for *purity*. Contaminants such as other crop seed, weed seed, and inert material (chaff, broken seed, soil, sand, gravel, and the like) lessen the value of the seed. Together, the purity, germination percentage, and germination rate are required to establish rates of sowing for the seeds under question.

Methods of germination testing are as follows. In a greenhouse, seeds may be sown on flats of sterile sand or peat moss. Tests vary in length from 10 days to 4 weeks, unless the seed is known to germinate slowly. Seeds may be placed between two blotters or on top of one blotter (Fig. 6–6). These blotters, held in trays, are placed in germinators with controlled optimal conditions of moisture, temperature, and light. Maximal care in handling is used to avoid conditions suitable for growth of disease microorganisms. Seeds may also be placed on rolled moist paper towels and put in a germinator. Seeds from woody trees or shrubs that need extensive preconditioning prior to germination are tested by excising the embryo, which is then tested directly for signs of germination. The chemical 2, 3, 5-triphenyltetrazolium chloride, which changes from colorless to red as a result of aerobic respiration, is also useful for germination testing. X-ray analysis can also be helpful for checking seed viability.

The timing for seed planting is dependent upon several variables. These include the number of days until maturity of the plant, the length of the growing season as determined by the average date of the last spring killing frost and the first fall killing frost, other climatic variables, the temperature preference of the plant, the desired date of harvest or sale of plants, the cold or heat hardiness limits of the plant, the number of days required for seed germination, and the timing of setting plants outdoors.

Fig. 6–6. Seeds being placed on a blotter that is overlaid on a heating grid. Seeds will then be tested for germination percentage and/or rate. U.S. Department of Agriculture photo.

Some of these factors apply only to seedlings started indoors or only to seedlings started outdoors. Recommended dates are usually available from the various state agricultural experiment stations or state extension services.

Disease control during sexual propagation is extremely important. Extensive losses of seeds, seedlings, and young plants can result from a group of diseases collectively referred to as *damping off*. Symptoms of this disease include seed decay or seedling rot prior to emergence, stem rot of the seedling at soil level and subsequent falling over, stem girdling and eventual death of the seedling or rot of the root system, and eventual death of young plants. The major fungi responsible for this disease are *Rhizoctonia solani* and *Pythium ultimum. Phytophthora* sp. and *Botrytis cinerea* may be involved to a lesser degree.

Pasteurization of the propagation media, discussed earlier in this chapter, will eliminate the microorganisms. Seeds should also be treated to prevent reintroduction of the microorganisms. Treatments include protectants, which prevent attack by microorganisms from the propagation media; chemical disinfestants, which eliminate microorganisms on the seed surface; and disinfectants that destroy organisms within the seed. Fungicides (Benomyl®, captan, chloranil, dichlone, ferbam, thiram, and zinc trichlorophenate) are used as protectants (sold under various trade names; see Chapter 10). Many fungicides come under federal and state regulations that must be observed by the user. Calcium hypochlorite at 1 to 2 percent strength for 5 to 30 minutes is used as a disinfestant, and hot water at 120° to 135°F (49° to 57°C) for 15 to 30 minutes serves as a disinfectant.

A number of environmental factors affect germination: water, temperature, gases, and light. These should be optimal for germination and early growth; often, optimal germination conditions are unfavorable for damping off. Insufficient moisture supply adversely affects the germination percentage, germination rate, and the rate of seedling emergence from the seedbed. In addition, as the moisture supply decreases, the actual concentration of soluble salts increases, which may be harmful if the concentration becomes high enough. Excess water can also be harmful because the aeration of the propagating media is reduced and damping off is favored; the buildup of algae and mosses is also favored. Water is usually supplied gently by hose, such as with a fogging nozzle or automatic misting systems. Alternately, bottom watering by placing the propagating container in a tub of water (do not submerge) is feasible for small numbers of containers. The reduction of excessive evaporation and water needs outdoors is helped by light mulching and indoors by the use of well-drained media with good water retention and the covering of flats or containers with glass or clear plastic wrap (gas permeable). Containers should be kept out of the sun and the cover removed when germination commences.

Minimal, maximal, and optimal temperatures have been designated for seed germination. These temperatures, however, are subject to variation, since they are influenced by diurnal cycles, seasonal cycles, and interaction with other environmental factors. Broad categories exist that are helpful in deciding which temperature regimes are to be provided. Cool-season crops (cabbage, broccoli, cauliflower, lettuce, radish, pea, spinach, pansy), vegetables and flowers that originated in the temperate climates, generally germinate poorly above 77°F (25°C). Warm-season crops (egg-plant, pepper, squash, melon, cucumber, dahlia, sweet corn, tomato, beans), originating from subtropical or tropical areas have a minimal germination temperature of 50° to 60°F (10° to 15°C). Many other cultivated plants have a temperature range for germination from as low as 40°F (4.5°C) to as high as 104°F (40°C). In many instances a night drop in temperature favors germination and seedling growth. Savings in energy costs also should be considered. The drop should be around 18°F (10°C). More detailed temperature requirements of many more species will be found in Part Three.

Oxygen, carbon dioxide, and in some instances ethylene can affect seed germination in the propagation medium. Oxygen is required for respiration of the germinating seed, and carbon dioxide is a product of that respiration. Anything that restricts oxygen influx and carbon dioxide efflux can be harmful to germination. A surface crust or excess water in the medium can hinder gaseous movement and availability. Ethylene is released by some seeds and may stimulate germination.

Light is required for the germination of some seeds (begonia, lettuce), inhibits the germination of others (phlox, portulaca, *Allium*), and is not required for still others. Some seeds respond to the photoperiod; birch (*Betula*) and the California poppy (*Eschscholzia californica*) are long and short day seeds, respectively. These light-dependent processes are con-trolled through phytochrome responses to light. Light requirement can be eliminated in some seeds by dry storage, chilling, alternating temperatures, chemical treatment, or phytohormone treatment. Light-sensitive seeds are usually tiny, and the best germination takes place at or near the soil surface, so that the seedlings sprout and commence photosynthesis quickly. Light-inhibited seeds often originate in dry desert environments. Light requirements of many seeds can be found in Part Three.

Low light intensities produce etiolated seedlings with low rates of photosynthesis. Optimal light intensity produces sturdy, vigorous plants that carry out higher rates of photosynthesis. However, very high light intensities can damage tender new seedlings. Therefore, some shading of certain seedlings may be necessary. If insufficient or no natural light is available, supplementary or completely artificial light may be used in some instances. Fluorescent lights are used to provide 1500 to 2500 or more footcandles for about 16 hours, followed by 8 hours of darkness. More details on the use of artificial light will be found in Chapter 8.

Indoor seedling production (Fig. 6-7) is carried out in greenhouses and other controlled environment facilities to produce plants for subse-quent transplanting to the field nursery and vegetable or flower garden or for use in the greenhouse or home. The bedding-plant industry is especially

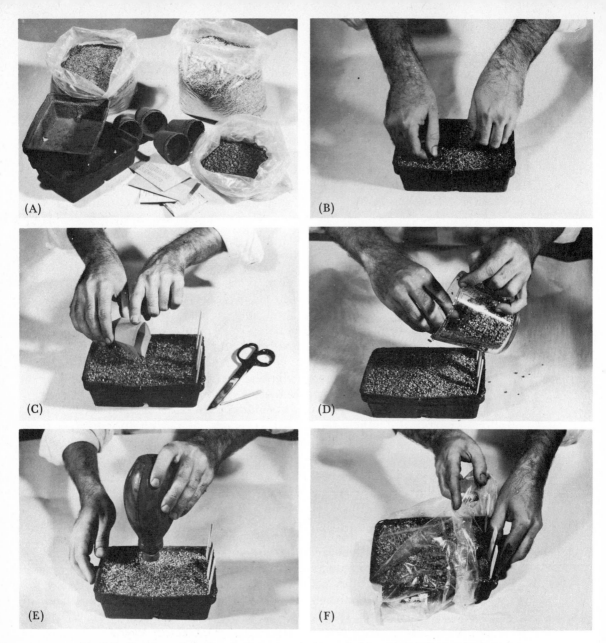

Fig. 6-7. Steps in seed starting. A. Have your starting materials ready: sterile containers, pasteurized propagation media, labels, and seeds. B. Place and level moistened media in container while pressing gently to minimize air spaces. C. Sow seeds in rows or even broadcast. D. Cover large seeds to required depth. Small seeds should be left uncovered and gently pressed into the media. E. Thoroughly moisten the media with a gentle, fine overhead spray or by bottom-watering. F. Place the container in a polyethylene bag and place in area with proper temperature for germination. A heated propagation mat may be needed. Seeds requiring light for germination should be placed under fluorescent lights or in indirect light (not direct sun). Others can be germinated in dark or dim areas. Remove the bag when seeds germinate. Water and fertilize as needed. These steps can be automated with machines, as is often done in large commercial operations. U.S. Department of Agriculture photos.

178

dependent on the production of seedlings indoors. Some container-grown stock is produced in this manner, too. Trees and shrubs for use in landscapes, reforestation, and as rootstocks are sometimes started indoors.

Containers of propagating media are leveled with a straight board and gently tamped (except for vermiculite where compaction destroys structure) to reduce the level to about ½ inch below the container top. The propagation media is thoroughly wetted prior to seeding it. Fine seeds are surface sown without any covering media, and medium to large seeds are covered two to three times their minimal diameter. Careful attention is paid to maintaining moisture, temperature, and light (if required). Bottom heating is supplied when needed. Misting or a glass or gas-permeable plastic cover (polyethelene) may be used to maintain moisture. If covered, they are kept out of direct sunlight. Covers are removed upon germination. After germination, temperatures are often lowered somewhat and light exposure increased to produce strong, vigorous seedlings. Fertilizer is provided if necessary. Surface dryness of the media is allowed once the root system is well-established. Seedlings are transplanted after they develop two or four true leaves.

Ferns (see reproduction cycle in Chapter 4) may be propagated from spores in a manner similar to that of seeds. Spores are surface sown and kept covered the same as for seeds. Propagation temperatures vary from 65 to 75°F (18° to 24°C), and bottom heat may be helpful. A mosslike growth, which consists of numerous prothallia, is produced; prothallia are removed by tweezers and transplanted to wider spacing. When these reach about ½ inch in size, tiny sporophyte plants are produced. With further transplanting these sporophytes subsequently develop into mature ferns.

Seeds of many annual, biennial, and some perennial plants (herbaceous and woody) are seeded directly outdoors. This method eliminates losses or slowed growth resulting from transplanting, but seed germination and density are more difficult to control. Often, thinning of seedlings is needed, which can be a considerable expense in commercial operations. Seed tapes, which consist of plastic tape with properly spaced seeds, are less wasteful of seed and can reduce or eliminate the need for thinning. Pelleted seed improves the handling of small seeds. Machines are also available for mechanized spacing of seeds.

Seedbed preparation is most important for seeding outdoors. Good contact between the seed and soil is needed for moisture supply, but pore space and good drainage are also required for adequate soil oxygen. Aggregate size is optimal if three-quarters of the particles range from 1 to 12 millimeters in diameter. Smaller particles, if too numerous, can eventually cause problems through the formation of surface crusts. Small areas are prepared with a spade, fork, or rototiller. Larger areas are done by plows or discs. Seedbeds should be worked 6 to 10 inches in depth. If moisture retention is poor, organic matter can be added or mulches used to conserve moisture. Fertilizer or limestone is added when need is indicated by soil tests.

Time of outdoor sowing is determined by the plant's temperature requirements, time to maturity, time of desired harvest, the length of the growing season, the date of the last killing frost in the spring, and the earliest time of soil workability. Your local agricultural experiment station or extension service should be consulted for the time of sowing.

Asexual Propagation

The principle behind asexual propagation is the potential for regeneration possessed by vegetative parts of plants (see Chapter 4). For example, adventitious roots arise from stem cuttings and, in turn, root cuttings can produce a new shoot system from adventitious buds. New shoots and roots can develop adventitiously from leaves.

Asexual propagation is important for growing cultivars that produce no seeds and preserving clones. It is possible to bypass certain undesirable characteristics or a particular developmental stage through asexual propagation of vegetative material from a later developmental stage. With some plants it may be more rapid and more economical than sexual propagation. Finally, the production of virus-free or mycoplasma-free plant material is dependent on propagation of heat-treated vegetative parts or certain other methods of asexual propagation.

Only the briefest generalizations are possible when discussing asexual propagation techniques. This results from the great differences that exist in the ease and methods of propagation among the very large numbers of plant species. Empirical trials are needed to establish propagation factors; these have been done for most horticultural plants of economic interest. Many others have yet to be investigated. A comprehensive book on plant propagation (see Further Reading) will provide the conditions for plants that have been examined. In addition, the asexual propagule for a large number of species can be found in Part Three.

PROPAGATION BY CUTTINGS

A part of a stem, root, or leaf can be cut from a stock plant, and this "cutting" can form roots and shoots under suitable environmental conditions. A low nitrogen-to-high carbohydrate ratio balance in the stock plant favors rooting in many cases, so fertilization prior to removal of cuttings should be avoided. Cuttings taken during the vegetative rather than the reproductive phase often root better.

New or adventitious roots and shoots develop from the cutting in three stages. Many cells can return to the meristematic condition (*dedifferentiation*) and initiate meristematic cells as root or shoot initials. In turn, these cells become root or shoot primordia, which grow and differentiate into roots and shoots. The sites of origin for adventitious roots in stem cuttings are usually within or near the vascular bundle. Sites of origin are varied for root or leaf cuttings.

Certain phytohormones and rooting cofactors, as well as synthetic root-promoting chemicals, favor the formation of adventitious roots. Stem cuttings are often dipped into synthetic root-promoting chemicals (0.4 percent indolebutyric acid) to enhance rooting. Fungicides, such as captan (25 percent) or Benomyl® (5 percent), are often mixed with indolebutyric acid as a precaution against fungal infection. Adventitious shoots are stimulated by treating root cuttings with cytokinins, and adventitious shoots and roots on leaf cuttings are enhanced by mixtures of auxins and cytokinins.

Cuttings are frequently rooted under mist and/or with bottom heat (75° to 80°F or 24° to 27°C), as described earlier under propagation aids. Some cuttings may be rooted directly outdoors. General air temperatures for rooting are 70° to 80°F (21° to 27°C) in the day and about 60°F (15°C) at night. Light intensity during rooting can be low, such as 150 to 200 footcandles, but should be higher once the cutting is rooted and growing. The light intensity will vary according to the plant species. Photoperiod appears to have some influence. The photoperiod under which the stock plant is grown, as well as during the rooting of the cutting, can influence the rooting and shoot development of some, but not all cuttings, depending on the plant species.

Cuttings (Fig. 6–8) can be classified as *stem cuttings, leaf cuttings, leaf-bud cuttings,* or *root cuttings.* Stem cuttings can be further divided into deciduous hardwood (currant, *Euonymus,* grape, *Hibiscus,* highbush blueberry), narrow-leaved evergreen (*Arborvitae, Cupressus,* Monterey pine, *Taxus*), semihardwood (*Allamanda, Azalea, Buxus, Clematis, Pittosporum, Pyracantha, Rhododendron*), softwood (*Acer, Cryptomeria, Hydrangea, Lantana, Viburnum*), herbaceous (*Begonia sempervirens, Impatiens, Zebrina*), and stem (cane) sections (*Dieffenbachia, Dracaena*).

Hardwood or narrow-leaved evergreen stem cuttings are taken from the wood of the previous or older season's growth during the dormant season. Semihardwood stem cuttings are taken during the summer from woody, broad-leaved evergreens or partially matured wood of deciduous plants. Softwood stem cuttings are derived from soft, succulent new spring growth (greenwood) of deciduous or evergreen species. Herbaceous stem cuttings come from the succulent tissue of herbaceous plants. Stem sections produce shoots from hidden buds, which can be removed and rooted, or the section can be left to produce shoots and adventitious roots.

Sometimes it may be desirable to take leafy cuttings during the pruning of woody plants or pinching of herbaceous plants. If propagation is not convenient at that time, cuttings can often be held in cold storage for several weeks. Cuttings are placed in polyethylene bags and stored at 31° to 40°F (-0.5° to 4.5°C). Low-pressure storage (1/30 atmosphere) at these temperatures extends the life of the cuttings. Examples of cuttings treated this way include azalea, privet, chrysanthemum, carnation, and geranium.

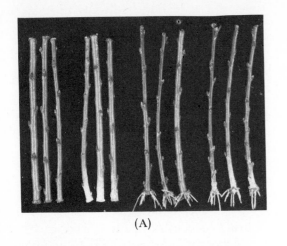

(A)

(B)

(C)

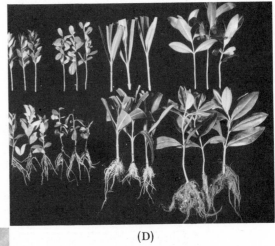

(D)

(E)

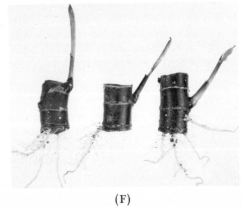

(F)

182

(G)

(H)

(I)

Fig. 6–8. Types of cuttings. A. Deciduous hardwood cuttings: peach. B. Narrow-leaved ever-green cuttings: *Juniperus.* C. Semi-hardwood cuttings: *Camillia.* D. Softwood cuttings, start-ing from left: *Myrtus, Pyracantha, Oleander, Veronica.* E. Herbaceous cutting: *Chrysanthe-mum, begonia* and *geranium.* F. Stem section cutting: *Dieffenbachia.* G. Leaf cuttings: top. *Saintpaulia,* bottom, *Peperomia.* H. Leaf bud cuttings: *Crassula argentea,* leaf (right) and leaf bud cuttings (left). Leaf cuttings of this plant will not produce a shoot. I. Root cut-tings: horseradish (left) and apple (right). All, except 6E, from Hudson T. Hartmann and Dale E. Kester, *Plant Propagation: Principles and Practices,* 3rd edition, © 1975, pp. 251, 280, 281, 282, 283, 285, 286, 287. Reprinted by permission of Prentice-Hall, Inc., Engle-wood Cliffs, New Jersey. No. 6E is U.S. Department of Agriculture photo now in National Archives (No. 16–N–22837).

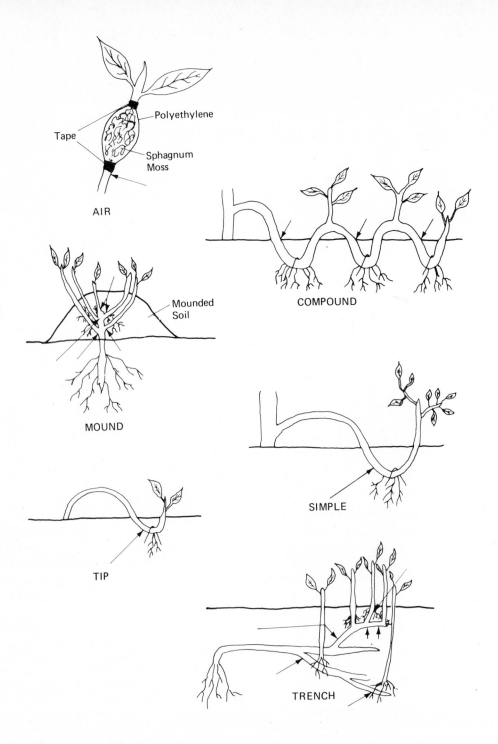

AIR

COMPOUND

MOUND

SIMPLE

TIP

TRENCH

Polyethylene

Tape

Sphagnum Moss

Mounded Soil

Fig. 6-9. Types of Layering. Area indicated by arrow is where shoot is severed after rooting is completed. This leaves mother plant and rooted daughter plants which are ready for planting elsewhere.

184

With leaf cuttings the intact leaf blade (*Echeveria, Gloxinia, Strepto-carpus, Sedum*), leaf sections with primary veins (*Begonia rex, Sanse-vieria*), and leaf blade plus petiole (*Cyperus, Saintpaulia, Tolmiea*) are used. Adventitious roots and an adventitious shoot are formed. *Leaf-bud cuttings* consist of the leaf blade, petiole, and a short piece of the stem with the attached axillary bud. Examples are *Crassula argentea* and *Camellia*. Root cuttings are sectioned roots, which are usually taken in late winter or early spring when the root's stored food is high (*Bouvardia, Cordyline, Daphne, Echites, Ligularia, Plumbago, Rubus*).

PROPAGATION BY LAYERING

Through layering, adventitious roots are formed on a stem that is still part of the parent plant. After roots are established, the stem is severed and planted. Unlike a cutting, there is no critical maintenance of the attached stem. Layering may be natural, as with *Forsythia* or trailing blackberries, or it may be artificially induced. Layering is simple and usually successful, but it is expensive on a large scale since it is not easily mechanized and hence requires much hand labor.

Layering (Fig. 6-9) can be induced by bending a stem downward, fastening it in place, and covering it with soil or a rooting medium. The darkening of the stem causes etiolation, which increases the formation of adventitious roots. If only the region of the stem near the current season's shoot is covered, this is termed *tip layering*. This form of layering is used for *Forsythia* and *Rubus* sp. When the submerged bend is 6 to 12 inches back from the tip, it is termed *simple layering*. This procedure is useful for hard-to-root shrubs, such as *Rhododendron, Jasminum,* and *Weigela*. If the stem is alternately covered and exposed along its length, such that each exposed section has at least one bud, it is termed *compound* or *serpentine layering*. Long, flexible vines (*Philodendron, Wisteria*) can be propagated this way. Other forms of ground layering include *mound (stool) layering* for clonal apple stocks and quince and *trench layering* for propagating difficult-to-root fruit tree rootstocks. Layering done on aerial stems is known as air layering, and it is useful for some tropical and subtropical shrubs, such as *Ficus, Codiaeum,* and *Coccoloba*.

PROPAGATION BY SPECIALIZED STEMS AND ROOTS

Other plants have growth habits or structures that lead to propagation naturally. Some of these are essentially natural forms of layering, although they can be accelerated, for example, by pinning the vegetative structure to the soil. Examples of the structures adapted to natural layering include runners, stolons, crowns, offsets, and suckers. These specialized stem structures were described in Chapter 2.

Fig. 6-10. *Chlorophytum* sp. produce runners which are easily rooted. Author photo.

Runners are specialized stems that grow horizontally along the ground, and their point of origin is from the axil of a leaf at the crown of the plant. The strawberry, *Chlorophytum* sp., *Episcia* sp., and *Saxifraga sarmentosa* (strawberry geranium) are examples of plants propagated from runners (Fig. 6-10). The daughter plants form at a runner node and are severed and dug after rooting. A similar form of specialized modified stem, the *stolon,* is produced by some plants. Aboveground prostrate or sprawling stems, such as with woody species like *Cornus stolonifera* or herbaceous species like Bermuda grass, are often considered to be stolons. However, some prefer to restrict the term stolon only to underground stolons, such as those found with mint (*Mentha* sp.) or those that develop into tubers, like the potato. Stolons after natural layering can be cut from the parent plant and planted elsewhere.

A *crown* is usually that part of the plant at ground level from which new shoots are produced. The crown is an arbitrary zone of demarcation between the shoots (aerial plant parts) and roots. With many plants the crown increases with age by the formation of new shoots and adventitious roots. Crowns in a sense develop through layering and, as such, provide a means for propagation. Propagation of crowns from herbaceous plants (*Chrysanthemum, Cyperus, Spathiphyllum, Maranta, Campanula*) and some woody shrubs (*Sorbaria, Viburnum, Spiraea*) is by crown division (Fig. 6–11). Many herbaceous plants require division every two or three years to prevent overcrowding anyway, and the divisions provide some increase of the plant. This method is more important to the noncommercial gardener, because of its limited return. Generally, smaller crowns are cut apart with a knife, whereas larger ones may require a hatchet or shovel. Spring- and summer-blooming plants are usually divided in the fall; those that bloom during late summer or fall are divided in the spring. Indoor potted plants are usually divided whenever necessary to prevent overcrowding.

Fig. 6-11. Propagation by crown division: *Hemerocallis*. From Hudson T. Hartmann and Dale E. Kester, *Plant Propagation: Principles and Practices,* 3rd edition, ©1975, p. 474. Reprinted by permission of Prentice-Hall, Inc., Englewood Cliffs, New Jersey.

An *offset,* sometimes called offshoot, is a lateral shoot or branch that arises from the base of the main stem. Bulbs also produce offset bulblets at their base. Often, offsets have a short, thick, rosettelike appearance. Offsets may also be lateral shoots arising from rhizomes. Rooted offsets can be cut close to the main stem and planted, or if rooting is insufficient, they can be treated as leafy stem cuttings. Many succulents (Fig. 6-12) and the pineapple can be propagated by offsets.

Suckers are shoots that arise belowground from an adventitious bud on a root. Shoots that arise from the vicinity of the crown are termed suckers, although by definition they are not, as they arise from stem and not root tissue. Technically, those shoots arising from an adventitious stem bud are watersprouts. Suckers may be severed after being rooted or treated as cuttings. The red raspberry produces true suckers.

Fig. 6-12. Offsets on *Hoodia gordonii* (left) and *Greenovia aurea* (right). Author photo.

(B)

(A)

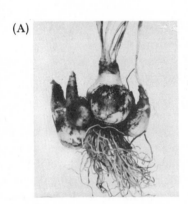

Fig. 6-13. A. Offsets produced by daffodil. B. Bulblets produced on belowground portion of stem of *Lilium*, which when mature are termed offsets. A, from Hudson T. Hartmann and Dale E. Kester, *Plant Propagation: Principles and Practices*, 3rd edition, © 1975, p. 483. Reprinted by permission of Prentice-Hall, Inc., Englewood Cliffs, New Jersey. B, American Society for Horticultural Science photo.

Certain specialized stems and roots are involved in food storage for the plant. These fleshy organs, already described in Chapter 2, include bulbs, corms, tubers, tuberous roots, rhizomes, and pseudobulbs. Since they contain buds, this enables the plant to survive dormancy and, after dormancy is broken, to produce shoots and to grow. As such, the organs are well suited to natural and induced propagation. Generally, these specialized organs are found in herbaceous, perennial plants that undergo dormancy as a result of warm-to-cold or wet-to-dry cycles found in temperate and tropical-to-subtropical areas, respectively. If the specialized organ is naturally detachable, the propagation method is termed *separation,* as opposed to *division* when the structure is cut into sections.

Bulblets develop from parent bulbs and at maturity are called *offsets* (Fig. 6–13). These can be propagated by separation and will eventually reach flowering size. Bulbils (aerial bulbs formed in the leaf axil as in certain lilies) may also be propagated in this manner. When it is desired to increase the production beyond offsets, three processes are used. Individual bulb scales are removed from the mother bulb, and under favorable growing conditions adventitious bulblets form at the base of the scales. This method, called *scaling,* is particularly useful for lilies (Fig. 6–14).

Fig. 6-14. Outer bulb scales of lilies can be removed and put in a container of damp redwood sawdust at 70° F. Bulblets arise (as in A) and in time they develop a root system (B). American Society for Horticultural Science photos.

(A)  (B)

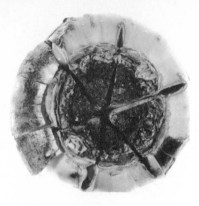

Fig. 6-15. Basal cuttage of hyacinth bulb. Observe development of bulblets at wounded area. From Hudson T. Hartmann and Dale E. Kester, *Plant Propagation: Principles and Practices,* 3rd edition, © 1975, p. 490. Reprinted by permission of Prentice-Hall, Inc., Englewood Cliffs, New Jersey.

Other bulbs, such as the hyacinth, can be propagated through *basal cuttage*, a form of wounding, which causes multiple development of adventitious bulbs (Fig. 6-15). Other bulbs may be propagated with *bulb cuttings,* such as *Hippeastrum* (Amaryllis). Bulbs are cut into portions that contain fractions of three or four scale segments and part of the basal plate (short, fleshy vertical stem axis). The cut areas are dusted with fungicide and cured at 70°F (21°C). They are subsequently treated as leaf cuttings. Bulblets appear and are grown to maturity.

Propagation of corms (gladiolus, crocus) is primarily through natural formation of new corms and cormels (Fig. 6-16). These are separated and cured at 95°F (35°C) for 1 week. Corms may also be sectioned. So that each section contains a bud. These sections will produce a new corm. Fungicide dusting is needed to prevent decay of the exposed tissue.

Fig. 6-16. *Gladiolus* showing formation of recent small cormels, new well-developed corms (upper right and left), and the old withered corm (bottom). Author photo.

Tubers can be propagated by planting the whole tuber or by cutting it into sections, each of which contains a bud or "eye" (Fig. 6-17). Cut sections are allowed to heal or *suberize* at 68°F (20°C) and 90 percent relative humidity for 2 or 3 days. Alternately, the cut surface can be treated with a fungicide. Examples of plants propagated by tubers include *Arisaema, Eranthis, Gloriosa, Solanum tuberosum* (Irish potato), *Caladium,* and *Helianthus tuberosus* (Jerusalem artichoke).

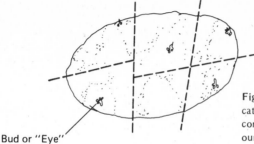

Bud or "Eye"

Fig. 6-17. A diagram of a potato tuber indicating cuts by dotted lines. Each section should contain one node and weigh about 1 to 2 ounces.

Tuberous roots can be propagated either from adventitious shoots or division. Sweet potatoes (*Ipomoea batatus*) can be forced to sprout at 81°F (27°C), and these sprouts will eventually form roots. The rooted shoots can be removed and planted. Division (Fig. 6-18) is used for tuberous root plants such as the dahlia. Other plants with tuberous roots are *Alstroemeria* and *Ophiopogon*.

Division is also favored for plants with a rhizome structure (Fig. 6-19). Rhizomes can be cut into sections, each of which contains a lateral bud, as is done with bananas, German iris, certain grasses, *Agapanthus*, and *Leucocrinum*.

Fig. 6-18. Division of the *Dahlia* tuberous root to include one adventitious shoot per section. U.S. Department of Agriculture photo.

Fig. 6-19. Rhizome propagation. A rhizome section from bamboo has produced adventious roots and a well-developed shoot. U.S. Department of Agriculture photo.

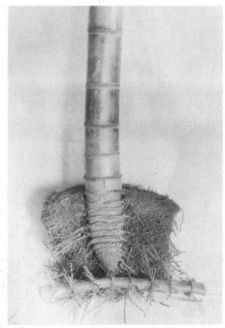

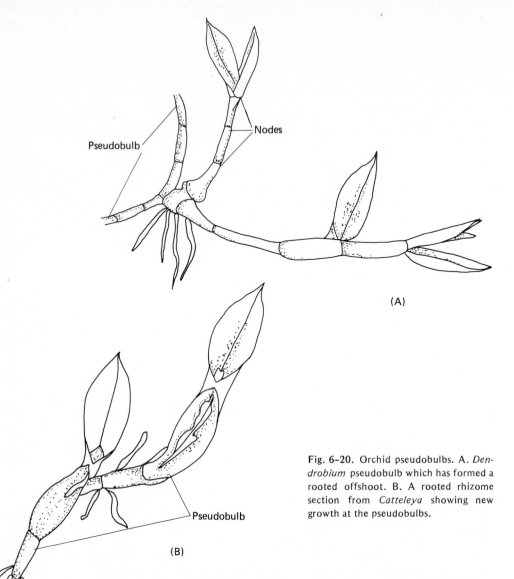

Pseudobulb

Nodes

(A)

Fig. 6-20. Orchid pseudobulbs. A. *Dendrobium* pseudobulb which has formed a rooted offshoot. B. A rooted rhizome section from *Catteleya* showing new growth at the pseudobulbs.

Pseudobulb

(B)

Pseudobulbs are produced by many orchids. Some pseudobulbs form rooted offshoots that can be cut and potted (Fig. 6-20). In others a rhizome with attached pseudobulbs can be planted in rooting media, and growth will begin at the pseudobulb base.

PROPAGATION BY GRAFTING AND BUDDING

Grafting is an ancient horticultural art whereby parts of plants are joined such that the tissues unite and growth follows as for one plant. The upper part of the graft union is the *scion*, and the lower is the *rootstock or understock,* or sometimes simply the *stock. Budding* is a special aspect of grafting in which the scion is a small piece of bark or wood with a single bud.

There are several reasons for the use of grafts. Grafting may be the method of economic feasibility for plants that are difficult to propagate by usual means, such as cuttings. Desirable cultivars might have poor root systems, and grafting would offer improvement. Grafting rootstocks can also offer increased tolerance toward unfavorable soils or plant pathogens in the soil. Other rootstocks may induce increased vigor or dwarfing in the scion. Grafting may also be useful for replacing undesirable cultivars on good, established trees, using the existing rootstock. This is termed top-grafting or top-budding, or both may be referred to as topworking. Certain forms of tree damage can also be repaired through grafts. Grafting is also useful for the creation of different plant forms, such as "tree" roses or combination fruit trees. Fruiting may also be improved by grafting a scion from a staminate plant onto a pistillate plant. Virus detection is also possible if a plant suspected of carrying a virus is grafted onto a susceptible plant.

Sometimes a third, different plant section may be inserted between the rootstock and the scion. This section is designated as an *intermediate stem section,* an *intermediate stock,* an *interstem,* or an *interstock.* This approach, called *double-working,* is used to overcome incompatibility between the scion and rootstock or to provide desired qualities not present in the conventional graft.

The most important part of grafting is the formation of a successful graft union, as illustrated in Figure 6–21. The required close matching of the callus-producing tissues of the cambium layer (see first step in Fig. 6–21) tends to restrict grafting to the dicotyledons in the angiosperms and to the gymnosperms, since both have a continuous tissue of vascular cambium between the phloem and xylem. Temperatures must be favorable for high cell activity 55° to 90°F (12.8° to 32°C), and high moisture is essential to prevent desiccation of the newly arising callus tissue. This is helped by waxing the graft union or wrapping it in a moist medium. Prompt waxing also helps to prevent infection by plant pathogens. The graft union must also be tightly held, such as by wedging, wrapping, nailing, or tying. This will prevent the dislocation of cellular growth, which would destroy the graft union. Finally, the scion and rootstock must be at the proper physiological stage. Generally, the scion is dormant and the rootstock is dormant or active, depending on the grafting method.

Not all plants can be united successfully, as grafting incompatibility causes failure of the graft union of many combinations. As the botanical relationship becomes more distant, the chances of a successful graft diminish. Grafting within a clone is assured, and grafting between clones within a species is almost always possible. The chances of successful grafting between species within a genus are less and are remote to nil for grafting between genera within a family and grafting between families.

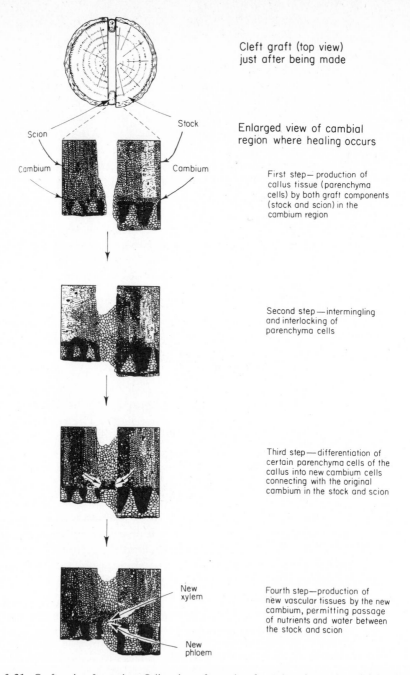

Cleft graft (top view)
just after being made

Scion

Stock

Cambium

Cambium

Enlarged view of cambial
region where healing occurs

First step — production of
callus tissue (parenchyma
cells) by both graft components
(stock and scion) in the
cambium region

Second step — intermingling
and interlocking of
parenchyma cells

Third step — differentiation of
certain parenchyma cells of the
callus into new cambium cells
connecting with the original
cambium in the stock and scion

New
xylem

New
phloem

Fourth step — production of
new vascular tissues by the new
cambium, permitting passage
of nutrients and water between
the stock and scion

Fig. 6-21. Graft union formation. Callus tissue formation from the scion and stock joins the two. Eventually a vascular cambium arises in the callus tissue and joins the scion and stock vascular cambium. New xylem and phloem are now produced by the joined vascular cambium. Normal development follows. From Hudson T. Hartmann and Dale E. Kester, *Plant Propagation: Principles and Practices,* 3rd edition, © 1975, p. 321. Reprinted by permission of Prentice-Hall, Inc., Englewood Cliffs, New Jersey.

194

GRAFTING

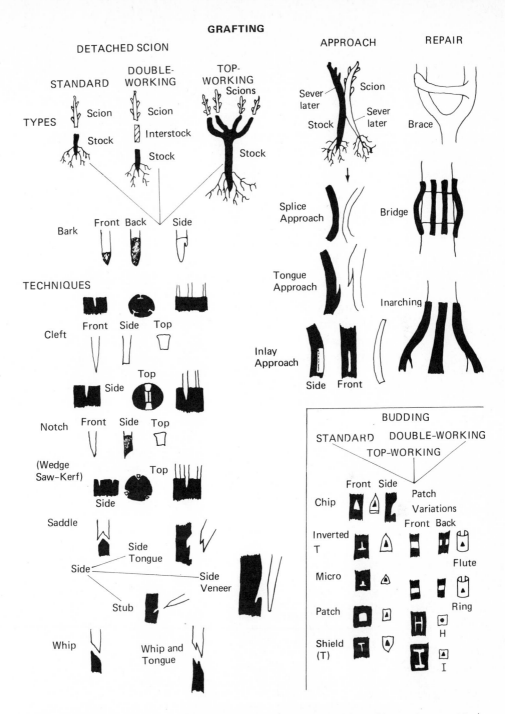

Fig. 6-22. Techniques used during detached scion (standard, double-working, and top-working), approach, and repair grafting plus budding are shown. Detached scion grafting and budding are used for trees that are relatively easily grafted, approach grafting for those that are difficult, and repair grafting for damaged trees with sufficient value to warrant saving them.

There are essentially two general forms of grafting: *detached scion* and *approach*. The first is the most commonly used; roots are present only on the rootstock and the scion is detached from the donor plant. The second method involves the grafting of two plants, both of which are self-sustaining. After the union is formed, the scion is severed from the donor plant. This method is employed with plants that are difficult to graft. There are numerous techniques of grafting and budding used to make either of these graft types. Some are shown in Figure 6-22.

PROPAGATION BY TISSUE CULTURE

Small pieces of plants can be used to develop new plants in an artificial medium under aseptic conditions; this process is *micropropagation* or *tissue culture*. The small parts of plants taken to start propagation are termed *explants*. Embryos, stems, shoot tips, apical meristems (frequently free of pathogens), leaf tissue sections, seeds, root tips, callus (parenchyma), single cells, and pollen grains are examples of explants.

The advantages of micropropagation are the rapid, vegetative propagation of plants and the production of pathogen-free plants. This is especially useful for plants with very slow normal reproduction, such as orchids, or for plants in which disease is transmitted through other propagation methods. In addition, it can be a more rapid and smaller-scale method for plant improvement. For example, the developing explant cells may be treated with mutagens, and the survivors cultured on to full-grown plants with altered genotypes. A number of ornamental and edible plants have been successfully micropropagated. Some of these plants have been recently micropropagated on a commercial level with favorable results (bromeliads, *Gerbera jamesonii*, ferns). Orchids have long been micropropagated by meristems and shoot tips.

Fig. 6-23. A plant successfully propagated by tissue culture. Photo from The Connecticut Agricultural Experiment Station, New Haven.

All procedures must be conducted under aseptic conditions. The medium is usually a semisolid agar with various additives to support and promote the development of the explant. Liquid media with aeration are also utilized. The various additives include minerals, sugar, vitamins, growth regulators, and other organic constituents. An example of micropropagation is shown in Figure 6-23.

Fertilization During Asexual And Sexual Propagation

Generally, the sexual propagation mixes covered earlier in this chapter or outdoor seedbeds have been amended with moderate levels of nutrients. These are usually sufficient to maintain the seedlings up to the transplanting stage. If a mixture or single component, such as milled sphagnum moss, is used without added fertilizer, or plants must be held in the propagation media or outdoor seedbed past transplanting time, it will be necessary to add some soluble fertilizer at levels recommended for seedlings. This may be added by hand watering or through automated means, such as a nutrient mist. Supplementary fertilization for outdoors and in plant-growth structures are covered in Chapters 7 and 8, respectively.

Most asexual propagation media contain little or no available nutrients, as opposed to outdoor propagation beds, which usually do. The vegetative propagules, such as cuttings or bulbs, are usually dependent upon stored tissue nutrients only, which are usually adequate to maintain them during the time of root formation. Under some conditions, nutrients may become self-limiting, and supplementary soluble fertilization will become necessary. Conditions include slow-rooting propagules, a delay in transplanting of the rooted propagule, and the fact that rooting of some species is improved by supplemental nutrients.

Interim Handling

Events following propagation, although not part of the process, seriously affect the successful establishment of the recently propagated plant. Rather than considering these events in terms of the more distantly removed management of the outdoor and indoor environments covered in the next chapters, they will be dealt with here. In addition, some of these events are germane to both chapters, such as the container plants that may be cultured both indoors and outdoors, so their inclusion here is more logical.

TRANSPLANTING AND HARDENING

Transplanting of seedlings, ferns, or asexually propagated materials is normally not a problem if root disturbance is kept minimal, unless environmental changes are associated with it (Fig. 6-24). Examples of

Fig. 6-24. Steps in seedling transplanting. A. Seedlings should be transplanted after developing two true leaves (left). Seedlings on right have gotten too large. B. Seedlings are carefully lifted with a spatula, knife blade, *etc.* such that root damage is minimal. C. Seedlings are carefully placed in a new container. After placement in the planting hole or slit, soil is gently firmed around the seedling. If individual containers are not used, adequate space for development should be provided. D. Proper development subsequently requires proper cultural care. Adequate fertilizer was used on the right, and none on the left. E. These eight week old seedlings are ready to be hardened off prior to outdoor planting. F. Improper timing produces transplants that are too large (or too small) for best results. Plant on left is a suitable size for garden use. U.S. Department of Agriculture Photos.

198

environmental changes would be transplanting indoor seedlings outdoors, outdoor foliage and flowering plants indoors, or transplanting cuttings from under mist to an unmisted site. For success in these and other cases, the plant material must be conditioned to accept the different conditions of the new environment.

Plants going from indoors to outdoors must be *hardened off*. Hardening involves lowering the temperature, withholding water, and a gradual change in the environment. This slows growth, and a carbohydrate reserve is built, which supplies the needed carbohydrates for rapid reestablishment of the rooting system. Coldframes (see Chapter 8) are used to harden plants for about 7 to 10 days prior to transplanting outdoors (Fig. 6-25).

Field-grown plants, such as recently propagated foliage and flowering plants, destined for indoor use must be conditioned or *acclimated* to the reduced light levels indoors. These may be transplanted into containers and placed in a lathhouse, where the development of shade leaves will allow it to survive in a lower light situation. Recent studies indicate that foliage plants (except for large-trunk-producing types) produced from the start under shade may be a better approach.

Plants propagated under mist must be hardened off to enable them to survive lower levels of relative humidity. Misting periods may be gradually decreased, and then the plants are transplanted; or the plants may be transplanted and placed in a humid, shaded area or another misting system operated at reduced levels.

Fig. 6-25. Hardening-off of transplants, such as in a coldframe, is essential when transplants are to experience moderate to severe environmental differences. Courtesy of National Garden Bureau, Inc.

(A)

Fig. 6-26. Containers for growing-on transplants. A and B. Large plastic pots and hanging baskets favored for indoor home plant use. Courtesy of Lockwood Products, Inc. C. Various containers suitable for commercial production. The outer right row of 3 plastic containers and the adjacent 3 metal containers are mainly used for container stock, whereas the others (plastic and clay) are useful for foliage/flowering plant production. Author photo. D. Old leaky buckets and other discarded large containers can be recycled for container plantings. U.S. Department of Agriculture photo.

(B)

(D)

(C)

Conditions that favor reduced transpiration after transplanting increase the chances of success. Favorable environmental factors include lower temperatures, lower light intensity, low or no wind speed, and high relative humidity. Best transplanting occurs during cloudy days, days of light rain, periods of no wind, and late afternoons. Plants should be watered well afterward. Shade can be provided after transplanting, if possible. A soluble fertilizer (starter solution) is sometimes added to aid the reestablishment of the plants, but not if soil moisture is low. The increase in soluble salts might damage the plant under conditions of low soil moisture.

In some instances recently rooted plants can be held in cold storage prior to transplanting. Cuttings rooted in late summer or early fall for spring planting can be protected over the winter if outdoors or allowed to grow on indoors. An alternative is to dig them and put them bared-rooted in polyethylene bags at 32° to 40°F (0° to 4.5°C). Reduced atmospheres (1/30 atm) seem to improve storage life.

CONTAINERS FOR GROWING-ON TRANSPLANTS

Transplanting may be done directly into outdoor sites or containers maintained indoors or outdoors. Handling at this stage is covered in Chapters 7 and 8. Containers covered in an earlier section in this chapter may be used. In addition, there are larger containers more suitable for container stock. These include plastic and metal containers (Fig. 6–26) in sizes from 1 quart to 15 gallons. These are available commercially or reclaimable from restaurants, canneries, and bakeries. They are quite useful for nursery plants and easily adapted to mechanized techniques.

GROWING MEDIA

Growing media are needed for transplants going into containers that will be placed in an indoor or outdoor site. Materials utilized to prepare the media and their pasteurization are similar to that described earlier for propagation media. A few mixtures can be prepared as follows. Many more can be found in some of the references cited in Further Reading.

The older growing media were based on soil. Their use has declined for a number of reasons. Among these are increasing scarcity of loam, variations in chemical and physical properties, and heavy weight. Mixtures for growing rooted cuttings or young seedlings contained:

1 to 2 parts sand
1 part organic matter
1 part loam soil

Organic matter could be peat moss, garden compost, shredded bark, or leaf mold. Container-grown nursery plants and potted foliage or blossoming plants not requiring a high humus content soil could be carried over in the following mixture:

> 1 part sand
> 1 part organic matter
> 2 parts loam soil

Potted foliage or blooming plants needing a humus-rich soil mixture were planted in the following mixture:

> 1 part sand
> 2 parts organic matter
> 1 part loam soil

Less variability was introduced with mixtures that did not include the soil loam. These mixtures consist of sand and peat moss or shredded bark or compost or leaf mold in varying proportions. One such example is that used for potted succulents:

> 1 part leaf mold
> 1 part sand

All these mixtures needed supplemental fertilization with time. Fertilizers could be added directly in the mixing process, if nutrient analyses indicated the need. The pH of the mixtures might be altered with lime, sulfur, or ammonium sulfate, depending on the pH requirements of the plants.

A well-known soil mixture for growing-on was developed a number of years ago at the John Innes Horticultural Institution in England by W. J. C. Lawrence and J. Newell (see Further Reading). The formulation, unless indicated otherwise, is by volume.

John Innes Potting Compost

> 2 parts sand
> 3 parts peat moss
> 7 parts loam
> 5 lb John Innes base per cubic yard of mixture
> 1 lb ground limestone per cubic yard of mixture

The base is a nutrient mixture composed of 2 parts by weight of hoof and horn meal at 1/8-inch grist with 13 percent nitrogen, 2 parts by weight of superphosphate of lime with 18 percent phosphoric acid, and 1 part by weight of potassium sulfate with 48 percent potash. The loam originally consisted of nutrient-enriched, composted strawy manure and pasture turf, but an effective substitute consists of soil from bare, fallow land.

Artificial growing media have been developed. These can be prepared on a large-scale basis, they provide a standardized mixture with little variation between batches, and they are more easily produced and maintained in a pathogen-free state. Two popular mixes are those developed at Cornell University by R. Sheldrake and J. W. Boodley and the University of California at Los Angeles by K. F. Baker and coworkers (see Further Reading). Many commercial variants of these artificial mixtures are also available. The Cornell Peat-Lite mixes do not require any decontamination. The Cornell Peat-Lite mix for container growing is as follows:

Peat-Lite Mix A (for 1 Cubic Yard)

11 bushels horticultural-grade vermiculite (nos. 2 or 4)
11 bushels shredded peat moss (at least 75% *Sphagnum* sp.)
5 lb ground limestone (best form is dolomitic)
1 lb superphosphate (20%) (best is powdered)
2 to 12 lb 5-10-5 fertilizer

Peat-Lite Mix B

Same as above, except substitute horticultural perlite for vermiculite

The amount of 5-10-5 fertilizer in mix A determines the length of time during which no supplemental fertilization is needed. Twelve pounds will supply nutrients for 5 to 6 weeks. Care must be observed to wet the peat moss during preparation. A nonionic wetting agent can aid the wetting of peat moss.

The University of California mixtures consist of varying amounts of sand, peat moss, and nutrients. The nutrients added differ, depending on whether the mixture is to be stored or utilized within 1 week of preparation. The mixtures are as follows:

Nursery Container-Grown Stock

75 percent sand (0.5 to 0.05 mm in diameter)
25 percent peat moss

Potted Plants

50 percent sand
50 percent peat moss

To each cubic yard the following fertilizer mixtures can be added.

For Storage

4 oz potassium sulfate
4 oz potassium nitrate
2 ½ lb single superphosphate
7 ½ lb dolomite lime
2 ½ lb calcium carbonate lime

2 ½ lb hoof and horn meal or blood meal (13 percent nitrogen)
4 oz potassium sulfate
4 oz potassium nitrate
2 ½ lb single superphosphate
7 ½ lb dolomite lime
2 ½ lb calcium carbonate lime

Be sure to wet the peat moss thoroughly. These complete mixtures can be steam or chemically pasteurized without any harmful effects.

One amendment recently introduced for growing media is a particulate gel (Viterra 2®) that can hold up to 130 times its weight in water. Its incorporation cuts the frequency of watering. Whether the economics will permit its uses in situations beyond hanging baskets and small potted plants remains to be seen.

FURTHER READING

Baker, K. F., *The U.C. System for Producing Healthy Container-Grown Plants,* Berkeley, Ca.: California Agricultural Experiment Station —Extension Service Manual 23, 1957

Ball, V., ed., *The Ball Red Book* (13th ed.) Chicago: George J. Ball Inc., 1975.

Bilkey, P. C., and others, "Micropropagation of African Violet from Petiole Cross Sections," *HortScience,* 13, no. 1 (1978), 37.

Boodley, J. W., and R. Sheldrake, Jr., *Cornell "Peat-Lite" Mixes for Container Growing.* Ithaca, N.Y.: Department of Floriculture and Ornamental Horticulture, Cornell University Mimeo Report, 1964.

Broome, O. C., and R. H. Zimmerman, "In Vitro Propagation of Blackberry," *HortScience,* 13, no. 2 (1978), 151.

Bunt, A. C., *Modern Potting Composts.* University Park, Pa.: Pennsylvania State University Press, 1976.

Chaturvedi, H. C., and others, "Shoot Apex Culture of *Bougainvillea glabra* 'Magnifica,'" *HortScience,* 13, no. 1 (1978), 36.

Copeland, L. O., *Principles of Seed Science and Technology.* Minneapolis: Burgess Publishing Co., 1976.

Garner, R. J., *The Grafter's Handbook* (4th ed.). London: Faber & Faber Ltd., 1979.

Hartmann, H. T., and D. E. Kester, *Plant Propagation* (3rd ed.). Englewood Cliffs, N.J.: Prentice-Hall, Inc., 1975.

Hawthorn, L. R., and L. H. Pollard, *Vegetable and Flower Seed Production*. New York: Blakiston Co., 1954.

Justice, O. L., and L. N. Bass, *Principles and Practices of Seed Storage*. Agriculture Handbook no. 506. Washington, D.C.: U.S. Department of Agriculture, 1978.

Kochba, J., and P. Spiegel-Roy, "Cell and Tissue Culture for Breeding and Developmental Studies of Citrus," *HortScience*, 12, no. 2 (1977), 110.

Koevary, K., and others, "Tissue Culture Propagation of Head Lettuce," *HortScience*, 13, no. 1 (1978), 39.

Lawrence, W. J. C., and J. Newell, *Seed and Potting Composts*. London: George Allen & Unwin Ltd., 1952.

Murashige, T., *Plant Propagation through Tissue Cultures*. Annual Review of Plant Physiology, Vol. 25. Palo Alto, Calif.: Annual Reviews, Inc., 1974.

——, and others, "Proceedings of the Symposia, Cell Culture and Tissue Culture," *HortScience*, 12, no. 2 (1977), 125.

Nehrling, A., and I. Nehrling, *Propagating House Plants*. Great Neck, N.Y.: Hearthside Press, Inc., 1971.

Reilly, Ann, *Park's Success with Seeds*. Greenwood, South Carolina: Geo. W. Park Seed Co., Inc., 1978.

Reinert, J., and Y. P. Bajaj, ed., *Applied and Fundamental Aspects of Plant Cell, Tissue, and Organ Culture*. New York: Springer-Verlag, New York, Inc., 1976.

Roberts, E. H., ed., *Viability of Seeds*. Syracuse, N.Y.: Syracuse University Press, 1972.

Schopmeyer, C. S., tech. coordinator, *Seeds of Woody Plants in the United States*. USDA Handbook no. 450. Washington, D.C.: U.S. Department of Agriculture, 1974.

Seeds. Yearbook of Agriculture. Washington, D.C.: U.S. Government Printing Office, 1961.

Sheldrake, R., Jr., and J. W. Boodley, *Commercial Production of Vegetable and Flower Plants*. Cornell Extension Bulletin 1056. Ithaca, N.Y.: Cornell University Press, 1965.

Welch, H. J., *Mist Propagation and Automatic Watering*. London: Faber & Faber Ltd., 1970.

Outdoor Management of the Plant Environment

The selection of an area for horticultural purposes, whether they are of the backyard or commercial variety, is of utmost importance in terms of environmental management. Any one of the several factors, for example, poor climate, shallow soil, shade from nearby trees and buildings, poor drainage, insufficient water resources, and pollutants, can reduce the horticultural productivity of a site.

Obviously, the climate and the horticultural crop(s) to be grown must be compatible. Factors such as the length of the growing season, seasonal temperature fluctuations, days to maturation, and relative hardiness of the plant material must be examined. The correlation between climate and crop(s) need not be as close for the home gardener as for the commercial horticulturist. With the latter, less than optimal conditions may mean financial disaster. The backyard horticulturist does have the option to alter the environment on a small-scale basis. Microclimates (see Chapter 5) must also be considered carefully. The localized effects of natural and artificial surface phenomena on frost dates, soil temperatures, air temperatures, wind movement, and pollution levels may improve or decrease the favorability of a site for horticultural production and/or landscape horticulture.

Other environmental factors besides climate must be considered. The soil should be deep, fertile, easy to work, and well drained. A subsoil layer of hardpan, gravel, or rock is a feature to avoid. Soil of less than ideal conditions can be improved through addition of fertilizers and soil conditioners, but the expense and labor must be weighed against the attributes or desirability of the site.

The site should have ample water of good quality, whether this be from rainfall or irrigation. Direct sunlight should be available all day long. As the available light decreases, the types of crops that can be grown and their favorable development decrease also. Shade from trees or buildings is to be avoided. Tree roots can also compete more successfully than smaller plants for nutrients and water.

Surface grades must be considered. A level site is acceptable if drainage is good, desirable with sandy or gravel-type soils in order to increase water infiltration and preferable to sites with grades of more than 3 percent. A southerly slope of not more than 1-1/2 percent is favorable for early crops. Grading modification between 0.5 to 1 percent are sometimes needed to improve drainage and reduce surface water accumulation, especially with fine-textured soils with poor structure.

There should be no low places where water can accumulate. Water from surrounding areas should not drain onto the site, and areas prone to flooding from streams are best avoided. Good drainage is essential. Drainage can be improved through the singular or combined use of agricultural tile, ditch work, breaking up hardpan subsoil plus incorporation of physical amendments, and modification of surface grades.

The site may also require protection against wind and animals. A windbreak, such as a hedge, a shelterbelt of trees, or a board fence, can provide wind protection. Fences can keep out stray or wild animals. In addition, the proximity and encroachment of suburbia must be considered by those involved in the commercial aspects of horticulture.

In today's age of energy shortages, the local availability and proximity of alternate energy sources is something that requires investigation. For example, a greenhouse industry in the Midwest about to have its natural gas curtailed in the dead of winter obviously has a need for alternate energy.

Pollution also plays a role in site selection. An area subject to atmospheric pollution, such as ozone or ethylene, would not necessarily be a good choice for commercial horticulture. A home gardener would not want to locate his garden in the leaching field of his septic tank, since he would run the risk of contamination of root crops by human pathogens.

DRAINAGE

Soil improvements usually deal with the physical condition and fertility of the soil. One physical condition that can be improved is drainage. Poorly drained soils are unsuitable for horticultural applications for many reasons, such as the decrease of soil oxygen caused by excess soil water or the slower warming of wet soils in the spring. Generally, some form of artificial drainage is required when the water table is within 3 feet of the surface, or when excess surface water cannot infiltrate fast enough to keep from seriously depleting soil oxygen.

Sites must be examined carefully to ascertain whether existing or potential drainage problems (will) occur, and, if so, whether the problems are associated with the surface, the root area, or the subroot zone area. This analysis will in effect determine the probable modification to alleviate the drainage problem. A further determination is necessary to ascertain whether the probable solution is actually possible. This determination is the *drainage capacity* of the soil, which is equal to the percent of total pore space minus the percent of pore space occupied by water at field capacity. The actual technique of this determination can be found in the texts suggested in Further Reading.

Soils with surface drainage problems are often level, fine-textured soils with poor structure found in areas of high rainfall. They usually have low drainage capacity, which eliminates the use of agricultural tile drainage in the subsoil. Modification of surface drainage includes slight grading (0.5 to 1 percent) and/or digging of surface drainage ditches (Fig. 7–1).

Drainage problems in the root area can be reduced by breaking up subsoil hardpan (if present) and incorporating physical amendments, such as 70 percent (by volume) sharp sand and 30 percent partially decayed organic matter, to improve aeration. Drainage problems in the subroot

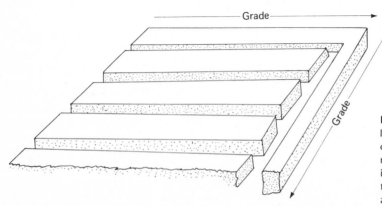

Fig. 7–1. Surface drainage problems can be reduced by a system of feeder drainage ditches terminating at a main drainage ditch, if grading is properly done such that gravity-produced flow is achieved.

Fig. 7-2. Drainage problems in the subroot area are minimized by installation of a plastic (shown here) or tile drainage system. U.S. Department of Agriculture/Soil Conservation Service photo by John B. Litchfield.

area (the soil above the root area has adequate drainage capacity) can be corrected by the installation of a tile drainage system (Fig. 7-2). The placement of the system and its depth depend upon several soil parameters, the types of plants to be grown, and whether it is economically feasible.

FERTILIZATION

An imbalance in nutrients in the soil can cause plant stresses that result in reduced growth and quality. Insufficient nutrients over a period of time produce typical plant-nutrient deficiency symptoms. These symptoms result from an actual deficiency of the nutrients or the presence of the nutrients in a chemical form unavailable to plants. The latter, as discussed in Chapter 5, is a function of the soil pH, cation-exchange capacity, nitrogen fixation, solubility properties, and mycorrhizae. Excess nutrients can also have detrimental effects, such as phytotoxicity or abnormal growth excesses. For example, excess boron would result in plant death, or excess nitrogen could cause luxuriant leaf growth at the expense of flowers and fruit.

By the time the appearance of nutrient deficiency or excess symptoms is noted, plant growth and probably yield have been reduced. At this stage it is not always likely that the correction of the nutrient deficiency will make up for the lost growth. These symptoms are not always easily diagnosed. For example, purple coloration on leaves may indicate a phosphorus deficiency, but in tomatoes it might also be indicative of a cold soil. Yellowing leaves may not be easily diagnosed. Nitrogen deficiency may appear as yellow-green leaf color on a normally dark green plant, potassium deficiency may appear first as yellow areas and these become dead and brown later, and iron deficiency (but also a magnesium deficiency or manganese deficiency or excess) may appear as yellowing leaves with green veins. It is obvious that nutrient deficiences are best avoided by early determination of soil fertility. If symptoms do appear, it may be necessary to resort to analyses of nutrient concentrations in the plant tissue coupled with soil analyses to determine the exact cause.

The actual nutrient requirement of horticultural crops is based on several parameters. They include soil diagnosis to determine the total nutrients, the available nutrients and the factors contributing to a nutrient's unavailability, and plant diagnosis to determine the actual amounts of nutrients absorbed by the plant. Together these are correlated to establish a relationship between plant development and the nutrient concentration of the plant tissue, as influenced by the levels of various nutrients in the soil. Standard curves derived in an empirical manner can be defined for each nutrient. These curves differ depending on the plant species. For example, the optimal range of nitrogen for a leaf crop, for example, spinach (*Spinacia oleracea*), would be excessive for a crop such as peppers (*Capsicum annuum*).

Soil diagnosis or testing services are usually available for varying moderate fees through your state agricultural experiment station or state extension service. Portable chemical test kits can be used by individuals, but they are not as reliable as the more extensive instrumental methods of analysis used by the state services or commercial soil-testing laboratories. However, they may be a valuable aid in indicating problems that can be confirmed or eliminated by a more comprehensive soil diagnosis. Analyses of nutrients in plant tissue (frequently leaves) are usually conducted in commercial or university laboratories.

In actual practice the horticulturist relies mainly on soil analyses before planting each crop. The nutrient needs of most horticultural crops have been established through the plant-tissue analyses. Soil analyses provide a base point, and the analyst can recommend the nutrients required to produce good results with the crops to be planted. Computers are increasingly being used to interpret the soil test results and to make fertilizer recommendations.

A key step in soil analysis is to obtain a soil sample that adequately represents the root zone area. For uniform mixtures of prepared propagation or growing media, a composite sample from two or three samplings is sufficient. For small areas (such as a garden), or larger areas up to 10 or 15 acres with uniform features (soil productivity, texture, color of topsoil, topography, tillage, drainage, and past management), a composite sample should be prepared from not less than 20 samplings. Each of the 20 samples should be a vertical slice or column of unfrozen soil of moderate moisture content taken to plow or spading depth (or deeper if the root zone will exceed this depth) and about 1.27 centimeters (0.5 inch) in cross section. If large areas show variation in the features mentioned, they should be divided into smaller areas with uniform features. Each smaller area should be represented by 20 samplings. The 20 samples should be mixed thoroughly in a clean container. This composite sample then provides the material for the soil test. About 0.5 to 1 pint of soil is required for the analyses. Subsoil samples taken at a depth of 46 to 61 centimeters (18 to 24 inches) can provide additional input for making lime recommendations.

Other information should also be supplied with the composite sample for the best fertilization recommendations. Desirable information includes the crop to be planted, the crop planted previously, the yield goal, previous liming and fertilization history, plow depth, and any unusual problems or conditions noted by the grower. One condition to note is temporary drought, which can temporarily increase soluble salt levels and depress pH (see Chapter 5). Since the pH and soluble salts will revert to predrought levels when the drought is ended, more lime than is needed could be falsely indicated.

The actual recommendation may be based on optimal plant response, or it may be based on the most economical plant response. This point may not matter much to the gardener, but it certainly does to the commercial horticulturist who does not wish to trade dollars for fertilizer. This situation results from the nonlinear response of yield versus fertilization cost (Fig. 7-3). As fertilizer applications (and hence costs) are increased to reach optimal plant response, the incremental yield increase realized decreases as optimal plant response is approached. In other words the dollar return measured against fertilization costs may be greater at some point short of the optimal plant response. The situation can be altered by factors such as density of planting, for a planting of high density may make higher applications of fertilizer more profitable than a planting of low density. Factors such as plant variety, insects, disease, weeds, inadequate water, and unfavorable temperatures can make the yield response less than optimal for a given amount of fertilizer.

Another point in favor of the prudent use of fertilizer is the accumulation of soluble salts with continuous, excessive use. These accumulations may appear as a white surface crust. Excessive salinity can cause foliar burn, retarded growth (and profits), and even death. Overfertilization should be avoided. Periodic irrigation may be required to reduce excessive salinity. Decreased fertilizer use also reduces pollution of watershed areas.

Fertilizers are composed of various nutrients in varying proportions. If a fertilizer contains the major plant nutrients, nitrogen, phosphorus, and potassium, it is termed a *complete fertilizer*. Secondary and micro-nutrient elements are added in varying amounts to *specialty fertilizers*. Alternately, they can be added separately to counteract a known deficiency. These elements are calcium and magnesium, usually applied as part of lime, and boron, copper, iron, manganese, molybdenum, sulfur, zinc, and chlorine, usually as the chloride form of one of the elements. The nutrient contents or the *analysis* or *grade* of most fertilizers is expressed as a set of three numbers, which express the percentage by weight of nitrogen, phosphorus (expressed as P_2O_5), and potassium (expressed as K_2O). To convert P_2O_5 into elemental phosphorus, multiply by 0.43, and to convert K_2O to potassium, multiply by 0.83. For the reverse, that is, element form to oxide, multiply by 2.33 and 1.20, respectively. Generally, fertilizers are expressed in terms of nitrogen and the two oxides for historical reasons. If the analysis is divided by the lowest common denominator, the result is known as the *fertilizer ratio*. To summarize, a 5–10–5 fertilizer contains by weight 5 percent nitrogen, 10 percent P_2O_5 (or 4.30 percent P), and 5 percent K_2O (or 4.15 percent K) and has a fertilizer ratio of 1–2–1.

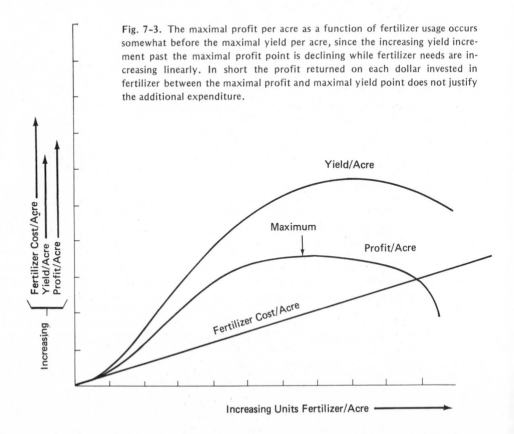

Fig. 7-3. The maximal profit per acre as a function of fertilizer usage occurs somewhat before the maximal yield per acre, since the increasing yield increment past the maximal profit point is declining while fertilizer needs are increasing linearly. In short the profit returned on each dollar invested in fertilizer between the maximal profit and maximal yield point does not justify the additional expenditure.

When the available nutrients in a fertilizer are more than or less than 30 percent, it is referred to as a high-analysis or low-analysis fertilizer, respectively. Examples of high-analysis fertilizers are ammonium polyphosphate (15–60–0) and urea ammonium phosphate (34–17–0). In terms of cost, the high-analysis fertilizers are a better buy, but their strength can cause fertilizer burns, so care must be used to apply only what is required.

Fertilizers can be derived from living organisms, *organic fertilizers,* or synthesized from chemicals and natural materials (such as rock phosphate), *inorganic fertilizers.* Examples of materials used to formulate each and their nutrient content are shown in Table 7–1. Chemical fertilizers are quickly available, but prone to leaching and fertilizer burning if used incorrectly. They are either formulated to be applied in a dry, granular form, or formulated to be soluble in water such that they can be applied in liquid form. Organic fertilizers have some of their nutrients in forms not usable by plants; however, the enzymic action of microorganisms converts these to usable nutrients. Organic fertilizers offer the advantage of nutrient availability over a period of time and less danger of leaching or burning. Organic fertilizers are available for dry or liquid application. Mixtures of organic and inorganic fertilizers, which offer the advantages of both, are becoming more common. An example of this is some lawn fertilizers, which contain inorganic nitrogen for quick use and organic nitrogen to extend nutrient availability, thus reducing the frequency of fertilizer application. Chemical fertilizers are also being produced in forms that extend nutrient availability and lessen the danger of leaching or burning. These include soluble nutrients coated with a resinous or membranous material or nutrients possessing reduced solubility. An example of a "synthetic organic" is ureaform. These synthetic organics tend to cost more than the straight inorganic forms. They are often incorporated into growing media to reduce the need for supplemental fertilization of container plants.

Nutrients removed by crops must be replaced if the soil fertility is to be maintained. Cover crops, such as legumes, returned to the soil cannot supply enough nutrients for many of today's high crop yields, so some form of fertilization is needed. A soil test will help determine what and how much nutrients, secondary and micronutrients, as well as lime and soil conditioners, are required. An application of a certain-analysis fertilizer will be recommended. This brings us to the timing of fertilizer application. Traditional times are just prior to or at the time of planting. This is still acceptable for the efficient use of fertilizer. However, applications of fertilizers in the fall may be desired, if the objective is to buy fertilizer at a lower price and to reduce the labor load at planting time. Fertilizers frequently cost more in the spring when the demand peaks and the supply decreases. The primary danger with fall fertilizing is the loss of nitrogen (mainly as nitrate) through leaching, especially where percolation losses of water are great. This can be minimized by using an ammonium form, such

TABLE 7-1. MATERIALS SUPPLYING FERTILIZER NUTRIENTS

Nitrogenous Materials	Nitrogen Content (% dry wt)
Anhydrous ammonia	82
Urea	46
Urea formaldehyde	38
Ammonium nitrate	33.5
Liquid ammonia	20–24
Diammonium phosphate	21
Ammonium sulfate	20.6
Sodium nitrate	16
Ammonium phosphate sulfate	16
Calcium nitrate	15.5
Potassium nitrate	14
Blood meal	10–14
Monoammonium phosphate	11
Magnesium ammonium phosphate	7–8
Cottonseed meal	6–7
Fish meal	5–7
Activated sewage sludge	5–6
Poultry manure	5
Steamed bone meal	2
Cattle manure	1–2

Phosphorus Sources	Phosphorus Content (P_2O_5)
Monocalcium phosphate	55
Diammonium phosphate	54
Monoammonium phosphate	48
Triple superphosphate	45
Magnesium ammonium phosphate	40
Steamed bone meal	23
Single superphosphate	20
Ammonium phosphate sulfate	20
Fish meal	1.5–6
Blood meal	1–5
Cottenseed meal	2–3

Potassium Sources	Potassium Content (K_2O)
Potassium chloride	50–60
Potassium sulfate	48–51
Potassium nitrate	46
Potassium frits	36
Potassium magnesium sulfate	22
Wood ashes (unleached)	4–10
Seaweed	4–5

as gaseous ammonia or an NH_4^+ salt, and applying it to a soil that is 10°C (50°F) or colder. Ammonium nitrogen can be applied in the fall to acid soils, since conversion to the leachable form, nitrate, is slow. Nitrogen fertilizers can also be applied safely in the fall where a cool-season crop is growing vigorously and will absorb the nitrogen or where nitrogen will be utilized by decomposition of crop residues, thus tying up nitrogen from loss by leaching. Fall applications of phosphorus pose no problem, and fall applications of potassium are only a problem on peat and muck soils where leaching losses are great.

One application of fertilizer during the growing season is often not enough. Losses of nutrients, especially nitrogen, occur by leaching; nutrients may also be tied up in unavailable form. Therefore, later application(s) may be required. Another problem is that the amount of nutrients required for an unproductive soil may be so large as to damage tender seedlings. Another choice, instead of additional applications, is to use a mixture of chemical and organic forms for quick availability and extended availability. Slow-release forms of fertilizer can also be used in place of the organic ones. Slow-release forms are particularly useful for container nursery stock in the field, where their addition reduces the frequency and labor of application. The economics of several applications of quickly available chemical fertilizers versus one application of the combined chemical and organic or chemical and slow-release fertilizers must be considered. The labor cost for the first is higher, but the cost of fertilizer is higher for the second.

There are several techniques for the application of fertilizer (Fig. 7-4). These include the following: *broadcast scattering,* the uniform scattering of fertilizer by hand or fertilizer spreader over the ground surface (suitable for gardens or small operations); *plow-sole,* the dropping by mostly mechanical means of fertilizer behind the plow in the opened furrow; *band placement,* the placement of fertilizer bands under or to the sides of the planted seeds; and *top* or *side* dressing, the application of fertilizer over the tops or sides, respectively, of growing crops. Soluble fertilizers may also be applied in irrigation water or in water during transplanting. The latter is termed a *starter solution.*

Fertilizers can also be applied through the foliage instead of the soil. This is termed *foliar nutrition* and is used to deal with special problems that cannot be solved readily through fertilization of the soil. Not all plants can tolerate foliar feeding; those that can generally have a heavy layer of cuticle wax, such as orchard trees. Situations where this approach is useful include the following. If a quick response is needed, such as with a sudden deficiency of a nutrient or a very rapid use of nutrients during a period of intense growth, foliar feeding responses are more rapid than conventional soil fertilization. The application of micronutrients as a foliar feeding is more effective than the application of chelated forms to soil, where in some cases soil reactions hinder the usefulness of the chelated form or the deepness of roots slows the effect. Foliar feeding

(A)

(B)

(C)

Fig. 7-4. Techniques for fertilizer application. A. While hand broadcasting of fertilizer requires no machinery, the use of a fertilizer (or lime) spreader facilitates the task and produces more even coverage. Fertilization of large areas requires equipment such as shown in B through D. B. Injection of nitrogen fertilizer (as gaseous NH_3) just before harrowing (equivalent to raking soil) of soil. C. Band placement of fertilizer in furrows prior to seeding operation. D. Plants being side-dressed with fertilizer. A is U.S. Department of Agriculture photo. Others are American Society for Horticultural Science photos.

(D)

may be more economical for some crops, such as forest trees, or serve as a useful alternative when soil fertilization is impractical, such as during a dry spell.

Decisions on the timing of initial and supplementary applications and placement of fertilizer are important if the fertilizer usage is to be efficient. These decisions depend upon several factors: the nutrient(s) needed, the crop and its changing nutrient requirements as a function of its development, the soil type and fertility levels, and the climate.

Some of the many points to keep in mind in regard to timing and placement are as follows. Nitrogen can be taken up as ammonium or nitrate. Ammonium can be held as an exchangeable, readily available form on clay and humus, but it is easily converted into nitrate by bacteria. Nitrates are easily lost by leaching. Therefore, levels of nitrogen cannot be maintained without supplementary additions. Nitrates are lost by leaching most readily in sandy soils. Nitrogen can become tied up and unavailable to plants in soils where active decomposition of organic matter is occurring until the carbon-to-nitrogen ratio is reduced to 12:1. Nitrogen availability can become reduced in cold and/or wet soils, such as in the spring, on account of unfavorable bacterial actions that decrease levels of nitrogen. The solubility and toxicity of nitrogen compounds limits the amounts that can be applied at any one time. Surface applications of nitrate forms are preferred over ammonium or amine forms (such as urea) on alkaline soils, since the soil lime and pH favors production of ammonia gas (NH_3) from the latter two and subsequent losses through volatilization. Ammonia should be applied at a 6-inch depth to maximize soil absorption and to minimize losses. Nitrates or ammonium salts can be applied at the surface, and the rain or irrigation water will leach them to root depths. Excess nitrogen can burn seedlings, cause vegetative development at the expense of flower and fruit in ornamental, fruit, and vegetable crops, or cause poor-quality fruit on trees. Excesses at the season's end may produce lush vegetative growth that is susceptible to winter injury.

Phosphorus mobility and availability are low, especially in strongly acid soils. It is important to build up soil levels of phosphorus prior to planting on account of limited availability, but its low mobility reduces the need for supplemental applications of phosphorus. Low mobility means the phosphorus must be placed close to the seed. Starter solutions containing 1500 parts per million of phosphorus are especially helpful with transplants and subsequent establishment of good root systems. Supplemental fertilization is often with a fertilizer lower in nitrogen and higher in phosphorus than the first applied fertilizer in order not to favor vegetative over reproductive development.

Potassium mobility is somewhere between phosphorus and nitrogen. The solubility and toxicity of potassium suggests placement not too close to seeds or plants. Injury is most likely with sandy soils. Potassium needs are less than nitrogen or phosphorus, so the need for supplemental applications is reduced.

SOIL CONDITIONERS

Organic matter plays an important role in maintaining good soil condition, and it also affects fertility, as was discussed in Chapter 5. An ideal loam may contain about 5 percent organic matter by weight. Continuous cropping reduces this amount of soil organic matter, perhaps up to as much as 2 percent of the total organic matter per year. Unless the organic matter is replaced by some form of soil conditioner in addition to fertilizer, the long-term economy of plant production will suffer. However, even a soil with sufficient organic matter can have poor soil conditions because of problems caused by tillage, cultivation, and erosion. These problems will be covered in later sections of this chapter.

The organic-matter content of a soil can be increased through crop residues, green-manure crops, cover crops, and organic amendments (Fig. 7-5). The most widely used method of maintaining a satisfactory level of organic matter, especially in soils utilized by commercial horticulturists, is the crop-residue approach. In general, soil fertility is maintained at high levels through the frequent applications of complete inorganic fertilizers, based in part upon frequent soil tests. Plantings are spaced to achieve high density, and the crop stubble is left after harvest. Then at planting time all field operations are done at one time: plowing, seeding, fertilizing, applying herbicides, and applying pesticides. This minimal tillage reduces the oxidative loss of organic matter and the problem of soil compaction.

Green manures are plants grown primarily to supply a steady release of nitrogen by plowing them under. These plants are usually legumes, such as vetch (*Vicia*), annual sweetclover (*Trifolium, Melilotus*), and *Crotalaria*. Rye (*Secale cereale*), a nonleguminous plant, is also used. Green manures do also increase the organic matter content of the soil, but they are less effective in increasing long-term organic matter in the soil than are crop residues. The soil nitrogen content must be watched, as the microorganisms will use soil nitrogen during decomposition of the green manure if the carbon-to-nitrogen ratio of the plowed under green manure exceeds 30:1 to 35:1. The large population of green-manure-induced microorganisms can attack the more resistant forms of soil organic matter, thus leading to the decline of organic matter over the long term.

Cover crops are either temporary or permanent. Temporary cover crops, such as grasses or legumes, prevent winter erosion of soil and are plowed under in the spring as green manure. Grasses or legumes can also be used as permanent cover crops. Cover crops of grass are frequently used on a permanent basis in orchards. Oxidation of soil organic matter is minimal because of the constant soil cover and lack of cultivation. Organic-matter levels are maintained by the constant degradation of grass roots.

Organic matter in the soil can be increased by the addition of organic amendments such as manures, compost, sewage sludges, and peat. A number of manures are available, such as cow, poultry, horse, and rabbit

(A)

Fig. 7-5. A. A cover group (Blando brome) is used between tree rows in this orchard. It is mowed, but allowed to set seed for the following year's cover crop. Since it is an annual grass, dying roots and shoots contribute to the soil organic matter content. There is no need for weed control by tillage, so soil compaction and oxidative loss of organic matter is minimal. U.S. Department of Agriculture/Soil Conservation Service photo by Clarence U. Finch. B. Soil organic matter can be increased by the addition of either liquid manure or sewage sludge. An effluent spreader shown here is used for application. U.S. Department of Agriculture/Soil Conservation Service photo.

(B)

manure. Besides their value as an organic amendment, they also contain about 1.3 to 5.0 percent nitrogen (dry weight basis), 0.3 to 1.9 percent phosphorus and 0.1 to 1.9 percent potassium. Wet manures average about 25 percent organic matter. Manures are somewhat difficult to handle and can easily lose nutrients through leaching and microbial activity. However, those who do use them recognize their value as a soil conditioner and fertilizer. Nitrogen and phosphorus (phosphorus content is low) from manures tend to have extended availability.

Compost, sometimes called synthetic manure, is a soil conditioner that requires some preparation prior to use. It is probably more useful on a smaller-scale basis, such as with gardens, as the other organic-matter sources are less labor intensive. Several misconceptions about compost still prevail, and the composting process is often made much more complex than is needed.

Accordingly, the composting procedure will be covered here. *Composting* of collected organic wastes starts when the indigenous microorganisms are subjected to favorable environmental conditions. These bacteria and fungi assimilate some carbon, nitrogen, phosphorus, and other nutrients made available by their enzymic activities during the process of aerobic decomposition. Heat is produced from these biological oxidations and much of it is retained, since organic matter acts as an insulator. The temperature rises up to 104°F (40°C), whereupon the population of mesophilic microorganisms is succeeded by thermophilic and thermotolerant microbes. Temperature increases up to 158°F (70°C) follow, during which time the rate of degradation accelerates. Decomposition eventually slows and the temperature returns to that of the ambient atmosphere. Changes induced by microbial degradation have converted the original wastes into a humuslike material. The volume and weight of the heap will have decreased by as much as 50 percent, and losses of carbon can be as high as 40 percent.

A simplified method for the construction of a compost pile is shown in Figure 7-6. The size of a pile is determined by the relationships among heat loss and surface area, heat production and volume, and oxygen content and compaction. While additives and limestone have been shown unnecessary for compost production, both carbonaceous and nitrogenous materials are needed during decomposition by microorganisms, which utilize about 30 parts of carbon per 1 part of nitrogen. In practice, the optimal carbon-to-nitrogen ratio is approximated by combining two parts of carbonaceous wastes to one part nitrogenous materials. The carbonaceous materials include dry leaves, hay, straw, sawdust, wood chips, chopped cornstalks, or other dry plant residues, and small amounts of shredded paper. Nitrogenous materials may be food wastes, manures, or fresh green plant residues, such as grass, weeds, and garden remains. Nitrogen content can also be increased by adding fertilizers. Dry leaves from deciduous trees, except for oak leaves, which are resistant to degradation, can be composted alone to form leaf mold, since their carbon-to-nitrogen ratio is sufficient to support microbial growth. The pile is kept moist to the touch, such as the feel of a wrung-out sponge.

Although not necessary for decomposition, shredding and aeration are often practiced. Aeration may be achieved by turning or mixing the outer edges with the center of the pile. If the oxygen content is too low, anaerobic decomposition results and is associated with foul odors and slower degradation. Shredding produces greater surface area, which increases susceptibility to microbial invasion and availability of oxygen. Extensive shredding and frequent aeration are the basis of rapid composting techniques.

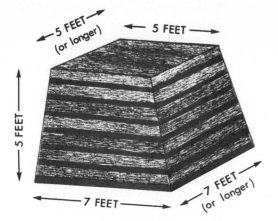

Fig. 7-6. Compost pile construction shown here assures good composting, as a proper carbon/nitrogen ratio will be achieved, compaction and hence anaerobic conditions will be minimized, but heat retention will be sufficient. The latter two are functions of surface area and volume. Photo by author from publication *The Biochemistry and Methodology of Composting*, Bulletin 754 of the Connecticut Agricultural Experiment Station, New Haven.

Generally, compost production in small containers such as cans or plastic bags is inefficient, because the container puts some constraints on volume and aeration. Production of compost under these conditions is slowed. Loose-fitting containers, 40 by 40 inches or larger and 5 feet in height, are usually preferred (Fig. 7-7).

Most plant and animal pathogens, weeds, and insects can be destroyed during composting if the temperature is maintained at 133°F (56°C) for a few days. More resistant pathogens require 149° to 158°F (65° to 70°C) for 3 weeks. Unless rigid control over temperature and turning (to thoroughly expose all material to these temperatures) is practiced, it is not advisable to use materials suspected of containing pathogens.

Fig. 7-7. A good compost container allows the temperature of the compost pile to reach its normal peak: 160°F. Courtesy of Rotocrop (USA) Inc.

Dry garden compost contains 1.0 to 3.5 percent nitrogen, 0.5 to 1.0 percent phosphorus, and 1.0 to 2.0 percent potassium. When added to soil, it improves aeration, soil tilth, and moisture retention. Compost piles are usually employed for small-scale use, such as in a garden. Large-scale needs for organic matter in soil can be met with sheet composting or green manuring.

Sewage sludge has value both as a fertilizer and soil conditioner. Digested sewage sludge, which is prepared by anaerobic digestion, contains 2.0 percent nitrogen, 0.6 percent phosphorus, and traces of potassium on a dry weight basis. Similarly, activated sewage sludge, which is prepared by aerobic decomposition, has about 5.0 percent nitrogen, 4 percent phosphorus, and traces of potassium. These materials have been used successfully, but have not been generally accepted because of possible risks involving human pathogens and phytotoxicity problems associated with heavy metals that contaminate many sewage sludges. Some municipalities are preparing an acceptable compost from sewage sludge and bark.

Available peats were discussed in Chapter 6. The best forms for use as a soil conditioner are moss peat, followed by reed-sedge peat. Peat is especially used as a soil amendment in landscape horticulture, where it is necessary to improve the water-holding capacity of coarse-textured soils and to improve the structure of fine-textured soils. It must be thoroughly wetted before incorporation into the soil. It will be necessary to add a complete fertilizer with it to compensate for the nutrients needed for its decay, as well as to replace the soil nutrients diluted by its incorporation into the soil. Some lime (see next section) may be needed to counter its acidic soil reaction.

SOIL pH

Soils can be naturally acid because of rain leaching lime downward or because they were formed from parent materials that were acid. Generally, acidic soils of the United States are found in areas of high rainfall, that is, the eastern half of the United States and the coastal Pacific Northwest (Fig. 7–8). Neutral-to-alkaline soils are found in the arid western and intermountain areas, where soil cations can accumulate because of limited leaching by rain. A soil analysis will be needed to determine the pH (see Chapter 5) of a specific soil. Horticultural practices also contribute to soil acidity, in that plant roots secrete hydrogen ions, and most sources of nitrogen (except sodium nitrate) in fertilizers acidify the soil. An indication of the acidity of many fertilizers is provided by instructions on how much lime is needed to neutralize them.

The pH preference of plants varies, but most horticultural crops grow reasonably well with a soil pH of 6.0 to 6.5. Exceptions include a number of succulents from arid regions, which prefer slightly alkaline soils from 7.5 to 8.0. A number of plants of the Heath family (*Ericaceae*), including

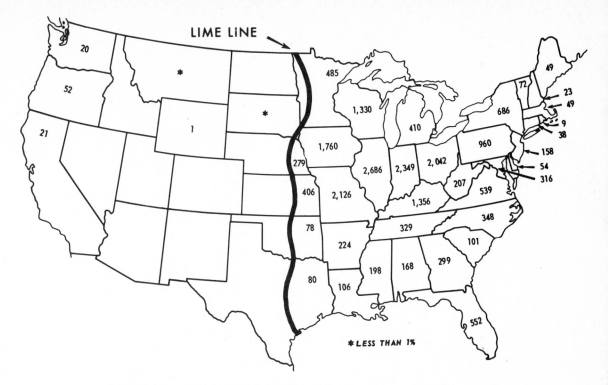

Fig. 7-8. Lime map of United States. The acidic soils and hence areas of heaviest lime use occur east of the lime line and to a lesser extent, the West coast. Numbers can be used as a rough guide to how many thousands of tons of lime used per year in each state. U.S. Department of Agriculture photo.

such plants as the gardenia, rhododendron, azalea, cranberry, camellia, blueberry, *Andromeda,* and mountain laurel, require a pH from 4.5 to 5.5.

Soils with pH values below 6.0 generally have decreasing availability of nutrients such as nitrogen, phosphorus, potassium, calcium, magnesium, and molybdenum (see Fig. 5-14). Other nutrients, manganese, aluminum, and boron, become increasingly available and cause phytotoxicity. Hence soils of this type need to have the pH increased for most garden crops. This is accomplished by the addition of lime. Common liming materials include ground limestone or calcic limestone ($CaCO_3$) and dolomitic limestone [$CaMg(CO_3)_2$]. Less common forms are quicklime (CaO) and hydrated lime [$Ca(OH)_2$], which must be handled with more caution. The mode of action of lime in acid soils is shown in Figure 7-9.

Lime has other values besides the regulation of soil pH. These include the addition of the essential element calcium, and magnesium if dolomitic forms are used, improvement of phosphorus availability, a beneficial regulation of potassium uptake, an increased availability of nitrogen as a result of accelerated decomposition of organic matter, and general improvement of physical conditions over a period of years.

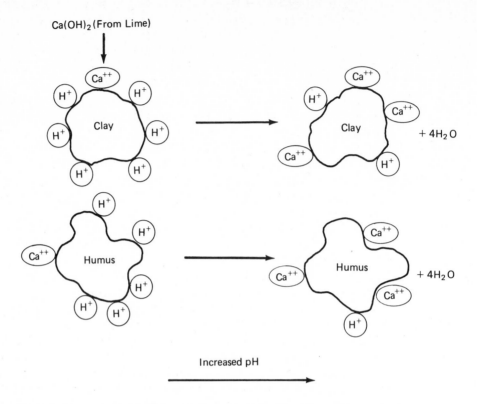

Ca(OH)$_2$ (From Lime)

+ 4H$_2$O

+ 4H$_2$O

Increased pH

Fig. 7-9. Lime reduces soil pH by replacing acid-causing H$^+$ on clay and humus particles. The released H$^+$ reacts with OH$^-$ ions to form neutral water, thus removing the contribution of H$^+$ to soil acidity.

The amount of lime to apply for the correction of acid pH is dependent upon several factors. These are the soil pH as determined by a soil analysis, the pH preference of the crop(s) to be grown, and the purity of the lime, which are obvious factors. The particle size of the lime is a consideration, since the finer the particle the faster the reaction. For example, limestone of 100 mesh or finer will react within a few months of application, 60 mesh in 1 to 2 years, and 20 to 60 mesh in 2 to 4 years. The solubility of the different forms of lime also affects the reaction. The higher the organic matter and clay in the soil, the greater the amount of lime that is needed to adjust the pH, since their buffering capcity resists changes in pH. The amount of lime is not the same per change in pH unit, but changes with different ranges of pH. For example, less lime is needed to go from pH 5.5 to 6.5 than from 4.5 to 5.5. It must be remembered that pH is basically a log function, so that the relative strength increases ten-fold for each unit change in pH.

The most efficient way to add lime is in small amounts yearly. However, the increased labor makes this a more costly approach. Besides the initial adjustment of pH, lime must be added afterward to replace losses due to leaching, removal by harvested crops, erosion, and acidifying properties of fertilizer. Spreaders are the most effective means of liming, and the lime should be thoroughly mixed into the soil. Lime is usually applied in the spring or fall as the soil is cultivated. Lime and all fertilizers can be safely applied together and mixed into the soil. If they are not mixed into the soil, such as on a lawn, fertilizers containing ammonia derivatives of nitrogen should not be applied with the lime, as ammonia formation and subsequent volatilization will waste much of the fertilizer. However, fertilizers not containing ammonia nitrogen can be applied concurrently, or ammonia derivatives could follow the lime safely after at least a 2-week interval.

If it is desired to acidify a soil, the most effective method is the application of sulfur. It is the least expensive and is long lasting. This might be desirable for growing acid-loving plants. For more rapid changes, aluminum or iron sulfate can be used. These are quick acting, but less effective over the long term than sulfur.

In summary, the maintenance of a desired soil pH is a yearly, ongoing process that necessitates yearly fall soil testing. This is necessary, in part, because of the influences of the environment (rainfall, plant materials, plant residues, soils), as well as cultural influences, including irrigation applications, mulching, and fertilizer applications. These influences cause soil pH to be unstable, and therefore there is a need for soil pH monitoring and modification.

Tillage

The working of soil through plowing, sowing, and cultivating afterward is termed *tillage*. Its primary purpose is the control of weeds. Secondary reasons include improving drainage and aeration by breaking up the soil, such as for preparing a seedbed. On small-scale situations, such as a garden, the soil is dug in the spring when the soil temperature and water content are correct (soil is ready when it does not pack into a hard ball, but crumbles readily). Sandy soils can be tilled anytime, but clay soils are better tilled in the fall to allow alternate freezing and thawing to improve soil structure. Digging is done with a spade, spading fork, or a mechanized tiller (Fig. 7–10). Subsequent cultivating is done with a hoe (Fig. 7–10) or cultivating fork. Larger operations use mechanized plows and cultivators (Fig. 7–11).

The need for tillage has been questioned since the development of herbicides for weed control. No-tillage systems have been developed and are successful with certain horticultural crops. Both approaches have their advantages and disadvantages, which will be discussed.

(A)

(B)

(C)

(D)

(E)

Tillage controls weeds and improves aeration and drainage of soils, especially fine-textured soils. The turning and burying of plant residues also helps control some diseases and insects. However, continual tillage decreases levels of organic matter in the soil by improving conditions for its oxidation. There is also deterioration of soil structure, since the continual use of heavy machinery causes soil compaction, and the plow causes pan formation where its bottom slides over and compresses the soil. This compaction and pan can inhibit aeration, water penetration, and root penetration. The use of a chisel plow can give temporary elimination of the compaction and pan problems, but the incorporation of soil amendments for long-range relief is needed (see drainage in this chapter). The extensive use of machinery is also high in energy costs.

These soil-structure problems are minimal on home-garden operations, since much of the work is done by hand or light machinery. On a larger scale, these problems can be lessened by cutting back on the number of cultivating steps or eliminating them entirely by mulching or chemical weed control.

In a no-tillage system, herbicides are used to kill weeds prior to planting and during the growing season. The earth is left covered by a mulch, which can be the residue of the previous crop, a cover crop grown as a mulch, or a crop grazed by animals. Planting and fertilization are done with a machine that slices through the mulch and soil, deposits fertilizer and seed, and then presses the soil over the seed.

Advantages include reduced losses of organic matter, a reduction of soil erosion, less destruction of soil structure, decreased labor, and less energy consumption. Disadvantages include the use of herbicides, which some consider objectionable, and increased losses from insect and rodent damage because of the attractiveness of the mulch.

At present, and undoubtedly for some time, the tillage system predominates in horticultural practices. The no-tillage system is used in some instances and undoubtedly has influenced a trend to reduced tillage systems. The no-tillage system is often objectionable to horticulturists on an esthetic basis.

Fig. 7-10. Tillage of garden plots (opposite page). A. Soil is initially turned over with a spade (shovel) or spading fork. Next the soil is worked further with a hoe to reduce sizes of soil "chunks" and to achieve some leveling. The final step is raking to produce a level seedbed free of debris and pulverized sufficiently for sowing seeds. B. Rotary tillers greatly facilitate the procedure in A, as they carry out the three steps in one operation. In addition they can be used for later cultivation to destroy weeds or used to incorporate organic matter in the soil. C. Appearance of seed beds prior to sowing seeds. D. After germination some thinning of seeds to achieve proper spacing is often required. E. Weeding is accomplished by cultivation with a hoe or more easily with the rotary tiller. A, C, D, E: Courtesy of National Garden Bureau, Inc. B, photo of TROY-BILT Roto Tiller courtesy of Garden Way Manufacturing Co. Inc., 102nd St. & 9th Ave., Troy, NY 12180.

Fig. 7-11. Soil tillage on a large scale uses the following equipment. A. Deep plowing is done with this high horsepower tractor. B. Plow used for plow depth of 12-14 inches. C. A disc is used to further work the soil after plowing. Soils that work very easily can sometimes be prepared with discing only in place of plowing/discing. D. Discs can be followed by ring rollers to both work and roughly level the soil. E. If leveling is critical, such as for furrow irrigation, a land-plane can be utilized for proper leveling. F and G. Final seed bed preparation consists of har-rowing (somewhat like raking) with a spring-tooth harrow (F) or drag harrow (G). This gives a leveling, pulverizing, and debris-removal effect. If the soil contains a lot of crop residues, the drag harrow is preferable as tangling is less than with the spring tooth harrow. H. Cultivation for weed control with a cultivation sled. I. Cultiva-tion with a toothed cultivating roller. All photos except C from American Society for Horticul-tural Science. C. U.S. Department of Agricul-ture.

(E.)

(F)

(G)

(H)

(I)

Irrigation

Much of the irrigation technology used for horticultural applications is derived from agricultural practices. The need for irrigation by horticulturists in the arid West is obvious. However, horticulturists in the East and South can benefit from the practice of irrigation. This results from the fact that the total rainfall may be adequate, but irregularities in the distribution pattern and timing may cause a number of drought days. Many failures with horticultural crops result from lack of or poor timing and quantity of irrigation.

The amount and frequency of irrigations can be determined from calculations or actual measurements. This is necessary so that water is not wasted and excess water does not harm plants. Calculations are used to determine the *consumptive use* of water by the plant, that is, the water lost through transpiration and evaporation. Consumptive use is affected by a number of variables, which include temperature, wind, humidity, sunshine levels, percentage of plant cover, root depth, the developmental stage of the plant, and even the available amount of moisture. Fortunately, a reasonable approximation of the consumptive use can be arrived at from average meterological data, empirical constants available for horticultural crops, estimates of the total plant cover, and measurements of evaporation determined with a Bellani black-plate atmometer. The difference between the consumptive use and the available moisture, the *water deficit,* is the amount of water that must be supplied.

The other approach is the direct measurement of soil moisture. Either a gravimetric procedure, which involves weighing a soil sample before and after oven drying, or instrumentation is utilized. The instruments required are a tensiometer and/or a Bouyoucos bridge. The tensiometer measures soil-moisture tension. The drying of soil increased the soil-moisture tension, which in turn causes a suctioning of water from the soil tensiometer that can be observed on the tensiometer gauge. Tensiometers are usually installed in several places and at varying depths in the zone of rooting. The Bouyoucos bridge measures the electrical resistance of a sorption block, such as gypsum, situated in the soil. The electrical resistance changes with the amount of absorbed water in the block, which is a direct function of the amount of soil moisture surrounding it. Generally, the amount of soil moisture should be supplemented by irrigation when the level of available water in the root zone reaches 50 percent of field capacity.

When the water is applied, it must be remembered that not all the irrigation water will be available for use by the plant. Some will be lost by evaporation of free capillary water, runoff, adherence to soil particles (hygroscopic water), or sinking below the root depth (gravitational water). The latter can be minimized by applying water only to the depth of rooting, which is known for most horticultural crops (see Chapter 5, soil moisture). Other points to consider are that soil fertility levels must be good for efficient water utilization and calculations and measurements are only a useful starting point. This must be tempered with experience, personal observations, and variations that prevail in localized sites.

230

Fig. 7-12. Types of surface irrigation. A. Furrow irrigation. Furrows are made by machines after the plowing, discing, harrowing, and landplanning are completed (see Fig. 7-11). B. Border irrigation. C. Corrugation irrigation. A is American Society for Horticultural Science photo. B and C are U.S. Department of Agriculture/Soil Conservation Service photos.

(C)

Certain soils in arid and semiarid regions, which are classified as saline, alkali, or saline-alkali, pose some special considerations in regard to irrigation. Saline soils have surface salt crusts. These can become productive if their drainage is either naturally good or improved by artificial means such that low-salt irrigation water will leach the salts below the root zone. Problems result when only high-salt irrigation water is available, or the water table is high. Alkali soils have a pH between 8.5 and 10.0. These soils require applications of sulfur to reduce pH and gypsum (calcium sulfate) to reduce the levels of exchangeable sodium. Saline-alkali soils have the problems of both saline and alkali soils and need all the preceding treatments. Horticultural crops also vary in their salt tolerance, which should be considered when selecting plants for these soils.

Water for irrigation purposes must be tested to ensure that it meets certain standards. Excesses of soluble salts, sodium, bicarbonate, boron, lithium, and other minor nutrients could cause phytotoxicity.

There are several irrigation systems: surface irrigation, sprinkler irrigation, subsurface irrigation, trickle or drip irrigation, and ooze irrigation. *Surface irrigation* is used in arid and semiarid areas with level surfaces. The four types of surface irrigation are *furrow, flood, corrugation,* and *border* irrigation (Fig. 7–12). Furrow irrigation, the oldest known, depends on water flowing by gravity from a main ditch down into furrows, on top of which are crops. Slopes must be less than 2 percent to avoid soil erosion. Some piping may be used to minimize water losses by seepage and evaporation. Water gates and siphons can be used to improve water distribution. In flood irrigation the water is distributed as a continuous sheet. Crops tolerant of excess water, such as rice or cranberries, are generally irrigated by the flood method. Corrugation irrigation can be used on slopes up to 5 percent. Many small furrows are used to guide rather than carry the water. Border or contour levee irrigation is used on gentle slopes. Narrow, leveled strips are enclosed by levees, and irrigation consists of flooding these areas.

Advantages of surface irrigation include a low initial investment (if land requires no grading), low energy consumption, and low water evaporation. Disadvantages include uneven distribution of water by gravity flow (costly grading may be required), and the weight of the water can reduce soil aeration and drainage.

Sprinkler irrigation consists of pumping water through pipelines and distributing it from rotary heads (Fig. 7–13). The result is a rainfall-like application. This system is popular for supplementing natural rainfall during periods of inadequacy, especially in the humid regions of the East and South. Advantages of this system include a reduction in land leveling, drainage problems, erosion, and the need for special skills. This results from the even, controlled rate of sprinkler irrigation. The initial investment, evaporation losses, and energy consumption are higher than those for surface irrigation, but the operational costs are less. In addition, winds may interfere with uniform distribution.

Fig. 7-13. Sprinkler irrigation lines in use. Courtesy of Johns-Manville Corp.

Subsurface irrigation depends upon an impervious layer below the soil surface. This will stop percolation losses and allow the maintenance of an artificial water table by pumping water through ditches and laterals. This irrigation system depends, therefore, on unusual soil conditions and is not widely utilized in the United States.

The previous irrigation systems are used mainly on large-scale horticultural operations in the field. The home gardener may use a hose and sprinkler during periods of inadequate rainfall, and this would be a simple form of sprinkler irrigation. The use of misting systems for propagation and automatic-watering sprinkler systems in the greenhouse is another form of sprinkler irrigation. However, other types of irrigation have become popular for greenhouse use, small nursery container operations, home gardens, and small field operations. They are known as *trickle* or *drip irrigation* and *ooze irrigation* (Fig. 7-14). With trickle irrigation, water is carried by polyethylene pipes into which are tapped smaller feeder lines that terminate with small emitters (valves). Water trickles from these valves, which can be shut off individually, or the feeder line can be removed and the hose plugged. These are particularly adapted to potted plants, nursery container stock, small vegetable and flower gardens, and orchard operations. *Ooze irrigation* employs polyethylene pipe for carrying water and the feed lines are porous polyethylene, which allows water to ooze through micron-sized holes. The reasonable cost and efficient water usage of these systems suggests their increased use in all horticultural operations.

(A)

Fig. 7-14. A. Drip irrigation is being used here on nursery container stock outdoors. American Society for Horticultural Science photo. B. Ooze irrigation: close-up showing oozing water. U.S. Department of Agriculture photo.

(B)

Mulching

The cultural practice of mulching has several advantages (Fig. 7-15). Mulches save on the labor of cultivation, since emerging and small weeds perish under their dark barrier. Therefore, they reduce tillage and the use of weed-control chemicals. Water is conserved, because mulches reduce the evaporation of soil moisture by lowering the soil temperature. Water absorption by a mulched soil is greater than by an unmulched soil, because of physical properties of the mulch and its prevention of the formation of impervious soil crusts. Consequently, soil losses from heavy washing and blowing are decreased. In effect, mulches are excellent conservation agents.

The insulating property of mulch averts drastic fluctuations of soil temperature, keeping the soil cooler in summer and warmer in winter. During the summer this improves both root growth and nutrient availability. Winter mulches reduce the risk of root damage. Protection with winter mulches will be covered further in the following section on protection. Soils are improved with organic mulches as lower layers decompose and become incorporated into the soil. At the end of the growing seasons, organic mulches can be tilled into the soil to further increase the organic-matter content. Finally, mulches impart a neat, trim look to gardens, reduce the incidence of mud-splashed flowers and vegetables after heavy rains, and decrease the frequency of vegetable rot caused by soil contact.

Introduction to Applied Horticultural Science / Part II

(A)

(B)

Fig. 7-15. A. Organic (straw) mulch used in the vegetable garden. Courtesy of National Garden Bureau, Inc. B. Black plastic mulch used with tomatoes. U.S. Department of Agriculture photo. C. Wood chip mulch around foundation plantings. U.S. Department of Agriculture photo now in National Archives (No. 16-N–34103).

(C)

Mulches do have some disadvantages. The application of a mulch requires labor of varying degrees, depending on the material selected. However, the labor saved from the decreased need of cultivation usually outweighs the initial input. Mulch and moisture around newly emerging seedlings or perennials can provide an ideal environment for diseases, such as damping-off or crown rot. The potential of disease is increased by long periods of rain. Diseases may overwinter in the mulch. Insects and rodents, who find it an attractive habitat, can cause plant damage. These problems can be minimized by avoiding direct contact between the plants and the mulch. Premature applications of organic and black plastic mulches may retard soil warming, and hence the growth of plants preferring warmer soils, such as tomatoes, peppers, and eggplants. An exception is clear plastic mulch, which will accelerate the warming of soil and produce earlier yields with warm season crops such as corn or muskmelons, but it supplies much less weed control than black plastic mulches, unless a herbicide is put down prior to the clear plastic. A too thick mulch or one prone to caking can impede the uptake of water and air by the soil, leading to possible plant damage. Mulches with a carbon-to-nitrogen ratio grater than 30:1 can steal nitrogen from the soil during decomposition, causing nitrogen deficiency in the mulched plant. Prior applications of nitrogen fertilizers can prevent this problem. Dry mulches that are combustible may be a fire hazard near buildings or in public places. Woody mulches near buildings can be a vector for termites.

A practical mulch should be easily obtained, inexpensive, and simple to apply. Availability and cost vary from region to region. Mulching materials may be found in yards, garden centers, lumberyards, sawmills, dairy farms, tree-service firms, breweries, and food-processing plants. A suggested depth is 2 to 4 inches, bearing in mind that too little will give limited weed control and too much will prevent air from reaching roots. Mulches should be applied prior to active weed growth and summer droughts or before the ground freezes if it is a winter mulch. For warm-season crops, such as tomatoes, the mulching should be delayed until blossoms appear. A list of mulching materials follows with specific emphasis on advantages and disadvantages.

Bark. Small pieces of bark are preferred over large chunks. Bark mulches vary, but all are attractive, durable, and suitable for foundation shrub plantings. Contact with wood framing is to be avoided, since bark can be a termite vector. The high carbon-to-nitrogen ratio of bark requires prior application of nitrogen fertilizer.

Buckwheat hulls. These are light, fluffy, and black. Since buckwheat hulls are prone to caking, they should be applied no deeper than 2 inches. Forceful watering will cause scattering of buckwheat hulls.

Cocoa shells. These are brown, light, easy to handle, and relatively noncumbustible. Cocoa shells have some value as a fertilizer and resist blowing in the wind. Their high potash content harms some plants, so they should not be applied to a depth greater than 2 inches.

Coffee grounds. These cake badly; a depth of 1 inch is recommended. Coffee grounds contain some nitrogen.

Compost. This is a good mulch, as it has fertilizer value and a soillike appearance. It is also a good organic amendment for tilling into the soil after the growing season ends.

Corn cobs. Ground corn cobs are a good mulch. Some find their light color objectionable. Other uses for gound corn cobs, such as in feeds and mash, tend to limit the supply for mulching.

Grass clippings. These contain nitrogen. Grass clippings cannot be applied thick when green, as they heat rapidly and form a dense mat that restricts the flow of air and water. They should be applied in thin layers and allowed to brown between each application.

Ground tobacco stems. These make a coarse, good mulch with some nitrogen value. Availability is limited, and they should not be used on plants susceptible to tobacco mosaic virus, since they can be a source of infection.

Leaves. Leaves are free, readily available in many areas, release some nutrients upon decomposition, and spread easily. However, they have a tendency to form a soggy, impenetrable mat. This problem can be overcome by mixing leaves with fluffy materials, such as hay or straw, or by shredding the leaves. Oak leaves are especially good for acid-loving plants, such as azaleas and rhododendrons.

Licorice roots. These are especially good on slopes, since they resist floating and blowing. Licorice roots also have an attractive appearance.

Newspaper. This is certainly readily available and economical, but somewhat difficult to apply. The high carbon-to-nitrogen ratio necessitates the prior application of nitrogen fertilizer. A good use for newspaper is as an undermulch; that is, place four or five sheets under a thin layer of an attractive, more expensive mulch.

Peanut shells. These are attractive and easy to apply. Peanut shells also contain nitrogen and are long lasting.

Peat moss. This is attractive, easy to handle, but somewhat expensive. Dry peat moss requires considerable time and water to become moist, so it should be applied only to a 3-inch or less depth and avoided in areas subject to drought. Its acidic pH makes it especially desirable for acid-loving plants.

Pine needles. These have an esthetic appeal and are not prone to forming a soggy mat as are leaves. They are especially good for acid-loving plants.

Polyethylene film. This is one of the few mulches that is readily available and economical enough to be used on larger-scale commercial applications. Polyethylene allows passage of gases such as nitrogen, oxygen, and carbon dioxide. Holes or slits facilitate the planting of seeds or plants and water entry. It can last several years if undamaged by machinery. Usually, it is used as black film. Clear film is sometimes used, but it offers limited weed control (unless herbicide is applied before mulching), since light passes through it. Earlier crops can be produced with the clear, and to a lesser degree, black plastic mulch because of the warming of the soil.

Straw, hay, salt-marsh hay. These materials are lightweight and easy to apply, but their appearance restricts their application mostly to vegetable gardens. They are used more frequently as a winter mulch for protection . They are not long lasting and frequently contain weed seeds.

Sawdust. Aged or partially rotted sawdust makes a satisfactory mulch that lasts a long time. Since it is prone to caking and has a high carbon-to-nitrogen ratio, apply it only 2 inches deep after adding nitrogen fertilizer to the soil.

Spent hops. These are resistant to blowing and have some nutrient value. However, when fresh they have some odor and may heat, thus causing some plant damage.

Stone. This is a very durable mulch with an attractive appearance. It is used often around trees or shrubs. Do not use crushed limestone around acid-loving plants.

Sugarcane. This is a good mulch when available in the crushed form. It has a moderately acidic pH, making it useful around acid-loving plants.

Tire fiber. This is an extremely durable and effective mulch.

Wood chips. Since these are moderately priced or free, attractive, readily available, and easy to apply, they make an excellent mulch. Their high carbon-to-nitrogen ratio requires an application of nitrogen fertilizer. Wood chips can last about 2 years. As with a bark mulch, one must consider that they can be a vector for termites.

PROTECTION FROM WIND AND SUN

In some areas the prevailing winds cause problems such as wind damage, snow accumulations, temperature drops, and soil erosion. These areas can occur near the seashore, level exposed areas, or on hilltops. The unsymmetrical shaping of isolated trees is a good indicator of wind problems that require modification. In the same or other areas, the sun may pose problems concerning temperature control within structures. Both problems may be modified through the judicious use of appropriate plant materials. The landscape horticulturist must choose plants not only on the basis of appeal, but also for their modifying influences upon the environment nearby.

Windbreaks (Fig. 7-16) are used to reduce or deflect excessive winds.

Fig. 7–16. Aerial shot of windbreaks used to protect crops and homes. U.S. Department of Agriculture photo.

Most of us are familiar with the moderating influences of windbreaks upon open spaces, such as the reduction of the windchill factor. Wind damage, snow accumulations, temperature drops, and soil erosion problems can be reduced by the proper placement of a windbreak. The effect of a windbreak upon horticultural and other buildings is important. Greenhouses and homes exposed to extensive prevailing winds require much greater energy input for heating than those not exposed to such winds. Unheated facilities can also benefit from wind protection, since their period of usefulness could be extended. Windbreaks can also be used to deflect wind to another location, such as the guiding of summer breezes through an outdoor patio. Proper placement of plants can also be used to reduce the speed of wind, rather than to deflect it elsewhere.

Windbreaks consist of natural materials, such as hedges or trees, or artificial materials, such as lath fencing used to control snowdrifts. If the prevailing winds are a winter problem, evergreen plantings would deflect them then, whereas deciduous trees would only reduce the wind speed somewhat during the winter. In general, windbreaks are useful for a distance that is ten to twelve times their height; the maximal protection occurs at a distance of five to six times the height of the barrier.

During the summer, the proper placement of a deciduous tree (Fig. 7–17) can shade the home and reduce energy consumption by an air conditioner or make an uncooled home more comfortable. During the winter, the nature of the deciduous tree would permit the sun's rays to warm the home. Alternately, a deciduous tree could shade and cool a small greenhouse in summer, yet detract minimally from needed winter light.

Fig. 7–17. Proper placement of trees and shrubs can make a home comfortable and more energy efficient. A high-crowned deciduous tree (A) on the west, a low-crowned deciduous tree on the east (B), and other deciduous trees to the south (D, C) shade homes in the summer. This shade cooling effect reduces the need for air conditioning, yet in the winter their deciduous nature allows the sun to warm the house when it is most needed. Evergreens around the house create a still air space, thus reducing loss of cool air from the house in the summer and heat loss in the winter.

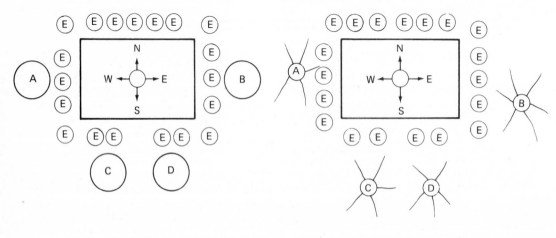

SUMMER   WINTER

During a hot summer, the shade of a tree over a patio is most welcome. In areas where sun is a year-round problem, the use of evergreen trees is preferable over deciduous ones.

WINTER PROTECTION

Plants in exposed locations, particularly in the northern and northeastern United States, may require some form of winter protection to improve their survival and development (Fig. 7–18). The protection is mainly against heavy winds, which can desiccate foliage; extremes of heat and light during winter, which increase problems of desiccation; alternate freezing and thawing, which could heave a plant from the ground; and protection against heavy snow, which can cause physical damage to the plant. The foremost plan against the winter threat is the optimal use of horticultural practices to build up and maintain good plant vigor.

Evergreen plants are susceptible to desiccation during the winter, especially when a combination of winds, bright sun, and moderate temperatures occurs. The leaves transpire, but the roots cannot take water from the frozen soil; hence leaf desiccation is the result. Protection includes the erection of burlap screens or the use of antitranspirant sprays.

Young, unestablished plants can be carried through the first winter by erecting a burlap square around them and filling in lightly with oak leaves. Young trees may require a wrapping of burlap or tree-wrapping tape around their trunks to prevent winter sun scald the first winter after being transplanted. Soil can be mounded around tender woody plants, such as roses, to protect crowns.

During the period of the year that includes late winter-to-early spring in the North, newly planted ground covers and newly planted or shallow-rooted herbaceous perennials and biennials are prone to heaving when soils (particularly fine-textured soils) undergo thawing and freezing sequences caused by succeeding periods of bright sunny days and cold nights. Applying a light, loose organic mulch to these plants after the ground freezes in midwinter helps to conserve the movement of heat and thereby maintain the soil in a constant frozen condition. This mulch is removed in early spring when air temperature fluctuations are not so extreme. Mulches can be pulled over low, tender plants for additional winter protection. The mulch must be loose and fluffy to ensure good air circulation. Winter mulches usually require some stabilization against winds.

Snow damage can be minimized by a lath or board structure over the tops of susceptible plants, by tying them together with cord, and by shaking the snow off before it freezes.

The labor of winter protection can be minimized by avoiding sites with exposure problems (for example, winds and/or falling snow from roofs) and avoiding plants that are apt to require winter protection.

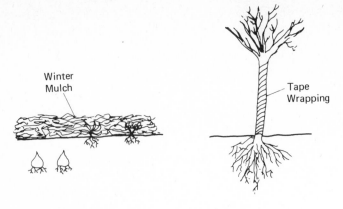

Winter
Mulch

Tape
Wrapping

FIRST WINTER PROTECTION FOR NEW PLANTS

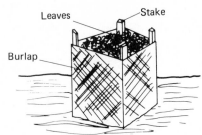

Leaves   Stake

Burlap

Fig. 7-18. Various forms of winter protection. Recent evidence questions the effectiveness of antitranspirant sprays to prevent winter desiccation. In view of this it might be wise to use both the spray and the burlap screen together, especially if the shrub is valued.

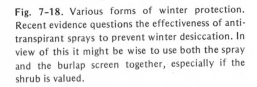

Burlap

Antitranspirant Spray

Cord

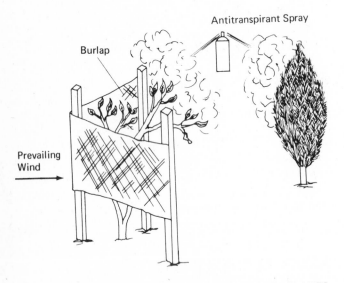

Prevailing
Wind

WINTER PROTECTION FOR ESTABLISHED PLANTS

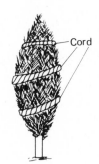

Board Shelter

Mounded
Soil

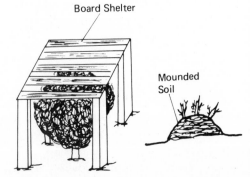

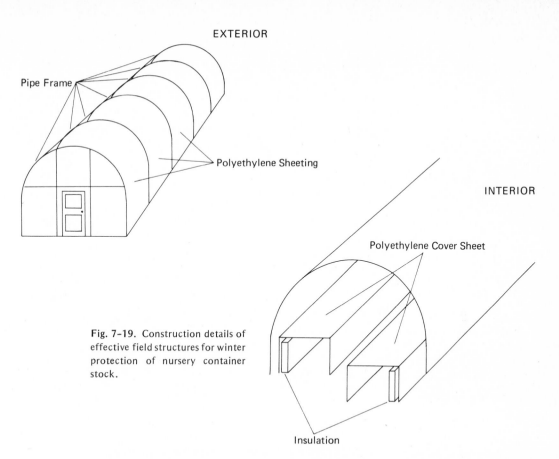

EXTERIOR

Pipe Frame

Polyethylene Sheeting

INTERIOR

Polyethylene Cover Sheet

Fig. 7-19. Construction details of effective field structures for winter protection of nursery container stock.

Insulation

The foregoing has been concerned with in-ground plantings. Container plantings can also be successfully stored and protected outdoors during the winter. Container and/or ball-and-burlap nursery stock are more susceptible to low-temperature root kill in a reduced soil volume as opposed to in-ground sites. Damage can also occur through bark splitting and desiccation.

Storage and protection should be started from mid-October through early November. The storage structure is usually erected right over the container plants at their growing sites, as the labor of plant movement is uneconomical. The structure usually consists of pipe hoops covered by a sheet of clear or milky polyethylene (Fig. 7-19). Supports may be needed in areas that experience heavy snowfalls. White polyethylene is suggested when early flowering or growth is disadvantageous and clear polyethylene when it is not. Containers should be irrigated prior to storage.

Plants with tender roots subject to root kill may require additional protection. This is usually a polyethylene blanket that is draped over the plants and tucked at the edges. Styrofoam sheets may be placed between the outermost containers and the polyethylene sheet. Desiccation injury is minimal with the polyethylene sheet.

FROST PROTECTION

If at all possible, tender plants should always be planted after the frost-free date. Plants susceptible to frosts at certain stages, such as the blossoming period of fruit trees when a frost would decrease fruit set, should be planted in microclimates such as thermal belts or near large bodies of water (see Chapter 5). A southern exposure may offer better frost protection than other exposures. The propagation and production of many herbaceous ornamentals is keyed to the variable known as the first frost-free date. It is a variable because of yearly seasonal fluctuations and because herbaceous ornamentals themselves vary with respect to their tolerance to frost and freezing temperatures. The bedding-plant industry must be able to calculate for each species, variety, and cultivar sufficient lead time so that plants will have time to reach "landscape size" at the time of sales or planting after the last spring frost date, yet still have time to mature before the fall frost.

There is always the possibility of the danger of a late, unexpected frost to tender plants recently planted or to established plants at a critical developmental stage. Here it may become necessary to protect against frost by conserving or adding heat. The gardener may protect against a light frost by laying newspapers, polyethylene, bushel baskets, or cloth over the threatened plants the evening before the frost. This acts to conserve the soil heat and keeps the plants from direct contact with cold air. Often the gardener may wish to risk the chance of frost to get an early start with tender plants such as tomatoes. Here a hot cap, a caplike container made of translucent paper, is set over the plants (Fig. 7–20). It acts like a greenhouse to increase the soil temperature and conserves heat at night. Plastic tunnels on wire horseshoe-shaped inserts are also utilized for this purpose. Hot caps and plastic tunnels are used on commercial scales for high-value horticultural crops.

At one time, smudge pots were used frequently to combat frost in citrus groves and fruit orchards (apple, peach, pear). However, air-quality standards are such as to make them illegal in many areas. Basically, they burned a cheap oil that gave some heat at tree level, and the black smoke cloud reduced the loss of heat by radiation. More efficient, clean-burning heaters are used now. These depend upon heat output and convection currents to reduce the danger of frost. Fuels are usually a good grade of oil or propane. Solid petroleum wax candles are also used.

During a radiation frost (see Chapter 5), the plant temperature and the air immediately adjacent may be cooler than that of the air above. Mixing of the air above by fans may increase the temperature around the plant and protect it from frost, if the radiation frost is not too severe. Fans and heaters together are even more effective (Fig. 7–20).

Water fogs can also reduce the danger of frost. When water turns to ice, energy, termed heat of fusion, is released. This may be enough to protect against a mild frost. The formation of a white frost on a plant can

Fig. 7-20. Frost protection. A. Hot cap used to protect early tomato plant. U.S. Department of Agriculture photo. B. Fans and heaters used to protect a young citrus orchard. American Society for Horticultural Science photo. C. A polyethylene bag over a wire frame provides frost protection and acts as a mini-greenhouse for that hopeful early garden planting. Author photo.

(A)

(B)

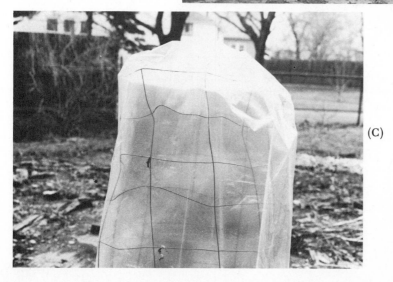

(C)

(A)

Fig. 7–21. Soil erosion through excessive water runoff is effectively controlled by terraces (A) and contour tillage (B). Both hold rainfall better and a more even water distribution results. U.S. Department of Agriculture photos.

(B)

be protection against frost damage if the air temperature does not decrease too much. Fog also provides a cover to reduce the heat lost through radiation. On occasion, plants touched lightly with frost may be saved by sprinkling them with water before sunrise.

A newer technique involves the insulation of the plant with a spray foam. The foam is nontoxic and usually disperses the following day.

Water and Soil Conservation

Water and soil need to be conserved, as the supply of each is not unlimited. The two are interrelated, since steps taken to conserve soil often aid in the conservation of water. Improper management of soil and water can lead to floods, droughts, erosion, and other conditions unsuitable for horticultural practice. Since this also affects agriculture, the consequences have a serious impact upon humans.

Water conservation is based upon the control of soil runoff and improvement of water absorption by the soil. In humid regions the aim is to minimize runoff and maximize water absorption. This holds true in arid regions where crops tolerant of dryness are grown. However, where the objective is to collect water for irrigation, the aim is to maximize runoff with minimal erosion.

On slopes, terracing and contour tillage are used to control runoff, which in turn reduces soil erosion (Fig. 7-21). Terraces and tillage are run at right angles to the slope. This acts to trap the rainfall until it can be soaked up by the soil.

Stubble left from crops is very effective at trapping snow, thus increasing storage water. Frozen soils can take up water when sufficient organic matter is present. The bulk of the water would be absorbed in the spring when the snow melted. Mulches increase the infiltration of rainwater and reduce soil erosion. Water is also lost by transpiration by plants. Sodded orchards are mowed closely to increase the amount of soil water available to the fruit trees, since less grass surface area means less transpiration.

Erosion of soil by wind and water can be decreased by providing a protective cover for the soil, especially at times when these problems are more apt to occur. Trees and grass are quite effective in preventing wind and soil erosion. Horticultural crops alternated with cover crops are also effective. Windbreaks can be helpful. Vegetation also increases water absorption by the soil by breaking the force of rain, which reduces soil particle scattering and in turn plugging of soil pores.

FURTHER READING

Fret, T. A., and E. M. Smith, "Woody Ornamental Winter Storage," *HortScience,* 13, no. 2, (1978), 139.

Havis, J. R., and R. D. Fitzgerald, *Winter Storage of Nursery Plants.* Amherst, Mass.: Cooperative Extension Service University of Massachusettes Publication No. 125, October, 1976.

McVickar, M., *Using Commercial Fertilizers* (3rd ed.). Danville, Ill.: Interstate Printers & Publishers, Inc. 1970.

Oliver, H., *Irrigation and Water Resources Engineering.* New York: Crane, Russak & Co., 1972.

Pearson, R. W., and F. Adams, eds., *Soil Acidity and Liming,* Agronomy Monograph Series No. 12. Madison, Wis.: American Society of Agronomy, 1967.

Poincelot, R. P., *The Biochemistry and Methodology of Composting.* New Haven Ct.: Connecticut Agricultural Experiment Station Bulletin 754, 1975.

Schilfgaarde, J. V., *Drainage for Agriculture.* Madison, Wis.: American Society of Agronomy, Inc., 1974.

Schwab, G. O., R. K, Frevert, T. W. Edminster, and K. K. Barnes, *Soil and Water Conservation Engineering.* New York: John Wiley & Sons, Inc., 1966.

Soil: The Yearbook of Agriculture. Washington, D.C.: U.S. Government Printing Office, 1957.

Tisdale, S. L., and W. L. Nelson, *Soil Fertility and Fertilizers.* New York: Macmillan, Inc., 1974.

Water: The Yearbook of Agriculture. Washington, D.C.: U.S. Government Printing Office, 1955.

Management of Controlled to Semicontrolled Plant Environments

A number of plant environments other than the outdoors are encountered by horticulturists. Some are partially dependent upon the natural environment, and others are entirely dependent upon artificial environments. Each has environmental management problems peculiar to itself. These alternate environments are a very important part of horticulture, and as such, will be considered separately in this chapter. Since these structures will often have combined water and electrical facilities, all installations should be by a qualified electrician, and ground-fault circuit interrupters are highly recommended to prevent possible electrocution.

Greenhouses

STRUCTURE

Greenhouses are an important plant-growth structure encountered on both the commercial and hobbyist levels of horticulture. Basically, greenhouses are either free-standing or lean-to (attached to another structure) types (Fig. 8–1). Free-standing greenhouses afford maximal light to every corner, but cost more to heat than lean-to greenhouses. One must balance light losses versus heat costs. Free-standing greenhouses are usually the choice of commercial horticulturists. Lean-to greenhouses are more apt to be the choice of home owners, who enjoy being able to walk directly into the greenhouse from their homes.

(A)

Fig. 8-1. Free standing glass and aluminum greenhouse (A), lean-to glass and aluminum greenhouse (B), and window unit glass and aluminum greenhouse (C). Courtesy of Lord & Burnham.

(B)

(C)

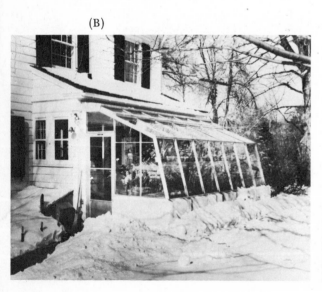

Introduction to Applied Horticultural Science | Part II

Fig. 8-2. A. The interior of a large commercial greenhouse constructed with fiberglass, as easily recognized by the rippled, translucent sidewall. Courtesy of Modine Manufacturing Co. B. An inexpensive polyethylene home greenhouse. U.S. Department of Agriculture photo. C. A commercial quonset type greenhouse covered with polyethylene plastic. William P. Raffone, Jr. photo.

There is a large variation in greenhouses in terms of structure. The framework usually consists of pressure-treated wood, steel, or extruded aluminum alloy. Heat loss through the metal framework is higher than through wood. Aluminum alloy comes closest to being maintenance free. Coverings consist of glass, fiberglass, and plastic films. Glass has the highest initial cost, but its long-term depreciation advantage still makes it a good choice. Fiberglass, once only good for several years exposure, is now available with a special coating that extends its greenhouse life to 20 years. Polyethylene films are much cheaper than fiberglass, but their breakdown upon exposure to ultraviolet light necessitates replacement once a season. The addition of ultraviolet inhibitors results in a plastic film that may last two or more seasons.

Many glass-covered houses still exist in the trade, but the trend today in commercial ventures is toward the plastic types, fiberglass and polyethylene film (Fig. 8-2), both because of lower capital investment and construction costs and the ease with which these types can be made more energy efficient. Home owners are increasingly attracted to the fiberglass type for similar reasons, although the esthetics of a quality glass greenhouse still appeals to some.

TEMPERATURE

Environmental control within a greenhouse involves the control of temperature, air exchange, light, humidity, and growing media. Temperature control is dependent upon regulation of heating and cooling. Some of the heat is supplied by solar radiation, which passes through the transparent glass and warms everything it contacts. Some of the absorbed heat is reradiated from the warmed objects as longer wavelengths, that is, infrared, which cannot pass back through the glass but is absorbed by carbon dioxide and water vapor in the greenhouse. This is commonly termed the *greenhouse effect*. Fiberglass is also effective, but polyethylene film is much less so. However, the film of water frequently found on the plastic film does absorb the infrared, so a greenhouse effect can occur even with polyethylene. The heat trapped by the greenhouse effect depends upon convection currents or fans for distribution.

However, on cloudy days and winter days, the heat produced by solar radiation may not be sufficient to maintain the desired temperature. Some form of artificial heat then becomes necessary. This heat source must have a constant high output with uniform distribution. This is necessary since heat losses through the glass or plastic and framework by conduction and through leaks is rapid. Cold spots may develop in parts because of factors such as cold winds blowing against a wall that increase heat loss in that area compared to other areas. Heat loss can be calculated on the basis of surface area and the rate of thermal transmission which varies according to the material involved. Heat-loss calculations for the structure, the desired running temperature of the greenhouse, heat losses if any by the heating system, expected external winds and temperatures, and the desired safety margin are factors that are needed to determine the number of Btu's that must be produced by the heating system (see Further Reading).

Numerous types of heating systems (Fig. 8-3) and fuels are available for greenhouses. Cost of installation and cost and availability of fuel must be considered. Hot water or steam is used with pipes fitted with fins to radiate heat and start convection currents; these systems are usually dependent on oil-, gas-, or coal-fired boilers. Circulating fans may be required for uniform heat distribution. Forced hot air may be distributed through ductwork. Oil-, gas-, or coal-fired furnaces can be used to produce

(A)

(B)

Fig. 8-3. A. Greenhouse heated by an overhead gas-fired forced (fan) hot air furnace. B. Overhead hot water heating unit. Hot water maintained by gas or oil energy. C. A combined overhead heating and ventilating unit that uses a polytube for distribution of heat and outside air. Heat can be from gas-fired hot air, or air heated by steam or hot water condensor maintained with oil or gas. D. Four inch steel fins on copper pipe with circulating hot water or steam used for heat distribution. Hot water from oil or gas-fired boiler. A, B, C, courtesy of Modine Manufacturing Co. D, Author photo

(C)

(D)

the hot air, or gas or electric fan-forced heaters can be utilized. All these systems are thermostatically controlled and often hooked into a temperature alarm system to warn of failure in the heating system. A backup system, even if a simple kerosene heater, can be useful in emergencies or in times of fuel shortages, which appear likely to be increasingly common today. Infrared heating is possible, but regulation and control problems limit its use. All electric heat is possible, but costs are prohibitive.

VENTILATION

Heating makes maximal use of solar radiation and secondary use of conventional heating. Conventional heating will be turned off when the environment reaches the set temperature, but solar radiation cannot be turned off as readily. Temperatures in an enclosed greenhouse can soar well over 40°C (100°F) in a short time on a sunny day. The simplest way to control this temperature rise is by ventilation. Sections of the greenhouse, usually along the ridge and sometimes on the sides, can be opened by a motorized drive (Fig. 8–4), which is thermostatically controlled and set to open the windows at a temperature several degrees above the set heat temperature.

Fig. 8–4. Ridge ventilating window opening is powered by electric motor (not visible) that has been activated by a thermostat. Author photo.

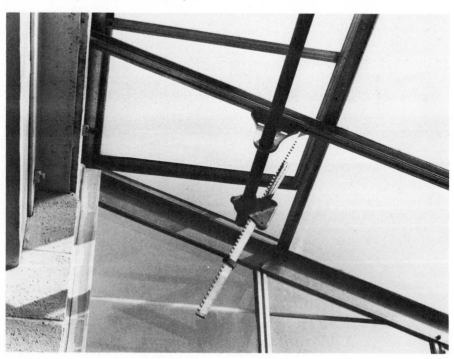

(A)

(B)

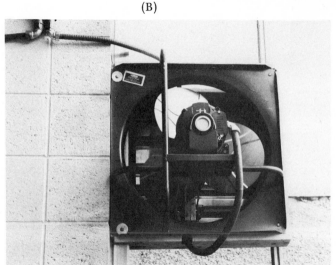

Fig. 8-5. A. A perforated polyethylene convection tube is connected to a fresh air intake fan with a motorized shutter. Control is by thermostat. Courtesy of Acme Engineering & Manufacturing Corp. This unit is often used in conjunction with an exhaust fan at the opposite end of the greenhouse (B, author photo). This would be a two phase ventilating system: phase one, air intake only; temperature continues to rise bringing phase two, air intake and exhaust together.

A more recent unit for this purpose in smaller greenhouses uses a heat-activated expanding chemical to operate the ventilation windows, which saves on electrical costs and is secure from failure. This type of ventilation system depends upon the rising of hot air and the entry of cool air to establish convection currents. Such a system can be improved with the use of air intakes and exhaust fans to exchange the air at a more rapid rate. Another type of ventilation makes use of an air intake fan and motorized exhaust shutter, which is regulated by a thermostat. A perforated polyethylene convection tube can be (Fig. 8-5) connected to the intake to improve air distribution and to reduce cold drafts in winter. In addition to heat control, ventilation also serves the useful purpose of replacing air that is depleted of carbon dioxide by plants.

(A)

(B)

Fig. 8-6. Some forms of greenhouse shading. A. A roll up shade with wooden slats which can be manually operated. Courtesy of Lord & Burnham. B. Green vinyl plastic shading which adheres to inside glass by capillary action and reduces light by about 65 percent. Note the end louvered window for supplementary ventilation. Author photo.

COOLING

During a hot summer, ventilation may prove inadequate to maintain the desired operating temperature. Shading of the greenhouse will provide additional cooling, as well as a reduction of summer light intensity (usually around 60 to 70 percent), which is often too intense for greenhouse plants. Shading (Fig. 8-6) may be provided by spraying shading compounds on the glass, or by the use of roller blinds, green vinyl translucent plastic, or vinyl mesh (Saran). Even under these conditions, ventilation cannot reduce the interior temperature below that found outside. Alternate means of cooling may be required, depending upon summer temperature conditions. Conventional air conditioning is too costly, but air cooling with fans and wet pads provides a reasonable alternative (Fig. 8-7).

Fig. 8-7. Wet pad evaporative cooling in a large fiberglass commercial greenhouse. A. Exterior showing opened sides for air intake through wet pads. B. Interior of same greenhouse showing wet pads. Courtesy of Lord & Burnham.

These wet-pad evaporative coolers will produce an interior temperature equivalent to an outdoor temperature measured in the shade. High-pressure mist systems are also used to reduce temperatures by evaporative cooling.

HUMIDITY

Other environmental aspects, such as humidity and light, can also be controlled in a greenhouse. Humidity can be supplied by steam humidifiers (Fig. 8–8) or cool-water humidifiers, which operate on an aerosol principle. Automation can be provided with a humidistat. Ventilation systems pose a problem, as humidity losses obviously occur when they are open.

LIGHT

Several factors determine the amount of visible radiant energy (light) in a greenhouse. These include variations of the sun's position with time of day and season, the location and orientation of the greenhouse, the slope of the roof, the shape of the greenhouse, and the amount of cloud cover. Because of these variables, the design of a greenhouse is at best a compromise in terms of maximizing the amount of visible radiant energy received per unit of time (radiant flux, commonly expressed as watts).

On a daily basis, the available radiant flux is highest at noon, with the start of the morning and the end of the afternoon being lowest. With seasonal changes, the higher the sun is above the horizon, the greater the radiant flux. The minimum occurs during winter. In addition, the day-length changes are such that the low angle of the winter sun and short daylength combined reduce further the radiant flux. These factors are also functions of latitude; southern latitudes have a greater radiant flux than northern latitudes in the United States during winter.

The radiant flux within the greenhouse is determined by the angle at which the light strikes the glass. The ideal angle of the glass to maximize transmission of light is a function of latitude and season. Since a greenhouse with a variably sloped roof is impractical, and the ideal roof slope is often such as to require an excessively high ridge and large area, the slope is compromised to provide a high (but not maximal) level of transmission, yet still to allow snow to slide off and condensation to run off rather than drip on plants. These roof slopes are usually 32° for greenhouses up to 25 feet wide and 26° for those over 25 feet.

The ideal shape for maximal light transmission as a function of hour of the day and season is the hemispherical dome. This shape is impractical for commercial operations. The quonset structure, seen frequently with houses of pipe half-hoops and polyethelene, is almost as good a shape as the dome. The other shapes seen in greenhouse construction are not quite as efficient in terms of light transmission.

(A)

Fig. 8-8. Two ways of providing humidity in a greenhouse. A. Cool water misting humidifier. Courtesy of Lord & Burnham. B. Steam humidifier. Author photo.

(B)

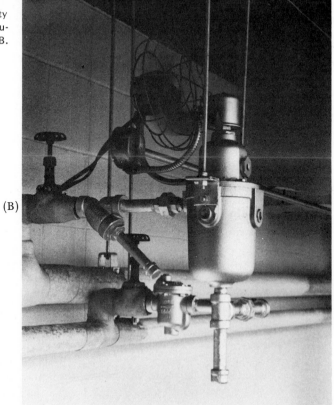

The directional alignment of the long axis of the greenhouse (*orientation*) should be such as to maximize radiant flux during the winter. At latitudes above 40°N, an east-to-west orientation is best (long axis faces south). Below this latitude as the distance to the equator decreases, a north-to-south orientation may be preferred. In terms of structure shape versus orientation, the shape has more potential for maximizing interior light.

During the summer, the radiant flux may be too high for most greenhouse plants. Light reduction can be provided with shading, as discussed previously with ventilation. Roller shades can be automated through a motorized drive and a photoelectric sensor. During the winter, increases in light intensity can be provided with artificial light sources, which can also be automated with photoelectric sensors.

For certain crops, flowering is subject to photoperiodism (see Chapter 5); thus it may become desirable to control the photoperiod. This could involve lengthening the day with artificial light, interruption of the night period with artificial light, or shortening the day with black shade cloth. Certain horticultural crops of high commercial value, such as poinsettias (*Euphorbia pulcherrima*) and chrysanthemums, are routinely brought to flowering for holidays or seasonal display through control of the photoperiod.

Chrysanthemums, a short-day plant, have optimal development of flower buds when the day length is about 12.5 hours. If the natural day is longer, the plants are covered with shade cloth, such as black polyethylene, black polypropylene, or black sateen cloth (64 by 104 mesh or smaller). No more than 2 to 3 footcandles should penetrate the shade cloth. Temperatures must be watched, since heat buildup is possible, especially during the summer. Some newer materials, such as black vinyl or polyester aluminized on one side, are available for dual purposes: shading and heat retention as a thermal blanket during the winter (see later discussion on fuel conservation).

The day can be lengthened artificially to promote flowering of long-day plants or to maintain the vegetative development of short-day plants. Since low light intensities are effective, it is economical. Incandescent or pink fluorescent bulbs, rich in red radiation, are quite effective for this purpose. Alternately, the night period can be interspersed with cyclic light for 20 percent of the time every 30 minutes for the duration of the additional hours of daylight needed. Brief light flashes of 4 seconds per minute are also effective. Since the overall light period is less, the operating cost is less than complete extension of the day length.

GROWING MEDIA

Media suitable for growing plants and techniques for pasteurization were described in Chapter 6. The various soil tests and tissue analyses described in Chapter 7 for outdoor cultivation are also applicable to growing media and greenhouse plants. These media must have their nutrient status checked routinely with soil tests. Tissue analyses are useful to confirm suspected nutrient deficiencies or to determine the levels of nutrients needed by greenhouse crops.

The pH level must be determined and maintained; ground limestone and sulfur are generally used to raise and lower pH in growing media, respectively. For most flower crops in the greenhouse, a pH of 5.8 to 7.0 is acceptable; for vegetables, 5.8 to 7.4; and for acid-loving crops (azalea, hydrangea, gardenia), 5.0 to 5.5.

Soluble-salt levels must be checked routinely, especially because of the small volumes of the growing media and the continuous use of fertilizer. Salts may be seen to accumulate as a white surface crust. Since excessive salinity can cause foliar burn, retarded growth, and even death, it is wise to avoid overfertilization, to monitor soluble-salt levels, and to use leaching if necessary to reduce salt levels.

WATERING

Watering is easily done by hand-held hoses with water breakers to reduce the force of the water. Fogging nozzles can be used for recently planted seeds and seedlings. Watering can also be automated, with spray irrigation, trickle or drip irrigation, ooze irrigation, or capillary mat irrigation (Fig. 8-9). These can be controlled with solenoid valves and time clocks or turned on manually when needed. Drip and capillary mat irrigation are primarily used for potted foliage and flowering plants; and ooze and spray irrigation are used for flower beds in greenhouses. Other uses for these systems were discussed in Chapter 7. Water temperature should be watched, especially in the winter, since plant development may be set back (Chapter 5).

FERTILIZATION

Application of nutrients is mostly in solution form. Numerous commercial forms of soluble fertilizers are available for liquid feeding. These formulations vary extensively in dry weight percentages of N, P_2O_5, and K_2O: 20-20-20, 30-10-10, 25-0-25, 10-6-4, and 9-45-15 are examples. Choices are made on the basis of horticultural crop and developmental stage. Many fertilizers also contain trace elements. The dilution and application schedules vary, but a dilute feeding during every watering or a stronger solution at weekly or biweekly intervals is common.

Soluble fertilizers can also be prepared to any desired nutrient analysis by the grower. Chemicals used include ammonium nitrate or urea (must be biuret free), monoammonium phosphate, potassium nitrate, calcium

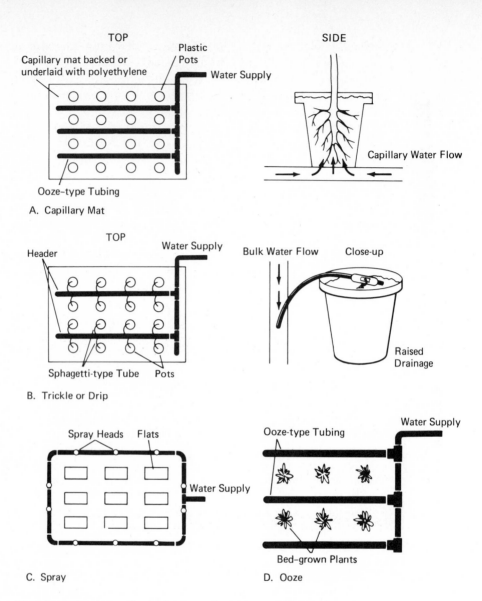

TOP

Capillary mat backed or
underlaid with polyethylene

Plastic
Pots

Water Supply

Ooze–type Tubing

A. Capillary Mat

SIDE

Capillary Water Flow

TOP

Header

Water Supply

Sphagetti–type Tube   Pots

B. Trickle or Drip

Bulk Water Flow   Close–up

Raised
Drainage

Spray Heads   Flats

Water Supply

C. Spray

Ooze–type Tubing

Water Supply

Bed–grown Plants

D. Ooze

Fig. 8-9. Various types of automated watering systems used in greenhouses.

nitrate, ammonium sulfate, and soluble trace elements if desired. Formulations are often richest in nitrogen, which is the most readily exhausted nutrient.

Liquid feeding is done during watering or sometimes through foliar sprays (see Chapter 7). For large-scale fertilization, a liquid fertilizer is usually injected into a watering system by the use of a proportioner (Fig. 8–10).

Controlled-release fertilizers greatly reduce the frequency of application and the risk of injury from excessive applications. (Fig. 8–10). How-

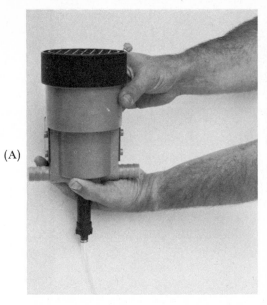

(A)

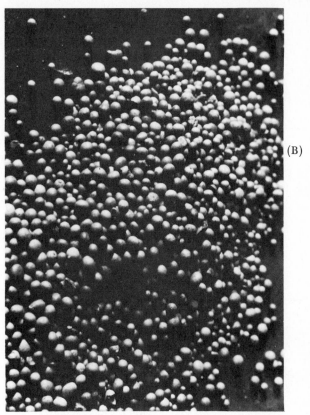

(B)

Fig. 8-10. Labor-saving methods which give controlled release of fertilizer. A. A proportioner which is hooked into an automatic watering system. It gives controlled liquid fertilization at desired dilutions. Courtesy of Profel Machine Tools, Inc., Morton Grove, Illinois. B. Controlled release fertilizer beads which supply fertilizer for 90 days. Author photo.

ever, they are more costly than soluble fertilizers. The slow release of nutrients over a 3- to 9-month period is regulated by either a membranous or resinous coating around granules of soluble fertilizer. Another approach is to use nutrients in chemical forms possessing reduced solubility or organic forms that require microbial breakdown to release nutrients in available form.

CARBON DIOXIDE ENRICHMENT

Another form of fertilization in an indirect sense is the enrichment of the air with carbon dioxide. Carbon dioxide, a substrate for photosynthesis that accounts for increases in dry weight, is normally present in the atmosphere at 0.03 percent or 300 ppm. In an enclosed environment with poor air circulation, such as a greenhouse, levels of carbon dioxide can be locally depleted in the immediate vicinity of the plant. Therefore, a fresh air supply and good circulation are important in maintaining normal levels of carbon dioxide. Increasing the level of carbon dioxide from 1000 up to 2400 ppm can, under some conditions, produce up to a 200 percent increase in photosynthesis. This in turn increases the dry weight production of the plants.

This increase can be realized for some, but not all, plants. Temperatures must be high (85°F or 29.5°C), and light levels must be saturating for photosynthesis. As the carbon dioxide levels increase, so does the level of light needed for saturation of photosynthesis. Plant leaves must not be shaded, so plant spacing must be adequate. The greenhouse must be well sealed.

Carbon dioxide may be derived from Dry Ice or compressed gas cylinders or by burning natural gas, propane, or fuel oil with fresh air. The latter fuels must be pure to prevent the release of toxic products. The gain in dry weight production must be weighed against the costs of the carbon dioxide supply and its control and distribution. The use of fuel systems tends to offset the costs somewhat as the heat may be used to warm the greenhouse. When the weather warms to temperatures where ventilation is needed for temperature control, the use of carbon dioxide becomes unrealistic.

AIR POLLUTION

Pollutants in the air (see Chapter 5) are as serious a concern for those managing a greenhouse as they are for those managing horticultural production outdoors. Crops grown in greenhouses in urban or industrial sites may become affected enough to force relocation of the greenhouse or a change to crops less sensitive to air pollution. Some air pollution can even arise from greenhouse operations. Sulfur dioxide and ethylene can be produced from oil- or coal-fired boilers used to heat the greenhouse. Other pollutants may be produced during careless or excessive use of wood preservatives, herbicides, insecticides (especially fumigants), or chemicals used for pasteurization. Mercury vapors may arise from certain paints, which should be avoided in a greenhouse.

AUTOMATION

All greenhouse functions can be automated. More recent automation, especially as practiced in European commercial greenhouses, can be handled from one preprogrammed console, which is equipped with various sensors to respond to environmental needs. Some advanced automation is now available for home greenhouses. Although not nearly as complex as commercial operations, it is quite helpful. For example, an automatic control system is available to tie in a wet pad evaporative cooler and the heating system. It can maintain a set temperature with no drift or overlap of heating and cooling operations. As labor costs increase, the use of automation will increase in greenhouse operations.

ENERGY CONSERVATION

Greenhouse operators, whether they are hobbyists or commercial types, have had to face an increasingly serious problem: the cost of energy. It appears that the availability and cost of fuel for heating will get worse before they get better, but some steps can be taken to lessen the severity of the problem.

Fuels must be examined closely. First, determine your fuel costs on the basis of cost per 100,000 Btu's. Another important fuel consideration is the heating cost per square foot over a year; that is, divide the total square feet of the greenhouse into the fuel cost per year. These two steps will provide an idea of the heating value per dollar and a cost basis for evaluating changes that reduce energy costs. Economy must be weighed against present and *future* availability. Heating systems that can burn an alternate fuel appear to be a good choice. One point to consider is where horticulture will stand on the priority scale if fuel is rationed. The Federal Energy Regulation Commission listing places greenhouse operations after hospitals, homes, and small businesses but ahead of big industry. Finally, some of the new heating systems are more efficient than older, existing systems. The installation cost versus projected fuel costs might make a new installation feasible.

Insulation offers considerable potential for the reduction of fuel consumption. The main thrusts with most existing greenhouses is an insulation system that forms a dead air space that reduces heat losses. Unfortunately, the production of new glass or fiberglass greenhouses with double glazing that is hermetically sealed is of limited feasibility because of the cost. Some greenhouses are being produced today with double glazing of acrylic plastic, polycarbonate, or fiberglass or with inflatable double polyethylene.

A double layer of polyethylene film (4 to 6 mil) can be put over the exterior of older glass or fiberglass greenhouses in a manner that allows normal operation of vents. It is kept inflated by a blower to produce a double bubble over the structure. Experiments indicate a possible reduction in heat loss of up to 50 percent. Savings are less (20 to 30 percent) when a single layer of polyethylene film is attached by stapling it on the inside of the house onto the wooden glazing bars. If the glazing bars are aluminum or steel, a bubble pack plastic (Fig. 8-11) is available that can be applied to wet glass and held in place by capillary action. Up to 20 percent light reduction is experienced, which might cause problems with crops requiring high light levels. One possible answer is the use of newer Mylar type plastic films that have much better light transmission than polyethylene.

Another form of insulation is the retractable heat sheet or, as it is sometimes called, thermal blanket (Fig. 8-12). These may cut fuel costs from 40 to 60 percent. Heat sheets are an insulating blanket of nonporous polyethylene or synthetic material, with a reflective bottom aluminum foil backing, which can be drawn by hand or by an automated, timed motor across the greenhouse at gutter height. The basic idea is to trap heat under the sheet, and thus to prevent the waste of heat on overhead space that does not contain plants. Care must be taken at temperatures below 0°F (–18°C), since heavy frost accumulations can occur and melt so slowly during the day that available light intensity becomes a problem. This problem is preventable with a tight-fitting heat sheet that prevents moisture leakage from under the sheet. In addition, heat sheets are not

Fig. 8-11. A bubble pack polyethylene insulation which sticks to the inside glass by capillary/adhesive action. Light reduction is 12 percent and heat loss by conduction can be cut 25 to 30 percent. Material shown here is AirCap,® Courtesy of Sealed Air Corporation.

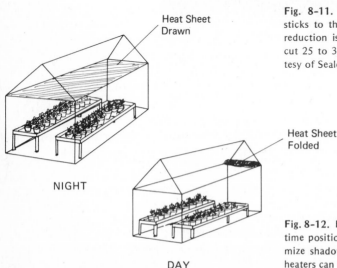

Heat Sheet Drawn

NIGHT

Heat Sheet Folded

DAY

Fig. 8-12. Principle of the retractable heat sheet. Folded daytime position should be at the north end (if possible) to minimize shadow. Support posts, hanging baskets, and overhead heaters can cause installation problems.

Introduction to Applied Horticultural Science / Part II

used during heavy snowfalls, since the melting of snow may be slowed enough such that structural collapse from snow weight becomes possible. When not in use, the heat sheet must be tightly folded to prevent light losses caused by large shadows. Heat-sheet materials and installations are being actively researched now and future improvements should be watched for. At present, they are being increasingly used in some commercial operations.

In older glass houses, injection of a sealant between overlapping glass panes (lapsealing) can give fuel savings of about 25 to 30 percent. Other minor (but effective collectively) insulation improvements include recaulking, north wall insulation with fiberglass aluminum-faced insulation (coldest wall and least light), installation of reflectorized insulation behind fin tubing on perimeter foundation walls, and insulation of heating pipes leading to the greenhouse. Windbreaks can reduce heat losses, which double in exposed areas from winds of zero to 15 miles per hour. During the zeal to conserve energy, it must be remembered not to seal the greenhouse completely airtight since oxygen is needed for people, plants, and open flames in heating and CO_2 is also needed for plants.

Solar energy may hold the answer for the long term. At present a major breakthrough is needed, such as an economical collector. Collectors constructed from plastics are being examined; these may be the answer. The cost of present solar energy systems is too high. Costs need to reach a level where the greenhouse operator could recover his investment in 3 to 5 years in fuel savings. One interesting possibility is a triple-layer polyethylene quonset structure misted between the inner two layers. The sun-heated water is stored in a sand bed that acts as the floor of the structure.

Another very promising approach to fuel conservation is split nighttime greenhouse temperatures. As seedlings, many plants require a uniform high temperature, but as they develop, their requirement becomes thermoperiodic, that is, optimal night temperatures are lower than day temperatures. How long is the optimal temperature required at night? Can the nighttime physiological processes be completed in a few hours at warm night temperatures and followed by colder temperatures without affecting development? The answer to the last question appears to be yes, based on recent research. Development of chrysanthemums, marigolds, petunias, and Easter lilies was normal when the normal optimal night temperature of 15°C (60°F) was maintained only to 11 P.M. and then allowed to drop to 7°C (45°F) until 6 A.M. To bring the soil temperature back quickly to the normal daytime temperatures, the plants were watered at 6 A.M. with warm water 21° to 24°C (70° to 75°F). Clearly, more research will be needed to test other plants, but this area could offer fuel savings to large-scale producers of certain horticultural crops. Questions to be answered also include the percentages of energy saved balanced against the energy needed to warm the water or to recover the daytime temperature from the lowered night temperature during periods of cloudy weather.

Hotbeds

Hotbeds can provide certain greenhouse functions, such as starting plants and rooting cuttings, at a modest cost (Fig. 8–13).

Essentially, hotbeds consist of a wooden, concrete, or cinderblock frame that is 18 inches high in the rear and 12 inches high in the front. The length and width are generally determined by the number of standard window sashes (3 feet by 6 feet) that are used to cover the frame. The hotbed is located in a sheltered, well-drained area, and it should be situated in a sunny, southern exposure. The bed generally consists of 6 inches of sand suitable for propagation.

Heating can be provided quite simply by a thermostatically controlled heating cable buried in the sand. Alternate heating can be provided by steam, hot water, or hot air forced through pipes. At one time, fermenting manure was used to supply heat for hotbeds. The soil temperature (heating cable) is usually maintained at 70° to 80°F (21° to 27°C) if it is to be used for rooting of cuttings. If the hotbed is to be used to start plants for the garden, air temperatures usually range from 55° to 70°F (13° to 21°C), depending on the requirements of the plants. Cooling is provided by manual ventilation or, more recently, by a mechanized window raiser that is operated by a heat-activated expanding chemical. Shading, as with greenhouses, is used to provide some cooling and a reduction in light intensity. Watering is usually done by hand.

Fig. 8–13. Construction details of an electrically heated hotbed. Manure can also be used to supply heat. A 24 inch layer of strawy manure, composed for 7–8 days, and then covered with six inches of soil can maintain a soil temperature 15°F. higher than ambient soil for up to 90 days. This was once the only heat source for hotbeds. U.S. Department of Agriculture photo.

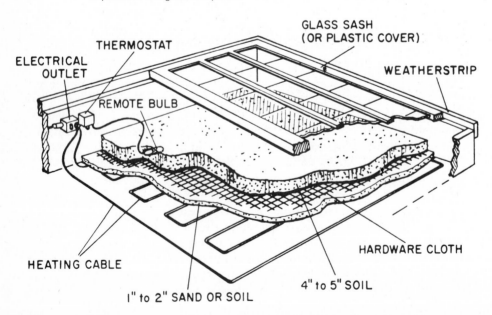

Introduction to Applied Horticultural Science / Part II

(A)

(B)

Fig. 8-14. A. Polyethylene covered coldframe. U.S. Department of Agriculture photo. B. Fiberglass covered coldframe. Courtesy of National Garden Bureau, Inc.

Coldframes

Coldframes are the same as hotbeds, but are not provided with artificial heat; they are warmed by solar energy through the greenhouse effect (Fig. 8-14). Cooling is by manual ventilation, a mechanized window opener, or shading.

Coldframes are used to harden off transplants raised in the greenhouse or hotbed, prior to setting them outdoors. The considerable change in temperature and light intensity in going from the greenhouse to outdoors directly is such that the plant is stressed. Death or at the least delayed development is the result without the intermediate environment provided by the coldframe, a halfway house. Coldframes can also be used to raise transplants of the more cold-tolerant crops in the spring, such as the cole crops or lettuce. They also provide a useful place to summer over house plants or to winter over plants that have borderline winter hardiness. For the latter purpose, the coldframe is covered with leaves or straw. A layer of polyethylene tacked over the sash, as in the greenhouse, will provide additional insulation.

Sun-Heated Pits

Sun-heated pits are enlarged coldframes sunk deeply into the earth to provide the dimensions of the greenhouse at a much reduced cost. They have a sloping glass roof that faces the south. Sunpits are heated by the sun through the greenhouse effect, and the surrounding earth is an effective insulator. Auxillary heaters might be used as an emergency measure on severe winter nights. On very cold nights an additional covering, such as blankets or straw mats, may be needed. Plants that tolerate cold temperatures, that is, night temperatures of $2°$ to $7°C$ ($35°$ to $45°F$), do especially well in sun-heated pits. As energy costs continue to rise, the use of sun-heated pits will most likely increase.

Lathhouses

Lathhouses consist of a simple frame of pipes or wood embedded in concrete (Fig. 8–15). The frame is covered across the top with thin wood strips about 2 inches wide and spaced to cut out from one-third to two-thirds of the natural light. Alternately, the top may be covered with a plastic woven fabric (Saran), which comes in different densities. Aluminum lathhouses are also available. The least expensive lathhouse consists of snow fencing attached to a frame.

Lathhouses provide shade and hence protection for nursery stock in containers. This is especially important in regions subject to high temperatures and light intensities during the summer. Other uses include protection of container plants after transplanting, acclimation of field-grown house plants to reduced light levels prior to selling them, the holding of shade plants, the further hardening of tender plants after the cold frame and prior to field planting, propagation beds, and holding plants for sale. Like the coldframe, the lathhouse may be considered a halfway house.

Light intensities can be controlled by adding or removing wooden laths or by the use of different densities of Saran. Temperatures are generally similar to those found in the shade. Watering frequency is reduced because the shade reduces the water losses incurred by evaporation of soil water or transpiration.

Introduction to Applied Horticultural Science / Part II

(A)

Fig. 8-15. Lathhouses. A. Commercial lathhouse covered with plastic woven fabric. American Society for Horticultural Science photo. B. Simple wooden home lathhouse. Courtesy of National Garden Bureau, Inc. C. Commercial wooden lathhouse. William P. Raffone, Jr. photo.

(B)

(C)

Cloche Structures

The basic principle behind cloches is to provide a greenhouse-like environment at the permanent site of the crops. Originally, cloches consisted of several sheets of glass held together by a wire framework. These units were fitted together end to end to provide a glass tunnel directly over the crop. The practice of using glass cloches has decreased because of material and labor costs. More recently, the use of polyethylene film on wire hoops, which can be applied by machine, has resulted in a revival of cloches. In the home garden, cloches can be as simple as a polyethylene bag over a frame or a gallon glass bottle with the bottom removed. A very economical, simple cloche consists of clear polyethylene film laid over ridge and furrow filled soil.

Cloches are used to produce early crops of vegetables and flowers, because air and soil temperatures within a cloche are higher than in the external environment. Additional protection against frosts and wind is offered. For example, the polyethylene over ridge and furrow has been used with sweet corn on truck farms to produce corn 2 to 3 weeks earlier than the normal market time. Such corn, of course, commands a premium price. The only control possible with cloches is complete removal when the season has progressed to where the internal temperatures are too high, or gradual perforation of the plastic types to allow ventilation.

The Home and Office

The home and office do not appear to be plant growth structures at first glance. However, one realizes after some thought that with today's increasing green consciousness and trends toward interior decorating with plants (Fig. 8–16) that these do qualify as controllable plant environments. The design, installation, and maintenance of ornamental plants for interior decoration in homes and commercial sites can be challenging and rewarding to those in landscape horticulture and floriculture.

Control is limited to a great degree by the dictates of creature comfort. Fortunately, the ideal conditions for health, comfort, and efficiency, that is, day to night temperatures of 68° to 70°F (20° to 21°C) to 60° to 65°F (16° to 18°C) with 50 percent relative humidity, are suitable for a wide range of foliage and flowering plants. Unfortunately, even with these compelling reasons and the additional problem of energy conservation, many homes and commercial buildings are still overheated and under-humidified. These conditions should be corrected prior to plant installations. The drop in night temperature not only conserves fuel, but is beneficial to plants. At night, plants respire and produce energy for growth. Part of this energy is thought to be wasted, and as the respiratory rates increase with increasing temperature, too high night temperatures result in a lower net photosynthesis for the day, and in turn a decrease in rates of plant development.

Fig. 8-16. A lighted foyer garden used for interior decoration. U.S. Department of Agriculture photo.

Temperatures within the home and office can be controlled to a limited extent through the use of microclimates. Window areas often experience a greater drop in night temperature than other parts of the room, since heat losses through the glass and curtain are usually higher than through the insulated wall. This temperature differential is often localized near the window, since air movement and disturbance become minimal in the house or office at night. A maximal-minimal thermometer placed near the window and monitored over a period of time will give an idea of the temperature range. In the day a southern window area will likely be warmer than a northern one, as it receives more sun. Sunporches are also cooler at night than other rooms, as heat loss is especially favored by the large glass area. If one checks with a thermometer, temperature differentials can be found to occur throughout a house or office. These depend on several factors, such as the distance of a room from the heating system, shading from surrounding buildings and trees, wind exposure, and efficiency of insulation.

Air conditioning does not pose a problem, since the temperature is maintained in a range suitable for many plants. As with heating, it is important to avoid a direct draft upon the plant. Drafts cause leaf desiccation and eventual plant deterioration. Problems found to occur in air-conditioned offices can usually be traced to the turning off of the air conditioning on a weekend or holiday, which creates a high temperature range unsuitable for plants.

Natural light in a home or office is often a limiting factor in the choice of plants for an indoor environment. Artificial light can be utilized to broaden the choice of plants; this will be discussed later. Generally, light intensity decreases in order for the following window exposures: south, east, west, and north. This assumes that the exposure is not shaded or obstructed by a large overhang, building, or tree. Light intensity also varies with the season. For example, a plant receiving favorable light in the winter at a southern window could receive too much light in the same window during the summer. A lightweight sheer curtain could be used to decrease the light intensity, or the plants could be located farther back from the window where the light intensity is less. A window or sunroom with a southern exposure is the most versatile, since the range of light intensity from directly at the window to a point farther back is suitable for a number of plants that vary in their light requirements.

Ideally, the relative humidity level should be maintained at 40 to 50 percent. With most indoor environments, this level cannot be maintained constantly without the help of a humidifier. If a humidifier is not available, one can take advantage of humid microclimates. Often the window area over a kitchen sink or in the bathroom has a higher level of humidity. Plants can also be grouped on wetted pebbles. The water level should be below the pot bottom to prevent capillary uptake, possible waterlogging of the soil, and salt accumulation on the surface of the growing media. The evaporation of water from the pebbles and transpiration from the grouped plants tends to create a humid microclimate.

Ventilation is important for plants in the home or office, especially in the winter when air movement is considerably less. The level of carbon dioxide in air is about 300 ppm, which can be less than optimal for plant development when other environmental parameters are optimal. However, in an enclosed environment, levels of carbon dioxide may rise as high as 600 ppm in the winter. These levels are beneficial for many plants if adequate ventilation is present. Otherwise, a localized depletion of carbon dioxide can occur in the immediate vicinity of the leaf. Ventilation is usually adequate in the spring, summer, and fall. To avoid ventilation problems in winter, plants should be located in traveled areas, and fresh air should be added to the environment, but never as a direct draft on the plant. Poor ventilation is often involved in disease problems, too, since localized increases in water vapor may occur in still air around the leaf and provide an ideal environment for plant pathogens.

Introduction to Applied Horticultural Science / Part II

Watering house plants is one of the hardest environmental parameters to control and explain. Any explanation is only a generalization; there is no substitute for cultural knowledge and experience. When watered, potted foliage and flowering plants are watered thoroughly until water flows from the drainage hole. Excess water is discarded from the saucer to prevent waterlogging of the soil. The time to rewater depends upon several parameters: temperature, relative humidity, light levels, rate of transpiration, soil properties, type of pot, and developmental stage of the plant. All these conditions interact differently, some in concert and others in opposition. Soil dryness, however, can be seen, felt, and even monitored with a moisture meter.

Water needs can vary considerably. Succulents, which come from arid regions, need much less water than tropical plants, some of which like a constant even soil moisture, whereas others like to have the top inch or two of soil dry out between waterings. Plants with hairlike, delicate fibrous roots and plants with wiry, thick roots are characteristic of the first and second group of tropical plants, respectively.

Artificial Lighting

Artificial light sources can be used in place of or to supplement natural light when there is insufficient natural light or to prevent the turning of plants toward the natural light in areas between the natural light and principal viewing point. Landscape architects should be consulted for effective placement of plant light sources.

Artificial light sources do not duplicate all the physical parameters of natural light, but they can affect developmental processes by providing the light wavelengths (spectral quality) and intensities most effective for their regulation. The artificial light sources by themselves can maintain photosynthesis and be used to manipulate the photoperiod. In conjunction with natural light, they can be used to increase photosynthesis and to alter the photoperiod. These aspects were covered in Chapter 5.

Artificial lighting for photosynthetic purposes is not used to any great extent on a commercial basis, except for plants used in interior decoration of commercial buildings and for greenhouse crops of high value, because of high electric costs. However, this use enjoys great popularity with amateur horticulturists, especially for potted foliage and flowering plants indoors. Control of photoperiod by artificial light is used extensively on a commercial scale (see lighting under greenhouses in this chapter).

Artificial light can affect photosynthesis by supplying radiation in the red and blue regions of the spectrum and photoperiodic responses by supplying red, far red, and blue. Different lamps vary in their spectral output. For example, incandescent light is richer in the red and far-red region than fluorescent light. Differences in lamps used for horticultural lighting are as follows.

Fluorescent lamps produce spectral emissions derived from fluorescent phosphors and low-pressure mercury vapor. The more common fluorescent light, called cool white, has the following energy output in the spectral regions given in decreasing order: yellow-green, blue, red, and far-red. This lamp is well-suited for human visual purposes and is also useful for photosynthetic purposes because of the blue and red emissions. By changing phosphors the energy output of a particular spectral region can be increased. Fluorescent lamps designed expressly for growing plants have phosphors that tend to increase the red and far-red emission at the expense of the yellow-green region. However, these lamps are not quite as energy efficient as the cool white. Expressed in terms of total energy, they produce about 75 percent as much as the cool white.

Plants can be grown quite satisfactorily under a 1 to 1 combination of ordinary cool white and warm white operated for 16 to 18 hours per day. Plant-growth lamps do offer some advantages, even though they cost more, since plants appear to flower a bit earlier and to become bushier and darker green when compared on an equal total-energy basis. Whether these esthetic improvements are worth the extra cost must be decided by the business or home owner. A setup of a simple fluorescent light arrangement for growing plants is shown in Figure 8–17. Lamps are commonly placed 6 to 9 inches above the plants.

Incandescent lamps are richer in the red and far-red spectral regions than fluorescent lamps. In addition, they are about one-third as energy efficient as fluorescent lamps, their lifetime is about fifteen times less, and their light distribution is not as linear. Their spectral output, however, makes them suitable for photoperiod control, since only low energy output is required. Fluorescent lamps are also available that produce more red than conventional fluorescent lamps. These, of course, are also useful for control of the photoperiod.

As would be expected from their complementary spectral emissions, a combination of fluorescent and incandescent bulbs is good for photosynthesis and control of photoperiod. A general combination is one or two 15-watt incandescent bulbs for every 80 watts of cool-white fluorescent lamps. With any incandescent bulb in the presence of plants, the heat output (infrared radiation) can be a problem in close, confined quarters with inadequate ventilation.

A third type of lamp is used in horticultural practice, the high-intensity discharge lamp (HID lamp). These overcome one disadvantage with fluorescent lamps, which at a distance of a few inches can supply only 8 to 10 percent of the light intensity provided by sunlight on a bright, sunny summer day. Light intensities with HID lamps can be as much as five times higher than with fluorescent lamps. These lamps are more costly than fluorescent or incandescent types. However, the following factors must be weighed. A number of fluorescent lamps must be used to give the same light intensity as one HID lamp; HID lamps have a long life, with some giving up to 24,000 hours when lit for five or more hours per start;

(A)

Fig. 8-17. A. A circular fluorescent lamp is used in this attractive plant growing arrangement. U.S. Department of Agriculture photo. B. A practical fluorescent lamp arrangement utilized for raising seedlings. Courtesy of National Garden Bureau, Inc.

(B)

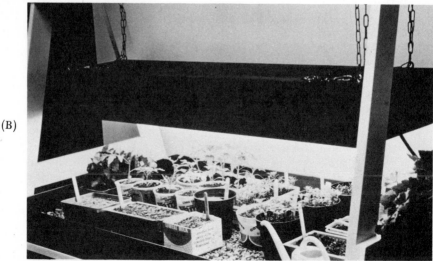

and some HID lamps have energy efficiencies that exceed those of fluorescent lamps. Collectively, these factors can make a HID lamp cost competitive over the long range.

High-intensity discharge lamps are available in three basic types: mercury vapor, metal halide, and sodium vapor lamps. Mercury vapor lamps, the oldest form of HID lamps, are available in clear and phosphor-coated forms (mercury-fluorescent lamp). The former has spectral emissions in the green and blue and the latter has, in addition, red. Mercury-fluorescent lamps have been used extensively in Europe for horticultural lighting.

Fig. 8-18. A self-ballasted mercury vapor phosphor coated lamp which can be used in a standard incandescent socket. Courtesy of Public Service Lamp Corp.

These lamps are generally smaller and have the lowest installation costs, the longest life expectancy, and the best light distribution pattern of all the HID lamps. Mercury-fluorescent lamps are also a good choice for home use (Fig. 8–18); they are used primarily with large trees in tubs as part of a practical decorative accent.

Metal halide lamps emit light in both the blue and red areas of the spectrum. A phosphor-coated form is even more effective for plant growth. Sodium vapor lamps emit most of their light in the red area. A combination of metal halide and sodium vapor lamps is very effective for plants.

Sodium vapor lamps come in low-pressure (LPS) and high-pressure (HPS) forms. The LPS is a *line* source of light (like a fluorescent tube) and is the most efficient of all the light sources described here. The HPS is a *point* source (such as an incandescent light) and is almost as efficient as the LPS; their light intensity does not drop rapidly with age, as do the fluorescent or mercury halide lamps, and their life is over 24,000 hours. Even though sodium vapor lamps cost more to install than fluorescent lamps, their efficiency makes them more economical in the long run. Their effectiveness is also better, so they are increasingly the choice in commercial operations for raising seedlings and supplementary lighting. Better growth is produced with LPS or HPS over cool-white fluorescent, and best growth when supplemented with incandescent light. Suggested installation heights are 5 feet or higher. Since these lamps produce some heat, a fan may be necessary to avoid heat damage to plants.

The best way to determine the number and proper placement of lamps, that is, spacing between lamps and height above plants, is to know the light requirements of the plants (see greenhouse and house plants in Chapter 13) and to measure the light intensity, following adjustments, until you reach the desired level.

Most plant light levels are still expressed in footcandles, so a meter to measure in these terms is suggested. As stated in Chapter 5, the light used by plants is more correctly measured in terms of radiant energy over those wavelengths utilized by plants (400 to 700 nanometers). Meters are available to take these readings, usually in terms of microwatts per square centimeter. Until this becomes more widely accepted, we will continue to see footcandles widely used in horticultural circles.

FURTHER READING

Bickford, E. D., and S. Dunn, *Lighting for Plant Growth*. Kent, Ohio: Kent State University Press, 1972.

Hanan, J. J., W. D. Holley, and K. L. Goldsberry, *Greenhouse Management*. New York: Springer-Verlag, New York, Inc., 1978.

Jensen, M. H., "Energy Alternatives and Conservation for Greenhouses." *HortScience,* 12, no. 1 (1977), 14.

Mastalerz, J. W., *The Greenhouse Environment*. New York: John Wiley & Sons, Inc., 1977.

Poincelot, R. P., *Gardening Indoors with House Plants*. Emmaus, Pa.: Rodale Press, Inc., 1974.

Taylor, K. S., and E. W. Gregg, *Winter Flowers in Greenhouse and Sun-heated Pit*. New York: Charles Scribner's Sons, 1969.

Walls, I. G., *The Complete Book of Greenhouse Gardening*. New York: Quadrangle/The New York Times Book Co., Inc., 1973.

Whitcomb, C. E., "A Self-Contained Solar-heated Greenhouse." *HortScience* 13, no. 1 (1978), 30.

Physical, Biological, and Chemical Control of Plants

The horticulturist is limited in his indirect efforts to control plant development through the manipulation of the plant's outdoor environment. Greater control becomes possible in plant-growth structures, as we have seen in the previous chapter. However, direct control of plant development through physical, biological, and chemical means is both feasible and highly satisfactory. Physical techniques are among the oldest forms of control, but they are still indispensable today. Biological and chemical controls are relatively new and are becoming increasingly sophisticated as our understanding of the science behind plant development grows. The control of plant development is an important tool of the horticulturist, which if properly mastered will pay worthwhile dividends.

The importance of the preceding horticultural practices to the optimization of physical, chemical, and biological control upon plant development cannot be stressed enough. Plants must have good vigor for best results. This depends upon proper identification of all environmental conditions in a site (site analysis). Once these conditions are known, the horticulturist can better select plants suited to that environment, whether it is to be left natural or modified where possible to maximize plant development. After the plantings are established, cultural practices are directed toward maintaining optimal plant development. Under these conditions the physical, biological and chemical practices of controlling plant development can be expected to yield their best results.

Physical Control   Control through physical means is used for a number of reasons. One may alter the size, shape, and direction of growth of plants through physical control for esthetic or practical reasons; this process is termed *training*. Control may also be necessary for plant health, such as the removal of a diseased or dead branch to prevent further plant damage. The quality of fruit, flowers, and vegetative parts and the quantity of fruit and flowers can be improved, such as by a thorough thinning out of unproductive wood to increase flowering and fruiting, the reduction of fruit-bearing branches to produce larger fruit, and the removal of fruit or flowers to increase the size of those left. The balance between growth and flowering can be regulated, such as the removal of older, mature wood to encourage the production of young wood for growth and/or flowering with some woody plants. Finally, it may be desirable to restrict growth, such as with foundation plants, to maintain a proper scale between the house and plantings. The above is accomplished by pruning.

PRUNING

Pruning is essentially the removal of plant parts. It is the primary means of physical control. Lesser techniques of physical control, such as wiring plant parts will be covered later. The pruning of plants is not a haphazard process; pruning must be carefully done based on a thorough knowledge of plant development if the consequences of pruning are to be predicted.

Assuming that the chapters on plant structures and plant development have been mastered, we will proceed to the effects of pruning. First, pruning of the top growth alters the balance between root and shoot, unless compensating root pruning is done. (Root pruning is generally done with container plants and field-grown nursery stock and will be considered later in this chapter.) The root area is untouched and continues to transport water, nutrients, and stored food upward. Since the shoot area is reduced, the upward flow is diverted in part to the buds below the cuts. These buds, which were dormant because of apical dominance, are capable of response now that the stem tips have been removed. This destroys the production of auxin from the meristem, and hence the inhibitory effect of auxin on lateral buds (apical dominance). These buds break dormancy, and the diverted water and nutrients bring about growth and subsequent branching. If, however, the pruning cuts removed lateral branches and not stem tips, branching is eliminated and strong growth of the stem tip is seen.

With pruning it is apparent that some photosynthetic production is lost. The new growth will compensate partially, but only after a lag time until the leaves are formed. This causes a temporary setback, but the remaining photosynthetic area and the stored carbohydrates in the untouched roots and older shoot parts carry the plant. Compensation for the lost part, therefore, is never 100 percent. The result of accumulative pruning is that of a dwarfing process.

Pruning of top growth appears to enhance juvenility in many plants, which results in vegetative growth. This is a consequence of the lower, more juvenile buds being forced into growth after pruning. The reasons for the enhanced degree of juvenility are not clear, but may lie in an alteration of phytohormone production or carbohydrate-to-nitrogen ratios. Root pruning, on the other hand, slows down vegetative growth and enhances flowering in many plants. The reason again is unclear, but may also be tied in with phytohormones and/or carbohydrates.

The response time to pruning will vary, depending on the time of season that the pruning is done. The correct time to prune varies with plant material, but some generalizations are possible. Deciduous trees are usually pruned in late winter and early spring, which makes it easier to see which branches should be removed and gives the new growth time to mature prior to winter. That time of year, when deciduous form is readily apparent, is advantageous both for normal, ongoing maintenance pruning and restoration and rejuvenation pruning of severely neglected deciduous trees and shrubs. Exceptions include *Acer* (maples) and *Betula* (birches), which bleed sap profusely at this time. These are usually not pruned until early summer when the sap runs more slowly. Mature fruit trees are pruned while dormant. Deciduous shrubs that bloom in the spring (*Forsythia, Syringa*) are pruned soon after flowering in order not to disturb flowering the following year. Later-blooming deciduous shrubs, that is, those that bloom on the current year's wood (*Hydrangea*), are pruned in late winter and early spring. Evergreens are pruned in early spring just prior to when growth commences or, if they are grown for flowers, right after flowering. Climbers are treated in the same manner as shrubs; their time of blooming determines when to prune them. These are generalizations, so a good pruning guide should be consulted to determine the correct time for pruning particular plants.

Some basic techniques and tools apply to all forms of pruning (Fig. 9-1). Reference to Figures 9-2 and 9-3 will aid the reader in understanding the following pruning terms. Pruning back growth to a bud is termed *heading back;* complete removal of selected growth back to a lateral, the trunk, or the crown is called *thinning out.* Heading back tends to produce a bushy appearance; thinning out results in a more open look. Of the two, heading back alters plant form more and can even spoil form (if excessive); thinning out tends to maintain and even improve plant form. The need for both pruning techniques varies with species, stage of development, and overall plant form. The slant and distance of a cut from the bud are important; examples of incorrect and correct cuts with pruning shears are shown in Figures 9-4 and 9-5. Cuts on larger branches with lopping shears would be similar. The position of the top bud is important. For example,

Introduction to Applied Horticultural Science / Part II

if the plant has an upright nature, the top bud should be an outside bud such that the branch has room to develop outward and keeps the center open. If you wanted to fill the center, you would use an inside bud. In older trees, top buds may dry out or be crowded out by lower ones, so the choice of top bud or its position may be for nought.

Watersprouts are rapidly growing, vertically growing, unbranched shoots produced from the main stem or head of a tree or shrub. *Suckers* are rapidly growing shoots produced from the plant roots (often incorrectly applied to watersprouts originating from stem tissue near the crown). Both are usually superfluous and are thinned out during maintenance or rejuvenation pruning. They may be selectively retained on occasion to fill out a sparse tree or shrub or to replace the loss of a nearby branch caused by disease or accident.

Fig. 9-1. Pruning tools. Top: pruning saw and leather scabbard. Bottom: left, hedge shears; center, anvil pruning shears; right, lopping shears, U.S. Department of Agriculture photo. Not shown is the extension pruning saw which has a telescoping handle for reaching higher branches.

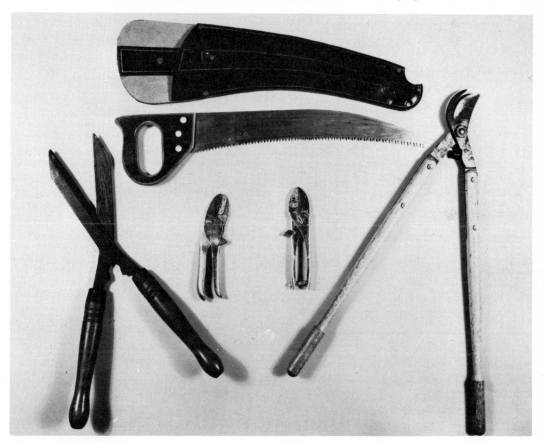

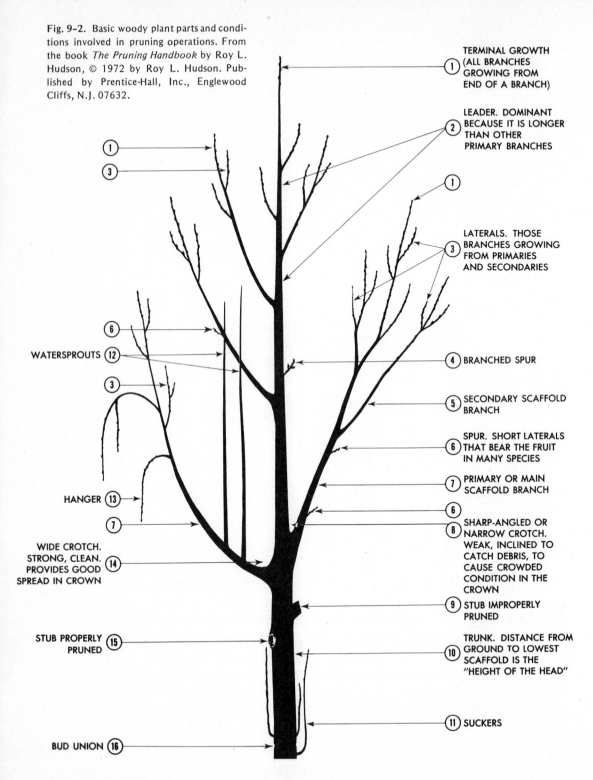

Fig. 9-2. Basic woody plant parts and conditions involved in pruning operations. From the book *The Pruning Handbook* by Roy L. Hudson, © 1972 by Roy L. Hudson. Published by Prentice-Hall, Inc., Englewood Cliffs, N.J. 07632.

(1) TERMINAL GROWTH (ALL BRANCHES GROWING FROM END OF A BRANCH)

(2) LEADER. DOMINANT BECAUSE IT IS LONGER THAN OTHER PRIMARY BRANCHES

(1)

(3) LATERALS. THOSE BRANCHES GROWING FROM PRIMARIES AND SECONDARIES

(4) BRANCHED SPUR

(5) SECONDARY SCAFFOLD BRANCH

(6) SPUR. SHORT LATERALS THAT BEAR THE FRUIT IN MANY SPECIES

(7) PRIMARY OR MAIN SCAFFOLD BRANCH

(6)

(8) SHARP-ANGLED OR NARROW CROTCH. WEAK, INCLINED TO CATCH DEBRIS, TO CAUSE CROWDED CONDITION IN THE CROWN

(9) STUB IMPROPERLY PRUNED

(10) TRUNK. DISTANCE FROM GROUND TO LOWEST SCAFFOLD IS THE "HEIGHT OF THE HEAD"

(11) SUCKERS

WATERSPROUTS (6) (12)

(3)

HANGER (13)

(7)

WIDE CROTCH. STRONG, CLEAN. PROVIDES GOOD SPREAD IN CROWN (14)

STUB PROPERLY PRUNED (15)

BUD UNION (16)

284

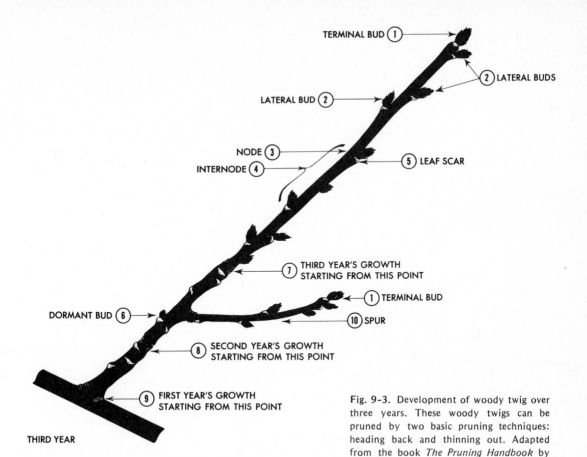

TERMINAL BUD ①

② LATERAL BUDS

LATERAL BUD ②

NODE ③
INTERNODE ④

⑤ LEAF SCAR

⑦ THIRD YEAR'S GROWTH
STARTING FROM THIS POINT

① TERMINAL BUD

DORMANT BUD ⑥

⑩ SPUR

⑧ SECOND YEAR'S GROWTH
STARTING FROM THIS POINT

⑨ FIRST YEAR'S GROWTH
STARTING FROM THIS POINT

THIRD YEAR

Fig. 9–3. Development of woody twig over three years. These woody twigs can be pruned by two basic pruning techniques: heading back and thinning out. Adapted from the book *The Pruning Handbook* by Roy L. Hudson, © 1972 by Roy L. Hudson. Published by Prentice-Hall, Inc., Englewood Cliffs, N.J. 07632.

SECOND YEAR

FIRST YEAR

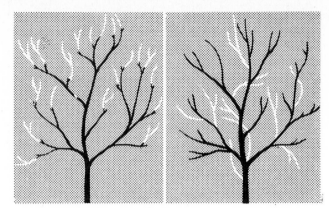

Fig. 9-3. Continued. Left: heading back, Right: thinning out.

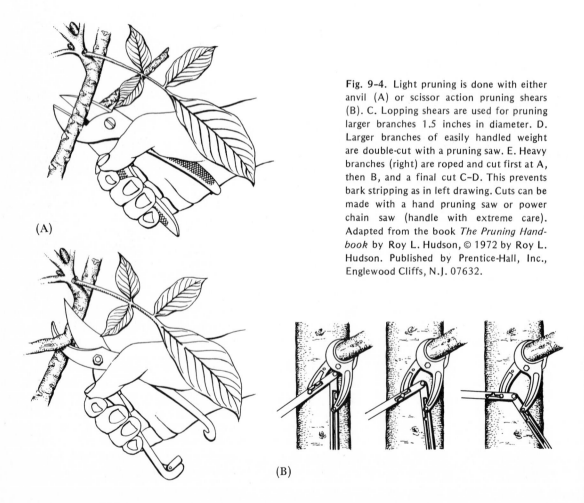

(A)

Fig. 9-4. Light pruning is done with either anvil (A) or scissor action pruning shears (B). C. Lopping shears are used for pruning larger branches 1.5 inches in diameter. D. Larger branches of easily handled weight are double-cut with a pruning saw. E. Heavy branches (right) are roped and cut first at A, then B, and a final cut C–D. This prevents bark stripping as in left drawing. Cuts can be made with a hand pruning saw or power chain saw (handle with extreme care). Adapted from the book *The Pruning Handbook* by Roy L. Hudson, © 1972 by Roy L. Hudson. Published by Prentice-Hall, Inc., Englewood Cliffs, N.J. 07632.

(B)

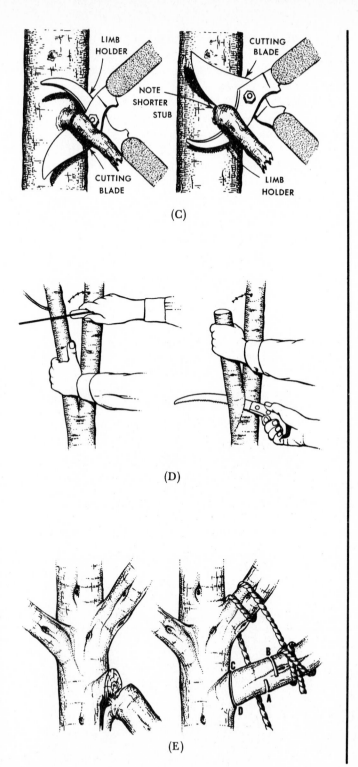

(C)

(D)

(E)

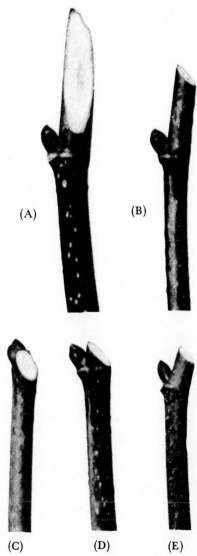

(A)　　　　(B)

(C)　　　(D)　　　(E)

Fig. 9-5. Correct and incorrect cuts in relation to buds on woody twigs. The correct cut is shown in D. The cut in E is also correctly used in Northern areas of extreme cold since it prevents bud desiccation. The cut shown in A results in excessive wounded area, twig dieback will result in B, and bud damage in C. From the book *The Pruning Handbook* by Roy L. Hudson, © 1972 by Roy L. Hudson. Published by Prentice-Hall, Inc., Englewood Cliffs, N.J. 07632.

287

Larger branches that can be handled on a weight basis, but are not easily supported with a free hand, are treated with a pruning saw as shown in Figure 9-4. Branches that are too large to be handled safely are thinned out as demonstrated in Figure 9-4. These double cuts are necessary to prevent splitting and bark stripping, which could lead to serious tree damage.

The final cut on the tree trunk after the removal of the branch is important in order to have correct healing and minimal chance of disease. A correct cut leaves no stub, but it is not absolutely flush with the trunk. This exposes more heartwood than is necessary, which leads to longer healing time. Instead a slight shoulder is left and cut on a slight angle to slant away from the trunk. This leaves a round to oval area of heartwood. The results of proper and improper cuts are seen in Figure 9-6. Cuts larger than 2 inches in diameter are painted with a wound compound. However, recent studies by the United States Forest Service indicate that the use of tree wound paint is more cosmetic than disease preventative. Healing of the wound and disease problems appeared to be statistically similar for

Fig. 9-6. A. The pruning cut was correctly done (round to oval) and callus tissue (light gray) is healing the wound. B. This pruning cut was incorrectly done (branch not undercut), leading to excessive wounding. This increases the chances for disease and lengthens the healing period. William P. Raffone, Jr. photos.

(A)

(B)

Introduction to Applied Horticultural Science | Part II

tree wounds that were painted and unpainted. The practice still persists, even though it may only be psychologically helpful. It may be useful initially to waterproof the wood, thus minimizing environmental suitability for pathogens, but yearly (impractical) applications would be needed to maintain waterproofing.

Caution should be exercised in pruning work, as well as with any horticultural activity where hand or power tools are utilized. When using an extension pruner for high branches, beware of power lines and falling branches. Unless roped properly, large branches can injure people or nearby structures. A large branch may miss you, but it can spring back upon impact and knock the ladder out from under you. Removal of large branches involving climbing, and perhaps most pruning of large trees, is best left to the professional arborist.

PRUNING TO TRAIN

Trees and shrubs will flower and fruit without pruning, but pruning allows the horticulturist to produce a well-balanced plant with a practical size and shape. It also produces a plant with strong, well-placed branches and a plant that is easier to maintain. This is particularly important to the commercial horticulturist, where fruit trees must be properly trained to allow mechanical harvesting.

Pruning to train a tree or shrub is started by the nurseryman and continued by the purchaser. The nurseryman generally sells young deciduous trees in the form of a single straight stem (*whip*) or a whip with a few short side branches a few inches in length (*feathers*). If the tree is grafted, it may be called a *maiden,* instead. These are sold bare rooted, are relatively inexpensive, and have maximal potential for training by the homeowner. Some will be found in containers or balled and burlapped. Whips and maidens can be trained further either at the nursery or by the homeowner into *standards* or *central leaders*. The former, used with medium-sized trees, has a clear stem for 6 feet, above which the framework is allowed to develop. When the tree reaches about 7½ feet, the tip is removed to promote branching. Smaller-growing trees are usually sold as half-standards (branching starts 3.5 to 4 feet above crown). The central-leader system is usually used with the larger trees. These have a clear stem for 6 to 8 feet, which allows for mowing underneath, and the leader is not pruned. Heading back is done on the other branches to produce a balanced head. The technique involved in the central-leader system is shown in Figure 9–7. Pruning may or may not be done to keep the center of the tree open. The above systems with medium to large deciduous trees allow for foot and vehicular traffic (street trees) or mechanized operations such as mowers (lawn trees) underneath. Other trees may be trained by pruning to maximize symmetry and ornamentation instead. These are usually destined to be specimen trees.

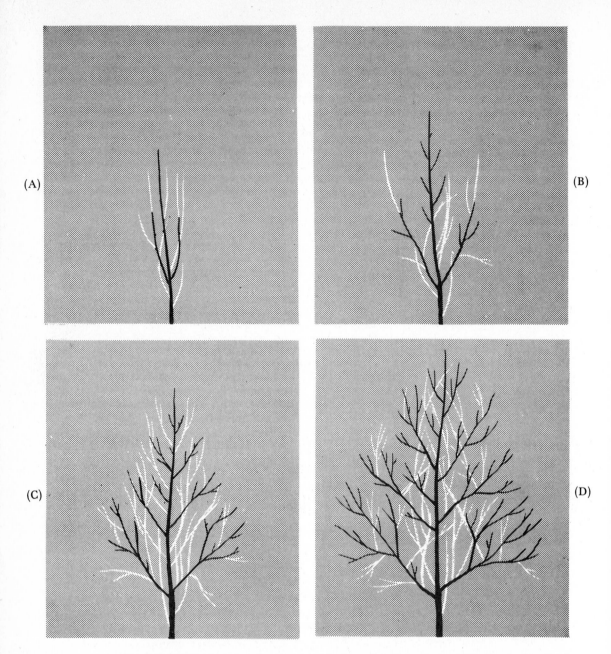

(A)

(B)

(C)

(D)

Fig. 9–7. Training technique for the central leader system. From the book *The Pruning Handbook* by Roy L. Hudson, © 1972, by Roy L. Hudson. Published by Prentice-Hall, Inc., Englewood Cliffs, N.J. 07632.

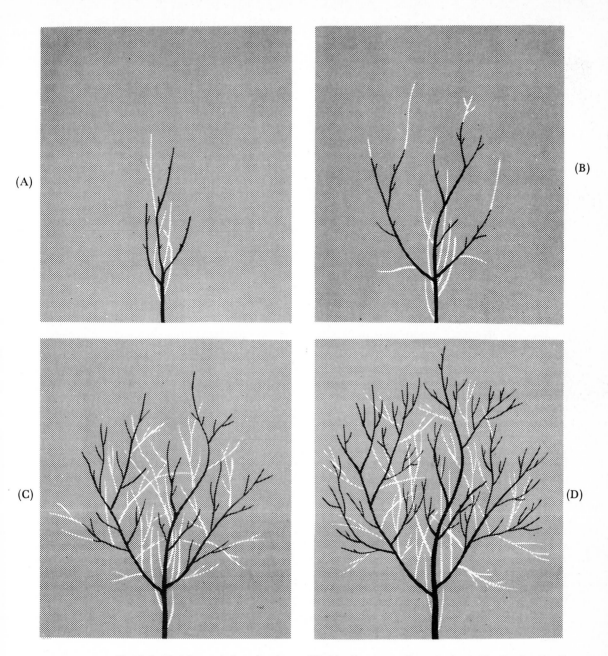

(A)

(B)

(C)

(D)

Fig. 9–8. Training technique for the modified leader system. From the book *The Pruning Handbook* by Roy L. Hudson, © 1972 by Roy L. Hudson. Published by Prentice-Hall, Inc., Englewood Cliffs, N.J. 07632.

Evergreen trees, conifers in particular, do not respond favorably to heavy pruning. Therefore, they are trained differently than deciduous trees. Training is minimal in the nursery, where the main thrust is to allow development of the natural form. Exceptions include evergreens that will be utilized as hedges (*Taxus, Tsuga*). Broad-leaved evergreen trees, such as *Arbutus,* are allowed to develop with a central leader, and side shoots are not removed. Surplus leaders are removed. Side branches are thinned out where crowded or crossed. Main branches are kept at 6 to 8 feet or higher. Long-needled conifers are usually only given light heading back of their terminal growth. If heading back is done beyond the new growth, such as candle growth on *Pinus* or *Picea,* twig dieback can occur. If dieback occurs on a branch, it is thinned out, not headed back as with deciduous trees. Unfortunately, such removal may spoil the symmetry. Surplus leaders are removed on conifers, except for certain shrublike ones (*Cephalotaxus*), especially if they grow away strongly. Conifers with fine or scalelike foliage, such as *Chamaecyparis, Thuja, Juniperus,* are sheared or tip-pinched partially on their new growth in order to train them into a desirable shape.

Young fruit trees are trained by three different systems: the central-leader, modified-leader, and open-center systems. The central-leader system, is the same as that shown in Figure 9–7 for ornamental trees. Once popular in orchards, it has fallen into disfavor because the tall, pyramidal shape makes spraying and picking fruit somewhat difficult. It still has an appeal to the homeowner who wants a fruit tree that is both ornamental in appearance and fruit bearing. The modified-leader system is the most popular at present (Fig. 9–8), since it produces a well-shaped tree of good structural strength suited for orchard conditions, that is, minimal shading of underlimbs and maximal fruit production. With open-center training, the branches all arise from one small area on a shorter, main trunk, and the center is primarily free of branches (Fig. 9–9). This type of form has a disadvantage in that the open center may be ideal for a water pocket and subsequent rot. The open-center form is also structurally weaker, with all the branches arising from a small area with narrow angles that withstand stress less than wide ones.

Although not hard and fast categories, the central-leader system is used for many nut trees and some tropical fruits. The modified-leader system is used with apples, apricot, cherry, olive, pear, persimmon, and plum. The open-center technique is used for almond, fig, nectarine, peach, Japanese plum, and quince.

Shrubs as a rule need less training than ornamental and fruit trees. Training produces a shrub with a well-balanced, evenly spaced framework. This is usually done after purchase from the nursery. Deciduous shrubs are initially pruned to three to five strong upright shoots, and side shoots are halved and left with two to three buds. The following year the stems are headed back about one third and crossed or crowded branches are thinned out. Less pruning is done with evergreen shrubs. Three strong shoots are selected and others are only lightly tipped. The following year crowded branches are thinned out.

Berries, such as blackberries and raspberries, are trained according to the growth patterns shown in the area that they are grown in. Where growth is limited, such as in the north and northeast parts of the United States, they are trained as unsupported bushes. On the Pacific Coast, growth of berries is so rampant that they are trained on trellises or stakes. Grapes are generally trained on wire and wood supports, trellises, and arbors. The latter are used mostly by the homeowner.

There are also some very specialized forms of training, which are rather labor intensive. They are utilized primarily for esthetic purposes and only to a limited degree on a commercial basis. These include topiary, espalier, cordon, and bonsai. *Topiary* is the training of a plant to grow as a certain form, such as an animal or geometric pattern. These were once the high point of formal gardens, but are seldom seen today. They do have some limited appeal on the house-plant scale, where one may encounter plants trained on shaped-wire pieces stuffed with sphagnum moss.

Fig. 9-9. Tree form produced by the open-center training system. Tree shown is peach. U.S. Department of Agriculture photo.

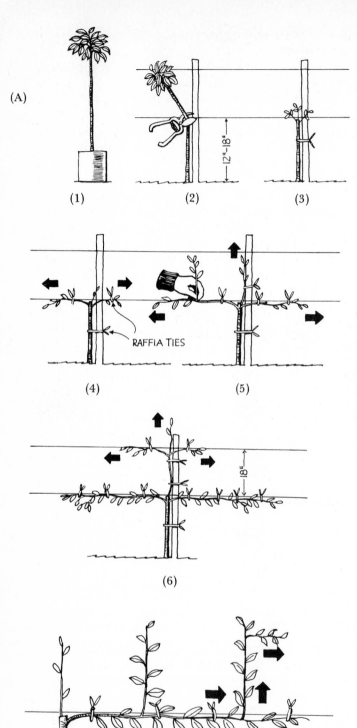

(A)

(1)   (2)   (3)

12"–18"

RAFFIA TIES

(4)   (5)

18"

(6)

(7)

Fig. 9–10. A. Steps in the espaliering technique. 1. Start with young tree. 2. Cut to height where first shoots wanted. 3. New shoots appear. 4. Select best and train along wires. 5. Leave center shoot and pinch off lateral shoots emerging on trained horizontal ones. 6. Select and train best lateral shoots on central shoot at second desired height. 7. Depending on type of espalier, various shoots on main trained branches can be retained and trained in various directions. Adapted from the book *The Pruning Handbook* by Roy L. Hudson, © 1972 by Roy L. Hudson. Published by Prentice-Hall, Inc., Englewood Cliffs, N.J. 07632.

(B)

Fig. 9–10. B. Fruit trees after completion of espalier training method. Courtesy of Stark Brothers Nurseries and Orchards Co.

Espaliers are fruit trees, shrubs, and vines that have been trained flat against a wall, trellis, or post and wire. If the espalier is restricted to one or two shoots trained in parallel or opposite directions, it is termed a *cordon.* An example of the espaliering technique is found in Figure 9–10. This technique is used commercially for fruit trees in Europe, to a limited extent in the United States, and worldwide for grapes. For grapes a specialized form of espalier based on the Kniffin system is used. Bonsai is the art of artificially dwarfing trees into miniatures through top pruning, root pruning, and wiring over a number of years.

Root pruning is an essential part of training practiced by nursery people. Without it the digging of field-grown nursery stock for transplanting and sale could not be done so successfully. Root pruning is a routine, annual cultural activity carried out by hand and also machines designed for that purpose. The techniques vary, but the simplest example is the forcing of a sharp spade into the ground in a circle around a woody plant about one year prior to moving the plant. Most nurseries use mechanized equipment to root prune alternate sides of a row in alternate years.

Fig. 9-11. Root ball of plant left unpruned and held too long in present size container. Some roots are dead as they do not have white coloration of healthy roots. Such a plant usually does poorly when transplated to its final outdoor site. Author photo.

This forces the development of fibrous roots near the base and facilitates the transplanting operation. Recovery is faster and losses minimal.

Root pruning should also be practiced on container nursery stock as follows. Container stock should be moved to bigger containers on time to avoid producing a constricted root system (Fig. 9-11). If constricted roots develop by oversight and remain uncorrected, the plant quality will be inferior. The outer, constricted, circular-growing roots must be pruned to eliminate this condition before the plant is moved on to another container. When the homeowner removes a container plant, he too must root prune if the plant is to be properly established in its new permanent environment. A sharp knife, hatchet, or spade is drawn through the outermost root area, such that it is partially quartered. This will produce outward root growth in the soil, and hence better establishment. If the root ball is not broken, root growth will usually remain constricted, and the plant will have a poor appearance and often will eventually die. Death occurs because the encircling roots girdle each other and the crown, resulting in disruption of food and water movement in the roots.

MAINTENANCE PRUNING FOR ESTABLISHED PLANTS

The preceding section has been concerned with the use of pruning for training of plant material. Much of the training occurs in the nursery. Once plants are trained and established, maintenance pruning, usually by the homeowner, becomes necessary. Additional pruning will be needed to restrict growth because of space limitations or deviations from form, to obtain a balance between vegetation and flowers and fruits, to maintain health, and to produce quality improvements of fruits and flowers. Some of these areas apply equally to the various plants we prune; others are handled somewhat differently depending upon the plant.

Pruning is but one aspect of the maintenance of plant health. Others will be covered in the following chapter. Weak branches, pests, and disease interfere with training and can hasten the death of a woody plant. In addition, the falling of large weakened branches can cause human injury and property damage. Diseased or dead wood should always be removed promptly back to healthy wood. Other branches that are potential sources of future health problems and/or detract from plant form should be removed before they become problems. These include watersprouts, suckers, crossed branches, horizontal forks of equal thickness, narrow-angled scaffold branches, and underbranches of a main branch. Suckers and watersprouts (Fig. 9-12) tend to grow rapidly and obliterate desirable plant form, are often soft growth, and may rub other branches if left too long. Suckers arising below a graft union must be removed, or the scion will be overtaken by the sucker growth. Under some conditions a watersprout arising above the graft union or on an ungrafted plant can be headed back and left to fill in an opening or to replace diseased material about to be or previously removed.

Fig. 9-12. The two straight upright slender branches in the center of this shrub are watersprouts. These should have been pruned out earlier before they detracted from the shrub form. They can still be removed. Author photo.

297

(A)

Fig. 9-13. Correct (A) and incorrect development (B) of scaffold branches. The tree in B is more prone to storm damage, more susceptible to rot in the crotch area, and has a less attractive form. Author photos.

(B)

298

If branches cross, one should be removed to prevent rubbing and bark damage. Horizontal forks tend to split under wind stress, especially if loaded with fruit, or if one branch is heavier than the other. One branch of the fork should be removed. Underbranches, as do suckers, arise from adventitious buds. They grow downward and then curve upward or straighten out horizontally. These too should be removed because of their structural weakness.

Narrow-angled scaffold branches are weak because of squeezed bark in the crotch and a lack of continuous cambium. They are also inclined to catch debris and to crowd the crown. Branches of this type should be thinned out at an early stage, before major tree surgery is required after a severe storm. Figure 9–13 shows incorrect and correct development of scaffold branches.

Under some conditions a tree may have progressed too far before corrective pruning was decided upon. This occurs if the tree is mature and the removal of narrow-angled scaffold branches requires major surgery or the removal of a weakened branch would leave a gaping hole. Under these conditions the scaffolds or weakened branches may be strengthened with bracing cables attached to bolts (Fig. 9–14). Stresses must be considered carefully if these braces are to work. This treatment may also be used to save a storm-damaged, irreplaceable tree of value.

Fig. 9-14. A tree with a weak crotch due to poor scaffold branch development. It has been strengthened with two brace bolts. U.S. Department of Agriculture photo.

299

After initial training some woody plants will need little or no pruning to maintain their form. Others will need a lot. Pruning will also be required to restrict plants whose mature size is larger than the desired size. This will vary from clipping the lawn to the pruning of foundation plants, ornamental trees, fruit trees, vines, and even house or greenhouse plants. Heading back and thinning out are the mainstays of size restriction and form retention. These techniques will also be used to maintain a balance between vegetative growth and flowering or fruiting. These maintenance pruning approaches will now be covered in a generalized way according to various plant categories.

On ornamental trees one should remove superfluous branches that develop down low or up near the tip. Branches that spoil the form are thinned out. Branches that must be removed because of size restrictions should be headed back to a vigorous side branch. Oldest branches should be cut out and younger wood trained into its place.

Pruning of established shrubs differs widely but is based upon the time of flower-bud development as related to flowering time and the age of the wood upon which the flower buds are developed. Spring-flowering deciduous shrubs bloom on last season's wood; that is, flower buds formed prior to dormancy. If left unpruned, these tend to flower less and less. Pruning is done shortly after flowering and consists of thinning out superfluous growth at the crown, such as old wood and young suckers plus watersprouts, and minimal heading back of growth that is too excessive in length or detracts from plant form. Other shoots are thinned out to open the center of the shrub. This tends to keep the shrub in good form and at a desirable size and promotes flowering.

Deciduous shrubs that bloom in the summer and autumn bloom on the current season's growth. These are dormant pruned. Pruning is the same as for the spring-flowering shrubs just covered, except for the time.

The removal of spent flowers (*dead heading*) is a time-consuming pruning practice. Where used, it is primarily with plants that have spent flower clusters that detract from plant esthetics or slow desired plant development because of seed and fruit formation. Shrubs of this type include *Syringa* and *Rhododendron*.

In mild climates, deciduous shrubs will often be subjected to insufficient chilling; therefore, buds will not start growth normally in the spring. Irregular leafing and flowering will result. Tip pinching of the first 2 or 3 inches during the summer will often improve leafing and blossoming the following year.

Broad-leaved evergreen shrubs generally require less pruning than deciduous materials. This is especially true of shrubs having a slow, even, compact growth, most of which originates from the terminal buds. This produces a domelike appearance with little interior growth. An example of this type is the *Rhododendron*. Branches that break from form are lightly headed back. Weak, diseased, dead, and crossed wood is thinned out. This light pruning can be done in early spring prior to growth; heavier

Introduction to Applied Horticultural Science / Part II

pruning should be delayed until after flowering. Winter-killed branches are headed back to healthy wood in the late spring after it is completely evident that they are dead. If desired, spent flowers are also removed then.

Fast-growing broad-leaved evergreens, such as *Berberis,* that follow the growth habits of the deciduous shrubs are pruned more heavily in the Pacific coastline and Deep South, where they are headed back and thinned out strongly. In other areas, pruning is less severe and not necessarily done on an annual basis. Pruning of those shrubs desired for flowers or berries is delayed until after their appearance; otherwise, pruning is done in early spring prior to active growth.

Coniferous evergreens usually reach the homeowner after minimal shaping and training at the nursery. Fine-foliaged conifers (*Chamaecyparis, Thuja, Juniperus*) are lightly sheared or tip-pinched to maintain shape, usually in the late spring or early summer when growth is completed. Evergreens used as hedges are more heavily sheared (*Taxus, Tsuga*). Pruning at this time will give maximal size restriction and minimal upkeep for the remainder of the season. Long-needled conifers can also be done then. They can be lightly headed back on their terminal growth. Older, dying branches in the lower portions should be thinned out (see training section on evergreens). Pruning in general should always maintain the inherent natural shape of the evergreen and be kept to the minimum.

Once trained, fruit trees are pruned to maintain maximal fruit yield. With all fruit trees, watersprouts, suckers, dead wood, and weak branches are removed. The trained form is also preserved. For example, with the modified-leader form the center is kept open and the leader is kept dominant. Strong young wood is selected through pruning to eventually replace the fruit-bearing wood, and spent fruit-bearing wood is removed. With many fruit trees the fruit is produced on spurs, which are kept productive through pruning to encourage a constant proportion of fruit-bearing spurs. Procedures vary according to the fruit. General pruning information for fruit trees is given in Part Three.

Grapes and berries are pruned to maintain the trained form and to encourage maximal fruiting wood. This varies from the complete removal of all canes that bore fruit right after harvest to only partial removal. Complete removal is practiced with brambles (raspberry, blackberry), which, though perennials, have canes with a biennial fruiting habit. Partial removal is used with grapes, where the fruit is produced on the current growth from buds set the previous season.

Flowering climbers and vines are generally pruned at a time determined by the age of the wood on which flowers are produced, as for flowering shrubs covered previously. In general, those that bloom on the current season's wood can be pruned during the winter. Those blossoming on the past season's wood are pruned after flowering is completed.

Pruning is also used to improve the quality of flowers and fruit. This involves the removal of some buds, flowers, or fruits to spread the development over those remaining. This increases the final size over that which would have developed if all were left. When used to increase the size of fruits, it is termed *thinning,* and *disbudding* when flower size increase is desired.

Biological Techniques

Grafting, the techniques of which have been covered in Chapter 6, is used for propagation, but it is also used for the biological control of plant growth. For example, modification of growth habits can be achieved through grafting. The choice of a rootstock can be used to dwarf the scion cultivar, such as with fruit trees. Others can be utilized to give the grafted plant exceptional vigor. Grafting is frequently done with certain highly pigmented forms of cacti, whose vigor is greatly enhanced by the choice of a proper rootstock. Other alterations of growth habits through grafting include the formation of tree roses and weeping birches or cherries. Rootstocks may also be chosen to allow the scion cultivar to tolerate unfavorable soil conditions, to resist diseases, to improve the size and quality of fruit, to encourage early bearing of fruit, and to replace rootstocks that are not cold hardy in a particular climatic zone.

Ringing and scoring are used to produce dwarfing and early fruiting or flowering. These treatments give only temporary effects and are not used to any great extent. Grafting to achieve the same results on a permanent basis is obviously a better choice. Ringing treatments remove the bark in a ¼-inch strip completely around the trunk. If it is done early in the season with grapes, current fruit production is heavier and the sugar content is higher. With fruit trees the setting of fruit buds is increased the following year. Scoring denotes the cutting of the bark around the trunk, but it is not removed. The effects result from phloem disruption, but are temporary since the phloem regenerates.

Shaking is another form of biological control, but a natural one that requires minimal output. The stresses of wind movement appear to strengthen plant stems and in some cases cause dwarfing. Unstaked and widely spaced container nursery stock requires no staking and less pruning when transplanted into the landscape than crowded, staked stock with minimal movement. The need to stake new trees, unless in a site exposed to excessive winds and hence shifting of the root ball, would appear to be minimal. Herbaceous material with easily broken stems and of tall height should still be staked. The area of natural stresses and their effect upon plant development still needs much research and should produce results of interest in the future.

Chemical Techniques

The growth and differentiation of plants is controlled extensively by endogenous chemical messengers, phytohormones, as discussed in Chapter 4. It would not be unreasonable to expect isolated phytohormones, synthesized phytohormones, and analogs, which are collectively termed plant-growth regulators, to alter plant development upon being applied to the plant. This has not proved to be as easy to accomplish as it appears. Advances have been made, and this area holds future promise to horticulturists who wish to modify plant development for the purposes of increasing products and reducing labor.

The major problem with achieving the desired results with plant-growth regulators undoubtedly lies with the complex interactions that take place between the various plant hormone systems. For example, the application of a plant-growth regulator may cause changes in the levels of an endogenous phytohormone. In turn, this might have no effect on producing the desired result, it might produce a result other than the one desired, or it might produce the desired result in conjunction with a good or bad side effect. Nevertheless, some results are possible, and they will be discussed later in this section.

For the plant-growth regulator to be effective, it must enter the plant. A number of possibilities exist. Uptake is possible through the roots, leaves, and stems. The degree of entry varies according to many things, such as the plant species, the part of the plant, the chemical nature of the plant-growth regulator, the temperature, the solubility of the plant-growth regulator, the relative humidity, and the nature of the carrier (solubilizer). In general, entry through the roots is easier than through the foliage.

After entry the fate of the plant-growth regulator is even more complex. Chemicals may be translocated, they may be nontranslocatable, they may be selectively translocated only by certain portions of the translocation system, they may be moved out of the plant, and their chemical structure may be unaltered or modified. Any environmental factors that affect translocation will also affect the movement of the plant-growth regulators.

In view of the complexities involved, it is sufficient for our purposes to say that entry of plant-growth regulators is possible. To obtain a chemical formulation that will enter the plant and produce a beneficial result is often a long, arduous task based on existing experimentation and much trial, error, and screening of formulations on large numbers of plants.

Application of a plant-growth regulator can bring about regulation by three means. The first involves the use of chemicals that alter endogenous phytohormone systems, the second uses chemicals that function as disruptive agents, and the third uses chemicals that bring about regulation through interactions between phytohormone systems.

Alteration of phytohormone systems can be achieved by applying an isolated phytohormone or an analog of it, which brings about regulation directly, or a chemical may be used to stimulate or inhibit the synthesis of an endogenous phytohormone. Disruptive agents disrupt the plant tissue and induce an altered physiological state, which includes changes in the phytohormone systems. Interaction reactions may be induced by applying two combined regulators to produce an enhanced effect, or by

GROWTH RETARDANTS

① ② ③ ④

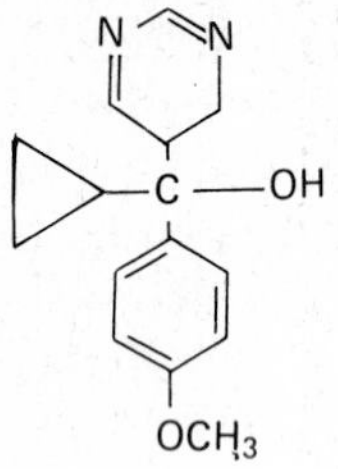

$[Cl - CH_2 - N^+ (CH_3)_3] \, Cl^-$

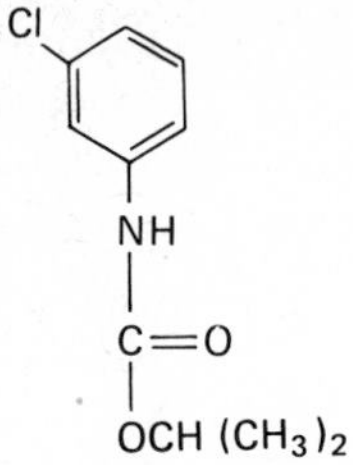

⑤

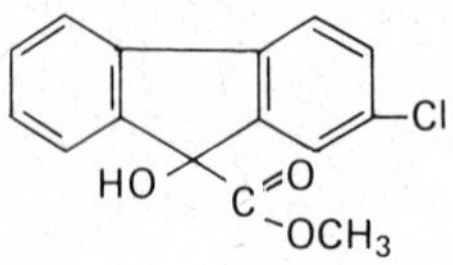

OCH_3 (compound 1)

HO, $C{\overset{O}{=}}$, OCH_3, Cl (compound 5)

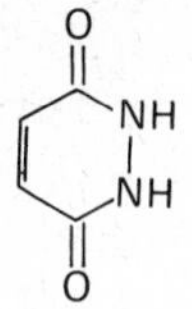

OH, NH, $C=O$, $OCH(CH_3)_2$, Cl (compound 4)

⑥

$HO-\overset{O}{C}-CH_2-CH_2-\overset{O}{C}-NH-N(CH_3)_2$

⑦

$\left[\; Cl \; CH_2-P{\underset{(CH_2)_3 CH_3}{\overset{(CH_2)_3 CH_3}{-(CH_2)_3 CH_3}}} \; \right] Cl$

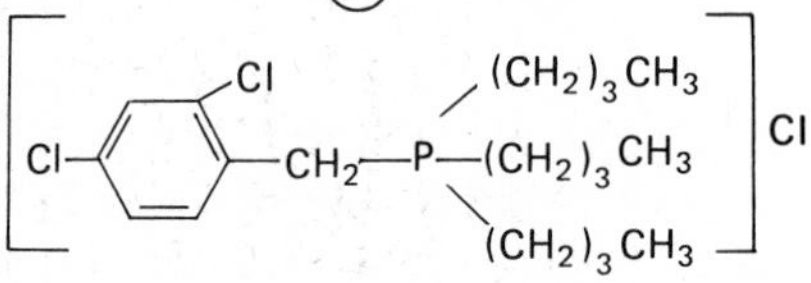

FRUIT THINNING AGENTS

① ② 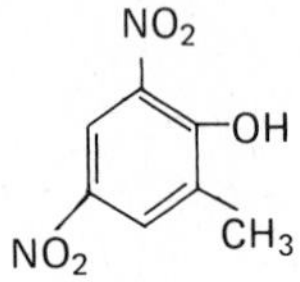 ③ 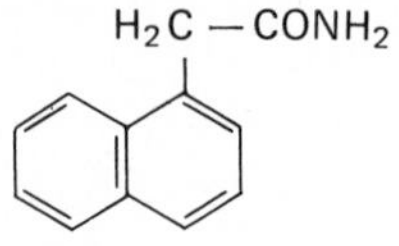 ④

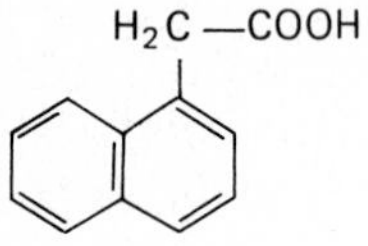

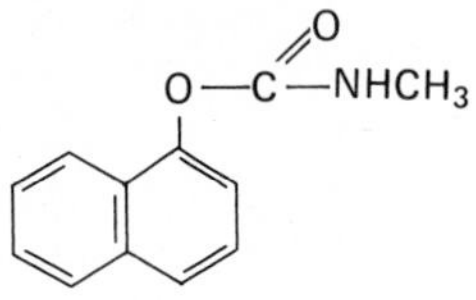

$H_2C-COOH$

NO_2, OH, CH_3, NO_2

$H_2C-CONH_2$

$O-\overset{O}{C}-NHCH_3$

ETHYLENE SOURCE

$Cl-(CH_2)_2-\overset{O}{P}-(OH)_2$

SCALD PREVENTATIVES

①

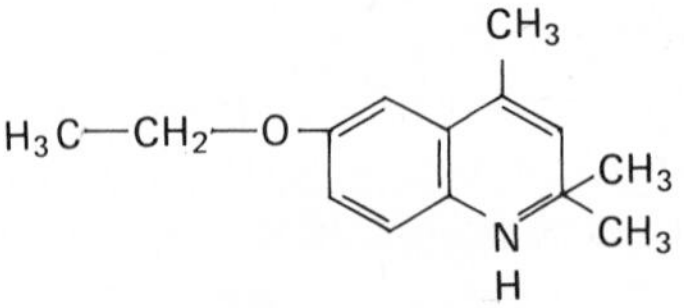

H_3C-CH_2-O, CH_3, CH_3, CH_3, N, H

②

$-NH-$

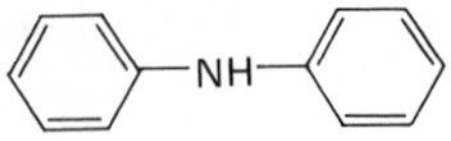

ABSCISSION AGENT

CH_3, O, OH, $CH-CH_2$, O, NH, O, H_3C

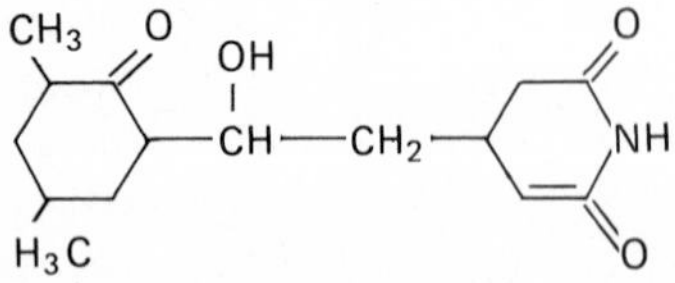

CHEMICAL PRUNER

$H_3C-(CH_2)_n-\overset{O}{C}-OCH_3$

$n = 7-11$

MISCELLANEOUS

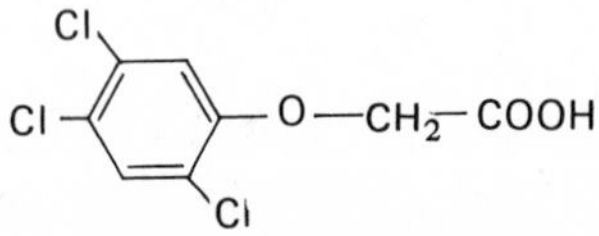

Cl, Cl, Cl, $O-CH_2-COOH$

applying one regulator that interacts with an endogenous phytohormone to produce the desired result. Examples of each class, respectively, include the use of ethylene to stimulate ripening of fruit, the use of cycloheximide to stimulate abscission of citrus fruits, and the application of auxin to pineapples to induce ethylene production, which in turn causes flowering.

Regulators and the responses of plants to them must be registered for use through the Environmental Protection Agency. In addition, the various states have laws controlling their use to varying degrees. The horticulturist must understand the laws governing plant-growth regulators (also insecticides, fungicides, herbicides, and other horticultural chemicals), obtain a license to use them when it is required by law, watch for changes in the regulations and lists of permissible chemicals, and observe all safety precautions.

Plant-growth regulators can be used with some horticultural crops to increase yield by stimulating flowering and fruiting. More important, however, is their use to modify quality and timing more effectively than other cultural techniques. If this use has the added advantage of reducing labor cost, so much the better. Some regulators are used specifically to reduce labor costs associated with high-value horticultural crops.

The structures of several plant-growth regulators are shown in Figure 9–15. There are a number of regulators, which can be grouped as natural phytohormones or according to function. Each will be treated separately. Trade names are capitalized and common names are not.

Fig. 9-15. Structures of growth regulators used on horticultural crops for various purposes.

Growth Retardants:
1. A-REST. 2. CYCOCEL. 3. MALEIC HYDRAZIDE. 4. SPROUT NIP. 5. CHLORFLURENOL. 6. ALAR. 7. CHLORPHONIUM.

Fruit Thinning Agents:
1. NAA. 2. DNOC. 3. NAD. 4. SEVIN.

Ethylene Sources:
ETHREL.

Scald Preventatives:
1. ETHOXYQUIN. 2. DPA.

Abscission Agent:
CYCLOHEXIMIDE.

Chemical Pruner:
OFF-SHOOT-O.

Miscellaneous:
SURE SET.

Refer to text for other trade/common and scientific names. Omission of trademarks and names is due to spacial limitations and simplified usage. It does not imply disrespect toward proprietary rights associated with trademarks nor an endorsement of any product.

Auxin (3-indolebutyric acid, Hormodin®) has been useful for improving the rooting of cuttings (Fig. 9–16). A formulation of it with a fungicide has been used to reduce the losses attributable to rotting of cuttings during rooting. Other combinations include a 1 : 1 mixture of indolebutyric acid and naphthaleneacetic acid, or the preceding two plus naphthalene acetamide (Rootone®). Indolebutyric acid is solubilized in a minimal volume of ethanol and diluted to a final concentration that ranges from 500 to 5000 ppm. Alternately, the solubilized indolebutyric acid may be blended with talc to produce a final strength of 0.1 to 1.0 percent. Empirical trials are necessary to determine the optimal concentration for rooting any particular species under a given set of conditions. Too much can inhibit bud development of shoots, and too little will not stimulate rooting. Softwood cuttings of many bedding plants respond favorably to the lower end of the range given. Generally, much of the cutting is immersed into the liquid; with powder formulations, only the base is dusted. Indolebutyric acid resists breakdown by sunlight and bacteria.

Indolebutyric acid is not effective in stimulating rooting of all plant species at any developmental stage. Some species that are difficult to root are believed to be lacking or limiting in naturally occurring rooting cofactors. In others it is believed that rooting inhibitors are present. Sometimes success results when the cuttings are leached to remove inhibitors. The identity of the rooting cofactors remains unknown.

Fig. 9–16. Chrysanthemum cuttings being dipped in 3-indolebutyric acid prior to rooting them. U.S. Department of Agriculture photo now in National Archives (16–N–22842).

Fig. 9-17. A. Loblolly pine on right was sprayed with 400 ppm of gibberellic acid at one year. Control is on left. Plants are now two years old and treated plant is taller and has a double thickness stem. B. Chrysanthemum on right was treated with gibberellic acid during the 4th-7th week of short days. It shows a higher stem and flower stalk and earlier blooming than the control on the left. U.S. Department of Agriculture photos.

GIBBERELLINS

Gibberellins, chemically 2, 4a, 7-trihydroxy-1-methyl-8-methylenegibb-3-ene-1, 10-carboxylic acid→ 1-4 lactone (gibberelic acid, Gibberellin, Gib Tabs®, Gibrel®, Gib-Sol®, Pro-Gibb®, Berelex®, Activol®, or Grocel®), have been utilized for achieving various plant responses (Fig. 9-17). A number of uses have been helpful to the horticulturist. Application of 20 to 40 ppm on 'Thompson Seedless' and other grapes causes cluster loosening and elongation and increases berry size. A concentration of 10 to 100 ppm at weekly intervals for up to 3 weeks will produce early flowering for many summer-flowering annuals. A concentration of 0.1 to 1.0 ppm is used to break dormancy in seed potatoes in lieu of low temperatures. Other functions include the induction of bolting and the increase of seed production with lettuce; the control of fruit maturity, fruit set, yield, rind color, and rind aging with citrus; the increase of yields and as a harvest aid with forced rhubarb and Italian prunes; the promotion of rapid emergence with beans, peas, and soybeans; and the extension of the harvest season for sweet cherries and artichokes. Concentrations vary and depend on several factors. The range is critical, and overdosage will produce plants that are horticulturally unacceptable. Aqueous solutions of gibberellins are unstable, so they should be freshly prepared prior to usage.

GROWTH RETARDANTS

Growth retardants are a horticulturally important group of plant growth regulators. These regulators slow growth, and many of them produce secondary effects of great value to the horticulturist (Fig. 9-18). Most growth retardants are dissolved in water, and a surfactant is added, unless one is present in the growth-retardant formulation. Surfactants usually consist of a detergent or a sulfonated alcohol and are added at a low concentration (0.1 to 0.2 percent). Growth retardants are applied as a foliar spray or a soil drench. The various growth retardants will be considered individually.

Butanedioic acid mono- (2,2-dimethylhydrazide), known as daminozide, Alar®, or B-Nine®, is used to reduce internode elongation, to induce heat, drought, and frost resistance, to produce darker green foliage and stronger stems, to inhibit undesirable stretching of transplants, and to produce earlier and multiple flowers. It is generally applied as a foliar spray to runoff at 2500 to 10,000 ppm; wetting agents are not added in this instance. It is used on a number of fruits, vegetables, and ornamentals. Most dicots respond, but monocots do not. It is used most frequently on chrysanthemums and bedding plants.

α-Cyclopropyl-α-(4-methoxyphenyl)-5-pyrimidinemethanol (ancymidol, A-Rest®, or Reducymol®) is used as a growth retardant on chrysanthemums, tulips, poinsettias, and lilies. It is applied as a soil drench. Also used as a growth retardant in the form of a soil drench are (2-chloroethyl) trimethylammonium chloride (chlormequat, Cycocel®, or Cycogan®) and tributyl-2, 4-dichlorsbenzylphosphonium chloride (chlorphonium, or Phosfon®). The first is used on red poinsettias and azaleas and the second on chrysanthemums and Easter lilies. Dilution rates vary with plant species, variety, and growing conditions.

Maleic hydrazide (chemically, 1,2-dihydro-3,6-pyridazinedione), known in the trade by about twenty names (Retard®, Slo-Gro®, Maintain-3®, to name a few), is a growth retardant with several uses. It is applied at rates of 0.75 to 3.0 pounds per acre in 50 to 100 gallons of water. It is used as a growth retardant on turf, as a chemical pruner on trees and ornamental shrubs, as a dormancy inducer for citrus, as a sprout inhibitor for stored onions and potatoes, and as a herbicide in some instances. Methyl-2-chloro-9-hydroxyfluorene-9-carboxylate (chlorflure-

Fig. 9-18. The growth retardant PHOSFON was used on the chrysanthemum at right when 8 weeks old. Control is on left. U.S. Department of Agriculture photo.

nol) is a growth retardant used in combination with maleic hydrazide on weed growth and turf. Isopropyl N-(3-chlorophenyl) carbamate (CIPC®, Chloro-IPC®, Sprout Nip®, or Spud-Nic®) is a growth retardant used to prevent sprouting of potatoes.

FRUIT-THINNING AGENTS

The regulators, 2-methyl-4, 6-dinitrophenol (DNOC, Elgetol) and 1-naphthyl-N-methylcarbamate (carbaryl, Sevin®) are used to thin apples. The first is applied at 0.66 to 1.5 pints per 100 gallons (825 to 1875 ppm) of water when 70 to 80 percent of the blossoms are open and the second at 1200 ppm 10 to 25 days after full bloom. Naphthalene acetamide (Amid-Thin W®) is applied at 25 to 50 ppm to thin apples and pears 19 to 25 days after bloom. Apples, pears, pineapples, and olives can be thinned and preharvest drop prevented by α-naphthaleneacetic acid (NAA®, Fruit Fax®, Trehold®) at 10 to 150 ppm. Since one of the above thinners is also a frequently used insecticide (Sevin®), care should be exercised to avoid following it with a thinner when used as an insecticide to prevent excessive thinning of fruit.

ETHYLENE SOURCES

The compound (2-chloroethyl) phosphonic acid (ethephon, Ethrel®, Florel®, or Cepha®) is used to release ethylene into plant tissues. It is used at various concentrations from 2.5 to 6.5 pints per acre. Responses include fruit loosening, promotion of uniform ripening, and increases of flower bud development for apples, fruit loosening and earlier ripening for cherries, earlier harvests for filberts, increased yields and uniform ripening for tomatoes, flower induction on pineapples; it improves harvest efficiency for blackberries and blueberries, hastens maturity and color intensity for cranberries, accelerates ripening of lemons and peppers, and loosens fruit and improves color for tangerines. Ethylene is also applied directly as a gas in a confined area at a concentration of 1000 ppm. It is used to degreen and ripen bananas, citrus, honeydew melons, pears, persimmons, pineapples, tomatoes, and walnuts.

SCALD PREVENTATIVES

Storage scald of apples can be prevented with the plant growth regulators 1,2-dihydro-6-ethoxy-2,2,4-trimethylquinoline (ethoxyquin, Stop Scald®, Santoquin®, or Nix-Scald®) and diphenylamine (Big Dipper® or Scaldip®). These are applied as a preharvest spray at 2700 and 2000 ppm, respectively, a few days prior to harvest. Fruit can also be dipped after harvest in a wax emulsion.

ABSCISSION AGENTS

Abscission agents are generally used to cause abscission for improving harvest efficiency. Fruit can be loosened for easier picking, or the plant may be defoliated for cleaner harvests. Citrus fruit can be loosened, but not dropped, with 3-[2-(3,5-dimethyl-2-oxocyclohexyl)-2-hydroxy-ethyl]-glutarimide, which is also called cycloheximide or Actidione®. It is applied 4 to 7 days before harvest at 0.33 pound in 500 gallons of water (about 80 ppm).

CHEMICAL PRUNERS

Methyl esters of C_9 to C_{13} fatty acids (Off-Shoot-O) are used as a chemical pruner on azaleas, *Cotoneaster,* juniper, *Ligustrum, Rhamnus,* and *Taxus.* The application is at 1.5 to 5 ounces per quart of water. No surfactant is added.

MISCELLANEOUS REGULATORS

Blossom set on tomatoes is improved with *p*-chlorophenoxyacetic acid (Fruitone® or Tomatotone®) up to 50 ppm. Deoxy derivatives of gibberellin (Pro-Gibb® 47) are sprayed on cucumbers at 50 ppm to develop male flowers on gynoecious cucumbers for purposes of seed production.

EXPERIMENTAL GROWTH REGULATORS

A number of plant-growth regulators that show promise for horticulturists are presently being used on an experimental basis. Some or all will undoubtedly be released on a commercial basis. The compound 6-benzylamino-9-(tetrahydropyran-2-yl)-9-(H)-purine (SD 8339, Accel) is being used on chrysanthemums, carnations, and roses to produce stockier, more branched forms. Another chemical, [(3-phenyl-1,2,4-thiadiazol-5 yl) thio] acetic acid, is a growth promoter that increases the number of branches on young apple trees in the first year of growth after grafting or budding. Suppression of foliar growth and an increase in yield with some species have been observed with *N*-4-methyl ([(1,1,1,-trifluoro-methyl) sulfonyl) amino] phenyl) acetamide (Fluoridamid, Sustar-2-S, or MBR-6033) on trials with turf, ornamentals, and soybeans. A multiple purpose regulator, 2,3-dihydro-5-6-diphenyl-1,4-oxathiin (UBI-P293), has been used to control tuber size with potatoes, shorten the height of ornamentals, delay flowering of peaches, chemically prune ornamentals, increase the sugar content of sugar beets, and as an antitranspirant. Ethane-dial dioxime (Glyoxime, Pik-Off) has shown a promise as an abscission agent on oranges.

FOREIGN PLANT-GROWTH REGULATORS

Some promising plant-growth regulators, unlike the preceding ones, are not cleared yet for use in the United States. These are primarily European products. They include the sodium salt of 2,3:4,6 bis-*O*-(1-methylethylidane)-α-L-xylo-2-hexulofuranosonic acid (dikegulac-sodium), 2-chloroethyltris (2′-methoxy-ethoxy)-silane (etacelasil) and *N*-*m*-tolyphthalamic acid (Tomaset®). Their respective purposes are as a chemical pruner and growth retardant for ornamentals, an abscission agent for olives, and a fruit- and flower-setting agent for tomatoes, eggplant, lima beans, cherries, and prunes.

FURTHER READING

Brown, G. E., *The Pruning of Trees, Shrubs, and Conifers.* Salem, N.H.: Faber & Faber, Inc., 1977.

Grounds, Roger, *The Complete Handbook of Pruning.* New York: Macmillan, Inc., 1973.

Hammett, K. R., *Fell's Guide to Plant Training, Pruning, and Tree Surgery.* New York: Frederick Fell Publishers, Inc., 1975.

Hudson, Roy L., *The Pruning Handbook* (3rd ed.). Englewood Cliffs, N.J.: Prentice-Hall, Inc., 1973.

Stutte, Charles A., *Plant Growth Regulators: Chemical Activity, Plant Responses, and Economical Potential.* Washington, D.C.: American Chemical Society, 1977.

Thomson, W. T., *Agricultural Chemicals, Book III: Fumigants, Growth Regulators, Repellents, and Rodenticides.* Fresno, Calif.: Thomson Publications, 1976 (Frequent revisions, a working publication of importance to the commercial horticulturist).

Plant Protection

Horticulturists are faced with a number of plant pests, which normally exist in the horticultural ecosystem. Failure to control these pests can result in consequences that vary in severity. These might only be intangible distress to the horticulturist or, worse, loss of professional status. A horticulturist in charge of commercial plantings that suffered financial loss because of failure to control plant pests or in charge of a landscape planting that became esthetically poor due to plant pests could quickly lose his or her professional status. Plant pests of concern include insects, mites and related creatures, weeds, bacteria, fungi, and viruses.

Control of plant pests, or *plant protection*, involves a number of steps. First, the horticulturist must recognize the possibility or existence of a problem. In this regard it is important to be knowledgeable with respect to plant susceptibility to attacks by plant pests and under what environmental conditions and during which season a plant is most susceptible. The horticulturist must also be capable of recognizing the presence of a plant pest at its initial appearance, or after it causes a plant response or *symptom,* and be able to identify the pest once it is discovered. Some pests can only be identified when they are considered in conjunction with the symptoms. Identification is followed by an assessment to determine the control required to eliminate the pest, the method of application, how much is needed, phytotoxicity problems, and the timing of application.

These aspects will be discussed in detail as they arise in the following sections.

If chemical control is chosen over other forms of control, the horticulturist must also consider the following. The label must be read carefully and the directions followed. Federal laws make it illegal to apply a plant-protection chemical in any manner except as stated on the label. The horticulturist must be licensed to use the chemicals if it is required by state law. Certain chemicals are now banned or allowed only on certain crops, as determined on a federal level, and further changes are possible on the state level. The horticulturist must know these restrictions.

The responsibility of the horticulturist does not end upon application of the chemical. One must know its toxicity, the pest developmental stages affected, rate of kill, spectrum of activity, and persistence in the environment. Finally, the horticulturist must be able to assess the effectiveness of the treatment and whether additional control is needed.

Diagnosis and Identification

The horticultural environment is not static, but is constantly changing. The magnitude of these changes can vary from minor to extreme. Plants, if they are to survive, must be capable of adjustment. On occasion, the plant may fail to adjust for various reasons. At this time the plant indicates its lack of adjustment through distress symptoms. The horticulturist can sometimes spot problems before they lead to distress symptoms; otherwise, the problem must be diagnosed and the causal agent identified once distress symptoms are evident.

Distress symptoms can arise from a number of sources, including artificial injuries, natural injuries, and plant pests. Often, interactions among these sources are possible. For example, on a tree an injury caused by a lawn mower (artificial) or a lightning strike (natural) may open a wound that gives access to a plant pest, such as an insect or disease. Therefore, all sources of injury to a plant must be considered when diagnosis of a problem is undertaken by the horticulturist.

Background information about the distressed plant is needed before a diagnosis of the problem can be made. First, the plant must be identified. Identification may only be needed at the genus level or the complexity of the problem may require identification to the species level. This may require specialized knowledge beyond that of the horticulturist, who must then resort to specialized texts or consultation with an appropriate expert. Once the host plant is known, other background information of importance includes the following. The age of the plant is important. For example, a tree so old as to be senescent may not be worth saving or impossible to save from a disease or insect attack. An idea of the vitality is helpful. This can often be determined from visual examination and comparison with other plants of the same species. A plant having poor vigor, everything else being equal, appears to be more susceptible to plant pests, or is at least more prone to severe damage or death when infestation occurs. The hardiness of the plant in question should be known. An extreme variation from the normal climate may distress a plant of borderline hardiness. It may be necessary to know the past history of the distressed plant, since pollution, drought, or waterlogged soil may not produce distress symptoms until after the causal agent has been corrected or alleviated.

Once this is accomplished, the distress symptoms are observed. The plant is examined for *signs* of a plant pest, such as the actual organism, a skeleton, or part of a pest. Diagnosis is made on the basis of combined symptoms, signs (if any), and the knowledge of what pests can be associated with the host plant in question.

If no signs of plant pests are present, it is likely that the distress symptoms resulted from artificial or natural injuries. The possibility does exist, however, that the plant pest has been eliminated by natural causes or moved on to another plant.

Confusion sometimes arises when a distress symptom is found that can be produced by a number of causes. For example, a wilt may be caused by bacteria, borers, drought, or even a lightning strike. Here careful observation is necessary to find another symptom or sign to eliminate false causes. When a symptom is caused by combined injuries and plant pests, the past history becomes very important. Sometimes injuries from artificial or natural causes may appear similar to those caused by plant pests.

If the symptom is traced to a plant pest, its identification becomes necessary. The horticulturist may have to resort to specialized texts, a plant pathologist, an entomologist, or even a taxonomist if available resources are insufficient. Once the pest is identified, control measures can be chosen. If the cause was artificial or natural injury, corrective measures can be taken.

Plant Pests of Animal Origin

Animal pests of the biotic environment that attack plants include nematodes, mites, insects, birds, rabbits, deer, mice, moles, other rodents, dogs, cats, humans, millipedes, slugs, and snails.

NEMATODES

Nematodes are roundworms, not insects, that live in soil and water (Fig. 10-1). Their size is about 1/64 to 1/8 inch, and they have a rather impermeable cuticle. They are classified in the phylum (equivalent to botanical division) Nemeta. Most are free living, but others are parasitic on plants and animals. Those that attack plants feed on the surface or interior of roots. A few are leaf feeders.

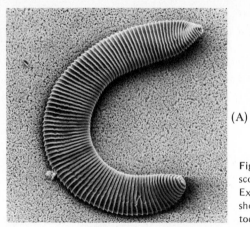

(A)

Fig. 10-1. A. A nematode as seen with a micro-scope. Courtesy of The Connecticut Agricultural Experiment Station, New Haven. B. These roots show cysts which contain eggs of the Golden nema-tode (*Heterodera rostochiensis*).

(B)

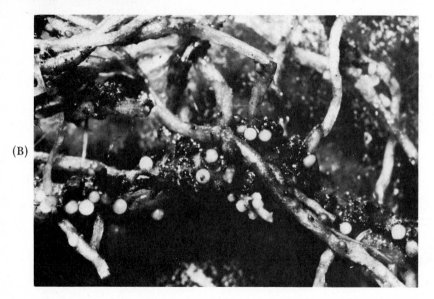

Fig. 10-2. (Below) Two-spotted spider mite (*Tetranychus urticae*), a very common horticultural pest. Courtesy of Union Carbide Corporation.

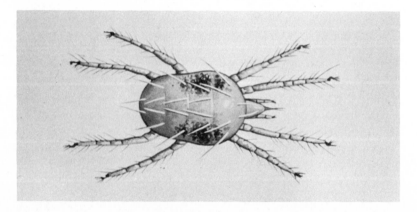

While feeding, nematodes may inadvertently introduce disease organisms that cause wilts or rots into the root system of the plant. Root-knot nematodes cause galls or swellings on the roots of many plant species. These growths block the flow of water and nutrients, producing a stunted, yellow, wilted plant. Meadow nematodes cause symptoms that could easily be mistaken for root rots. Cyst and sting nematodes cause stunted plants with wilted, curled foliage.

ACARIDS

The animal class Arachnida includes spiders, mites, scorpions, and ticks. These differ from the true insects in the class Insecta (Hexapoda). The arachnids have four pairs of legs, as opposed to three pairs for insects. They also lack antennae, wings, true jaws, and compound eyes. In addition, they have only two body regions with no separate head; insects have three body regions, one of which is a separate head. Of the arachnids, the mites, members of the order Acarina, are a serious plant pest (Fig. 10-2). Mites, along with ticks, are also referred to as acarids. The most common are the spider mite, broad mite, and the cyclamen mite.

Mites are small, usually less than 1/100 inch long. They damage plants by their feeding action, the sucking of plant sap. They can cause extensive damage since they are prolific, and chemicals used to control insects often eliminate the competitors and enemies of mites, leading to explosive increases of mites. Many insecticides are combined with miticides to prevent such occurrences.

Symptoms of mite damage are as follows. Some species cause fine yellow, gray, or white flecking or stippling of the foliage; the whole leaf becomes chlorotic with severe damage and eventually drops off. Webbing may be seen to cover the plant with large populations of certain mites. Other mites cause russetting of leaves, gall production, bud injury, and blistering. Host plants include a very wide range of horticultural material throughout the United States, with some mites being very host specific and others being more far ranging.

INSECTS

Insects are members of the same phylum, Arthropoda, as are mites. The class Insecta (Hexapoda) contains the true insects, which differ from mites and other arachnids as discussed in the preceding section.

The species in Insecta constitute the bulk of the plant pests of animal origin. Some insects are pests well adapted to plant destruction because of their feeding habits, small size, protective coloration in some cases, the ability to fly, and their prolific nature.

(A)

(B)

Fig. 10-3. Stages of insect metamorphosis with the Japanese beetle (*Popillia japonica*). A. Eggs hatch in the soil to produce a white grub (above). Eventually the grub enters a resting stage (pupa) shown left. B. The adult form which emerged from the pupa. Both the grub and adult form are horiticultural crop pests. U.S. Department of Agriculture photos.

318

Insect life cycles are such that we can consider two groups of insects, those that undergo complete metamorphosis and those that undergo gradual metamorphosis. The former group has distinctly different states, which give the insect different forms ideally suited for feeding on plants and for reproduction; these are of obvious disadvantage to the horticulturist. In gradual metamorphosis, the change between stages is gradual; hence the distinction between stages is not obvious. An example of insect metamorphosis is shown in Figure 10-3.

Insecta contains a number of orders with both plant pests and beneficial insects:

Coleoptera: weevils and beetles; complete metamorphosis and have chewing mouth parts.

Collembola: springtails; wingless, without metamorphosis, and have chewing mouth parts.

Dermoptera: earwigs; simple metamorphosis and have chewing mouth parts.

Diptera: flies; complete metamorphosis, with chewing or reduced mouth parts in larvae and sponging in adults.

Hemiptera: true bugs; gradual metamorphosis and have piercing-sucking mouth parts.

Homoptera: whiteflies, aphids, leafhoppers, scale, mealybugs, planthoppers, spittlebugs, and psyllids; gradual metamorphosis and have piercing-sucking mouth parts.

Hymenoptera: ants, wasps, bees, and sawflies; complete metamorphosis, with chewing or reduced mouth parts in larvae and chewing-lapping in adults.

Isoptera: termites; gradual metamorphosis and have chewing mouth parts.

Lepidoptera: moths, skippers, and butterflies; complete metamorphosis, with chewing mouth parts for larvae and siphoning for adults.

Neuroptera: antlions and lacewings; complete metamorphosis and have chewing mouth parts.

Orthoptera: grasshoppers, crickets, katydids, male crickets. mantids, and walking sticks; undergo gradual metamorphosis and have chewing mouth parts.

Thysanoptera: thrips; simple metamorphosis and have rasping-sucking mouth parts.

There are at least 700,000 species in Insecta and very few plants are completely safe from these insects. Damage runs the gamut from little chewed holes in leaves to complete destruction of the plant. A generalized relationship between these symptoms and various source insects is shown in Figure 10-4. In addition, some insects are vectors for certain plant diseases, which will be covered in a later section.

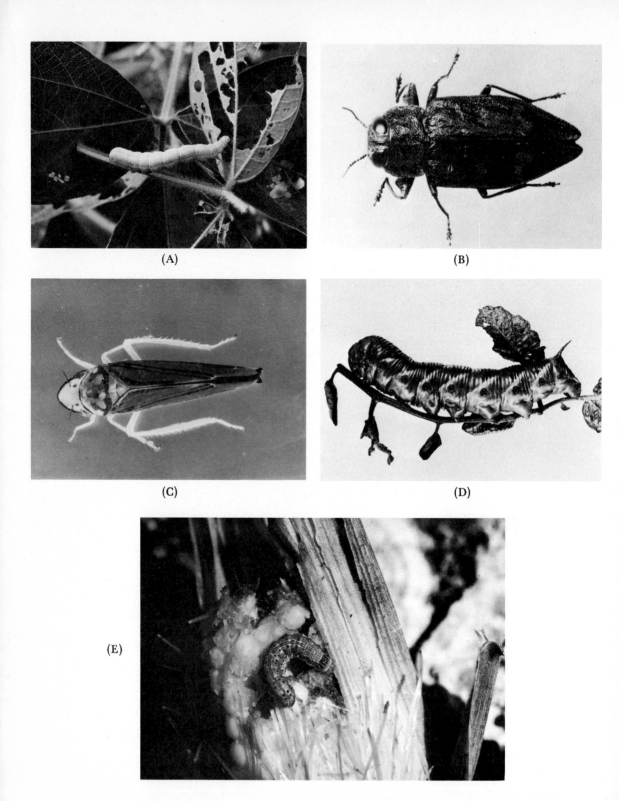

(A)

(B)

(C)

(D)

(E)

320

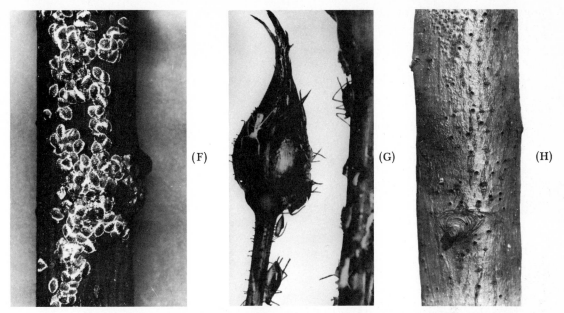

(F) (G) (H)

Fig. 10-4. A number of insect pests and some aspects of plant damage caused by insects. A. Cabbage looper (*Trichoplusia ni*) on soybean. It is a serious pest on cole crops. B. Flatheaded apple tree borer (*Chrysobothris femorata*) which attacks a wide range of deciduous fruit and shade trees during the grub stage. C. Leafhopper (*Graphocephala coccinea*) commonly found on garden flowers. These suck plant juices and many are vectors of viral diseases. D. Tomato hornworm (*Manduca quinquemaculata*) can strip foliage and small fruits quickly from tomato plants. E. Corn earworm (*Heliothis zea*), one of the worst sweet corn pests. F. European elm scale (*Gossyparia spuria*). One of many sucking scales, most of which are serious crop pests. G. Aphids, a soft bodied sucking insects. These are widespread and can be vectors for certain viral and bacterial diseases. H. Tree trunk showing extensive borer damage. I. Mealybugs, essentially soft scales, are sucking insects which are serious greenhouse and garden pests. A–F, H are U.S. Department of Agriculture photos, G is U.S. Department of Agriculture photo now in National Archives (54-M-5020), and I is author photo.

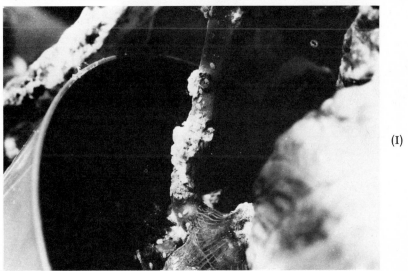

(I)

LARGER ANIMAL PESTS

Some larger animals can occasionally cause extensive damage to horticultural crops. These include birds, rabbits, various rodents, deer, cats, dogs, and humans. Birds can be a plant pest when they eat newly planted seeds and ripe fruits such as berries, cherries, and grapes. Rabbits and other rodents damage young crops by eating them and often girdle fruit trees when they eat the bark during the winter. Deer can also damage fruit trees by eating the bark during the winter. Dogs and cats damage plants by digging activities, sharpening (cats only) of claws on bark, and the use of plants as an outdoor toilet. Humans damage plants through ignorance, carelessness, and vandalism (Fig. 10-5). Damage by the larger animal pests can reduce the esthetic qualities of plants, increase their susceptibility to attacks by other plant pests, or kill plants.

Fig. 10-5. A street maple planted at a school bus stop and vandalized by having the terminal leader broken off. This forced sprouting at lower shoot area and has produced an unattractive form for a street tree. If not corrected by pruning, the tree will become worthless. Author photo.

MISCELLANEOUS ANIMALS

Millipedes, slugs, and snails damage plants in the garden and greenhouse. Damage varies from holes caused by eating of plant tissues up to partial or total loss of herbaceous plants. During foraging, slugs and snails leave a slimy trail, which can be used to identify their presence.

Plant Pests of Plant Origin

A number of plant organisms (or plant and nonanimal organisms depending on which classification system is used) interfere with the successful cultivation of horticultural crops. These include viruses, mycoplasmalike organisms, rickettsialike organisms, bacteria, fungi, dodder, mistletoe, algae, moss, weeds, strangling vines, and strangling roots. These agents compete with plants, and some cause diseases. The microbial agents are known as *pathogens,* and the resultant disease is termed a pathogenic disease. This is to distinguish it from nonpathogenic diseases resulting from mechanical, environmental, or animal causes. Many prefer to view nonpathogenic diseases as injuries, rather than diseases.

VIRUSES

The virus is the smallest pathogen, with dimensions in the millimicron range. As such, a virus can only be visualized with the aid of an electron microscope. Viruses usually consist of a sheath of protein surrounding a nucleic acid core. They are not capable of reproduction except within the cells of a living host. Inside the cell they are dependent on the metabolism of the cell, where they direct part of the cell's protein synthesis toward their own reproduction.

Plants of horticultural value affected by viruses include the tomato, cucumber, squash, potato, corn, bean, sugar beet, strawberry, raspberry, apple, peach, and many ornamentals. For a given virus, the host range may be broad or as narrow as a few cultivars within a species. The plant response to viruses, that is, the symptoms (Fig. 10-6), provides a basis for division into two viral groups, the *yellows* and *mosaics.* The basic symptom of the former is yellowing of leaves; mosaic results in a patchwork pattern of normal green and light green to yellow areas in the leaves. Other virus symptoms include spotting, dwarfing, leaf curl, leaf roll, leaf wrinkle, excessive branching, and gall formation. These symptoms result in part from destruction of chlorophyll and, possibly, disruption of hormonal levels and interference with translocation.

Viruses may be transmitted in several ways among plants. The yellows group can be transmitted by leafhoppers and the mosaics group by aphids. Other insects with sucking mouth parts, such as nematodes, may also be

(A)

(B)

(C)

Fig. 10-6. Plants with symptoms of viral diseases. A. Healthy dahlia leaf on left compared to one on right infected with dahlia mosaic virus. B. Squash leaves deformed and mottled by cucumber mosaic virus. C. Curly top virus on tomato. U.S. Department of Agriculture photos.

transmission agents, or *vectors*. Transmission can occur when a viral-infected plant is grafted onto a healthy plant; in fact, this is used as a test to detect suspected viruses in plants where symptoms are few or lacking. The healthy portion of the graft is a highly susceptible stock, which will show symptoms rapidly. Some can be transmitted by hand, such as tobacco mosaic virus, which also infects tomatoes. Mechanical transmission is also possible, such as rubbing leaves or rubbing viral-infected plant sap into a healthy leaf.

MYCOPLASMALIKE AND RICKETTSIALIKE ORGANISMS

Mycoplasmalike organisms consist of a membrane enclosing a living protoplasm, and, as such, the organism has no constant shape. Mycoplasmalike organisms are usually found in the phloem tissue, where they utilize food from the host, disrupt translocation, and sometimes cause hormonal imbalance. Rickettsialike organisms are highly modified bacteria. Both of these organisms produce symptoms that resemble viral diseases; in fact, many of these diseases were formerly thought to be viral in nature.

Mycoplasmalike organisms are now known to cause aster yellows in strawberries, certain vegetables, and ornamentals; blueberry stunt disease, peach X-disease, elm phloem necrosis, and pear decline. The last two are spread by the elm leaf hopper (*Scaphoedeus luteolus*) and pear psylla (*Psylla pyricola*), respectively. Rickettsialike organisms are associated with Phony disease of peach and Pierce's disease of grapes.

BACTERIA

Bacteria are single-celled organisms larger than viruses; unlike viruses, they can be seen with a light microscope. Some bacteria are plant pathogenic. These can enter plants through natural openings or through wounded tissue. Insects with sucking mouth parts can also transmit bacteria. For example, the cucumber wilt bacteria is transmitted by the striped cucumber beetle.

Diseases of bacterial origin produce symptoms that often resemble symptoms caused by fungal attacks. Symptoms include rots induced by enzymic degradation, wilts caused by bacterial blockage of the vascular system, gall formation, and tissue destruction at various plant parts. Examples of bacterial disease include potato blackleg, Stewart's disease of early sweet corn, crown gall on fruit trees, berry bushes, and roses, fireblight on apple, pear, and quince, cucumber wilt (Fig. 10-7), and shothole disease of stone-fruit trees. Water-soaked spots on leaves of house plants are often of bacterial origin.

Fig. 10-7. Cucumber on right is healthy, but one on left is infected with cucumber wilt (*Erwinia tracheiphila*). This disease is carried by the cucumber beetle (*Diabrotica undecimpunctata* and *Acalymma vittata*). Author photo.

325

FUNGI

Fungi are multicelled organisms with a threadlike growth pattern; they account for more parasitic plant diseases than viruses or bacteria. Fungal entrance is through natural openings, wounds, or even directly through the cuticle by enzymic attack. Fungal spores are readily dispersed by air and water movement. Insects, such as the Dutch elm beetle, can spread diseases (Fig. 10-8).

Symptoms of fungal diseases are conspicuous and are the basis of simple categories of fungal infections: wilt, canker, blight, mildew, rot, leaf spot, rust, scab, and smut (Fig. 10-9 on pages 328-329). Other symptoms are rolling of leaves, fruit deformation, fruit and blossom drop, dwarfing, chlorophyll destruction, tumorlike tissues, and tissue destruction. The various fungal disease types and some symptoms are shown in Table 10-1. Fungal diseases are widespread in the plant material used by horticulturists. All plant parts and all developmental stages are susceptible to fungal attack.

Fig. 10-8. This American elm (*Ulmus americana*) is showing severe symptoms of Dutch elm disease (*Ceratocystis ulmi*). The disease is carried by the Dutch elm beetle (*Scolytus multistriatus*). U.S. Department of Agriculture photo.

Introduction to Applied Horticultural Science / Part II

TABLE 10-1. FUNGAL PLANT DISEASES

Group	General Symptoms
Anthracnose	Sunken spots on leaves and fruit; black, shriveled leaves
Blight	Sudden spotting or drying of leaves, flowers; no noticeable wilting
Canker	Initial well-defined lesions; later enlargement to wounds or girdle areas; some oozing
Damping-off	Seed or seedling rot in soil; stem rot and falling over; stem girdled and seedling stunted
Leaf spot	Spots with clearly defined margins, usually of a different color than interior
Mildew	Powdery coating, usually white
Mold	Colored fungal patches (white, pinkish, black)
Rot	Soft pulpy tissue, usually brown or black
Rust	Red or rusty brown spore patches
Scab	Dark, raised spots
Smut	Large masses of black spores
Wilt	Foliage yellowing; collapse of plant

DODDER, MISTLETOE, ALGAE, AND MOSS

Dodder is a plant lacking chlorophyll that is parasitic on some plants of horticultural value, such as the chrysanthemum. Mistletoe is a familiar parasitic plant found high in trees. Both are debilitating to the vigor of the host plant and may cause eventual death. In addition, dodder may be a vector for certain viral diseases.

Algae and moss are simple plants, which can be undesirable in certain horticultural situations. Green growths of algae on clay pots or in water bodies are unsightly. Greenhouse floors and moist garden paths can be slippery and dangerous when coated with algae. Moss may give an undesirable appearance in lawns or other areas.

WEEDS

Weeds are broadly defined as any plant that grows where it is unwanted. Naturally, this could even include plants that do not fit the typical image invoked by the word "weed." In actual practice, weeds are usually considered to be native or naturalized plants that compete aggressively with plant material grown for horticultural purposes.

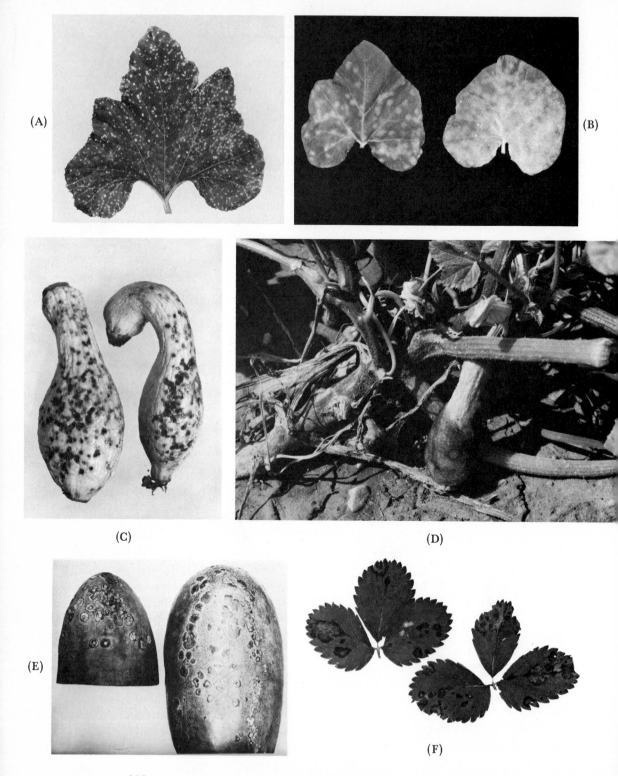

(A)

(B)

(C)

(D)

(E)

(F)

328

(G)

(H)

Fig. 10-9. Some common fungal diseases of plants and their symptoms. A. Squash leaf infected with downy mildew (*Pseudoperonospora cubensis*). B. Squash leaf infected with powdery mildew (*Erysiphe cichoracearum*). C. Squash fruit infected with scab (*Cladosporium cucumerinum*). D. Squash fruit infected with rot (*Choanephora cucurbitarum*). E. Watermelon infected with anthracnose (*Colletotrichum lagenarium*). F. Strawberry leaves infected with leaf blight (*Dendrophoma obscurans*) G. Potato tuber and leaves infected with scab (*Streptomyces scabies*). H. Rose leaf with blackspot (*Diplocarpon rosae*) I. Grape leaf with black rot (*Guignardia bidwellii*). Fruit also rots. J. Maple dieback observed on street maple. Probable result of fungal disease(s) setting in after environmental stressing of tree, such as by pollution or poor aeration and minimal water due to pavement surrounding tree. A recent disease seriously infecting urban street plantings of maple; exact cause and organisms not completely elucidated. Photos A–G, I are from U.S. Department of Agriculture, H is U.S. Department of Agriculture photo now in National Archives (54-M–5010), and J is Author photo.

(I)

(J)

Weeds are highly successful as competitors; in fact, they usually win out in competition with horticultural material unless steps are taken to control these plant pests. The numbers of seeds produced by weeds are overwhelming. For example, the weed purslane (*Portulaca oleracea*) can produce about 190,000 seeds in one growing season. These seeds are also extremely viable for many years in a natural or induced dormant stage. When this dormancy is broken, such as by tillage of soil, which exposes them to favorable conditions, they can germinate for many years after they were produced. Many weed seeds will retain viability for 10, 20, and some even 40 years when buried in soil.

Their growth habits often contribute to their aggressiveness. Seed germination and seedling growth are often rapid, and subsequent growth is often extensive. Furthermore, weeds are more tolerant of extremes of heat, cold, and drought as well as plant pests, than are many horticultural plants, which seem to have lost some of these characteristics over years of intensive breeding. Some weeds have long taproots that must be completely removed; otherwise, regrowth will occur. A well-known example of this type is the dandelion (*Taraxacum officinale*). Others have vigorous rootstocks that can spread underground and send up new plants at many locations. Quack grass (*Agropyron repens*) and western sage (*Artemisia ludoviciana*) are examples of this type. In addition, many weeds possess a fibrous surface root system well suited to moisture and nutrient removal.

Weed control is practiced for a number of reasons. The most serious reason is their detrimental effect on crops. In the competition for light, water, and soil nutrients, weeds can seriously weaken or kill horticultural crops. This obviously reduces the crop's economic and esthetic values. In addition, weeds may be alternate hosts for insects and plant diseases that attack horticultural material. They are esthetically unpleasing in lawns, ornamental plantings, vegetable gardens, and roadsides. Some weeds pose health hazards because they are dermal irritants, such as Pacific poison oak or poison ivy (*Rhus diversiloba* and *R. radicans*), or allergens like ragweed (*Ambrosia artemisiifolia*) and goldenrod (*Solidago sp.*).

Types of Natural Controls

Natural controls is a large category that includes all forms of control other than that by chemicals which do not occur naturally and must be synthesized. Natural controls are not new, but the interest in and use of natural controls is on the increase, especially in view of the increasing concern about damage to our environment.

The advantages of natural controls are that they are much less of a threat to the balance of nature and our environment, and some are self-perpetuating once established. Some disadvantages are that natural controls are not always capable of instant results, and the level of control is less than complete in many cases. These disadvantages are readily accepted by organic gardeners and many noncommercial horticulturists, but by few commercial horticulturists. The latter group's reluctance stems from the fact that the disadvantages cut into the profit margin, and many consumers are reluctant to accept less than perfect appearance for horticultural products.

As natural controls become more readily available and more refined, they will become increasingly adopted. An effective compromise is that of *integrated control,* which makes use of natural controls wherever feasible and chemical control when no other system will work. The chemical control chosen should be the one least apt to damage the environment, and the harsh, persistent chemicals should be chosen only as a last resort.

Natural controls include the following: mechanical, physical, cultural, biological, physiological, and genetic controls. These will be discussed first in terms of general approach and later in terms of specific application against groups of plant pests.

MECHANICAL AND PHYSICAL CONTROL

Once a plant pest is identified, a number of control measures are possible. Mechanical methods can be used. These include hand picking and destruction of larger insect pests. Traps with sex attractants or sticky baits or electric black-light traps are helpful in trapping flying insects. Sticky bands around trees can trap caterpillars. A cardboard or newspaper band around new transplants can prevent damage from some soil insects. Fences can keep small animals away. Screening may be used to keep birds or insects away from fruit or mice from gnawing tree bark. In some instances, small bags can be placed around apples or pears to prevent insect damage. These techniques will also control some plant pathogens, that is, to the degree that they stop insects that are vectors for certain plant diseases.

Hot-water dips at 100° to 122°F (43° to 50°C) for 5 to 25 minutes have been employed to destroy plant insects and mites. Heat treatments from 100° to 135°F (38° to 57°C) for varying times have been used to destroy bacteria, fungi, and viruses. Obviously, this approach is only successful with plants, seeds, and bulbs that can withstand these temperatures. Forceful syringing or misting with water can be used on plants that can withstand the force to remove spider mites, mealybugs, and aphids from house and greenhouse plants.

Metallic strips or edging may be helpful in preventing the entry of weeds or turf into areas such as shrubbery borders or gardens. Mulches (covered in Chapter 7) control weeds through their action as a physical barrier. Heat treatments, such as pasteurization, eliminate weed seeds and pathogens in soil or propagation media as discussed in Chapter 6. Overburning of land and spot flaming have been utilized to destroy weeds.

CULTURAL CONTROL

A number of cultural techniques are practiced in the battle against plant pests. Plants suspected of harboring plant pathogens or pests should not be planted. Any plant debris should be promptly removed; otherwise, it may provide a breeding area or overwintering material for insects and disease organisms. Plants with uncontrollable plant pest infestations should be destroyed. These practices are termed *sanitation.* *Surgery* involves the removal of infected plant parts and *roguing* is the removal of diseased plants or seeds. These practices may prevent the spread of insects and disease.

Plant diseases and insects can be unwittingly transmitted through shipments of plants. Federal and state regulations exist to control the import of plants and sometimes the shipments of certain plants across state lines in order to prevent such occurrences. These regulations forbid the entry of certain plants and require inspections of others and/or phytosanitary certificates.

Removal of weeds through hand picking, hoeing, and mulching (see Chapter 7) in and around the garden eliminates sources of diseases and insects that utilize these plants as alternate hosts. Rotation of crops may disrupt the life cycles of diseases and insects by eliminating their food supply for a season. Timing a crop in regard to sowing or transplanting may be used to avoid peak periods of insect infestations or the time of laying of eggs.

Other cultural practices include the planting of insect- and disease-resistant cultivars whenever possible. Soil pH may be altered to provide an unfavorable environment for certain disease-causing fungi. Humidity levels and temperatures may be varied in greenhouses to provide less favorable environments for disease microorganisms. Air circulation in greenhouses can be used to speed drying of foliage in order not to provide a good environment for disease initiation.

Certain plants may be grown to discourage plant pests. The African and French varieties of marigolds have a root exudate that repels nematodes. This is an example of a repellent plant. Mixed crops may be less susceptible to a particular insect than would a monocrop of the preferred host plant. Other plants may be preferred by insects. These may be planted as trap plants to draw insects away from what would normally be a host plant in lieu of the preferred species.

BIOLOGICAL CONTROL

Biological control is based on the natural antagonism and competition that exists among organisms. For example, many insects that attack plants have natural enemies. These enemies are *predators,* which prey on other insects for food, and *parasites,* which lay eggs in or on other insects that then serve as a source of food for the young. Predators, parasites, and other insects that serve a useful purpose in the horticultural and agricultural community are collectively termed *beneficial insects.* These insects unfortunately are often casualties when chemical control is used.

Other forms of biological control are the use of bacterial diseases to infect insects, the use of viral diseases to infect bacteria, the use of insects to eat weeds, and the introduction of sterile male insects to reduce harmful insect populations. The use of botanical or naturally occurring insecticides is another.

PHYSIOLOGICAL CONTROL

Slight alterations of the plant's physiology are possible through such actions as fertilization and pruning. Sometimes these physiological alterations can enable a plant to resist some plant diseases; this is the basis of *physiological control.* Fertilizer can be applied to cause excessive growth, and the plant may outgrow the disease. Alternately, the fertilization may be such to cause slow growth, which would have an adverse effect upon those diseases that thrive in vigorously growing plant tissue.

GENETIC CONTROL

Some plants have the ability to resist insects and diseases to varying degrees. This natural or *genetic resistance* may be complete for a particular pest, that is, the plant has *immunity,* or the resistance may only be *partial resistance.* A complete lack of resistance is termed *susceptibility.* Resistance may take another form called *tolerance;* the plant becomes infected and even injured, but is able to function. The basis of *genetic control* is the introduction through plant breeding of resistances into plants that do not possess natural resistance. Horticulturists should strive to keep abreast of developments in this area and to specify the use of resistant plants wherever possible.

NEMATODES AND NATURAL CONTROLS

Natural controls of nematodes include the following. Do not buy or transplant plants showing symptoms of nematode infestation. Infested plants should be destroyed. Small batches of soil can be steam pasteurized to eliminate nematodes (see Chapter 6). Use cultivars that are known to have some degree of nematode resistance, such as sweet potato 'Nemagold'; use rootstocks, such as Nemagard rootstock for peach, that are nematode resistant to graft trees that will go into nematode-infested areas.

Fig. 10-10. Exudates from the roots of African or French marigolds repel soil nematodes. Best results are with mass plantings of marigolds followed by crops in same area. Companion planting of marigolds/crops were not effective according to recent research. Courtesy of National Garden Bureau, Inc.

Rotate susceptible crops with resistant crops to reduce nematode populations. Exudates from the roots of African or French marigolds have nematode-repellent properties (Fig. 10–10). Organic amendments, such as compost, help reduce nematode populations. Some fungi appear to be biological controls for nematodes. The success of compost in the reduction of nematodes may lie with its ability to increase fungal populations, some of which control nematodes. Someday, when identification and culturing of nematode-attacking fungi are complete, we may be able to inoculate soil with these fungal cultures. Hot-water treatments of 65 minutes at 110°F (43°C) have been used to free sweet potatoes of nematodes prior to planting them.

MITES AND NATURAL CONTROLS

Natural applied controls, especially useful for plants in the home and greenhouse, include soapy water sprays and hot-water dips. Soap sprays consist of a tablespoon or two of a flaked pure soap dissolved in a gallon of room-temperature or slightly warmer water. The hot-water dip consists of immersion in water at 112°F (44°C) for 15 to 20 minutes; it has been used to control cyclamen mite infestations of African violets. Sprays of light- to medium-weight petroleum oils have been used as dormant and summer oils to control mites on shrubs and trees. These oils will be covered in more detail in the following section. Warm, dry conditions favor mites, so an increase in relative humidity and temperature reduction may reduce their numbers. Forceful blasts of water, such as with a fogging nozzle, can be used to wash mites off foliage (Fig. 10-11).

Insect predators of the plant pest mites include lady beetles (*Stethorus* spp.), predatory mites from the family Phytoseiidae, some thrips, and an anthocorid bug. No insect parasites of mites are known. These natural controls are effective, but usually after the infestation peaks. Chemical controls applied at the peak of infestation usually eliminate these beneficial insects. For the least damage to plant and beneficial insects, chemical controls should be utilized before the peak of infestation. Present research with predatory mites may lead to an effective natural control in the future.

Fig. 10-11. Mites can often be removed from house plants by forceful rinsing with water. U.S. Department of Agriculture photo now in National Archives (16-N-30557).

INSECTS AND NATURAL CONTROLS

A number of natural controls are used against insects. Numerous predators and parasites attack insects that damage plants. Some of these have been successfully introduced into the horticultural world, and others are still in the laboratory testing stage. A number of both are shown in Table 10-2. These predators and parasites may be other insects (Fig. 10-12), bacteria, fungi, or viruses.

TABLE 10-2. BIOLOGICAL SUPPRESSANTS OF INSECT PESTS

Predators	Prey	Parasites	Hosts
Insects		*Bacteria*	
Vedalia beetle (*Rodolia cardinalis*)	Cottoncushion scale (*Icerya purchasi*)	*Bacillus thuringiensis*	Cabbage looper (*Trichoplusia ni*) and other larvae of moths, butterflies
Green lacewing (*Chrysopa carnea*)	Aphids (*Aphis* spp.)		
Cryptolaemus montrouzieri	Citrus mealybug (*Planococcus citri*)	Milky spore disease (*Bacillus popillae, B. lentimorbus*)	Japanese beetle
Lady beetle (*Coleomegilla maculata*)	Aphids		
Praying mantis (*Tenodera aridifolia*)	Various insects	*Fungi*	
European ground beetle (*Calosoma sycophanta*)	Gypsy moth *Porthetria dispar*	*Beauveria* spp.	Pine sawfly
		Metarrhizium spp.	Wireworm
Invertebrates		*Insects*	
Spider (*Xystichus sp.*)	Sawfly (*Pristiphora* spp.)	*Apanteles congregatus*	Tomato hornworm (*Darapsa myron*)
		Exenterus amictorus	Pine sawfly
Nematodes		*Diplostichus lophyri*	
Neoaplectana glaseri	Japanese beetle (*Popillia japonica*)	*Ooencyrtus* spp.	Elm spanworm (*Ennomos* spp.)
N. carpocapsae	Codling moth (*Laspeyresia pomonella*)	*Chelonus annulipes*	European corn borer (*Ostrinia nubilalis*)
		Brachymeria intermedia	Gypsy moth
Vertebrates		*Encarsia formosa*	Whitefly (*Trialeurodes vaporariorum*)
Birds	Various insects	*Itoplectis conquisitor*	Numerous species of butterfly caterpillars
Toads	Various insects		
Deer mouse (*Peromyscus maniculatus*)	Gypsy moth	*Agathis pumila*	Larch casebearer (*Coleophora laricella*)
	Pine sawfly (*Diprion Similis*)	*Chrysocharis laricinellae*	
		Trioxys pallidus	Walnut aphid (*Chromoaphis juglandicola*)
		Aphytis holoxanthus	Florida red scale (*Chrysomphalus aonidum*)
		Aphelinus mali	Wooly apple aphid (*Eriosoma lanigerum*)
		Tiphia vernalis	Japanese beetle
		Macrocentrus ancylivorus	European corn borer
		Viruses	
		Nuclear polyhedrosis types	Numerous insects

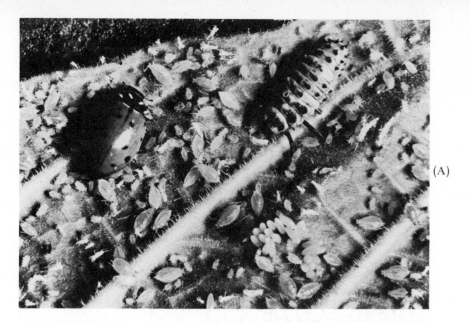

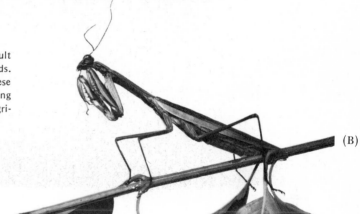

Fig. 10-12. A. Lady beetle larva and adult (*Hippodamia convergens*) feeding on aphids. B. Another beneficial insect, the Chinese mantis (*Tenodera aridifolia sinensis*), laying in wait for prey. U.S. Department of Agriculture photos.

Other controls include the use of botanical or naturally derived insecticides such as pyrethrin, rotenone, and dormant oils (2 to 7 percent). The first two are botanicals (moderately toxic) from plants and are used to control caterpillars, beetles, aphids, mites, leafhoppers, thrips, weevils, Japanese beetles, flea beetles, loopers, and cabbage worms. The last, natural petroleum oils (relatively nontoxic), are used on dormant woody plants to control insects that winter over, such as scale. Lighter-weight oils or summer oils have been used on nondormant plants to control scale, mites, mealybugs, whiteflies, aphids, pear psylla, and various insect eggs. A recent area of promising insect controls is the use of insect-growth regulators to interfere with the insects' development. One such regulator is 2-propynyl(2E,4E)-3,7,11-trimethyl-2,4-dodecadienoate (Enstar 5E®, slightly toxic). Insect pheromones (sex attractants) are used in traps to lure insects, like the spruce bark beetle (*Ips typographus*), to their death.

Soil

Fig. 10-13. Newspaper collar around tomato plant buried about 1-2 inches in soil foils cutworms.

Large insects, such as the tomato hornworm, can be picked off by hand and destroyed. Cardboard or paper collars can be put at the soil line of such plants as tomatoes, peppers, and eggplants to discourage cutworms (Fig. 10-13). Sticky bands can be put on trees to trap climbing caterpillars. Boards painted yellow and coated with SAE 90 motor oil trap whiteflies and psylla. Black-light or other traps may be used against flying insects (Fig. 10-14). Weeds should be eliminated to remove breeding areas for some insects. Crops should be rotated to disrupt feeding and breeding cycles of insects. Bags may be placed around large fruits such as apples to prevent damage (Fig. 10-15). Disease-carrying aphids are repulsed by aluminum foil (Fig. 10-16).

Fig. 10-14. A Japanese beetle trap. Bait contains geraniol, an attractant, and kerosene or oil is used in the bottom to hold and smother the beetles. Author photo.

Fig. 10-15. Waxed bags can be placed around fruits prone to heavy insect attack, such as apples. These should be placed as quickly as possible after fruit set if they are to be effective. The method has moderate to good success, but is useful only on a small scale level.

Fig. 10-16. A. Aluminum mulch around these squash reflects light and causes disorientation for the flying aphid, thus preventing aphids from landing on these plants. This minimizes aphid damage and chances of infectious diseases carried by aphids. Courtesy of National Garden Bureau, Inc. B. Netting is used here to prevent birds from eating ripening strawberries. U.S. Department of Agriculture photo.

Resistant plants should be planted wherever possible. For example, the butternut squash is more resistant to squash vine borer than summer squashes. 'Mammoth Red Rock' cabbage is less attractive to cabbage loopers than green cultivars.

OTHER ANIMAL PESTS AND NATURAL CONTROLS

Bird control must be balanced against the harmful insects they eat. Natural harmless controls include nets for small fruit trees and berries (Fig. 10-16), cheesecloth for new seedbeds, scarecrows, and noise makers (carbide cannon, intermittent siren, bird distress calls). Often bird damage is tolerated and compensated for by planting extra material.

Natural controls for rabbits, mice and other rodents, and deer include fences, screening around tree trunks, and harmless traps. Dogs and cats may be discouraged by fences or twigs placed over new seed beds. Containers of stale beer are used to attract and drown slugs or snails. Unfortunately, there is limited control for humans who damage plants through vandalism and ignorance. Education appears to be one answer.

VIRUSES AND RELATED ORGANISMS: NATURAL CONTROLS

The natural control of viruses consists mainly of controlling insects that are viral vectors, since the cure of viral-infected plants is generally not achieved. Virus-infected plants should be destroyed to prevent further contamination. Some viruses can be transmitted through seeds or asexual propagation of plants, so methods have been developed to maintain virus-free stock. The shoot apex is often virus free, and these can be propagated by aseptic methods of micropropagation (see Chapter 6) or by grafting onto disease-free rootstock seedlings. Heat treatments from $110°$ to $135°F$ ($43.5°$ to $57°C$) for 30 minutes to 4 hours or at about ($38°C$) $100°F$ for 2 to 4 weeks, depending on the plant or seed, will often free plants of viruses. A combined heat treatment and shoot apex approach is used for plants where neither method alone is satisfactory. Plants known to be virus prone, such as the strawberry, should be purchased as certified virus free. Research and breeding of viral resistant plants is in progress and should be watched for future developments. Some viral resistant plants are available (Table 10–3).

Natural control of microplasmlike and rickettsialike organisms is unlikely at present. The treatments to produce virus-free stock are often effective against these organisms. At present, limited genetic resistance is possible (Table 10–3).

BACTERIA AND FUNGI: NATURAL CONTROLS

Diseased plants should be destroyed, and diseased portions of plants removed. Garden and field sanitation should be practiced; plant residues and peripheral weeds should be removed since they may be substitute hosts for some fungal pathogens. Rotation of crops will discourage infection from some fungi. Maintain plant vigor. Avoid overfertilization, since in some cases like fire blight (*Erwinia amylovora*), this can make the plant more susceptible. Whenever possible, one should use cultivars with known resistance to prevailing fungal diseases (Table 10–3). For example, cultivars of tomatoes resistant to verticillium and fusarium wilts can be grown in areas where these diseases are a problem. Known alternate hosts for fungi that have life cycles which require two hosts should not be grown in proximity; for example, the apple-cedar rust needs both plants to complete its life cycle. Insects known to carry fungal or bacterial diseases should be controlled. Soils and propagation media should be pasteurized to reduce soil-borne bacterial and fungal diseases (see Chapter 6).

TABLE 10-3. EXAMPLES OF DISEASE-RESISTANT CULTIVARS

Crop	Cultivar[a]	Disease
Apple	'Prima,' 'Prisilla,' 'Sir Prize'	Scab
Apricot	'Harcot'	Bacterial spot, brown rot
Aster	'Madeleine'	Aster wilt
Blackberry	'Cheyenne'	Anthracnose, orange rust
Bluegrass	'Fylking'	Stripe smut, helminthosporium leaf spot, melting out, stem rust
Cabbage	'Golden Acre'	Yellows
Corn, sweet	'Bi-Queen'	Stewart's wilt, northern and southern leaf blight
Crabapple, flowering	'Adams,' 'Pink Spires,' 'Winter Gold'	Fire blight, cedarapple rust, powdery mildew, scab
Cucumber	'M and M Hybrid'	Mosaic, mildew
	'Poinsett'	Anthracnose, angular leaf spot
	'Progress No. 9'	Fusarium wilt
	'Wisconsin SMR 18'	Scab, mosaic
Geranium	'Marian'	Botrytis
Grape	'Glenora'	Phylloxera
	'Veeblanc'	Downy, powdery mildew
	'Welder'	Pierce's disease (rickettsialike organism)
Lettuce	'Florida 74'	Lettuce mosaic virus, bidens mottle virus
Pea	'Corvallis'	Enation virus, pea streak virus
	'Green Arrow'	Downy mildew
Peach	'Havis'	Bacterial spot, brown rot
Pear	'Aristocrat'	Fireblight
	'Old Home'	Pear decline (mycoplasmalike organism)
Pecan	'Schley'	Scab
Pepper	'Bell Boy Hybrid'	Tobacco mosaic virus
Raspberry, purple	'Brandywine'	Spur blight, powdery mildew, anthracnose
Rose	'Bonavista,' 'Elmira,' 'Moncton,' 'Sidney'	Powdery mildew, blackspot
Rye grass, perennial	'Pennfine'	Snowmold, helminthosporium leaf spot, melting out, stem rust
Spinach	'Melody Hybrid'	Downy mildew, mosaic
Strawberry	'Cataldo'	June yellows
	'Delite'	Verticillium wilt
	'Redchief'	Red stele
Sweet potato	'Carver'	Fusarium wilt
Tomato	'Floramerica'	Fusarium wilt, verticillium wilt, several others
Watermelon	'Sweet Favorite'	Anthracnose, fusarium wilt

[a]Some are commercially available, others will be in the future, and others may only be utilized in breeding programs.

PARASITIC OR LOWER PLANTS: NATURAL CONTROLS

Removal by hand is the natural control for dodder and mistletoe. Algae can be scrubbed away. Moss can be raked from a lawn. Since soil conditions of low fertility and/or acidity are associated with moss, these should be corrected to prevent reoccurrence.

WEEDS AND NATURAL CONTROLS

Natural controls of weeds consist of the following. The oldest form of natural weed control is hand removal. This is laborious, but works well for many annual and biennial weeds; perennial weeds and others with rootstocks of multiple shoot habit are apt to reoccur since some roots may be left after hand pulling. The use of hand hoes gives the same results, but with somewhat less work. Power cultivators achieve the same results with far less effort. However, the problems of tillage to reduce weeds, that is, increased losses of organic matter, soil compaction, and pan formation (see Chapter 7), must be considered.

Mulches (discussed in Chapter 7) are also effective weed controls, especially in gardens. Black polyethylene appears to be the only feasible mulch for large-scale weed control in commerical operations. Optimal horticultural practices to produce maximal growth in minimal time can sometimes favor crop competition over weeds. Edging strips may be used to prevent grass from spreading into flower and shrub borders (Fig. 10–17). Geese can be utilized to control weed grasses among crops that are not palatable to them. Another promising area is the utilization of insect pests specific for troublesome weeds. Soil pasteurization eliminates weed seeds in propagation media (see Chapter 6).

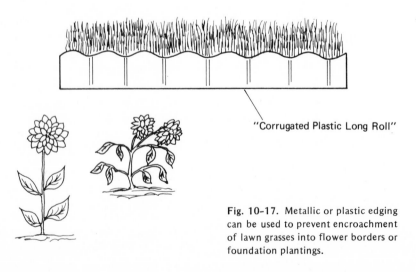

"Corrugated Plastic Long Roll"

Fig. 10-17. Metallic or plastic edging can be used to prevent encroachment of lawn grasses into flower borders or foundation plantings.

By far, the area of control most used by horticulturists is chemical control. Chemicals used to control plant pests through toxic action are termed *pesticides*. These chemicals usually kill the pest, but in some cases they may only be sickened. This toxicity is not necessarily for all life stages of the plant pest, but is often most effective at the most vulnerable stage in the life cycle of the pest. Chemicals that discourage plant pests from attacking plants because of some property that makes the chemical disagreeable to the pest are termed *repellents*.

Pesticides are usually divided into categories on the basis of which pest they control. These are insecticides, miticides (acaricides), nematicides, bactericides, fungicides, rodenticides, and herbicides. Specific examples of each will be discussed in a later section. Pesticides can also be divided into two broad categories; systemics and nonsystemics. *Systemics* are water-soluble chemicals, which can be absorbed by plant leaves and/or roots, and then translocated through the plant. The plant itself is then toxic to the attacking organism. *Nonsystemics* coat the plant and affect the organism upon contact or subsequent uptake.

Pesticides to a large degree show some degree of selectivity. For example, a miticide would be most specific for mites and would not show any fungicidal activity. Insecticides might only be useful for a small group of insects. However, certain pesticides are designed to eliminate a wide range of plant pests. These are termed *wide-spectrum* or *general-purpose* pesticides. They may be a single chemical that shows toxicity to a wide range of organisms, or they may be combinations of chemicals, such as a mixture with an insecticide, fungicide, and miticide. Unfortunately, pesticides are usually not selective enough to spare beneficial organisms.

When combining two or more pesticides to achieve multiple control, one must have information on pesticide compatibility. If they are not compatible, plant damage or reduced efficiency of one or more components may result. Pesticides should not be mixed unless such information is available.

Pesticide Application

The form in which pesticides are sold is termed the *formulation*. The formulation may be ready for direct application, or it may require dilution prior to application. The form that the formulation takes is determined by several factors: storage, handling, method of application, effectiveness, and safety. Formulations may be applied as sprays, dusts, granulars, aerosols, fumigants, or encapsulated materials.

SPRAYS

Spraying is the most prevalent mode of application. About 75 percent of all pesticides are applied as sprays. Formulations sold for use as a spray

(A)

(B)

(C)

(D)

(E)

Fig. 10–18. Various types of spray equipment. A. Hand operated pump type compression sprayer for small operations. B. Larger volume sprayer for bigger garden areas, shrubs, small trees where hand pump is utilized to build up air compression to force out spray. C. Back pack type of compression sprayer. D. and E. Larger units where compression is built up by gasoline engines. These would be used for large gardens, lawns, trees and small orchards. Courtesy of H. D. Hudson Manufacturing Company. Still larger sprayers are used by orchardists and large scale commercial horticultural growers (see text).

344

include water-miscible liquids, emulsible concentrates, wettable powders, water-soluble powders, flowable suspensions of ground pesticide in water, oil solutions, and ultralow-volume concentrates.

The bulk of the pesticides sold for use as a spray are in the form of emulsible concentrates. These are concentrated oil solutions of pesticides with added emulsifier, which allows the oil to form a milky-white emulsion when mixed with water. These emulsions are usually stable for several days and require no agitation when applied as a spray.

Wettable powders are pesticide dusts with a wetting agent added to facilitate mixing with water. The carrier in the pesticide dust is usually clay or talc, which will sink, so the solution will require agitation during spraying. Water-miscible liquids and water-soluble powders form solutions readily with water and require no agitation during application.

Pesticides not readily water or oil soluble are sold as wet dusts containing talc or clay, which can be suspended in water and sprayed with agitation. Oil solutions are diluted with oil and sprayed. Ultralow-volume concentrates are high-strength pesticides used directly for spraying without dilution to eliminate large spray volumes.

Sprays sometimes fail to give good coverage of plant leaves because of the cutin layer, which is water repellent. *Wetting agents* or *spreaders* are added to give more uniform coverage, since their properties are such that they break the surface tension of the droplets sprayed on the leaves. These are usually detergents or sulfonated alcohols. *Stickers* are chemicals added to increase the amount of pesticide that sticks, as well as the length of time that it sticks. Sprays are most effective as a thin, complete coat. Overspraying is indicated by dripping or runoff.

Spray equipment of many types is available (Fig. 10–18). Orchardists use a sprayer with a bank of nozzles, which uses air blasts from a rotating fan as the carrier for the spray. Air-driven sprayers avoid the large water volumes needed as a carrier; these sprayers are ideal for the ultralow-volume pesticide concentrates. Arborists use a high-pressure, high-volume sprayer with a fire-hose type of nozzle. Boom-type sprayers are often used on row crops. Smaller sprayers are available as tanks on wheels, backpack tanks, and hand-held tanks. Many of them rely on hand-operated force pumps for the compressed air needed to deliver the spray. The larger ones use a gasoline-driven engine to produce the compressed air. Siphon sprayers that connect to water hoses and use water pressure to dilute and deliver the spray are also available. A backflush valve should be used with these to prevent accidental entry of pesticide into the water supply.

DUSTS

Dusts are simple to formulate and easy to apply. Generally, dusts are not available as pure pesticide but contain an inert diluent such as clay or talc. However, dusts are less effective and economical than sprays, because

Fig. 10–19. A small hand operated duster. Power driven units are also used. Courtesy of H. D. Hudson Manufacturing Company.

coverage is often incomplete owing to a poor rate of deposit on plant material. Retention may not be as good as with a spray.

Power dusters use gasoline-driven fans to distribute the dust. Smaller power units are available as backpack units. Hand-operated units are also available (Fig. 10–19).

GRANULAR AND ENCAPSULATED MATERIALS

Granular pesticides are formulated as small pellets of clay, which range in size from 20 to 80 mesh. Drift problems are much less than with dusts. Systemic pesticides are available in granular form. Granulars are usually applied with a hand-operated rotary type of distributer. They can also be applied directly in the drill at planting time.

Encapsulated pesticides use tiny beads of polyvinyl or other plastics to surround the pesticide. These are slow-release forms. They are a relatively new form of pesticide.

Introduction to Applied Horticultural Science / Part II

AEROSOLS, FOGS, AND FUMIGANTS

In an aerosol bomb, which is designed for convenience, the pesticide is sprayed through the release of a compressed gaseous carrier. These are useful for small applications, such as with house plants. A fog effect is created by heat volatilization of kerosene or other oil carrier containing a pesticide. Fogs are not usually used for horticultural purposes; they are used mostly for mosquito and fly abatement. Fumigants are pesticides that consist of small volatile molecules that exist as gases above $40°F$ ($4°C$). Often they have a density greater than air. They are highly penetrating, which accounts for their use in soils and greenhouses. They are released as a fumigant by heating or, in some cases, by wetting.

Pesticide Problems

Pesticides solve or lessen the severity of many of the plant-pest problems encountered by the horticulturist. However, their use should not be indiscriminate in view of the many problems associated with pesticides.

SAFETY

Pesticides are often toxic to humans and, unless handled properly, can cause illness or death to humans who apply pesticides, come into contact with pesticides before or after application, or eat horticultural crops contaminated with pesticides. Pesticides can also cause soil and water pollution, since some have a long residual life and move through the soil and water cycles. Accordingly, some federal regulation of pesticides came about in the early 1970s. Horticulturists using pesticides must be aware of the laws concerning their use.

The provisions of the Federal Environmental Pesticide Control Act are in brief as follows. Pesticide use other than stated on the label is prohibited. Pesticides are classified as general or restricted use. Restricted pesticides can only be applied by a person certified to do so by the state. Certification is possible in several different categories. Certification is usually recognized by the issue of a license after the applicant has demonstrated proficiency in the following areas: label and labeling comprehension, safety, pesticides, pests, environment, equipment, methods of application, and laws and regulations. All violations are punishable by heavy fines and/or imprisonment. Other aspects of the law cover the manufacturer and registration procedures for pesticides.

There are a number of safety precautions to be observed when working with pesticides. When purchasing the pesticide, be certain the label is intact and the information therein is up to date. Try not to buy more than is needed for the current season. While transporting the pesticide, use containers that are intact and be alert for leaks of liquid or fumes. If pesticides

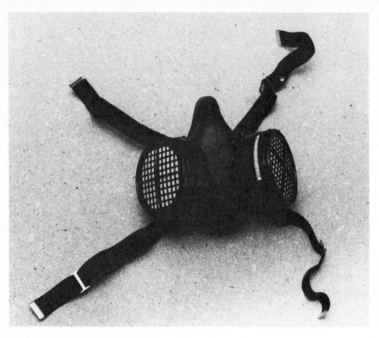

Fig. 10-20. A cartridge type horticultural respirator used with spray or dust pesticides when toxicity of pesticide warrants it. Author photo.

must be stored, be sure to store them in a secure, locked area away from foodstuffs or water supplies. The storage area should be dry, well ventilated, out of direct sunlight, and kept above freezing. The pesticide storage area should be clearly identified as such. Outdated material or damaged containers should be removed.

Directions should be carefully followed when preparing pesticides for use. Do not smoke or eat while working with pesticides. Mix them in a well-ventilated area away from winds or water supplies. If some is spilled on you, wash immediately with plenty of soap and water. Remove contaminated clothing promptly.

Wear appropriate protective clothing and equipment, such as breathing masks, when the toxicity of the pesticide warrants it (Fig. 10-20). This is especially important during application of the pesticide. Use clean, reliable equipment for application and do not use herbicide sprayers to apply insecticides and the like. Follow rates of application, and do not apply pesticides on windy days. Avoid walking in clouds of spray or dust, and keep pesticides away from the body. Try not to knowingly spray when beneficial or pollinating insects are most active; notify nearby beekeepers in advance of any spraying.

After pesticides are applied, observe the required waiting period before reentry or harvest. Be sure the area is clearly posted. All equipment should be thoroughly cleaned after use. Clothing should be changed afterward, and the operator should bathe completely with soap and water. If an accident occurs, notify a physician at once and have all the label information available in case it is needed for purposes of an antidote.

Empty containers should be rinsed and damaged to prevent reuse for any purpose. Containers and unused pesticides should be disposed of in a proper manner. Contact your state extension service or environmental protection agency to determine the recommended means of disposal.

These safety precautions should be carefully observed at all times. Avoid complacency and shortcuts. The risk of injury, death, or lawsuits is high otherwise.

APPLICATION PROBLEMS

The timing of pesticide applications is critical; otherwise, the plant pest may not be controlled or the horticultural crop will be contaminated when harvested. The pesticide should be applied before symptoms become very noticeable, that is, not when the plant pest has overrun the plant material. If the pesticide is only effective at a particular life cycle of the organism, it should be applied only then. Complete coverage is sometimes difficult because of the plant density or cutin layers. Pruning or chemical spreaders and stickers may be helpful. Rain or wind may also hinder the effectiveness of a pesticide.

PHYTOTOXICITY PROBLEMS

Sometimes a plant will show leaf damage after the utilization of a pesticide. The damage appears as marginal burn, spotting, or chlorosis. Sometimes distortion or abnormal growth is observed. These are the symptoms of *phytotoxicity*. The younger leaves and blossoms, or new growth, are most likely to be damaged. In cases like this, the cure may be worse than the original problem.

In many cases it is known that certain pesticides will damage specific plants; the pesticide is phytotoxic to that plant. The horticulturist must be aware of these cases. It is possible to minimize the phytotoxic effect of any pesticide. Apply pesticides in the cooler parts of the day, preferably early morning, which allows the foliage and pesticide droplets to dry before the temperatures go over 85°F (28°C). High temperatures increase phytotoxicity problems. The temperature of the water can also be a factor; water should be tepid, not cold or hot. If a pesticide is available as both a wettable powder and emulsifiable concentrate, the wettable powders are usually less phytotoxic.

Sometimes the carrier, such as the organic solvent in which the pesticide is dissolved, can cause phytotoxicity. Sometimes pesticides may show no evidence of phytotoxicity until they are used as a combination spray. Several other factors that influence phytotoxicity are the concentration of the pesticide, the developmental stage of the plant, and the health of the plant.

In many cases the full limits of phytotoxicity of pesticides are unknown or poorly characterized. Whenever possible, the horticulturist should test the pesticide on a few plants under the particular conditions that will prevail during subsequent spray applications. If phytotoxic symptoms appear, another pesticide should be chosen.

CONTAMINATION PROBLEMS

The toxicity of pesticides is clearly known. There are several ways of expressing toxicity, but the most common is the LD_{50}, which is the median lethal dose of toxicant (given as milligrams per kilogram of body weight) fatal to 50 percent of the test animals (usually white albino rats).

On the basis of LD_{50} values, pesticides can be listed in terms of decreasing toxicity to the spray or dust applicator. When a choice is available, select the pesticide with the highest LD_{50}. The LD_{50} values are expressed as acute oral and dermal figures. Both are important, but the dermal value is probably more useful to the horticulturist, who is more apt to contact pesticides dermally rather than orally. Another expression used is LC_{50}, which is similar, except that it is based on fatality through inhalation and is given in terms of milligrams per liter.

In the pesticide tables in this chapter, an indication of their relative toxicity will be given based on the following. Highly toxic compounds have oral LD_{50} values of 0 to 50 milligrams per kilogram, dermal LD_{50} values of 0 to 200 milligrams per kilogram, and inhalation LC_{50} values of 0 to 2000 milligrams per liter. Moderately toxic compounds have respective values of over 50 to 500, over 200 to 2000, and over 2000 to 20,000. The respective LD_{50} oral and dermal values for slightly toxic compounds are over 500 to 5000 and over 2000 to 20,000, and for relatively nontoxic compounds, over 5000 and over 20,000, respectively. The probable oral lethal doses for these categories in decreasing order of toxicity are a few drops up to 1 teaspoon, 1 teaspoon to 1 ounce, 1 ounce to 1 pint or 1 pound, and over 1 pint or 1 pound. All appropriate safety precautions must be taken. It should also be remembered that what is considered slightly toxic or relatively nontoxic may turn out to be tomorrow's carcinogen. Some pesticides are suspect even now and may be withdrawn from the market.

Based on their toxicity, some level of pesticide, usually expressed in parts per million, will be allowed in horticultural food crops. Pesticides obviously leave a residual material on the plant. The residual lifetime will vary, depending on several factors, such as chemical stability, water solubility, temperature, and microbial susceptibility. Some pesticides will break down quickly, whereas others will remain relatively unchanged for long periods of time. Given a choice, the horticulturist should chose the pesticide with the least residual lifetime.

The longer the residual lifetime of a pesticide, the more likely it is to appear in the environment as a pollutant. The residual pesticide on the plant can be passed through the food chain to humans, the residual pesticide can wash into the soil from rain, and sprays or dusts will land on the soil. Once in the soil, the pesticide may enter the water table and end up in the food chain.

CONTROL PROBLEMS

Constant dependence upon the same pesticide may lead to less effectual control, since the plant pest may develop genetic resistance. For this reason, it may be better to alternate different pesticides capable of controlling a plant pest.

Another problem is the possible disturbance of the ecosystem. Pests of minor importance may emerge after the elimination of the major pest by the use of pesticides. The pesticide may be toxic to the beneficial organisms that controlled the minor pest, and the killing of the major pest as well has removed all the competition. The horticulturist must be observant and move quickly to control new outbreaks, either by using another form of plant protection to control the new pest or by switching to a form of plant protection that controls both pests.

Pesticides and Plant Pests

This section contains several tables of pesticides. A large number of pesticides are cited for the following reasons. Availability will vary according to region, supply, and restrictions. Restrictions are imposed by the Federal Environmental Protection Agency (EPA). When the tables were prepared, the pesticides listed were not banned outright by the EPA. However, some were under consideration for possible banning or for use under restricted conditions. Others were on waiting lists to be considered similarly. Restrictions are such that some may be purchased by nonlicensed people and used according to their labels and others only by licensed persons tested by the individual states. Further restrictions on availability and licensed usage beyond those of the EPA vary by state. It is hoped that, out of the many, a few will meet the requirements dictated in the various situations nationwide. The author and/or publisher assumes no responsibility or obligation for the correctness, legality, or current EPA and state status of the listed pesticides. It is the horticulturists' responsibility to determine such status through local agencies such as the state agricultural experiment station, the state extension service, and state agencies regulating pesticide usage (such as environment protection agencies). All precautions should be observed, and the previous section on pesticide safety should be learned well. The omission or inclusion of any pesticide does not imply any rating in regard to its merits.

Fig. 10-21. Nematicides are usually volatile and applied as soil fumigants. This mechanized unit applies first a plastic mulch over the soil and then a nematicide under it. The plastic prevents atmospheric loss of nematicide. The plastic can be removed later or used as a mulch. American Society for Horticultural Science photo.

NEMATICIDES

Nematodes have a rather impermeable cuticle, so chemicals with high penetration properties are used (Fig. 10-21). There are four groups of nematicides: halogenated hydrocarbons, organophosphates, isothiocyanates, and carbamates. The most widely used are the halogenated hydrocarbons, which, because of their volatile properties, can be used as soil fumigants. Examples of halogenated hydrocarbons used as nematicides in the form of fumigants are shown in Table 10-4. Nematicides other than the halogenated hydrocarbons include the fumigants shown in Table 10-4. Many nematicides also serve other purposes, such as herbicides, insecticides, rodenticides, or fungicides. Sometimes the nematicides shown in

Table 10-4 are not used alone, but in combinations of two, such as methyl bromide plus chloropicrin or Telone plus chloropicrin. Some of the nematicides are phytotoxic, so they are applied as preplant materials.

A number of compounds of a nonfumigant nature are also used to control nematodes. These compounds are used as combined insecticide-nematicides, and as such will be covered in the later section on insecticides. Briefly, by trade name they include the contact insecticide-nematicides Dasanit®, Mocap®, Sarolex®, Nematicide VC-13®, Nemacur®, and Furadan®, and the systemic insecticide-nematicides Lannate®, and Temik®.

TABLE 10-4. NEMATICIDES

Halogenated Hydrocarbons	Toxicity[a]
Chloropicrin[b] (trichloronitromethane)	HT
D-D (1, 3-dichloropropene 1, 2-dichloropropane)	MT
Dorlone® (79% telone, 19% ethylene dibromide)	MT
Ethylene dibromide (1, 2-dibromethane)	MT
Methyl bromide (bromomethane)	HT
Telone® (1, 3-dichloropropene)	MT
Telone C® (85% Telone + 15% chloropicrin)	HT

Nonhalogenated Compounds	
Carbon bisulfide (carbon disulfide)	HT
Vapam® (sodium N-methyldithiocarbamate dihydrate)	ST
Vorlex® (20% methyl isothiocyanate, 80% mixture of dichloropropanes and dichloropropenes)	MT

[a]HT, highly toxic; MT, moderately toxic; ST, slightly toxic. See text under contamination problems, this chapter, for LD_{50} ranges.
[b]Popular trade name or common name.

MITICIDES (ACARICIDES) AND INSECTICIDES

Pesticides applied at the peak of mite or insect infestation usually eliminate beneficial parasites and predators. For the least damage to plant and beneficial organisms, chemical controls should be utilized before the peak of infestation or in an integrated pest-management program based on coordinated natural and chemical controls. Chemical controls against mites are listed in Table 10-5; those against insects include the nonsystemic insecticides (Table 10-6), the systemic insecticides (Table 10-7), and combined nonsystemic pesticides (Table 10-8). The nonsystemics kill by acting as a stomach poison, which is effective against chewing insects, and/or a contact poison, which is effective against both chewing and sucking insects. The systemics are primarily effective against sucking insects, although some of the systemics do act also as contact poisons.

TABLE 10–5. ACARICIDES

Cyclo Compounds	Toxicity[a]
Fenbutatin[b] [hexakis (β, β-dimethyl phenethyl)-distannoxane] oxide	ST
Pentac® (bis [pentachloro-2, 4-cyclopentadien-1-yl]	ST
Plictran® (tricyclohexyltin hydroxide)	ST
Propargite® [2-(*p-tert*-butylphenoxy) cyclohexyl-2-propropynyl sulfite] or Omite®	ST

Diphenyl Compounds	
Acarin® or Kelthane® [1, 1-bis(*p*-chlorophenyl)-2, 2, 2-trichloroethanol]	ST
Chlorobenzilate (ethyl-4, 4'-dichlorobenzilate)	ST
Chloropropylate (isopropyl-4, 4'-dichlorobenzilate)	ST
Tedion® or Tetradifon (*p*-chlorophenyl-2, 4, 5-trichlorophenyl sulfone)	NT

[a]ST, slightly toxic; NT, relatively nontoxic. See text, this chapter, under contamination problems for LD_{50} range.
[b]Popular trade name or common name.

TABLE 10–6. NONSYSTEMIC INSECTICIDES

Cyclo Compounds	Toxicity[a]
Endrin[b] (hexachloroepoxyoctahydro-endo, endo-dimethanonaphthalene)	HT
Hexachlor (1, 2, 3, 4, 5, 6-hexachlorocyclohexane)	MT
Lindane (γ-1, 2, 3, 4, 5, 6-hexachlorocyclohexane)	MT

Carbamates	
Carbaryl, or Sevin® (1-naphthyl methylcarbamate)	MT
Propoxur or Baygon® [2-(1-methylethoxy) phenol methylcarbamate]	MT

Diphenyl and Nonphosphate Compounds	
Methoxychlor [2, 2 bis(*p*-methoxyphenyl)-1, 1, 1-trichloroethane]	NT
Perthane® [1, 1-bis(*p*-ethylphenyl)-2, 2-dichloroethane]	NT

Inorganic Compounds	
Acid Lead Arsenate (acid or basic lead arsenate)	MT
Calcium Arsenate (tricalcium arsenate)	HT

Organic Phosphates	
Aspon® (*0, 0, 0, 0-tetra-n*-propyl dithiopyrophosphate)	ST
Chlorpyrifos [*0, 0*-diethyl-*0*-(3, 5, 6-trichloro-2-pyridyl) phosphorothioate]	MT
Fonopos (*0*-ethyl *S*-phenyl ethylphosphorodithioate)	HT
Terbufos (*S-tert*-butylthiomethyl-*0-0*-diethyl phosphorodithioate)	HT
Tetrachlorvinopos [2-chloro-1-(2, 4, 5-trichlorophenyl) vinyl dimethyl phosphate]	ST
Trichlorofon [dimethyl(2, 2, 2-trichloro-1-hydroxyethyl) phosphonate]	MT

TABLE 10–6. CONTINUED

Synthetic Analogues of Botanical Insecticides

Allethrin (*dl*-2-allyl-4-hydroxy-3-methyl-2-cyclopenten-1-one ester of *dl cistrans*-chrysanthemum monocarboxylic acid)	ST
Camphechlor (octachlorocamphene)	ST
Resmethrin [(5-benzyl-3-furyl)methyl 2, 2-dimethyl-3-(2-methylpropenyl) cyclopropanecarboxylate]	ST

[a]HT, highly toxic; MT, moderately toxic; ST, slightly toxic; NT, relatively nontoxic. See text, this chapter, under contamination problems for LD_{50} values.
[b]Popular trade name or common name.

TABLE 10–7. SYSTEMICS WITH INSECTICIDAL AND OTHER PESTICIDAL ACTIVITY

Insecticides	Toxicity[a]
Monocrotophos[b] (dimethyl phosphate of 3-hydroxy-*N*-methyl-*cis*-crotonamide)	HT
Insecticides-Acaricides	
Acephate or Orthene® (*O, S*-dimethyl acetylphosphoramidothioate)	ST
Cygon® [*O, O*-dimethyl-*S*-(*N*-methylcarbamoyl methyl) phosphorodithioate)	HT
Demeton or Systox® [*O, O*-diethyl-*0*-2-(ethylthio) ethyl phosphorothioate mixture with *0, 0*-diethyl-*S*-2-(ethylthio) ethyl phosphorothioate]	HT
Di-syston® [*O, O*-diethyl-(*S*-2-(ethylthio) ethyl) phosphorodithioate]	HT
Metasystox-R® [*S*-(2-(ethylsulfinyl) ethyl) *O,O*-dimethyl phosphorothioate]	MT
Methamidophos (*O, S*-dimethyl phosphoramidothioate)	HT
Mevinphos (2-carbomethoxy-1-methylvinyl dimethyl phosphate, alpha isomer)	HT
Insecticides-Acaricides-Nematicides	
Aldicarb or Temik® [2-methyl-2-(methylthio) propionaldehyde-*O*-(methylcarbamoyl) oxime]	HT
Oxamyl or Vydate® [methyl-*N'*, *N'*-dimethyl-*N*-((methylcarbamoyl) oxy)-1-thio-oxamimdate]	HT
Insecticides-Nematicides	
Ethoprop or Mocap® (*O*-ethyl-*S, S*-dipropyl phosphorodithioate)	MT
Furadan® (2, 3-dihydro-2, 2-dimethyl-7-benzofuranyl methylcarbamate)	HT
Methomyl or Lannate® (*S*-methyl-*N*-(methylcarbamoyl) oxy) thioacetimidate]	HT
Nemacur® [ethyl-3-methyl-4-(methyl thio) phenyl (1-methylethyl) phosphoramidate]	HT

[a]HT, highly toxic; MT, moderately toxic; ST, slightly toxic. See text, this chapter, under contamination problems for LD_{50} values.
[b]Popular trade name or common name.

TABLE 10–8. COMBINED NONSYSTEMIC INSECTICIDES

Insecticides-Acaricides	Toxicity[a]
Dialifor[b] [O, O-diethyl-S-(s-chloro-1phthalimidoethyl) phosphorodithioate]	HT
Diazinon [O-O-diethyl-O-(2-isopropyl-6-methyl-5-pyrimidinyl phosphorothioate]	MT
Dicrotophos (dimethyl phosphate ester with 3-hydroxy-N, N-dimethyl-cis-crotonamide)	HT
Dioxathion [2, 3-p-dioxanedithiol-S, S-bis (O, O-diethyl phosphorodithioate]	HT
Endosulfan (6, 7, 8, 9, 10, 10-hexachloro-1, 5, 5a, 6, 9, 9a-hexahydro-6, 9-methano-2, 4, 3-benzodioxathiepin-3-oxide)	HT
EPN (O-ethyl-O-p-nitrophenyl thionobenzenephosphonate)	HT
Ethion (O, O, O',-O'-tetraethyl-S, S'-methylene bisphosphorodithioate)	MT
Fenthion [O, O-dimethyl-O-(3-methyl-4-(methylthio) phenyl) phosphorothioate]	MT
Formetanate [N, N-dimethyl-N'-(3(((methylamino) carbonyl) oxy) phenyl) methanimidamide]	HT
Guthion® [O, O-dimethyl-S-(4-oxo-1, 2, 3-benzotriazin-3) (4 H)-ylmethyl) phosphorodithioate]	HT
Malathion (O, O-dimethyl phosphorodithioate ester of diethyl mercaptosuccinate)	ST
Methyl Parathion (O, O-dimethyl-O-p-nitrophenyl phosphorothioate)	HT
Naled (dimethyl-1, 2-dibromo-2, 2-dichloroethyl phosphate)	MT
Parathion (O, O-diethyl-O-p-nitrophenyl phosphorothioate)	HT
Phosalone [O, O-diethyl-S-((6-chloro-2-oxobenzoxazolin-3-yl) methyl) phosphorodithioate]	MT
Phosphamidon [2-chloro-2-(diethylcarbamoyl)-1-methyl vinyl)-dimethyl phosphate]	HT
Phosmet [N-(mercaptomethyl) phthalimide-S-(O, O-dimethyl phosphorodithioate]	MT
Supracide® [O, O-dimethyl phosphorodithioate, S-ester with 4-(mercaptomethyl)-2-methoxy-1, 3, 4-thiadiazolin-5-one]	HT
Tepp (tetraethyl pyrophosphate)	HT
Vapona® (dimethyl-2, 2-dichlorovinyl phosphate)	MT

Insecticides-Nematicides	
Dasanit® [O, O-diethyl-O-(4-(methylsulfinyl) phenyl) phosphorothioate]	HT
VC 13 Nemacide® (O, O-diethyl-O-2, 4-dichlorophenyl phosphorothioate)	MT

Insecticides-Acaricides-Ovicides	
Chlordimeform [N'-(4-chloro-o-tolyl)-N, N-dimethyl formamidine (monohydrochloride)]	MT
Trithion® [S-((p-chlorophenylthio) methyl) O, O-diethyl phosphorodithioate]	HT

Insecticide-Acaricide-Fungicide	
Chinomethionat (6-methyl-quinoxaline-2, 3-dithiolcyclocarbonate)	ST

[a]HT, highly toxic; MT, moderately toxic; ST, slightly toxic. See text, this chapter, under contamination problems for LD_{50} values.
[b]Popular trade name or common name.

TABLE 10-9. FUNGICIDES

Antibiotics	Toxicity[a]
Cycloheximide[b] [3(2-(3, 5-dimethyl-2-oxocyclohexyl)-2-hydroxyethyl)-glutarimide]	HT

Carbamates	
Ferbam (ferric dimethyldithiocarbamate)	ST
Maneb (manganous ethylenebisdithiocarbamate)	NT
Zineb (zinc ethylenebisdithiocarbamate)	NT
Ziram (zinc dimethyldithiocarbamate)	ST

Inorganics	
Bordeaux (copper sulfate–calcium hydroxide mixture)	ST
Caddy (cadmium chloride)	ST
C-O-C-S (copper oxychloride sulfate)	ST
Coprantol® (copper oxychloride)	ST
Cupravit Blue (copper hydroxide)	HT
Cuprocide (cuprous oxide)	MT
Lime sulfur (calcium polysulfide)	ST
Malachite (basic copper carbonate)	MT
Sulfuron (sulfur)	NT
Zinc coposil (basic cupric zinc sulfate complex)	ST

Metal Organics	
Cadminate® (cadmium succinate)	ST
Fentin Hydroxide (triphenyltin hydroxide)	MT

Miscellaneous Organics	
Anilazine [4, 6-dichloro-N-(2-chlorophenyl)-1, 3, 5-triazin-2-amine]	ST
Benomyl® or Benlate® [methyl-1-(butylcarbamoyl)-2-benzimidazolecarbamate]	NT
Botran® (2, 6-dichloro-4-nitroaniline)	ST
Captan [N-(trichloromethylthio)-4-cyclohexene-1, 2-dicarboximide]	NT
Chlorothalonil (tetrachloroisophthalonitrile)	NT
Difolatan® [cis-N-(1, 1, 2, 2-tetrachloroethylthio)-4-cyclohexene-1, 2-dicarboximide]	NT
Dodine (N-dodecylguanidine acetate)	ST
Folpet [N-(trichloromethylthio) phthalimide]	NT
Karathane® (2, 4-dinitro-6-octylphenyl crotonate and 2, 6-dinitro-4-octylphenyl crotonate)	ST
Oxycarboxin (5, 6-dihydro-2-methyl-1, 4-oxathiin-3-carboxanilide-4, 4-dioxide)	ST
Pentaphenate (sodium pentachlorophenate)	MT
Piperalin [3-(2-methylpiperidino) propyl-3, 4-dichlorobenzoate]	ST
Sanoside [hexachlorobenzene)	NT
Terrachlor® (pentachloronitrobenzene)	NT
Terrazole® (5-ethoxy-3-trichloromethyl-1, 2, 4-thiadiazole)	ST
Tersan-SP® (1, 4-dichloro-2, 5-dimethoxybenzene)	NT
Thiram (tetramethylthiuram disulfide)	ST

[a]HT, highly toxic; MT, moderately toxic; ST, slightly toxic; NT, relatively nontoxic. See text, this chapter, under contamination problems for LD_{50} values.
[b]Popular trade name or common name.

OTHER ANIMAL PESTICIDES

Chemical repellents against birds include 9, 10-anthraquinone (anthraquinon, slightly toxic) and 4-aminopyridine (Avitrol®, highly toxic). Other chemicals are used to kill birds or as birth-control agents. Chemical repellents against rabbits, other rodents, and deer include tetramethylthiuram disulfide (Thiram), also used as a fungicide and bone tar oil, (Magic Circle Repellent). These are slightly toxic. The bone tar oil repellent is also used to repel dogs and cats. Many of the chemical insecticides will control millipedes.

Molluscicides for snails and slugs include 4-(methylthio)-3,5-xylyl methylcarbamate (Methiocarb) and metaldehyde (Antimilace). These are moderately and slightly toxic, respectively.

MICROORGANISMS AND PESTICIDES

The control of viruses by pesticides is not possible at present. Indirect control is achieved by the use of pesticides (Tables 10-6 through 10-8) against insects known to be viral carriers.

Microplasmalike and rickettsialike organisms are inhibited by antibiotics such as oxytetracycline. Some measure of control has resulted from the injection and translocation of antibiotics within infected woody plants.

Bacterial diseases have been controlled by the use of antibiotic sprays such as streptomycin (Agri-Strep, Agrimycin 17, Phytomycin). This is relatively nontoxic. Chemicals can be used to prevent fungal infections and to fight those already present. Examples of fungicides used on plants and seeds are shown in Table 10-9 on page 357.

ALGAECIDES

Algae can be eliminated through chemical control, such as with copper sulfate penthahydrate and triphenyltin acetate.

HERBICIDES

Chemicals used to kill weeds are termed herbicides. These may be divided into several groups based on their functions. Some herbicides are *nonselective;* that is, they destroy any and all plants they touch. These may be useful for treating propagation media or for clearing areas of unwanted vegetation, such as roadsides, railroad right-of-ways, sidewalks, driveways, or areas about to be utilized for horticultural purposes for the first time. Others may be *selective* and kill some plants (weeds), but not

(A)

Fig. 10-22. A. This machine is preparing soil for planting and applying a preplanting treatment of herbicide during the same operation. U.S. Department of Agriculture/Soil Conservation Service photo by Jerry D. Schwien. B. Herbicide is being applied here as a postemergence treatment. U.S. Department of Agriculture photo by Murray Lemmon.

(B)

others (horticultural crops). Selective herbicides would be useful for weed control in lawns, orchards, or monocrops such as corn. However, selective herbicides are not as effective for mixed horticultural crops, because the various crops will show different susceptibility to being damaged. Selectivity may be brought about by physical means, such as avoiding the spraying of crops, or by physiological means, that is, by differences in contact or uptake of the herbicide. These may be possible if crops have heavier cuticles, less exposed growing points, different metabolism, less surface area, or deeper roots that limit herbicide effectiveness.

Some herbicides destroy weeds only on areas of *contact;* others destroy even untouched areas since they are absorbed and *translocated* throughout the plant. Herbicides (Fig. 10–22) can be applied prior to planting crops (*preplanting treatment*), after crops are planted but before they emerge (*preemergence treatment*), or directly to growing crops (*postemergence treatment*). Whatever treatment is used, care must be exercised to avoid drift of herbicide to susceptible plants. Separate sprayers must be used for herbicides to avoid accidental destruction of plant material. Some herbicides in horticultural use are shown in Table 10–10.

TABLE 10–10. HERBICIDES

Benzoic and Phthalic Compounds	Toxicity[a]
Alanap®[b] (*N*-1-naphthylphthalamic acid)	ST
Chloramben[b] (3-amino-2, 5-dichlorobenzoic acid)	ST
Dacthal® (dimethyl tetrachloroterephthalate)	ST
Dicamba (3, 6-dichloro-*O*-anisic acid)	ST
Dichlobenil® (2, 6-dichlorobenzonitrile)	ST
2, 3, 6-TBA (2, 3, 6-trichlorobenzoic acid)	ST

Carbamates	
Chloropropham (isopropyl-3-chlorophenylcarbamate)	ST
CDEC (2-chlorallyl diethyldithiocarbamate)	ST
Di Allate (*S*-2, 3-dichloroallyl diisopropyl thiocarbamate)	MT
Eptam® (*S*-ethyl dipropylthiocarbamate)	ST
Pebulate (*S*-propyl butylethylthiocarbamate)	ST
Propham (isopropyl-*N*-phenylcarbamate)	NT

Dinitro Anilines, Amides, and Anilides	
Bensulide [*S*-(*O, O*-diisopropyl phosphorodithioate ester of *N*-(2-mercaptoethyl) benzenesulfonamide]	ST
Butralin [4-(1, 1-dimethylethyl)-*N*-(1-methylpropyl)-2, 6-dinitrobenzenamine]	ST
Dinitramine (N^3, N^3-diethyl-2, 4-dinitro-6-trifluromethyl-1, 3-phenylenediamine)	ST
Diphenamid (*N, N*-dimethyl-2, 2-diphenylacetamide)	ST
Napropamide [2-(α-naphthoxy)-*N, N*-diethyl propionamide]	ST
Prufluralin [*N*(cyclopropylmethyl)-α, α, α-trifluro-2-dinitro-*N*-propyl-*p*-toluidine]	NT
Randox® (*N, N*-diallyl-2-chloroacetamide)	ST
Trifluralin® (α, α, α-trifluoro-2, 6-dinitro-*N, N*-dipropyl-*p*-toluidine)	ST

TABLE10–10. CONTINUED

Heterocyclic Nitrogen Derivatives

Ametryne [2-(ethylamino)-4-(isopropylamino)-6-(methylthio)-S-triazine)	ST
Aminotriazole (3-amino-1, 2, 4-triazole)	ST
Atrazine (2-chloro-4-ethylamino-6-isopropylamino-s-triazine)	ST
Diquat [6, 7-dihydrodipyridol (1, 2-a:2′, 1′-c pyrazidinium dibromide]	MT
Paraquat (1 : 1′-dimethyl-4, 4′-bipyridylium dichloride)	MT
Prometon (2-methoxy-4, 6-bis (isopropylamino)-S-triazine)	ST
Simazine [2-chloro-4, 6-bis (ethylamino)-S-triazine]	ST

Metal Organics and Inorganics

Ammate® (ammonium sulfamate)	ST
Atlacide (sodium chlorate)	ST
Borascu (sodium borate)	ST
Cacodylic Acid (dimethylarsenic acid)	ST
Cama (calcium acid methylarsenate)	ST
DSMA (disodium methanearsonate)	ST
MSMA (monosodium acid methanearsonate)	ST

Phenoxy Compounds

Agritox (2-methyl-4-chlorophenoxyacetic acid)	ST
Bifenox [methyl-5-(2, 4-dichlorophenoxy)-2-nitro benzoate]	NT
2, 4-D (2, 4-dichlorophenoxyacetic acid)	MT
MCPB [4-(2-methyl-4-chlorophenoxy) butyric acid]	ST

Urea Compounds

Bromacil (5-bromo-3-sec-butyl-6-methyluracil)	NT
Chloroxuron® [3-(p (p-chlorophenoxy) phenyl)-1, 1-dimethylurea]	ST
Diuron [3-(3, 4-dichlorophenyl)-1, 1-dimethylurea]	ST
Linuron® [3-(3, 4-dichlorophenyl)-1-methoxy-1-methylurea]	ST
Siduron® [1-(2-methylcyclohexyl)-3-phenylurea]	NT
Terbacil (3-tert-butyl-5-chloro-6-methyluracil)	ST

Miscellaneous Compounds

Acrolein (2-propenal)	HT
Dalapon® (2, 2-dichloropropionic acid)	NT
Dinoseb® (2-sec-butyl-4, 6-dinitrophenol)	MT
Glyphosate [N-(phosphonomethyl) glycine]	ST
Nitrofen® (2, 4-dichlorophenyl-p-nitrophenyl ether)	ST
Petroleum oils (weed oils)	NT
Pronamide [3, 5-dichloro-N-(1, 1-dimethyl-2-propyl)-benzamide]	NT

[a]HT, highly toxic; MT, moderately toxic; ST, slightly toxic; NT, relatively nontoxic. See text, this chapter, under contamination problems for LD_{50} values.
[b]Popular trade name or common name.

FURTHER READING

Anderson, W. P., *Weed Science: Principles.* St. Paul, Minn.: West Publishing Co., 1977.

Andus, L. J., ed., *Herbicides: Physiology, Biochemistry, Ecology,* Vols. 1 and 2. New York: Academic Press, Inc., 1976.

Baker, K. F., and J. R. Cook, *Biological Control of Plant Pathogens.* San Francisco: W. H. Freeman & Company Publishers, 1974.

Coppel, H. C., and J. W. Mertins, *Biological Insect Pest Suppression.* New York: Springer-Verlag New York, Inc., 1977.

Crafts, A. S., *Modern Weed Control.* Berkeley: University of California Press, 1975.

Johnson, W. T., and H. H. Lyon, *Insects That Feed on Trees and Shrubs.* Ithaca, N.Y.: Comstock Publishing Associates, 1976.

Kado, C. I., and H. O. Agrawal, *Principles and Techniques in Plant Virology.* New York: Van Nostrand Reinhold Co., 1972.

Klingman, R. L., and F. M. Ashton, *Weed Science: Principles and Practices.* New York: John Wiley & Sons, Inc., 1975.

Metcalf, R. L., and W. Luckmann, ed., *Introduction to Insect Pest Management.* New York: John Wiley & Sons, Inc., 1975.

Pfadt, R. E., *Fundamentals of Applied Entomology.* New York: Macmillan, Inc., 1971.

Pirone, P. P., *Diseases and Pests of Ornamental Plants* (5th ed.). New York: John Wiley & Sons, Inc., 1978.

Roberts, D. A., and C. W. Boothroyd, *Fundamentals of Plant Pathology.* San Francisco: W. H. Freeman & Company Publishers, 1972.

Scher, H. B., ed., *Controlled Release Pesticides.* Washington, D.C.: American Chemical Society, 1977.

Smith, K. M., *Plant Viruses.* New York: Halsted Press, 1976.

Thompson, W. T., *Agricultural Chemicals, Book I Insecticides; Book II Herbicides; Book III Fumigants, Growth Regulators, Repellents, and Rodenticides; Book IV Fungicides.* Fresno, Calif.: Thomson Publications, 1976/1977 revisions (Frequent revisions, working publications important to commercial horticulturists).

U.S. Department of Agriculture, *Biological Agents for Pest Control: Status and Prospects.* Washington, D.C.: Superintendent of Documents, 1978.

Ware, G. W., *Pesticides, An Auto-Tutorial Approach.* San Francisco: W. H. Freeman & Company Publishers, 1975.

Welch, S. M., *The Design of Biological Monitoring Systems for Pest Control.* New York: Halsted Press, 1979.

Westcott, C., *The Gardener's Bug Book* (rev. ed.). Garden City, N.Y.: Doubleday & Co., Inc., 1979.

　, *Plant Disease Handbook* (3rd ed.). New York: Van Nostrand Reinhold Co., 1971.

Westwood, M. N., *Viruses in Fruit Crops. HortScience,* 12, no. 5 (1977). 463.

Yepsen, R. B., *Organic Plant Protection,* Emmaus, Pa: Rodale Press, Inc., 1976.

Horticultural Crop Improvement

Horticultural productivity and quality has been improved ever since the cultivation of horticultural crops by humans began. The largest part of this improvement was a result of selection, or the propagation of plants that showed some improvement over others. With the advent of plant genetics, a scientific basis for plant improvement was established (see Chapter 4). The resulting technology allowed the plant breeder to manipulate plants such that the production of plants with more desirable features, of which there was now some choice, could be accelerated, rather than depending on the random appearance of different individuals with features that may or may not have been desirable.

Improvements in horticultural crops are not necessarily confined to the production of higher yielding vegetables and fruits. They may extend to include uniform ripening for purposes of mechanical harvesting, increased vitamin and/or protein content, better taste, improved keeping and processing qualities, or resistance to insects and disease. With ornamentals it may be desired to obtain esthetic improvements such as larger or double flowers, improved form, or even improved foliage appearance. Dwarfing may be desired to minimize pruning and space requirements of ornamentals and fruit trees. Lawn grasses may be improved in tolerance to shade and abuse. Resistance to insects and diseases would be a desirable trait for all horticultural crops, since it minimizes pesticide needs. Other desired characteristics might be increased tolerance toward heat, cold, or drought, more efficient absorption of soil minerals, insensitivity to photoperiod, more adaptability to mechanical harvesting, or a shorter period to the production of flowers, fruits, or vegetables. Many aspects of horticultural productivity and quality can be realized through plant improvement directed by the plant breeder. Those aspects resulting in increased marketability receive more attention for obvious reasons.

Sources of Gene Pools for Plant Improvement

Often a wide diversity of genotypes exists for many cultivated crops. Some are well known, and these are usually presently or recently in cultivation. Others have been neglected or fallen out of favor, and still others may only remain in isolated areas of the world. This is especially true for the primitive ancestors of our cultivated crops, which were discarded as more favored genotypes were produced by natural selection.

The more diverse the genetic background, the more scope the plant breeder has to work with. The more primitive ancestors or closely related, but uncultivated, species may possess genes for qualities the plant breeder wishes to incorporate into today's cultivated plants. Accordingly, the plant breeder must assemble a collection of these plants. Some are more readily obtainable than others. For example, the lineage of the strawberry and the garden dahlia are reasonably well known, because of the economic value of the strawberry and the enthusiasm of the dahlia specialist and collector. Other horticultural crops, especially those of limited economic value, have lineages that are known to a very limited degree. Most horticultural crops of economic value usually have existing collections of genotypes maintained by governments in developed nations. Collections of genotypes for vegetables may especially increase in the future, since the National Academy of Sciences (USA) has pointed out that many vegetables run the risk of genetic vulnerability toward plant pest epidemics, because of limited diversity in the gene pool of cultivars widely grown.

Often the New Crops Research Branch of the Agricultural Research Service and its Regional Plant Introduction Stations (U.S. Department of Agriculture) are useful places to track down existing genotypes. Specialized plant societies, particularly for ornamental plants, are another source. Sometimes historical societies are involved in genotype preservation. For example, through the creation of a historical orchard the Sturbridge Village group in Massachusetts has preserved many of the old apple cultivars no longer grown. Commercial seed companies, botanical gardens, arboreta, and plant research centers are also sources. Sometimes it becomes necessary to go to the geographic center of origin to locate normally unobtainable genotypes. These centers of origin are often a source of a rich gene pool. For example, recent plant expeditions to New Guinea have uncovered about 300 species of rhododendrons, which should provide a rich gene pool for those plant breeders interested in hybridizing the rhododendron.

Natural and Artificial Occurrences of Plant Variability

Mutations of plants can occur by chance or be artificially induced to produce an altered plant (sport) of horticultural value. Sports arise from a change on the chromosome or gene level. The nucleotide sequence of the gene may be rearranged, a portion of a chromosome may be relocated within itself or to another nonhomologous chromosome, a chromosomal portion may be lost or duplicated, or entire chromosome sets may be duplicated. Spontaneous mutations can also occur in tissues that do not give rise to sex cells (somatic tissue). These somatic mutations, or *chimeras,* are sometimes stable enough to be propagated asexually. Both the sport and chimera have been propagated asexually as sources of improved plants. Examples of sports are the pink-fleshed grapefruit and the seedless 'Washington Navel' orange. Examples of chimeras are the thornless blackberry and many variegated plants. Occasionally, a natural hybrid between two species or even two different genera may arise as a result of chance cross-pollination by insects or wind. Even if the hybrid is self sterile, it may backcross with one or both parents, thus still introducing new genetic variability.

The horticulturist is ever alert for these natural occurrences of variability and selects altered forms that have horticultural value for purposes of propagation. However, the plant breeder has no control over these accidental events. Increases in hereditary variation can also be artificially induced through the use of radiation such as X-rays or chemical mutagens like colchicine. These procedures are useful for developing new characteristics in horticultural crops.

Often, desired characteristics can be found scattered throughout varieties, cultivars, and species. It may be possible to cross these plants to produce offspring with desirable characteristics from both parents. This can occur naturally through pollination by insects, but work by the plant breeder under controlled conditions produces the bulk of today's acceptable hybrids.

Techniques of Plant Breeders

HYBRIDIZATION

Controlled hand pollination is a basic technique used by plant breeders (Fig. 11-1). The technique depends on flower structure and pollination characteristics of the different species, but it is possible to state some general principles.

Contamination by unwanted pollen must be minimized. Alcohol can be used to wash the hands and instruments prior to collecting and trans-

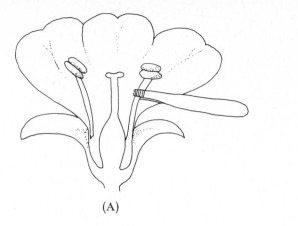

(A)   (B)

Fig. 11-1. Steps in hand pollination. A. Removal of stamens from a flower.
These will serve as a pollen source, or if flower to receive pollen is self fertile,
stamens are removed (emasculation) to prevent self-pollination. B. A piece of
soda straw is used to protect a stigma which has not become receptive yet. C.
Pollen is transferred to a receptive stigma via a soft camel hair brush. D. Pollinated flower is bagged to eliminate additional pollination and properly labeled.

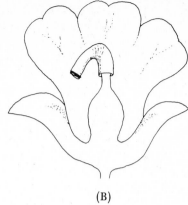

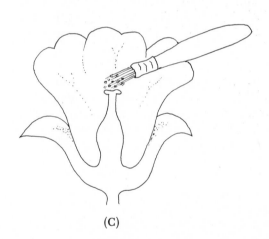

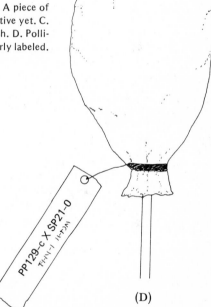

(C)   (D)

ferring pollen. Anthers should be taken from flower buds just prior to
bloom and before they open or *dehisce* or, if they are dehisced anthers,
only from caged plants or bagged flowers. These precautions will minimize
unwanted contamination by foreign pollen.

Pollen is taken from removed anthers. This can be done indoors to
prevent contamination by foreign pollen. Anthers can be separated from
about-to-open flower buds with a gentle pull by forceps, by pressing the
flower between the fingers, or by scraping the bud across a wire screen.
These anthers are dried in a warm room on paper until they can be seen to
dehisce under a magnifying glass. The pollen can be sieved to remove

impurities. Alternately, pollen may be collected when ripe on a pocket knife blade directly from flowers that were caged or bagged to prevent contamination.

If the seed parent is not ready to receive the pollen because of different blooming times or environmental effects, it is sometimes necessary to synchronize the flowering of the seed and pollen parents or to store the pollen. This is relatively easy, but success depends upon the plant groups involved, since two types of pollen exist. One has two nuclei (*binucleate*) and the other three (*trinucleate*). These two pollen types require different storage conditions. The binucleate type stores well at 10 to 50 percent relative humidity and 0°C (32°F) or lower; the trinucleate type is better stored at higher relative humidity and temperature just above freezing. Binucleate pollen appears to be common for plant families containing ornamental plants. Trinucleate pollen has been reported in the Cactaceae, Compositae, Geraniaceae, and Caryophyllaceae. Some families, such as Campanulaceae and Labiatae, have both kinds.

If the seed parent is self-fertile, the stamens should be removed with a fine-pointed forceps or scissors just before the flower opens (usually late bud stage) to prevent self-pollination. This process is called *emasculation*. It may be necessary to remove some flower parts (petals, sepals) in order to remove the stamens; care must be taken that only parts known not to affect the pistil are removed. For example, the pistil may dry upon removal of the calyx in some species. The stigma will need protection at this time if it has not become receptive (usually fully expanded and often sticky). This will prevent accidental foreign pollination, since pollen can remain alive for a while after landing on the stigma. The stigma can be protected in several ways. The petals, if large enough, can be tied together or the stigma can be covered with aluminum foil, a length of soda straw with the tip bent over, or half of a gelatin capsule. If the flower is self-sterile, it need not be emasculated, but only enclosed to prevent insect or wind pollination. Whole plants can be caged with screens, or large numbers of plants can be protected in insect-proof greenhouses.

Once the stigma is receptive, the pollen may be applied with a camel hair brush, the fingertip, the anther itself, a pencil eraser, a pocket knife, or a glass rod. Care must be used not to damage the stigma. If the two parents differ greatly in the length of the style, the long-styled parent should be the pollen source, since pollen from short-styled plants is sometimes not able to grow down a long style. Earliest possible pollination will also give the pollen tube maximal time to grow down the style. Then the seed parent flower should be immediately enclosed in a paper or closely woven cloth bag or caged to eliminate the chance of further pollination by insects or wind. The flower or flower branch should also be labeled with seed parent and pollen parent identification, along with the date. Bags may have to be removed with some species prior to complete seed ripening to prevent damage by excessive moisture. Records should be kept and observations recorded.

ARTIFICIAL MUTAGENS

Polyploids occur naturally (see Chapter 4), and some can be produced artificially. Some plants having polyploid forms are raspberry, apple, pear, cherry, plum, strawberry, cabbage, potato, aster, daylily, rose, chrysanthemum, delphinium, iris, narcissus, and marigold. The chemical mutagen, colchicine, is used to double the chromosome number. Colchicine is an alkaloid derived from the autumn crocus (*Colchicum autumnale*). With colchicine treatments, some of the plants derived from diploids are tetraploids. Occasionally, a polyploid will be produced with a more than double the number of chromosomes, such as an octaploid from a colchicine-treated diploid. Technically, these various polyploids (triploids, tetraploids, and so on) are termed euploids (see Chapter 4).

The technique is simple, but often large numbers of seeds or plants must be treated to ensure production of a desirable polyploid. Seeds are germinated on water-moistened filter paper in a closed petri dish. When the cotyledons expand, the seedlings are placed upside down in another petri dish containing an aqueous colchicine solution. This is done to avoid colchicine treatment of the roots. Since colchicine is a poison, it should be handled with care. The concentration of colchicine varies from as low as 0.01 percent to as high as 1.0 percent. The concentration varies based upon the nature and maturity of the seedlings, as well as the length of exposure. A good starting point is 24 hours with 0.5 percent colchicine solution. The colchicine-treated seedlings should be placed under fluorescent lights for around 24 hours and then transferred to a propagation media.

Larger seedlings or plants are treated by applying the colchicine solution directly to the growing tip. A wetting agent may be necessary if the growing tip has young leaves with a very waxy cuticle. The length of treatment varies. With some plants it may be necessary to keep the growing tip in contact with colchicine solution for several days. Starting with plants rather than seeds offers the advantage of knowing the starting genotype, whereas seeds can sometimes produce variable genotypes.

About 5 percent of the surviving plants will exhibit evidence of tetraploidy. This includes thicker, broader leaves and flower parts. Other signs are slower growth or larger cell, pollen, and stomate size; however, an actual count of the chromosomes is required to verify tetraploidy. If no tetraploids result, it may be necessary to increase the colchicine concentration and/or exposure time. Plants distorted in appearance are sometimes produced. These usually have an unbalanced chromosomal number (aneuploidy; see Chapter 4) and are discarded.

Flowers, fruits, leaves, and stems of some tetraploids, being larger, are often more attractive or desirable. The production of tetraploids by colchicine is also useful for other reasons, such as overcoming hybrid sterility or for crossing with diploids to produce triploids. From Chapter 4 we know that euploids can be either autopolyploids (direct increase of entire genomes within a species, ABC → ABCABC) or allopolyploids (multiples of genomes derived from two different species, ABCA′B′C′→. ABCABCA′B′C′A′B′C′). Autopolyploids are usually sterile, but allopolyploids are fertile. If a hybrid is sterile from a lack of chromosomal pairing at meiosis, the production of a tetraploid from the sterile hybrid will produce a fertile allopolyploid, which is a double diploid hybrid (amphidiploid). This is possible because each chromosome will have an identical pairing partner in meiosis. These fertile tetraploids also often breed true from seed. If the tetraploid and diploid of the same species are crossed, the resulting autopolyploid is a triploid that is sterile. This is only possible with a few plants (banana, citrus, watermelon, marigold), since most tetraploid-diploid crosses abort due to endosperm failure. Triploids are often valued for their increased vigor or flower and fruit size. Their sterility often means an extended blooming period, since there is no seed production to terminate flowering (Fig. 11–2). Other triploids produce seedless fruits such as watermelon and citrus, which are valued for food purposes (Fig. 11–3).

X-rays, gamma rays, and chemical mutagens other than colchicine have been used by plant breeders as a means of inducing mutations. Fewer useful genetic traits have been produced with these techniques in comparison to the wide genetic variability available naturally in the higher plants plus those produced with colchicine. For plants the X-ray dosage is usually from 35 to 400 roentgen units. Neutrons and alpha particles have been used also. These mutagens have been used with limited success, and most have been used with agricultural and not horticultural crops.

Fig. 11–2. Triploid marigolds have an extended blooming period compared to normal diploid marigolds. This results from their sterility and resultant failure to produce seeds. Marigold shown is triploid 'Orange Nugget,' courtesy of Burpee Seeds.

Fig. 11–3. A triploid watermelon, 'Triple Sweet.' Because of sterility, seeds do not complete their development. They remain soft and hence edible. Courtesy of Burpee Seeds.

HYBRID VIGOR

Heterosis or hybrid vigor is an important tool for the plant breeder, who can use it to restore the vigor lost through continual inbreeding (self-pollination) of a line of plants (inbred line). The original line was heterozygous and normally cross-pollinated. Continued inbreeding produces homozygosity (genetic consistency) for desired qualities, but at the expense of vigor. If two inbred lines are cross-pollinated, the F_1 generation often (not always) can show an increase in vigor (hybrid vigor), yet still retain the desirable characteristics existing in the two parental inbred lines. The effects of hybrid vigor may be expressed as increased growth rates, increased size, increased yield, earlier germination, uniformity, and resistance to unfavorable environmental factors.

The most popular use of hybrid vigor by plant breeders is the single cross (the crossing of two inbred lines). The resulting seed, once known to produce an F_1 generation with hybrid vigor and a desirable phenotype, can be marketed depending on the ease of emasculation and pollination (see controlled pollination systems, this chapter). Pollination control is needed to ensure hybrid seed unmixed with inbred line seed.

Hybrid vigor has been most successful with sweet corn. It has also been found and utilized to varying degrees in many other horticultural crops, such as Brussels sprouts, cabbage, carrot, eggplant, onion, pepper, tomato, ageratum, marigold, petunia, snapdragon, and zinnia. Hybrid vigor is being increasingly used in the commercial production of horticultural crop seed.

BACKCROSSING

Backcrossing is a technique used by the breeder to accumulate desirable genes in a rapid manner. A wild species may possess a characteristic that is highly desirable (nonrecurrent parent), such as insect or disease resistance, but have many characteristics that are horticulturally undesirable. It is crossed with a cultivated species possessing many desirable horticultural characteristics (recurrent parent), but which is highly susceptible to insects or disease. The resulting F_1 hybrid is selected for the desirable characteristic and then is crossed back with the recurrent parent, and the best improved F_1 hybrid of this cross is in turn crossed with the recurrent parent. This process is repeated until the desirable characteristic is bred in, and the many desirable characteristics of the recurrent parent are retained.

If the characteristic that is desirable is recessive, it will not show in the hybrid. The breeder must resort to self-pollinations to select the plant(s) in the next generation that possess the desired characteristic; these plants can then be used for the backcross. Obviously, the technique of backcrossing, when dealing with a desirable gene that is recessive, will easily work with plants that can be self-pollinated, but can still be used with cross-pollinating plants if they are bagged or isolated some other way and then artificially self-pollinated.

Regardless of whether the desired characteristic is dominant or recessive, the backcross procedure works well with cross- or self-pollinating plants. The desired plant could be obtained strictly through selecting during a selfing program after hybridization, but the use of the backcross produces the desired plant in fewer generations. For best results the desired character should be controlled by a single or few genes and be easily inherited. Best results have been observed with increasing insect and disease resistance or height changes.

Selection

The testing process leading to the final selection is time-consuming and a laborious part of the plant breeder's task.

Since the creation of new cultigens by the plant breeder and the adaptation of existing plants to new areas or, needs are the plant breeder's primary effort, much of the breeder's success rests with proper selection. This follows from the fact that any adaptation stems from changes in the genotype, which are recognized through selection of those phenotypes showing the desired new characteristics.

NATURAL VERSUS ARTIFICIAL SELECTION

Artificial selection differs from natural selection in several aspects. Artificial selection is carried out under the direction of the breeder, unlike unsupervised natural selection in the wild. For example, in the artificial selection of insect-resistant plants, the breeder may grow plants under natural infestation in a field or in a temperature-controlled greenhouse with the specific insect introduced on the plants. Artificial selection is faster, favorable for genotypes that would not survive in nature, such as ornamental triploids, and more specific for desired alterations. Natural selection is slower and favors those altered genotypes with enhanced chances of survival. These may or may not be desirable for human purposes. Artificial selection occurs with controlled crossing of a relatively small number of selected plants, in contrast to the uncontrolled random crosses among relatively large numbers of chance plants.

METHODS OF SELECTION

The success of the plant breeder in achieving genetic gains through selection ultimately rests on the ability to choose desirable genotypes, as determined by the ability to select those phenotypes showing alterations in characteristic(s). This involves perception and assessment of relative merits. A thorough knowledge of the morphological and physiological characteristics of the crop is required. Often, selection becomes difficult because the phenotype may be greatly influenced by environmental factors. Some view selection as an art in which the breeder with intuition does best.

Individual plants can be selected on the basis of individual merit; this is individual or single plant selection. If the seeds of several or more individual selections are pooled and subsequently bred, it is termed mass selection.

SELECTION IN SELF-POLLINATING PLANTS

Single plant selection with self-pollinating plants is essentially the selection of a large number of superior phenotypes from a genetically variable population of a horticultural crop. These are each self-pollinated, and the resulting seeds of each plant are propagated. These progeny, often in different environments, are selected for best lines, and their selfed-progeny in turn are selected, until no observable differences are found in the remaining lines. The remaining lines are compared then with established cultivars for two or more seasons until a single, pure line is retained. This type of selection is also called pure-line selection.

Characters determined by one or a few genes, such as disease resistance, are more easily selected in the preceding manner than are characters inherited quantitatively (see Chapter 4). Characters selected as described become fixed through continual selfing in homozygous progeny, that is, a pure line.

Mass selection with self-pollinated crops consists of culling or roguing out undesirable phenotypes. Care must be exercised not to overselect off-types; otherwise, the number of lines is reduced and the genetic base is narrowed. The resulting mass-selected variety consists of several closely related lines (not one pure line), which imparts maximal adaptability to the crop in terms of environmental variability. Mass selection can be used to retain the best characteristics of a crop in terms of adaptability, but in less time than with pure-line selection. Pure-line selection can be used to further improve mass-selected lines.

In established self-pollinating crops, individual and mass selection for variability has been utilized nearly to its fullest. Plant breeders use mass selection now mostly for the propagation of established cultivars (both old and new) as a means of preserving identity and variability.

The most widely used selection technique by plant breeders for self-pollinated crops is *pedigree selection* after hybridization. After hybridization of two parents having characteristics the breeder desires to see in one plant, the F_1 generation is examined, and obvious offtypes are removed. In the F_2 generation the best individuals are selected. In the F_3 generation the progeny (*family*) derived from the individual F_2 selections are examined, and the best individuals are selected. This is continued through the F_4 and F_5 generations.

In the F_6 generation the selection changes from individual to family. With *family selection* the plant breeder judges the phenotypes of the individual and its siblings collectively. Family selection involves the choice or rejection of entire families with one or both parents in common, since individual selection within the family is no longer possible due to homozygosity. If they have both parents in common, they are full sibs and have at least 50 percent of their segregating genes in common. If the sibs have one parent in common, they are half-sibs with at least 25 percent of their segregating genes in common. From this we may conclude that the more closely related two individuals are, the more reliably we may consider the evaluation of one to be representative of the other. For this reason, family selection is quite useful with inbreeding, in which one expects a high genetic relationship.

In the F_7 generation the best families (now from replicates) are selected. At this point, variability is greatly reduced, and we use the term line or selection to denote reduced variability. The best lines are chosen and sown in replicates through the F_{10} generation. In the F_{11} and F_{12} generations, strip tests and seed increase of the final chosen line are undertaken. This entire procedure of pedigree selection may take 10 to 16 years.

Another type of selection for handling hybrids is *bulk selection*. Instead of retention of individuals in the F_2 generation, undesirable types are rogued out and the entire seed crop is retained. This process is continued, and individual selection is not attempted until the F_5 generation at the earliest or as late as the F_8 generation. In addition to roguing out, natural selection will help somewhat to reduce inferior types. Growing the generations in different environments will help to ensure that survivors adapted to that environment become predominant. After the individual selections are made, they are essentially treated as the lines were in pedigree selection. Bulk and pedigree selection are also used in combination.

In summary, the pedigree technique permits early removal of offtypes, allows evaluation over several years and hence different conditions, and achieves rapid homozygosity. Bulk selection is usually the choice when extensive numbers of plants from complex or several crosses are involved, and natural and/or artificial selection is such that large numbers of offtypes can be quickly eliminated.

SELECTION WITH CROSS-POLLINATING PLANTS

Single plant selection with cross-pollinated plants is possible, but to a more limited extent than with self-pollinated plants. Continued inbreeding leads to elimination of undesirable recessive genes and genetic consistency, but also a loss of vigor, fertility, and productivity. Inbreeding and selection of the best phenotypes is practiced over fewer generations, and then crossing between phenotypes may be necessary to introduce hybrid vigor. Single plant selection opportunities are essentially nil when the plant in question is self-incompatible.

Mass selection with cross-pollinated plants consists of selecting desirable phenotypes and pooling their seed. It is most effective for easily observed or measured characteristics; its greatest contribution is in the development of cultivars for special purposes or adaption of cultivars to new areas. With established crops, its effectiveness decreases.

Progeny selection can be used to choose between apparent improvements in the phenotype that are environmentally and not genetically induced. Seeds from individual plants selected from mass selection are not pooled, but evaluated under different environmental conditions. Plants are then selected on the basis of performance, and each may become a progeny line. The pooling of several progeny lines is termed *line breeding*.

Recurrent selection is widely used with cross-pollinated plants. In its basic form it consists of selecting superior phenotypes from a cross-pollinated population. These are then self-pollinated, and the following generations derived from these selfed selections are intercrossed in all possible combinations. Further selection and intercrossing follow.

A more complex method is as follows. From two separate, genetically distinct populations, selections with desirable characteristics are made. These two selected groups are self-pollinated and also cross-pollinated with a random sample from the original opposing population (not from the population selected from). The selfed seed is kept in reserve. The crossed progeny from the two sources are evaluated in replicated field plots. The parentage of superior phenotypes among the crossed progeny is noted. From the selfed seed reserve, the parentage that has demonstrated superiority, based on the field trials on crossed progeny, is chosen. These are sown in separate blocks and each intercrossed among themselves. The seeds from each group are sown to produce two groups of plants, which may be cross-pollinated to produce a final seed source, or the whole cycle may be repeated with these seeds.

SELECTION AND QUANTITATIVE INHERITANCE

Selection as outlined previously and in conjunction with hybrid vigor or backcrossing obviously works best with characteristics that are easily observed or measured, such as tall or short and red or white flowers. Other characteristics, such as fruit size and color, resistance to drought or cold, or time to maturity are regulated by several genes (quantitative inheritance, Chapter 4) and are not subject to easily discerned differences. Selections for these can only be chosen by biometrical analysis of numerical quantities derived from the plants in question. Details are beyond the general treatment here, but most plant breeders find themselves dealing with this concept at some time.

TISSUE CULTURE SELECTION

The newest selection technique for the plant breeder holds great promise in terms of reduced labor and costs, as well as less need for test land. This is based on the aseptic method of propagation covered in Chapter 6. Basically, single cells of plants are maintained in tissue culture. One such petri dish of cells could be equivalent to a field containing tens of thousands of plants. These cells are grown in the haploid state whenever possible, because it is easier to select mutants from them than from diploid cells. In a haploid cell the one copy of a mutant gene can be clearly expressed, without interference from the other gene copy present in a diploid cell.

The cells can be treated with mutagens, such as radiation or colchicine, and then placed in the presence of screening agents for which the breeder wishes to alter the plant. For example, the screening agent may be a fungal toxin produced by a plant pathogen; the cells would be subjected to mutagens and then the fungal toxin added. Those that survive might possibly be converted into plants that are resistant to the plant disease caused by that fungus. These haploid survivors sometimes change spontaneously to diploids, or colchicine may be added to increase diploid production. The diploids, of course, have identical gene copies and, as such, are instant inbreds that will breed true when self-pollinated. This assumes that the diploid cells can be induced to develop roots and a shoot upon the addition of phytohormones and growth regulators. Limited production of mutated plants has been achieved to date. Techniques for regeneration of whole plants from cells are lacking for most horticultural crops at present, and the development of selective conditions in which only the desired mutants survive is also limited.

Another possibility with cultured cells is the creation of hybrids by cell fusion. This would be especially valuable for producing hybrids that cannot be obtained by sexual means because of incompatibility. Cells of two different tobacco species have been stripped of their cell walls by enzymic treatment and the resulting protoplasts fused together. The fused cell generated into a plant by the techniques mentioned previously for cell cultures. The resulting plant was a hybrid identical to the hybrid between the two species produced by conventional means.

Although exciting, this plant breeding tool will not replace, but only complement, classical plant breeding techniques. It will reduce the work load and offer some approaches that cannot be handled with classical techniques. However, there will always be a need for field selection.

Mass Seed Production

The plant breeder's work is not over once a selection has been made. The selection, if it is to be of horticultural value, must be reproduced on a large scale. If the plant can be propagated asexually, the production of an unaltered genotype is straightforward. If one wishes to maintain the genetic purity of a cultivar propagated by seed, this variability must be controlled. The degree of difficulty in controlling genetic variation varies according to the pollination mode of the cultivar. Regardless of the pollination mode, offtype plants should be removed, especially before pollination. Weeds should be controlled. Testing must be conducted to assure genetic purity and to detect any genetic drift that may occur if conditions controlling pollinations should fail to any degree or if environmental pressures over time favor adaptability changes.

SEEDS FROM SELF-POLLINATED PLANTS

In plants that are only self-pollinated, the pollen arises only from the same flower, from a different flower on the same plant, or from different plants of the same clone. Self-pollinated plants are largely homozygous; consequently, their offspring are also homozygous and have the characteristics of their parents. Cultivars of these plants can be preserved quite easily, even in the vicinity of closely related cultivars. Self-pollinated plants that can also accept pollen from a different plant or clone can produce seed that is a result of cross-pollination. Maintenance of these cultivars requires the prevention of cross-pollination.

Isolation is the method of restricting unwanted cross-pollination by wind or insect vectors, as well as preventing accidental mechanical mixing of harvested seeds. The distance required depends upon the species, the type of pollination, and the desired degree of seed certification for genetic purity. The distance for production of seed from self-pollinated plants need not be as great as with cross-pollinated plants. Minimal spacing is 10 to 15 feet for self-pollinated plants.

SEEDS FROM CROSS-POLLINATED PLANTS

Seeds produced from cross-pollinated plants are variable, since cultivars of this type are often heterozygous. We have seen that heterozygosity is important to the plant breeder as a source of new characteristics. Heterozygous plants produce plants with segregated characteristics (such as tall versus dwarf), which through continued self-pollination can segregate into nearly true breeding lines. Selection to a standard and prevention of cross-pollination from unwanted sources can produce a cultivar that, although not necessarily homozygous, has its variability under control. Cross-pollination can be restricted by isolation, as discussed under seeds from self-pollinated plants. Distances vary from 165 to 5280 feet, depending on species and desired degree of seed certification.

HYBRID SEEDS

Continual self-pollination of normally cross-pollinated plants produces an inbred line that has phenotypic uniformity, but often at the expense of vigor. This may be restored by breeding with another inbred line (single cross) to produce a hybrid cultivar with uniform phenotype, but also hybrid vigor. A desirable hybrid may be realized from other crosses, such as between two species. Other hybridization is between two single crosses (double cross), an inbred line and an open pollinated cultivar (top cross), or between a single cross and an inbred line (three-way cross). However, these latter hybridizations are used for few horticultural crops of value. Hybrid cultivars cannot be used for seed because of extremely variable progeny, so the hybrids must be continually reproduced from the parent inbred lines.

A problem arises in hybrid production on a large-scale since the mode of cross-pollination without any self-pollination is needed to produce pure hybrid seed.

Emasculation and hand pollination, which are used to produce hybrids during the selection process, are of limited value in terms of commercial hybrid production. The prohibitive labor cost makes it feasible only for high-value seed such as begonia, melon, petunia, pansy, tomato, or snapdragon. Commercial production of hybrids requires some system of sterility that favors cross-pollination over self-pollination.

The removal of male flowers of monoecious species followed by natural pollination by the other parent, such as the detasseling of sweet corn, reduces labor and obviously enforces cross-pollination and not self-pollination. However, there still is considerable labor involved in detasseling those sets of parents chosen to be the seed producers.

Fortunately for the plant breeder, a genetic factor for male sterility exists in many crop plants. This factor is responsible for inhibition of pollen production or nonviability of produced pollen. It may arise from a cytoplasmic factor (*cytoplasmic male sterility*) or a genetic factor (*genetic male sterility*). Male sterility has been observed in beet, carrot, corn, cucumber, lima bean, melon, onion, potato, and tomato. Plant breeders have been able to breed male sterility into a number of hybrid cultivars and have worked out the methods to maintain the male-sterile parental line. The most useful system involves sterility produced from both cytoplasm and nuclear gene activity. This approach has been used with horticultural crops, such as onion and sweet corn, to prevent self-pollination. Future applications appear promising.

Cabbage hybrids have been produced because of the factor of self-incompatibility. This factor results from the inability of the pollen tube to grow properly in the style of a flower on the same plant or clone (Fig. 11-4). The pollen will grow normally on another plant. Hybrids can be easily produced with dioecious plants (male and female flowers on separate plants), such as spinach.

Fig. 11-4. Self-incompatibility. Pollen from same plant or clone fails to produce pollen tube (left) in time to reach receptive ovule, thus preventing self pollination. Same pollen on another genetically similar but not identical cultivar produces pollen tube in sufficient time to allow fertilization (right).

Certain monoecious plants, such as the cucumber, tend to produce female flowers predominately at the first nodes, and male flowers appear later. Breeders have utilized this trait with cucumbers to produce *gynoecious* plants, that is, plants that produce mostly female flowers. If the other inbred line is a standard monoecious type, hybrid seed is easily produced.

Growth regulators have been used to shift sex expression in certain monoecious plants, such as the cucurbits (squash, cucumber, and melon). Gynoecious cucumbers treated with 1000 ppm of gibberellin three times a week from the expansion of the first true leaf can be induced to produce male flowers. Treatment of one parent and not the other makes it easy to produce hybrid seed.

SEEDS FROM TREES AND SHRUBS

Clones that are heavily heterozygous, such as many herbaceous and woody perennials, are not ideally suited for seed production because of extreme seedling variability. However, depending on the intended purpose, seeds may be desired over vegetative propagation. In that case, some control may be achieved by following recommended seed-selection techniques. Seed should be gathered from areas of comparable environment and especially from trees, shrubs, and herbaceous perennials showing a desired phenotype. Progeny testing, as discussed under selection, may be required for seed evaluation from these sources.

SEEDS BY APOMIXIS

Seed may also arise in some plants through an asexual process, apomixis, which provides a means to assure uniformity. Plants, which can produce seed both by self- and cross-pollination or by both apomixis and normal pollination, can be maintained pure with some effort, but the predominate form of seed production must be known. Some citrus cultivars produce apomictic seed, which is used to raise uniform, vigorous rootstocks for grafting. Kentucky bluegrass (*Poa pratensis*) can produce a mixture of apomictic and sexually produced seed.

Plant Variety Protection Act

Recent legislation has improved the prospects for those who produce new plants. Since it may take 10 to 12 years to go from initial crosses through selection and final introduction of a cultivar, some form of protection to give exclusive propagation rights to the seed producer is desirable. The Plant Variety Protection Act (1970) does just that. The U.S. Department of Agriculture can issue a Plant Variety Protection Certificate for the seed production of the cultivar, which is good for 17 years. To obtain such protection, the cultivar must meet certain criteria to prove that it is indeed novel. Vegetatively propagated materials may be patented based on 1930 plant patent legislation. A few horticultural crops are exempt from these rules: potato, cucumber, carrot, tomato, pepper, and celery.

The protection and enhanced marketability associated with new improved plants have led to an increasingly larger number of introductions. Hybrid seed is commercially available for about 40 different flowers grown in the garden and greenhouse and for about 20 different vegetables. In addition, a great number of new, improved plants are sold in propagating forms other than seed. These include ornamental shrubs and vines, fruit trees, nut trees, cane and vine fruits, bulbs, corms, and tubers. These offerings would be much less without plant patents.

All-American Selections

The ultimate honor for a flower or vegetable hybrid is to be chosen as a medal winner in the All-American Selections or All-American Rose Selections. Besides the honor, the financial return on such a hybrid is usually good because of public awareness of the selections. The competition for this annual award is intense among plant breeders from seed companies, botanical gardens, government research stations, and universities.

Over the years hybrids of a number of flowers and vegetables have been chosen. A few recent winners are shown in Figure 11-5.

Fig. 11-5. (facing page) Recent All-America Selections and All-America Rose Selections.
A. Dianthus 'Snowfire' hybrid.
B. Nicotiana 'Nicki-Red' hybrid.
C. Marigold 'Queen Sophia.'
D. Pepper 'Dutch Treat.'
E. Cucumber 'Saladin' hybrid.
F. Rose 'Grandiflora.'
A–E, courtesy of All-America Selections.
F, courtesy of All-America Rose Selections.

(A)

(B)

(C)

(D)

(E)

(F)

FURTHER READING

Allard, R. W., *Principles of Plant Breeding.* New York: John Wiley & Sons, Inc., 1960.

Briggs, F. N., and P. F. Knowles, *Introduction to Plant Breeding.* New York: Van Nostrand Reinhold Co., 1967.

Brooklyn Botanic Garden. *Breeding Plants for Home and Garden.* Brooklyn N.Y.: Brooklyn Botanic Garden Handbook Series, Vol. 30, 1974.

Darlington, C. D., *Chromosome Botany and the Origins of Cultivated Plants* (2nd ed.), London: George Allen & Unwin Ltd., 1963.

Darlington, C. D., and A. P. Wylie. *Chromosome Atlas of Flowering Plants.* London: George Allen & Unwin Ltd., 1955.

Frankel, O. H., and E. Bennett, eds., *Genetic Resources in Plants—Their Exploration and Conservation.* Philadelphia: F. A. Davis Co., 1970.

Lawrence, W. J. C., *Practical Plant Breeding* (rev. 3rd ed.). London: George Allen & Unwin Ltd., 1951.

McGregor, S. E., *Insect Pollination of Cultivated Crop Plants.* Agriculture Handbook No. 496. Washington, D.C.: U.S. Department of Agriculture, 1976.

National Academy of Sciences (USA), *Genetic Vulnerability of Major Crops.* Washington, D.C.: National Academy of Sciences Printing and Publishing Offices, 1972.

Nelson, R. R., *Breeding Plants for Disease Resistance: Concepts and Applications.* University Park, Pa.: Pennsylvania State University Press, 1973.

North, C., *Plant Breeding and Genetics in Horticulture.* New York: Halsted Press, 1979.

Staff. *Polyploidy and Induced Mutations in Plant Breeding.* New York: Unipub, 1974.

Harvesting and Postharvest Preparation

The last event in the horticultural process is the harvest, which is of special concern to commercial horticulturists. Some ornamentals may not be harvested, unless the appreciation of their esthetic values can be called a "harvest." Other ornamentals will be harvested as cut flowers. Edible products will be harvested for immediate consumption, sales, or preservation. These horticultural products will begin to deteriorate after harvest because of respirational and microbial activity. The well-rounded horticulturist will be able to predict harvest dates and will know what treatments are necessary after the harvest.

Days to Harvest

Climate may be related to periodic events in plant (and animal) life. The science dealing with this relationship is termed *phenology*. Phenological data for horticultural crops include dates of planting, germination, budding, flowering, ripening, and harvest. These dates are determined by climatic factors prior to and at the time of the event. These data can be presented in a map on which an *isophene* connects phenological events that take place on the same date (see frost-date maps in Part Three for isophenes).

A useful phenological concept for the prediction of harvest time is based on degree-days. One degree-day is defined as a day when the mean daily temperature is one degree above the minimal temperature for growth (*zero temperature*) of a plant. Zero temperatures vary according to the crop considered. For example, pea, a cool-season crop, has a zero temperature of 38°F (4°C); sweet corn, a warm-season crop, has a zero temperature of 50°F (10°C); potatoes, 44.5°F (7°C); and most deciduous fruit and nut trees, 41°F (5°C). A day when the mean daily temperature was 48°F (9°C) would provide 5, 0, 2, and 4 growing degree-days, respectively, for peas, sweet corn, potatoes, and deciduous fruit or nut trees. From accumulative research through various regions of the United States, it has been determined how many degree-days (sometimes referred to as heat units) are needed for various horticultural crops to reach the harvest stage.

There are some problems in this approach. It cannot differentiate between seasonal temperatures, such as a cold spring–hot summer versus a warm spring–cool summer, between differences in soil–air temperatures, or among unusual extremes of temperature during the growing season or during the diurnal variation of temperature. Certain developmental stages of the crops may also be more temperature sensitive than others or dependent upon photoperiod. One such crop with developmental stages of differing temperature sensitivity and dependence on photoperiod is the pea. The earliest possible plantings of peas may require 45 percent more growing degree-days to harvest than a planting made 5 weeks later because of increases in the mean daily temperature and photoperiod. Unless the horticulturist were aware of this, the prediction of harvest dates and timing for processing could be considerably wrong.

Additional refinements of the heat-unit concept to correct for photoperiod and effects of temperature change upon developmental stages are possible. The degree-days can be multiplied by the number of daylight hours to give a degree-day-length unit. A further refinement is to measure the rate of growth at the various temperatures in the growth range and to multiply the number of hours of each day at each temperature by the relative growth rate of the plant at that temperature. This concept is that of positive degree-days or growing degree days.

The concept of negative or dormancy degree-days must also be considered for perennial deciduous fruit and nut crops, which have minimal chilling requirements to break dormancy (indicated when known for fruits and nuts in Chapter 15). This type of data can be used to predict the time of bud swell and to assess the danger of late spring frosts. Sprinkling of fruit trees near bud swell cools by evaporation and reduces accumulation of degree days; bloom can be delayed up to 2 weeks. Growing degree-days can be used to predict the harvest date from the time of bud swell.

The use of phenological data, especially both growing and dormancy degree-days, is feasible. Best results are obtained for those horticultural crops for which development is not closely dependent upon photoperiod or differences in temperature sensitivity. Heat units can be useful for even these crops, if appropriate corrections are made. Heat units are best tempered with a working knowledge of your local area's climate and microclimate.

Careful utilization of phenological data makes it possible to establish planting and harvesting schedules that are most suitable for the local climate, labor availability, and fresh market processing demands. It may also be possible to schedule crops to avoid seasonal weather phenomena, maximal insect and disease activity, or maximal need for irrigation.

The importance of phenological data to harvesting and food processing should not be underestimated. If a harvest glut occurs in a limited time period, overtime labor and the use of more mechanized equipment than necessary often result. Crops may even spoil before they are picked. On the processing end, the factory's capacity may be exceeded, or it may have to be larger than is really efficient, overtime may be necessary, and some crops may spoil or be at less than optimal quality when processed because of time delays. The efficient use of labor and equipment at planting and harvest time therefore makes it expedient to make successive sowings of a crop to spread the harvest period based on phenological data. By doing so, the harvesting, processing, and marketing labor and equipment requirements are not overwhelming in a short time span, equipment and labor are used more efficiently over a longer period each season, and processing capacities need not be as large. This is especially useful for crops whose duration of harvest period is very short for quality maintenance. Successive harvesting of one crop can be realized by making several plantings of one variety at spaced time intervals and/or making one planting of several varieties with different maturation times. Different crops may be planted on dates determined by phenological data to ensure a smooth transition, with adequate time between crop changeovers for harvesting and processing. If adverse climatic factors cause a deviation from the predicted harvest date, all fields within the processing plant region should be affected similarly. Therefore, the sequence of harvest is only shifted, and thus the processing plant operation is affected minimally.

A number of other criteria are useful for establishing the time of harvest, which is influenced by factors such as the genetic constitution of the variety, the planting date, and the environmental conditions during the growing season. These criteria vary with the crop being considered, but include the number of days from germination, setting out plants, and flowering, skin color, degree of softness, sugar levels, sugar-acid ratios, abscission zone formation, differences in sound obtained from tapping, overall appearance, and the taste test. These criteria will be discussed in relation to specific crops in Part Three.

(A)

Figure 12–1. Hand harvesting. A. Much of the lettuce is harvested by hand in the United States. Mechanized machinery is available and only slowly being adopted because of high cost, but increasing labor unrest may speed up its acceptance. B. Apples are mainly harvested by hand. However, as dwarf trees grow in hedgerows and trellis systems increasingly replace conventional orchards of standard size trees, harvesting machinery will be increasingly used. Above photos from U.S. Department of Agriculture. C and D. An experimental citrus picker which makes harvesting easier and less costly. American Society For Horticultural Science photo.

(B)

(C)

(D)

**Harvest
Techniques**

The oldest harvesting technique is hand harvesting, which is still practiced (Fig. 12-1). The home horticulturist makes extensive use of this approach for his vegetable, fruit, and nut crops. Commercial operations also use this technique when mechanized operations are not available or are bypassed in favor of pick-your-own operations, or the cost and/or feasibility of the equipment is questionable. At one time commercial operations did not have suitable mechanized harvesting equipment for easily bruised crops, such as the bramble fruits, or the equipment was of limited feasibility because of variations in the time of ripening. However, harvesting machinery is being developed even for easily bruised fruit (Fig. 12-2), and plant breeders have aided mecha-

(A)

Figure 12-2. A. A mechanized harvester for blueberries, which are easily bruised. U.S. Department of Agriculture photo. B. Hand harvesting of blueberries with a blueberry rake is much slower. U.S. Department of Agriculture photo now in National Archives (33-S-23044C).

(B)

nized harvesting by developing crop varieties with uniform ripening habits, compact growth, less easily bruised fruits, more easily detached fruits or vegetables, or an angle of hang better suited for mechanized removal. In addition, more and more growth regulators are used as mechanized harvest aides to promote uniform ripening, color, and abscission. Harvesting equipment usually saves time and cuts the harvesting cost considerably. Many agricultural engineering departments in universities, the U.S. Department of Agriculture, and industrial research laboratories are actively involved in developing new harvesting machinery.

One type of mechanized harvester consists of motor-driven tree and bush shakers with catching belts and some storage capacity. Earlier models caused shaking damage, such as to the bark, but this has been minimized in more recent models. Possible problems associated with this type of mechanized harvesting are soil compaction, injury to lower fruit-bearing branches (such as spur apples) by falling fruits, and possible damage to the root system. Shaker harvesters have worked fairly well with cherries and most nuts.

More recent mechanized harvesters for various fruit trees make use of finger-type pickers, slappers, paddles, and impacter rods. These often have decelerators, such as numerous rubber strips, to slow fruit fall and hence minimize damage. Some of these newer machines are designed to be used with dwarf fruit trees or orchards trained to the hedgerow system or trellis system (see Chapter 15). Other low-growing fruits and vegetables are harvested with what amounts to mechanized slappers. Mechanized harvesters are constantly being improved, and new ones are being developed. In some instances, these machines can harvest crops with less damage than occurs by hand harvesting. One aim for the future is to develop a multiple picking harvester capable of being adjusted to more than one crop.

Dry seeds from fruits that dehisce or shatter readily at maturity can be mechanically harvested. Generally, the equipment is based on a revolving cylinder that acts as a beater to loosen seeds and a screening unit to separate seeds from debris. This procedure works fine with many flower crops (pansy, petunia, delphinium) and vegetables (cole crops, onion). Seeds from indehiscent fruits (grasses, corn) are harvested with a combine. Seeds from fleshy fruits (pepper, tomato, eggplant, cucumber, squash) are harvested with machines that macerate the fruit.

The mechanization of harvesting varies as to degree and the number of steps involved. For example, many apple orchards are still picked by hand because the size and spacing of large trees necessitates large harvesting machinery that is costly and somewhat difficult to handle. Experimental harvesters designed for compact trees, hedgerows, and trellis systems are more feasible and harvest fruits with minimal bruising. These can be expected to increasingly replace hand labor in many orchards. However, the bulk of tree and bush fruits intended for the fresh market are still hand harvested. Some operations may be handled in a partially mechanized manner, for example, the hand picking of fruit or vegetables followed by

Introduction to Applied Horticultural Science | Part II

(A)

Figure 12-3. Tomatoes being harvested for processing. A. Machine cuts tomato plant from field and it is conveyed to a shaker. Field sorters pick out vines and unripe fruit. Most of the tomatoes are ripe, since varieties used for processing have uniform ripening. B. Same harvester about to load sorted ripe tomatoes into bins along side. American Society for Horticultural Science photos. C. Field of harvested onions. Harvest and piling operation is mechanized. Onions are left to air dry. U.S. Department of Agriculture photo by Robert Kane.

(B)

(C)

mechanized grading and packing. Crops such as apples, broccoli, cabbages, cauliflower, peppers, and pears would be handled in this manner. Crops intended for processing are extensively harvested by mechanical means (Fig. 12-3) because of the savings in labor dollars and the lesser importance of crop bruises. These operations might be completely mechanized in one location; peas utilized for canning may be mechanically picked, vined, and shelled right in the field. Crops harvested this way include pea, beet, bean, carrot, lima bean, onion, potato, radish, sweet potato, tomato, spinach, sweet corn, and nuts.

Postharvest Changes and Their Prevention

When a horticultural crop is harvested, it is still a living organism, and life processes continue to produce changes in the picked crop. These changes constitute the science of *postharvest physiology*. The biochemical and physiological changes may enhance or detract from the appeal of the crop. The appeal of the crop is determined by three factors: quality, appearance, and condition. Quality is used in reference to flavor and texture, as determined by taste, smell, and mouth-tongue touch. The appearance refers to the visual impression of the crop, whereas the condition refers to the degree of departure from a physiological norm or the degree of bruising or disease.

Postharvest alterations are numerous and include water loss, the conversions of starch to sugar and sugar to starch, flavor changes, color changes, toughening, vitamin gain and loss, sprouting, rooting, softening, and decay. These processes may improve or detract from the quality, flavor, and texture of the crop. For example, the loss of water by transpiration and evaporation is bad for apples, but certainly useful in producing raisins and prunes from grapes and plums, respectively. Most people are familiar with the loss of quality associated with the rapid enzymatic conversion of sugar to starch that occurs with harvested sweet corn, and the reverse reaction, starch to sugar, that takes place in potatoes. However, the latter reaction results in the improvement of picked bananas and pears, and is one of the reasons that pears and bananas are the few fruit crops that can be picked when immature and ripened to a high-quality product.

On the whole, postharvest changes produce more detrimental results than beneficial ones. If the crops are harvested for subsequent processing, the time between harvesting and processing should be minimal, so postharvest alterations are not a serious problem. However, crops destined for fresh market sales must be handled in a manner that will minimize the detrimental effects and maximize any beneficial effects, if possible.

No one set of conditions will achieve this end, because postharvest alterations are a function of the crop type, and each crop varies as to the optimal air temperature, relative humidity, and concentrations of oxygen and carbon dioxide required for stabilization. Some form of air circulation is usually beneficial in reducing disease problems and removing any released gases that might accelerate harmful physiological and biochemical changes. The level of disease pathogens should also be kept to a minimum.

The application of postharvest preventative measures will also vary as to when they should be applied. For example, if a postharvest change is both rapid and seriously detrimental, some treatment might be required from the moment the fruit is picked. This would be a problem with strawberries, whose postharvest longevity is 7 to 10 days, but not with lemons, which have a postharvest longevity of several months. This treatment might be a temporary refrigerated storage area to hold the freshly picked crop until enough is accumulated. Then it might have to be shipped to a grading and packing area in a refrigerated truck. The grading and packing area might even need to be temperature controlled, yet the comfort of the workers must be considered. For example, apples are often graded and packed in areas where the temperature is kept between 55° to 60°F (13° to 15.5°C), and hot-air blowers or radiant heaters are directed at the feet of the workers to keep them comfortable.

The most important aspect of postharvest treatments is in storage facilities designed for long-term crop storage, so that the crop can be sold at a reasonable price over an extended period of time. The best storage conditions slow the life processes but avoid tissue death, which produces gross deterioration and drastic differences in the quality, appearance, and condition of the stored crop.

Although conditions vary, it is possible to make some generalizations. Fruits, nuts, and vegetables intended for storage should be free from mechanical, insect, or disease injuries and at the proper stage of maturity. Storage for the less critical crops, in terms of better postharvest longevity, usually consists of common (unrefrigerated) storage, which lacks precise temperature and humidity control. Examples of common storage include insulated storage houses, outdoor cellars, and mounds. Root crops such as rutabaga and carrots can be stored in such facilities, as well as other crops with good keeping qualities such as winter squash. More reliable storage consists of cold (refrigerated) storage with precise control of temperature and humidity. An improvement upon this form of storage is the addition of controlled atmosphere (CA storage) by which the concentration of oxygen and carbon dioxide are kept at an optimal level. Low temperature decreases the rate of respiration, the controlled humidity level slows water losses through transpiration and evaporation, and the controlled atmosphere brings about an overall reduction in the rates of chemical reactions occurring in the stored materials. The overall change in reactions is reflected in the inhibition of respiration and ethylene production. Apples and pears are examples of fruit stored under controlled atmosphere, temperature, and humidity conditions.

Controlled atmosphere conditions are usually 5 percent oxygen and 1 to 3 percent carbon dioxide. A compensating factor for the higher cost of controlled atmospheres is the fact that optimal temperatures are often several degrees higher if a controlled atmosphere is present. Not only is a savings in refrigeration cost realized, but physiological disorders sometimes associated with the lower temperatures are minimized.

Storage temperatures and humidities, of course, differ for each fruit and vegetable, but some generalizations are possible. A number of vegetables can be held near 32°F (0°C). These include asparagus, beet, broccoli, Brussels sprouts, cabbage, carrot, cauliflower, celery, garlic, leek, lettuce, lima bean, onion, pea, spinach, sweet corn, and turnip. The relative humidity is usually maintained at 90 to 95 percent, except for garlic, leek, onion, and pea, for which it is kept at 70 to 75, 85 to 90, 70 to 75, and 85 to 90 percent, respectively. Other storage temperatures for vegetables are 45°F (7°C) for pepper and snap bean (85 to 90 percent relative humidity), 45° to 50°F (7 to 10°C) for cucumber and eggplant (90 to 95 percent relative humidity), 50°F (10°C) for okra (85 to 95 percent relative humidity), 38° to 40°F (3.5° to 4.5°C) for potato (85 to 90 percent relative humidity), 50° to 55°F (10° to 13°C) for winter squash and pumpkin (70 to 75 percent relative humidity), 40° to 50°F (4.5° to 10°C) for summer squash (85 to 95 percent relative humidity) and ripe tomato (85 to 90 percent relative humidity), 55° to 60°F (13° to 15.5°C) for sweet potato (80 to 95 percent relative humidity), and 36° to 40°F (2° to 4.5°C) for watermelon (80 to 85 percent relative humidity).

Most temperate-zone fruits can be held at 32° to 41°F (0° to 5°C). Subtropical and tropical fruits would show chilling injury if held at these temperatures for a prolonged period of time. Bananas, for example, will not tolerate temperatures below 53°F (12°C). Relative humidities are usually kept at 85 to 90 percent. Nuts can be stored at 34° to 45°F (1° to 7°C) at 65 to 70 percent relative humidity.

Seed storage varies. Seeds only being held from one season to the next are usually held in open storage without temperature or moisture control. Better storage conditions are to hold the seeds at 40° to 50°F (4.5° to 10°C) at relative humidities of 3 to 8 percent.

Premarketing Operations

Premarketing operations include washing, trimming, waxing, curing, precooling, grading, prepackaging, and shipping.

WASHING, TRIMMING, WAXING, AND CURING

Washing is especially needed for root crops, since they have adhering soil particles that reduce consumer appeal and may increase the disease problem. Fruits are often washed or brushed to enhance appearance. Trimming (Fig. 12-4) is used to cut back discolored leaves or to cut back green tops for vegetables such as beet, carrot, celery, lettuce, radish, rutabaga, spinach, and turnip. Nursery plants and cut flowers are often trimmed

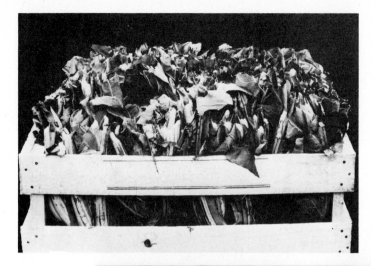

Figure 12-4. Harvested celery which has been trimmed to improve appearance. U.S. Department of Agriculture photo.

Figure 12-5. Oranges emerging from a washer and waxer on a conveyor belt. American Society for Horticultural Science photo.

prior to sale. Waxing is used to improve appearance and to slow moisture loss, which causes shriveling. Waxing has been employed to varying extents on cucumber, muskmelon, pepper, tomato, rutabaga, turnip, and citrus fruits (Fig. 12-5). Some roots, such as dahlia, are waxed for improved storage. Dormant deciduous nursery stock is frequently waxed. Curing is used to heal cuts or bruises incurred during harvesting of crops such as potatoes and sweet potatoes and to prevent rot of bulbs during storage through removal of water on the outer scales. Potatoes are cured by applying heat with humidity control to minimize water losses but to expedite healing; bulb treatment is mostly heat control to remove water.

PRECOOLING

Precooling is used for rapid removal of heat from recently harvested fruits and vegetables. This permits the grower to harvest at maximal maturity and yet be assured that the crop will have maximal quality when

Figure 12-6. Lettuce being placed in a vacuum cooler. American Society for Horticultural Science photo.

it reaches the consumer. Precooling slows the rate of respiration in the fruit or vegetable, slows wilting and shriveling by decreasing water loss, and inhibits the growth of decay caused by microorganisms. Precooling is an energy-requiring process; its cost can be minimized by harvesting at night and in the early morning, when the temperature of the crop is lowest.

Major precooling methods include hydrocooling, contact icing, vacuum cooling (Fig. 12-6), and air cooling. In hydrocooling, cold or iced water flows through the packed containers, absorbing the heat from the vegetables and fruits. Crops that are suitable for hydrocooling include asparagus, beet, broccoli, carrot, cauliflower, celery, muskmelon, pea, peach, radish, summer squash, and sweet corn. Contact icing is simply the placing of crushed ice in or around the packed produce. This technique has declined in use.

Vacuum cooling cools the crop through the rapid evaporation of water under reduced pressure, that is, evaporative cooling. Cabbage, lettuce, and spinach are suitable for vacuum cooling. Air cooling is simply cold air blown over the crops. This technique is used on bean, berries, cucumber, eggplant, pepper, and tomato.

After precooling the crop, it must be kept cool in transit by shipment in refrigerated trucks, railroad cars, and cargo holds in planes. If stored, cold storage is used, and refrigerated display cases are preferred.

GRADING

Uniformity in quality, size, shape, color, condition, and ripeness is usually preferred (Fig. 12-7). This is for reasons of consumer appeal, freedom from disease, and ease of handling by automated machinery. To achieve this end, fruits and vegetables are often graded according to standard grades, which form a basis of trade. The requirements of grading vary depending on factors such as whether the crop is destined for the fresh market or for processing. Grades have been established by federal law and are strongly suggested; however, they are not enforced legally unless the federal grades are used to describe the produce. Government grades include such designations as Fancy or Extra Fancy, no. 1, no. 2, and so on. Fruits, vegetables, and nuts not meeting grading standards are usually processed or employed in products using lesser grades and off sizes.

(A)

Figure 12-7. A. Apples are being graded by hand on the basis of color, bruises, and other defects. U.S. Department of Agriculture photo. B. Grading of oranges by size is being done here by mechanized equipment with rollers spaced at increasing distances. When fruit reaches spacing comparable to its size, it falls through. American Society for Horticultural Science photo.

(B)

PACKAGING

At least half of all vegetables and fruits are packaged in retail units at some point prior to reaching the retail store. This type of packaging is called *prepackaging* (Fig. 12-8). Products are placed in bags of transparent plastic film or in trays and cartons covered over by transparent film. Transparent film is usually made of polyethylene, which can be made to have a differential permeability toward oxygen and carbon dioxide. Often a controlled atmosphere can be reached in these containers covered in polyethylene, which consists of less oxygen and more carbon dioxide than is present in normal air. Apples store particularly well in this modified atmosphere. The retention of water vapor is also improved with plastic film, which is usually beneficial for the quality of the fruits and vegetables. Paper and mesh bags are also used for prepackaging products such as onions and gladioli. Trays and cartons used for the prepackaging unit are usually made of Styrofoam or wood pulp. The master containers for the prepackaged units are usually made of paperboard. Wooden crates and bushel baskets are seen less and less. Master containers are designed for easy stacking, shipping, and minimal bruising.

Figure 12-8. Prepackaged cauliflower. Heads are covered with plastic wrap and shipped in carton like that shown. Heads can be removed at the market and a pricing label stuck to the plastic wrap. U.S. Department of Agriculture photo.

SHIPPING

Fruits and vegetables may be shipped a number of ways. Large trucks with refrigerated containers, which may or may not have a controlled atmosphere, are used extensively. Refrigerated railroad cars are also used. Crops are also airfreighted by large cargo planes. Cargo boats are another source of transportation. Loading and unloading are often facilitated by the use of wooden pallets and forklifts.

Preservation

Vegetables, fruits, and nuts destined for fresh consumption can be made available over a longer time by storage, which has been discussed earlier. However, vegetables, fruits, and nuts destined for processing can be preserved by a number of techniques. These include canning, freezing, dehydration, pickling, fermentation, sugar concentrates, and chemical preservatives.

The types of food processing utilized for various vegetables, fruits, and nuts are shown in Tables 12-1 through 12-3. The use of fruit, vegetable, and nut is strictly in terms of food terminology in these listings, not in botanical terms, since they are processed for food. All, of course, can be eaten fresh directly or held in storage besides being processed. The banana, for example, is more likely to be eaten fresh than dehydrated. Those crops not found in the tables are only consumed fresh for reasons of poor transportation, or they are easily bruised, or they process poorly. Examples are fruits such as the avocado, nuts such as the Brazil nut and macademia, and vegetables such as lettuce, kohlrabi, Swiss chard, and endive. A number of crops are also processed into oil, such as the olive, coconut, and almond, and into starch, such as the potato, cassava, sweet potato, and yam. Many crops are also processed directly, through canning, freezing, and so on. Waste products from some crops are processed into nonfood products, such as fiber, perfume oils, medicinals, and gums.

Crops grown for processing usually cost less per unit area of land or per ton. This is because, unlike crops grown for the fresh market, appearance is not of major concern. However, size, quality, and uniformity are important. An extended harvest through successive plantings or the use of varieties with different maturation dates is needed to maintain a constant supply. This enables the processing factory to operate at an even flow over a reasonable period of time.

TABLE 12-1. KINDS OF PROCESSING FOR VEGETABLE CROPS

	Canning	Freezing	Dehydration	Pickling
Asparagus	+	+	−	−
Bean	+	+	+	+
Beet	+	−	−	+
Broccoli	−	+	−	−
Brussels sprouts	−	+	−	−
Cabbage	−	−	−	+
Carrot	+	+	+	−
Cauliflower	−	+	−	+
Celery	−	−	+	−
Cucumber	−	−	−	+
Eggplant	−	−	−	+
Garlic	−	−	+	−
Leek	−	−	+	−
Lima bean	+	+	+	−
Okra	+	−	+	−
Olive	−	−	−	+
Onion	−	−	+	+
Parsley	−	−	+	−
Pea	+	+	−	−
Pepper	−	−	+	+
Potato	−	+	+	−
Pumpkin	+	−	−	−
Rutabaga	−	+	−	−
Spinach	+	+	−	−
Summer squash	−	+	−	−
Sweet corn	+	+	−	−
Sweet potato	+	−	+	−
Tomato	+	−	+	+
Turnip	−	+	−	−

TABLE 12-2. KINDS OF PROCESSING FOR FRUITS

Apple	Juice, cider, sauce, slices
Apricot	Canning, dehydration, sugar concentrates
Banana	Dehydration
Blackberry	Freezing, sugar concentrates
Blueberry	Freezing, canning
Cherry	Canning
Citrus	
Orange	Juice, canning, sugar concentrates
Lemon	Juice
Lime	Juice
Grapefruit	Juice, canning
Cranberry	Juice, canning, freezing
Currants	Dehydration, sugar concentrates
Date	Dehydration, confections
Fig	Dehydration, canning
Gooseberry	Sugar concentrates
Grape	Fermentation, dehydration, juice, sugar concentrates
Guava	Juice, sugar concentrates
Mango	Sugar concentrates
Muskmelon	Freezing
Papaya	Sugar concentrates, juice
Passion fruit	Juice
Peach	Canning, dehydration, freezing, sugar concentrates
Pear	Canning
Pineapple	Canning, sugar concentrates
Plum	Dehydration, canning
Raspberry	Freezing, sugar concentrates
Rhubarb	Freezing
Strawberry	Freezing, sugar concentrates
Watermelon	Freezing

TABLE 12-3. KINDS OF PROCESSING AND USES FOR NUTS

Almond	Confections, baked goods
Cashew	Roasted
Chestnut	Boiled, roasted, steamed
Coconut	Dehydrated
Filbert	Roasted, confections
Hickory nut	Confections, baked goods
Peanut	Salted, roasted, confections
Pecan	Confections, baked goods
Pine nut	Salted, roasted
Pistachio	Salted, roasted
Walnut	Confections, baked goods

Introduction to Applied Horticultural Science / Part II

(A)

(B)

Figure 12-9. A. Commercial processing of peaches. These are about to be sealed and heat processed in a pressure cooker. B. Home canning equipment used in the boiling water method. C. Hot packing peaches prior to processing in a boiling water bath. U.S. Department of Agriculture photos.

(C)

CANNING

Vegetables and fruits selected for canning must have as high quality and condition as possible, since the canning process lowers quality to some degree. The reason for the reduction in quality rests with the need for temperatures high enough to stop enzymic degradation of the food and to destroy any pathogens harmful to man or the keeping qualities of the canned products. The canning containers, metal cans for commercial processing and glass jars for home and commercial processing (Fig. 12-9), must also be sterile when the product is added and sealed tightly. This prevents any reinfection of the canned products by pathogens.

Generally, there are two canning processes, a boiling water bath and a pressure processing boiling water bath. The former develops a temperature of 212°F (100°C) and the latter, 240°F (112°C) or higher. Acid crops are processed at the lower temperature, since the threat of food poisoning from the toxin produced by *Clostridium botulinum* (lower pH limit for growth is 4.5) is negligible. However, low- to medium-acid crops (pH 7.0 to 4.5) can support the growth of this organism, so they are processed at the higher temperature. The acid to high-acid crops have a pH below 4.5 and include most fruits, berries, tomatoes, and any pickled or fermented crops. Low-acid crops have a higher pH and include most vegetables. It is strongly recommended that home-canned low-acid crops be boiled for 15 minutes before eating in order to destroy any *C. botulinum* toxin that might be present by chance. Commercial canning temperatures are high enough to eliminate any doubt. Storage life of canned goods varies according to temperature, but a shelf life of 3 to 5 years in the temperate climates is not unusual.

FREEZING

Freezing stops the growth of microorganisms and slows or stops enzymic activities that lead to deterioration of quality. Some form of heat treatment, such as exposure to boiling water for a few minutes (blanching), is usually done prior to freezing the crop. This inactivates enzymes that would cause some flavor and color losses even in a frozen product.

After blanching, products are cooled rapidly, such as by contact with cold water, to prevent further cooking that might lead to losses in quality. After cooling, the products are packed in plastic bags or waxed paper and cardboard containers to prevent dehydration during freezing or storage; otherwise, freezer burn occurs. Losses of quality are associated with freezer burn. Best results are obtained with rapid as opposed to slow freezing, since less cellular damage is produced from the smaller ice crystals produced during rapid freezing. Large crystals disrupt cells, producing an unacceptable soft, pulpy texture in the thawed food. Freezing is accomplished by exposure to liquified gases (air, carbon dioxide, nitrogen), blasts of cold air, immersion in a cooling liquid, or contact with refrigerated metal plates or grills. Optimal freezing temperatures are between $-20°$ to $-40°F$ ($-29°$ and $-40°C$). These are only realized in commercial operations. Once frozen, products are stored at $-26°$ to $32°F$ ($-12°$ to $0°C$). Freezer life varies according to the product, but a shelf life of 1 year is common.

DEHYDRATION

Removal of water, or dehydration, causes physiological disruptions in the cells, which inactivate enzymes that normally cause deterioration of quality. The low moisture level and high sugar concentration also prevent the growth of microorganisms involved in food decay.

The oldest form of food preservation was probably dehydration brought about by exposure to sun in dry areas. Today, sun drying is still used, but hot-air drying (Fig. 12-10) offers the conveniences of not being climate dependent, more rapid drying, improved quality, few problems with pathogens, and requiring less space. The one advantage of sun drying is the lower cost, since some form of energy is expended to supply hot air. A recent form of dehydration is called freeze-drying; the produce is quick frozen and maintained in a vacuum until all the moisture is removed. It appears to produce a somewhat higher quality product, but its use is not as widespread as hot-air drying.

Figure 12-10. Dehydration of vegetables has reduced the amount shown at top to that at the bottom. Courtesy of National Garden Bureau, Inc.

Fruits and vegetables can be dehydrated at temperatures near 140°F (60°C). Vegetables are usually blanched prior to dehydration to inactivate enzymes. Sulfur dioxide treatments are often used prior to dehydration as an additional aid to preservation. Fruits are dehydrated to 15 to 25 percent moisture content and vegetables to 4 percent since they have lower sugar contents. Dried products should be stored at low temperatures and low humidities.

PICKLING AND FERMENTATION

Under certain conditions some microorganisms can have the desirable effect of preserving, rather than spoiling, fruits and vegetables. The trick is to control the environmental conditions to allow fermentation through the enzymic, anaerobic decomposition of carbohydrates by beneficial microbial growth. Fermentation can produce alcohols through alcoholic fermentation and acids by partial oxidation. Once the levels of alcohol or acid reach a critical concentration, the growth of the microorganisms that produced the alcohol or acid is suppressed. These conditions also inhibit the growth of unfavorable organisms. Wine is an example of alcoholic fermentation with grapes, and partial oxidation of apple cider can be used to produce vinegar. If salt is used to suppress all but the favorable organisms, the combined salting-fermentation is called pickling.

Once pickling is complete, the salt content is reduced by leaching it from the pickled product. The pickled product is usually canned for long-term storage, since the enzymes remaining after microbial growth is stopped can cause undesirable deterioration. Pickling and fermentation are important processes for preserving some vegetables and fruits (see Tables 12-1 and 12-2).

CONCENTRATED MOIST FOODS

Fruits, which have an acidic pH, can be concentrated to 65 percent or more solids, or if the pH is only slightly acidic, to over 70 percent solids. This high solids content, usually much of which is sugar, and the low moisture provide conditions unsuitable for microbial spoilage. Mild heat treatments are used to extend the preservation life.

The making of jelly, jam, or preserves is based on this principle (Fig. 12-11). Sugar is added to the fruit to achieve a high carbohydrate level in the final concentrated solids. Usually, high quality but unattractive fruit, such as bruised fruit, is utilized. Jellies are made only from the juice of fruit, jams from combined juice and fruit pulp, and preserves from juice and whole fruits. Fruit butter, such as apple butter, is a smooth, semisolid prepared from high concentrations of mashed, sieved fruit pulp. Marmalades are made from the combined juice and peels of citrus fruit. These all have a gellike texture resulting from natural or added fruit pectins. The final product after heat treatment is either covered by paraffin on the exposed top portion in the jar or vacuum sealed. This prevents moisture loss, oxidation, and mold growth.

High levels of sugar may be produced in fruit by soaking in changes of sugar solutions of increasing concentrations. The fruit can be washed and dried and is sold as candied fruit. If a thin glazing of sugar is added as a coating, the final product is called glacéed fruit.

Figure 12-11. Pouring of hot jelly processed at home. U.S. Department of Agriculture photo.

MISCELLANEOUS PRESERVATION

Juices with or without some pulp can be extracted from fruits and vegetables and be preserved by canning or freezing. Some juice finds its way into jelly processing. Pulp and fruit parts, either left from other forms of processing or made deliberately, appear in processed products such as ice cream, confections, or baked goods. Nuts may be salted and roasted for preservation or be processed in ice cream, candy, or baked goods.

CHEMICAL PRESERVATION

Certain chemicals present naturally in foods inhibit the growth of undesirable microorganisms or help to forestall deterioration of quality. Sugars, salt, vinegar, and alcohol fall into this group. Other chemicals that are not naturally present in fruits, nuts, or vegetables (or perhaps found in a very few) are added for the same reasons. These are classed as antioxidants, stabilizers, emulsifiers, bleaching agents, neutralizers, firming agents, and acidulants.

Specific chemicals used for these purposes include benzoic acid, vanillic acid esters, propionates, glycols, monochloroacetic acid, dehydroacetic acid, sulfur dioxide, and sulfites. Sulfur dioxide, for example, is used as a preservative with dehydrated fruits, and sulfites are used in wine fermentation. The use of chemical preservatives is regulated by law.

**Cut-Flower
Preservation**

The preservation of cut flowers is dependent upon maintenance of turgidity, a supply of respirable substrates, and prevention of damaging effects caused by ethylene and microorganisms.

Ethylene promotes senescence, and microorganisms and their released substances can plug flower stems. Physiological plugging of the stem is also possible; air bubbles may also cause stem plugging.

For maximal cut-flower life these factors must be controlled both during storage of cut flowers and during vase life. Controls vary with species and time of harvest, that is whether flowers are fully open, partially open, or in the tight bud stage. Some generalizations are possible.

Controlled atmospheric storage improves the longevity of some cut flowers, but its cost and certain problems, such as petal color changes, do not justify its use to any great extent. The most widely used storage is low-temperature dry storage. The temperature is 31°F (-0.56°C), and the cut flowers are packed in moisture-proof (but not air tight) containers. Cut-flower stems are placed in water after removal from storage prior to shipment. A water temperature of 100° to 110°F (38° to 43°C) appears to be best. Shipping is usually by airfreight or trucks, which are unrefrigerated except for long-distance shipping.

Preservative solutions are used to prolong vase life and to stimulate opening of flowers picked in the tight bud stage. The chemicals used vary, but the preservative should consist of a sugar, a bacteriocide, a heavy metal, or an acidifying agent. The sugar is usually sucrose or glucose (2 to 10 percent weight by volume), which is a respirable substrate. The bacteriocide acts to prevent stem plugging; a common choice is 8-hydroxyquinoline sulfate (200 to 1000 ppm) or 8-hydroxyquinoline citrate (10 to 100 ppm). These compounds also inhibit physiological stem plugging, which is only partially attributed to their acidifying properties. One commercial perservative is Floralife and 8-hydroxyquinoline citrate is marketed as Oxine citrate. Silver nitrate (about 1000 ppm) has been used to stabilize flower petal color, and it also appears to prevent stem plugging by counteracting the effects of bacterial excretions. A heavy metal to stabilize petal color seems unneeded if the pH is 4.0 to 4.8. Chemicals are also added to inhibit ethylene production, such as L-2-amino-4-(2-aminoethoxy)-*trans*-3-butenoic acid (0.068 mM). Fungicide sprays have also been used to avert disease problems. Benomyl (200 ppm) has been utilized.

FURTHER READING

Bibliography of Agriculture. Phenology entries extensive in this yearly publication; excellent source for tracking down such data. Macmillan Information, N.Y.

Introduction to Applied Horticultural Science / Part II

Chang, Jen-Hu., *Climate and Agriculture*. Chicago: Aldine Publishing Co., 1968.

Desrosier, N. W., *Technology of Food Preservation* (3rd ed.). Westport, CT: Avi Publishing Co., 1970.

Haard, N. F. and D. K. Salunkhe, *Postharvest Biology and Handling of Fruits and Vegetables*. Westport, CT: Avi Publishing Co., 1975.

Hanson, L. P., *Commercial Processing of Fruits*. Park Ridge, N.J.: Noyes Data Corp., 1976.

Luh, B. and J. G. Woodroof, *Commercial Vegetable Processing*. Westport, CT: Avi Publishing Co., 1975.

McWilliams, M. and H. Paine, *Modern Food Preservation*. Fullerton, CA: Plycon Press, 1977.

Neild, R. E. and J. O. Young, "An Agroclimatological Procedure for Determining and Evaluating Time and Length of Harvesting Season for Processing Tomatoes." *Proc. Amer. Soc. Hort. Sci.* 89, 549 (1966).

Sacher, J. A., *Senescence and Postharvest Physiology*. Annual Review of *Plant Physiology* Vol. 24, 197, 1973.

Thornthwaite, C. W., "Climate in Relation to Planting and Irrigation of Vegetable Crops" *Publ. Climatol.* 5 No. 5, 1952.

Thornthwaite, C. W. and J. R. Mather, "Climate in Relation to Crops." *Meteorol. Monographs* 2 No. 8, 1, 1954.

United States Department of Agriculture, *Climate and Man. Yearbook of Agriculture*. Washington, D.C.: 1941.

Waggoner, P. E. (ed.), "Agricultural Meteorology." *Meteorol. Monographs* 6, 28, 1965.

Wang, Jen-Yu, *Agricultural Meteorology* (3rd ed.). San Jose, CA: Milieu Information Service, 1972.

Woodroof, J. G., *Tree Nuts: Production, Processing, Products*. Vol. 1 and 2. Westport, CT: Avi Publishing Co., 1967.

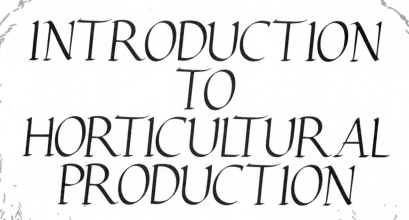

part three

INTRODUCTION TO HORTICULTURAL PRODUCTION

The science and technology of horticulture must be mastered, but unless they are put into practice, the practitioner is an incomplete horticulturist. Many commercial and noncommercial avenues are open, and these will be covered in following chapters. The satisfaction of having a horticultural career or a recreational and hobby interest can be experienced in any or all the horticultural divisions: ornamental horticulture, olericulture, and pomology. The rewards are there, whether they be beauty experienced in the flower garden, the good taste of fresh fruits and vegetables, the satisfied feeling experienced upon viewing one's landscaping efforts, or the financial reward of a horticultural career. Germane to all this is the production of horticultural crops, and special interest will be directed toward that topic in the chapters that follow.

Some of my greatest joys have been derived from being a horticulturist: working with tropical plants in my greenhouse on a snowy day, the first flowering bulbs of spring, the beauty of trees framed by the sky, the first bloom of a newly obtained species, the good conversation with other horticulturists. I could go on and on, but the fact that you are using this book leads me to suspect that you have experienced or hope to experience these and many other exciting aspects of horticulture.

Ornamental Horticulture

Ornamental horticulture, practiced by 43 million American households in 1979, can be divided into three areas: floriculture, nursery production, and landscape horticulture. Seed production of ornamentals is listed as a fourth area by some. Each area will be considered on its own merit as a viable, working concept, but the blurred distinction sometimes introduced by commercial consolidation should be remembered. Nonornamental plants (fruit trees and vegetable plants) may even be present in some operations. One concept (an essential part of this chapter) still maintains its distinctness: the plants propagated, grown, and sold.

Floriculture

Floriculture is concerned with the production and wholesale and/or retail distribution of cut flowers, potted foliage plants, greenery, and flowering herbaceous plants (Fig. 13-1). The latter category would include potted or container flowering plants utilized as house plants or for seasonal color, annual and biennial plants used as bedding plants, herbaceous perennials used in permanent plantings, ferns, and bulbs (both true bulbs and others such as corms and tuberous roots). Much of the floricultural production is directed to seasonal mar-

(A)

(B)

(C)

(D)

Figure 13-1. A few floricultural products. A. Wax begonias, ever popular as flowering bedding plants and pot plants. Shown here are mixed 'Scarlet and Pink Sensation'. Courtesy of George J. Ball, Inc. B. Marigolds in compressed peat bedding pack. These are popular flowering bedding plants. Courtesy of Jiffy Products of America. C. Coleus, popular foliage bedding and pot plants. Cultivars shown are mixed 'Saber Clown' (foreground) and 'Saber Pineapple' (background). D. Seed geraniums, a flowering bedding plant of recent popularity. 'Scarlet Flash' shown here. C and D courtesy of Pan-American Seed Company.

kets: cut flowers and potted flowering and foliage plants for special holidays; bedding plants, perennials, and potted flowering plants for the spring garden and outdoor summer living; bulbs and perennials for fall planting; and foliage and flowering plants for winter living indoors.

In terms of value, floriculture is the most significant part of ornamental horticulture. The wholesale value of ornamentals was $824 million in 1970. Of this, floriculture comprised $485 million. The retail value is about three times as great. The wholesale breakdown was $229,944,000 for cut flowers, $125,826,000 for flowering potted plants, $38,376,000 for potted foliage plants, $81,697,000 for propagation stock, $44,824,000 for bedding flowers, and $8,826,000 for greenery.

Of the cut flowers in 1970, carnations, chrysanthemums (pompom and standard), roses, and gladioli accounted for $194,875,000. In 1975 the five cut flowers accounted for $193,264,000, which is a substantial decrease in light of inflation. Little increase was seen in 1976 ($206,245,000) and 1979 ($231,000,000). On the other side, the wholesale value of potted foliage plants soared to $184,898,000 in 1975, reached $235,768,000 in 1976, and $299,000,000 in 1978. Flowering bedding plants reached $62,022,000 in 1976 and about $97 million in 1978. Cut flowers plus potted seasonal flowering plants (chrysanthemum, poinsettia, geranium, lily, and hydrangea) totaled $346,759,000 in 1976 and $404,000,000 in 1978. The wholesale value of floricultural ornamentals increased to about $700 million in 1976 and about $1 billion in 1978, with most of this increase from gains for potted flowering and foliage plants. In 1979 the total retail value of floricultural goods and services was $4 billion.

Floricultural operations tend to be dominated by greenhouse operations, where bedding plants and potted flowering and foliage plants are raised (Fig. 13–2). Some cut flowers are raised there, but many are also cultured outdoors. The production of cut flowers and foliage plants has tended toward centralization in the 1970s. California accounts for at least 45 percent of the wholesale production (1978) of cut flowers, and Florida less than 20 percent. Florida and California also account for about 41 percent and 29 percent, respectively, of the wholesale production (1978) of foliage plants. Cut-flower production is more economically feasible in these areas because of lower tax rates, lower labor costs, and a more favorable climate. Even the cost of air freight makes these cut flowers competitive with those grown locally in other areas.

Climate is a major factor in foliage plant production. Energy costs are minimal in Florida and California in regard to foliage plant production, as opposed to the ever-increasing energy costs of operating greenhouses in the north for foliage plant production. Many foliage plants can be grown directly outdoors, either in full sun or under shade.

Potted flowering plants (poinsettias, Easter lilies, chrysanthemums) and flowering bedding plants have resisted the centralization trend and remain more regionalized in production. Regional production of these persists probably because of high shipping costs and the heavy emphasis on controlled production (such as photoperiod) to ensure production mainly at holidays and proper regional spring planting time, as opposed to year round sales of cut flowers and foliage plants. A dependence on centralized production in conjunction with a shipping delay or strike would be economic disaster for those relying heavily on holiday or spring sales.

Figure 13-2. Greenhouses for raising flowering foliage plants for bedding and indoor plantings are a mainstay of the floricultural industry. William P. Raffone, Jr. photo.

Commercially, there are two areas of floriculture: (1) a business dealing exclusively with floriculture, or (2) a business that is a portion of an overall horticultural conglomerate. There is generally a production industry that mass produces floricultural crops for the wholesale market and a retail outlet for sales of plants and floral arrangements. The latter might be a retail florist or a portion of a garden shop. In some instances both the wholesale and retail portions are owned by a larger industry, which may not even be involved in horticultural sales, but has acquired the operation for diversification. Most wholesale and retail operations are controlled by single proprietors or families. However, wholesale growing operations are increasing in size, and a trend toward partnerships and corporate ownership will be evident in the future. Another future trend to watch is the inroad of imported cut flowers on American production. Carnations are already seriously affected, and the chrysanthemum is not far behind. In 1978 the amounts imported were 41 percent for carnations, 38 percent for pompom chrysanthemums, and 13 percent for standard chrysanthemums.

Production Nursery

The production nursery is concerned mainly with the production and sales (wholesale) of woody plants such as trees, shrubs, and vines, herbaceous plants used as ground covers, and turf grass sod. The woody material can be deciduous and/or evergreen, and often goes beyond ornamentals in that fruit trees may be included. Perennial vegetables are also produced by some. Fruit trees and perennial vegetables will be covered in later chapters.

(A)

(B)

Figure 13-3. A. Much of the nursery stock in production nurseries is grown in containers. B. Balled and burlapped stock is still produced in production nurseries, but is on the decrease. William P. Raffone, Jr. photos.

Of the $824 million wholesale value of ornamental crops in 1970, about $289 million was accounted for by production nursery crops. The biggest portion of this was $245,542,000 for woody and herbaceous plants, of which the bulk was trees and shrubs. Sod wholesale value was about $43 million in 1970 and $100 million in 1976. About 25 percent of the wholesale value (woody and herbaceous plants excluding sod) was concentrated in California. Other important production states included Connecticut, New York, New Jersey, Pennsylvania, Ohio, Illinois, Iowa, Florida, Tennessee, Texas, and Oregon. These each accounted for 3 percent or more of the sales. New York and Michigan were the leading wholesale producers of northern sod, and Florida of southern sod.

Nurseries are divided into wholesale, retail, and mail-order types. Wholesale nurseries usually specialize in large-scale growing of a few crops, with certain firms noted for their specialities (for example, roses, broad-leaved evergreens, or bulbs). Often the specialities are determined by the climate where the nursery is situated. Some of the crop is sold in the form of *lining-out stock,* which is purchased by other wholesale, mail-order, or retail nurseries and grown on to marketable size, which may require up to several years. Other wholesale nurseries sell finished products directly to the retailer. Many retailers have stopped growing nursery stock and act only as a distributor between the wholesalers and the consumer. Trends are toward container-grown nursery stock because of less labor, better acceptance by consumers, quicker recovery after transplanting and year-round marketability (Fig. 13–3). Balled and burlapped (B&B) stock is still used, but the labor cost is reducing the use of this form (Fig. 13–3). Bare-root stock is used for small deciduous plants, but losses can run higher with this form because of failure to control temperature and humidity at all storage sites.

| Landscape Horticulture | Landscape horticulture can be broken into *landscape design, landscape construction,* and *landscape maintenance.* Landscape design is concerned with the *design* (site planning) of ornamental plantings (see later section); landscape construction is involved with the *total preparation of the site* for the planting, which may be for a school, home, shopping mall, housing development, business, industrial complex, church, and so on (Fig. 13–4). This includes the *procurement of the necessary plants,* trees, shrubs, sod, vines, and so on; the *modification* of the site which may involve grading, soil modification, walls, terraces, pools, paved surfaces, drainage systems, and excavation, *plant installation,* and *site maintenance* until construction is complete. Landscape maintenance involves *maintaining established ornamental plantings,* which includes fertilization, pruning, mowing, and plant protection against insects, disease, and weeds. Established ornamental plantings |

are mainly lawns, shrubs, trees, and gardens of private homes, schools, churches, commercial establishments, parks, golf courses, cemeteries, and the like.

The above services may be provided from several sources. Landscape architects or designers (see Careers) are involved in landscape design, landscape contractors in landscape construction, and landscape maintenance firms in landscape maintenance. In some instances a retail nursery (see Production nurseries) may attempt to provide design, construction, and even maintenance services in addition to retailing ornamentals. Combined operations of this sort are referred to as landscape nurseries.

Landscape horticulture is a billion dollar plus industry. Gross receipts alone for the service aspects, that is, exclusive of landscape contractors, approached the billion dollar mark in 1976. These horticultural services ranked second only to animal husbandry services in all the agricultural services. More than 50 percent of the firms providing landscape horticultural services were concentrated in five states: Pennsylvania, New York, California, Ohio, and Florida. The majority were owned by individual owners.

Design

Landscape design and architecture are core elements of ornamental horticulture and, to a lesser extent, the other areas of horticulture. Design is intimately associated with humans, land, plants, structures, and the environment. Its aim is to balance the requirements of each, but yet present an aesthetically pleasing image. The final product must be economical, functional, beautiful, safe, durable, and have a positive impact upon the environment. Good design has a profound effect on the environment. Bad design does not have a positive impact, and may even be harmful if physical alterations are not done with foresight. The landscape designer or architect should work with and not separately from the building architect.

The concepts involved are quite extensive; indeed, they can be found in texts devoted entirely to landscape concepts. The area of concern to us is only the relationships possible among ornamental plants, horticultural structures (walls, fountains, pools, and the like), and the site, and how to integrate them with nonhorticultural structures (home, church, school, industrial building). Only minimal detail is provided.

The basic elements of design are color, form, line, and texture. Plant *color* varies considerably, and all parts must be considered: foliage, stems, bark, flowers, and fruit. Variations in color with season must be considered. A few examples include the dark green, seasonal color of a deciduous tree versus the blue-green of certain evergreens, or the reddish bark apparent in the winter with certain deciduous woody plants. Colors involved with inanimate nearby structures (walls, pools, fountains, paving, walks) must also be weighed.

(A)

(B)

(C)

(D)

(E)

(F)

(G)

Figure 13-4. Landscape nurseries are involved in simple though complex horticultural projects as shown here. A. Paved garden walkway showing imaginative use of common materials. B. Patio deck which gives the illusion of floating over the fountain pond. Not only good design, but good landscape engineering is evident here. C. Even a child's play area can be made to be pleasing in the overall landscape. Photos A–C from American Society for Horticultural Science. D. Landscaping does much to enhance the aesthetic appeal of this home. U.S. Department of Agriculture photo now in National Archives (16-N-31746). E. Penthouse terrace looking over New York City. F. Riverside Mall in California. G. Industrial headquarters. Shopping and working areas can be more enjoyable with the aid of landscape nurseries. Photos E–G, U.S. Department of Agriculture.

Plant *form* is essentially the visual impact of the total mass of a plant or, collectively, of a group of plants. A low, spreading plant and an upright, narrow plant have horizontal and vertical forms, respectively. A long hedge composed of numerous plants has a horizontal form. A large evergreen might be said to have a conical form with some species or a columnar form with others.

Collective plant masses or horticultural structures may be used to guide or direct eye movement; this constitutes the concept of *line.* A hedge or walk or wall may be used to direct the eye toward a focal point, such as a specimen tree or an attractive fountain.

The visual impact of the surfaces of plants and horticultural surfaces denotes *texture.* Grass and pea gravel have a fine texture, whereas a large-leafed tree would have a coarse texture.

These design basics must be artistically blended when plantings, horticultural structures, and buildings are considered as an overall picture. Plantings and structures should be mixed to produce *variety,* instead of a monotonous sameness. Some *repetition,* however, is necessary to bring order into the overall image. Skillful use of plantings and structures can be used for *emphasis* of a focal point. The overall arrangement must be considered for *balance,* whether it be exact on either side of an axis, as in a formal garden, or a bit asymmetrical to give an informal appeal.

In design, the environmental requirements of the planting must also be known. Failure of plantings would seriously harm the reputation of the landscape designer or landscape architect. The mature form and heights of plants must be known to maintain the proper *scale* of plantings with time. A planting that in time obscures the facade of a building and is not amenable to pruning obviously distorts scale.

The effects of microenvironment, especially in urban situations, must not be neglected. On the whole, the environment may appear suitable for a planting, but the microclimate may be sufficiently adverse to cause failure of the planting. One example in an urban situation will suffice (see Chapters 5 and 7 for others). Landscape installations in high-density urban sites are often under the detrimental influences of microclimates. This can be manifested by daily late summer-early fall temperature fluctuations. During this season, in particular, air temperatures can be halved in a matter of minutes. Late afternoon sun, perhaps intensified by building proximity (particularly buildings with reflecting surfaces), can produce very high desiccating air temperatures (120°F plus). Moments later the sun is shielded by buildings, and air temperatures often drop suddenly (60° F, plus or minus). This can be detrimental to plant development. Microclimates in these sites often influence air temperatures in spring and/or fall to the extent that "growing seasons" are extended. This influence can result in late season growth that is unable to harden off and therefore is very susceptible to cold temperature injury in the form of twig dieback. This environmental influence may also produce early season growth that can be very susceptible to cold injury produced by early spring nocturnal freezes.

Design as stated before is a unifying concept in ornamental horticulture, which involves close cooperation with activities in other areas, especially site preparation and subsequent maintenance. Necessary modifications of the site for purposes of design and to provide a satisfactory environment (grading, drainage, and so on) can only be achieved through cooperation between the designer and the landscape contractor. A mistake could cause loss of plants (improper drainage) or an incorrect design (misplacement of wall or a plant). Close attention to planting is critical, as improper planting may mean an unrealized final design. Maintenance personnel must know what the designer had in mind. Formal pruning of a woody plant could destroy the informal effect the designer had intended for the landscape.

Ornamental Seeds

The production of ornamental seeds, exclusive of grass seed, is the smallest in terms of wholesale value; it accounts for at most about 5 percent of the total wholesale value of ornamentals. However, its value to homeowners and floriculturists makes it very important; garden seeds alone retail yearly at $125 million.

Plant growing and seed harvesting are important aspects of ornamental seed production. Seed processing includes milling, cleaning, packaging, and storing. Much of this was covered in the chapter on propagation. California is the leading state in the production of ornamental seeds. Much seed production is under contract with seed houses. As with nurseries, there are wholesale, retail, and mail-order parts to the seed industry.

The wholesale value of grass seed in 1976 was estimated at $150 million for clean field seed at the farm level. Oregon, followed by Missouri, is the leading producer of grass seed.

Careers in Ornamental Horticulture

There are a large number of career positions associated in whole or part with ornamental horticulture. A brief coverage follows, which is not necessarily all inclusive, and is not meant to imply that 100 percent of a person's activity will be concerned with some aspect of ornamental horticulture. For more detail consult Further Reading under G. M. Kessler.

FLORICULTURE

The production area in floriculture involves four basic positions: production superintendent, grower, marketing manager, and an inventory controller. The production supervisor coordinates production and oversees the growers, who are actually responsible for the production of floral crops. Scheduling of crop production for correct timing and maximal profit is handled by the inventory controller. The marketing manager controls the handling of the harvested flowers and the sales (wholesale) and deliveries that follow.

The wholesale commission florist is an intermediate firm between the production and retail aspects. There is usually an overall manager, buyer, sales manager, and salesmen, whose positions are self-explanatory. This group usually handles cut flowers and hard goods.

The retail florist is the main outlet for retail sales of floricultural crops and products, but garden centers also offer a fair share. Some products are also sold by supermarkets and discount stores. The florist shop generally has a manager, often the owner, sales personnel, and floral designers. There may be a managerial position associated with the floricultural portion of the garden center. Generally, little growing, other than holding plants until sold, is involved in the retail end.

PRODUCTION NURSERY

Positions in the production nursery include the manager, propagator, inventory controller, field foreman, field superintendent, sales manager, shipping foreman and traffic manager, salesmen, and a broker.

The main manager is responsible for the overall operation of the nursery. The propagator is responsible for propagation to produce marketable stock and is usually in charge of a propagation crew. The field foreman oversees the crews that bring the recently propagated stock to marketable size. The field superintendent is actually responsible for all production stages, from time of propagation through reaching marketable size and must make many decisions involved with fertilization, pest control, and the like.

The inventory controller coordinates order requests with stock. The sales manager, salesmen, and shipping foreman and traffic manager are self-explanatory positions. The broker (usually self-employed) is a middleman between retail outlets, such as garden centers, and the production nursery.

LANDSCAPE HORTICULTURE

Landscape architects have a degree in landscape architecture and are required to be licensed in most states. Landscape designers are usually unlicensed and do not have a degree in landscape architecture, but do have sufficient training and education. Both specialize in landscape design either as self-employed individuals or as employees of firms involved in landscape horticulture.

Landscape contractors are usually self-employed and direct the men and equipment needed to provide the wide range of landscape construction activities described previously.

The landscape nursery operation is run overall by a manager. Landscape designs for homes and small buildings are usually handled by a landscape designer working on commission for the nursery, and larger projects (industrial complexes, housing developments, schools) are mainly handled by a landscape architect, who is self-employed in many cases. The construction superintendent corrdinates construction jobs, crews, plant material, and construction equipment. The construction foreman oversees construction activity. Services are sold by a salesman, who may double as a landscape designer.

Some specialists may be employed by the landscape nurseries, or they may depend on subcontractors. The latter include landscape contractors involved with site preparation, site modification, and construction of horticultural structures. Other specialists may be required for irrigation installations and lawn construction. These services must be provided by trained people, otherwise the reputation of the nursery may suffer.

Landscape maintenance firms are run by a manager, who may also oversee assignments in small firms. Larger firms employ a superintendent of operations for the latter task. The crew foreman oversees the work crews, and services are sold by a salesman.

Firms may specialize in one maintenance activity or several. Some of the specialities are as follows. One speciality is grounds maintenance for private individuals, public and private institutions, businesses, cemeteries, arboreta, parks, botanical gardens, and others. This service may include planting and transplanting, and even plant improvement, depending on whether it is involved with landscape contractors or with a speciality firm. When plant improvement is included, the speciality is more correctly named *plantation maintenance;* this form is more apt to be found at arboreta, botanical gardens, and private estates. Another speciality is arboriculture, which deals with maintenance and management of woody plants. Turf care is another common speciality. Chemical control with pesticides, growth regulators, and fertilizers is still another.

GARDEN CENTERS AND RETAIL NURSERIES

Garden centers and retail nurseries are basically retail outlets for production nursery crops, floricultural crops, nonornamental crops, horticultural hard goods such as fertilizers and tools, and sometimes horticultural services such as designing. Development and maintenance of the home landscape by the homeowner are made possible by such outlets, which may be individually owned, a partnership, an incorporated chain, or part of a horticultural or other conglomerate. They generally have a manager and buyer, and sometimes a designer and plant doctor. Although these operations are not completely involved in ornamental horticulture, it usually constitutes a large share of the business.

ARBORETA AND BOTANICAL AND HORTICULTURAL GARDENS

Horticultural institutions are involved with plant collections (not completely ornamental plants) and require people knowledgeable in ornamental horticulture. Some of these positions are the director of the institution, a superintendent of horticultural operations, a propagator, a curator (involved with planning, obtaining, labeling, and so on, of plant material), a greenhouse manager, a librarian, writers, educational director, and various researchers (see education and research).

EDUCATION AND RESEARCH

Teachers of ornamental horticulture (and other areas) are needed in high schools, vocation training or skill centers, community colleges, colleges, and universities. Cooperative extension service agents, dealing in whole or part with ornamental horticulture, are found in all states. They are educational consultants operating on a governmental level and dealing with homeowners, commercial horticulturists, and others in need of their services. Private self-employed consultants may also provide similar services. Writers and lecturers (usually self-employed) may provide ornamental horticultural information in magazines, books, newspapers, radio, television, and lectures.

Researchers dealing with various specialities within or associated with ornamental horticulture can be found in universities, state agricultural experiment stations, federal research laboratories, industry, arboreta, and botanical gardens.

SEED PRODUCTION

Commercial seed firms dealing with ornamentals need plant breeders, propagators, growers, sales managers, market specialists, and researchers.

HORTICULTURAL THERAPISTS

Horticultural therapists work with physically or mentally handicapped people. Much of the activity is involved with ornamental horticulture.

**Plants in
Ornamental
Horticulture:
Annuals**

Annuals are plants that pass through the vegetative and reproductive cycles and senescence in one growing season (Fig. 13–5). Plants that pass through the vegetative and reproductive cycles but not senescence, but do die as a result of killing frosts, are also treated as annuals. Annuals are the mainstay of the flower garden, since they give a greater effect for less money and labor than other flowers, and their exciting, long-term flowering makes them ideal choices for summer color in the landscape.

Accordingly, they are valued both by the home gardener and commercial horticulturists, who have established the bedding plant industry because of their appeal. Over 25 million flats of flowering bedding plants were sold in 1978. Annuals sold in this manner include ageratum, alyssum, aster, calendula, celosia, coleus, dahlia, dianthus, marigold, pansy, petunia, phlox, portulaca, salvia, seed geranium, snapdragon, verbena, vinca, viola, and zinnia. These are the flowering bedding plants that succeed in a sunny to partly sunny location. Those adaptable to shadier areas include begonia and impatiens.

There are many more annuals than these. Good seed houses sell at least 150 annuals, and there are many varieties of each of these to choose from. Surprisingly, many of these are seldom seen in gardens. There is no excuse for being bored with annuals.

The growing treatment of annuals provides a basis for two groups: (1) those that are sown in the place where they are expected to bloom, and (2) those that should be started early in a hotbed or greenhouse in order to provide sufficient time for blooming during the growing season. Some of those sown directly in place, if sown earlier indoors, can be expected to have an extended blooming period because of the earlier start. However, those that are fall short-day plants will not benefit from an early start.

Scheduling of propagation dates (see Table 13–1) is of extreme importance to the floriculturist involved in the production of annual bedding plants. Such plants must attain a size sufficient to assure flowering at or close to the correct marketing and planting time range.

Uses for annuals include edging purposes, colorful massing, quick screening, cut flowers, foundation plantings, window and porch boxes, and patio containers (Fig. 13–6). Uses are determined by such factors as seasonal time of bloom, height, frost tenderness, response to hot or cool weather, appearance, color, disease resistance, insect susceptibility, and lasting qualities as cut flowers. Extensive information on many (but not all) annuals is found in Table 13–1.[*] Most should not be started outdoors until all danger of frost is past (Fig 13–7). Some annuals are hardy and will tolerate a light frost. Those started indoors are started 6 to 10 weeks prior to setting out in the garden, depending on their speed of development. Starting seeds indoors and transplanting were discussed in Chapter

[*]Tables are at the end of the chapter.

(A)

(B)

(C)

Figure 13–5. A few of the hundreds of available annuals. A. Marigolds, front: 'Red Nugget' Hybrid^{TM}, back: 'Gold Lady' Hybrid. Courtesy of Burpee Seeds. B. Begonia 'Glamour White'. Courtesy of Pan-American Seed Company. C. Zinnia, 'Peter Pan Gold'. Courtesy of All-America Selections. D. Petunia 'Chiffon Magic'. Courtesy of Pan-American Seed Company. E. Cosmos. F. Celosia. G. Verbena. H. Bells of Ireland (*Molucella laevis*) I. Sunflower (*Helianthus annus*). E–I. courtesy of National Garden Bureau, Inc.

(D)

(E)

(F)

(G)

(H)

(I)

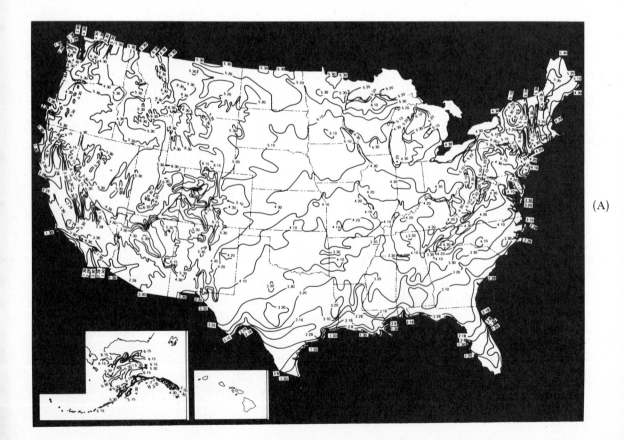

Figure 13-6. A. Annuals used for colorful massing in a flower border. U.S. Department of Agriculture photo. B. Annuals used in containers to enhance balcony. Courtesy of National Garden Bureau, Inc.

6. Depth to sow seeds varies and is indicated on seed packets. A good rule of thumb is to cover a seed two to three times its thickness, and very fine seeds such as petunia or begonia should not be covered, but lightly pressed into the soil. Seeds that require light for germination should be pressed into the media and not covered with growing media, paper, or anything opaque during germination.

A word or two about the annuals in Table 13-1 is in order. The majority are not true annuals, but perennials or biennials treated as annuals. If they are used in areas where they are winter hardy, their perennial habit would be observed. These perennials and biennials can be expected to bloom the first year, if they are given an early enough start. This is accomplished by starting them indoors if the growing season in your area is too short or if earlier blooming is desired. Alternately, if their seeds are of sufficient hardiness, they can be started early outdoors directly after the danger of heavy spring frost is past or planted in late fall just before the ground freezes. Because of the extensive variation of growing conditions in the Unites States, it would be wise to check with your local agricultural

Figure 13-7. A. (page 428) Mean date of last spring frost. B. (below) Mean date of first fall frost. These two dates determine the length of the growing season for annuals. U.S. Department of Agriculture photos.

(B)

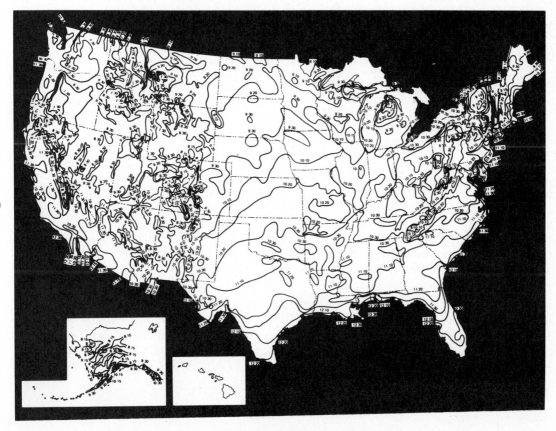

extension service for exact treatment in your area. Not all the annuals given in Table 13-1 will be suitable for your area either, and this should be checked with a local source. For example, some of the annuals would succeed well in California but not New England. In addition, some of these annuals may self-sow in your area. Finally, where a genus is indicated, it should be borne in mind that not all the species will necessarily be in cultivation.

Biennials

Biennials are plants that require two growing seasons to complete their life cycles. Generally, the first year is vegetative in habit, and the second reproductive. Many biennial ornamental plants are treated as annuals by giving them an early start indoors or by seeding in late fall or early spring outdoors, as mentioned in the preceding section on annuals. Some perennials are also so short-lived as to be regarded as biennials. Biennials are less popular than annuals, which give quick returns for minimal labor, and perennials, with their more intensive care, but lasting returns.

Examples of biennials include the following ornamentals: *Alcea rosea* (hollyhock), of which an "annual" strain exists: *Campanula medium* (Canterbury bells); *Daucus carota* (Queen-Anne's lace), *Digitalis purpurea* (foxglove), *Lunaria annua* (honesty), *Verbascum* (mullein), and *Viola* X *Wittrockiana* (pansy). Pansies are one of the more popular biennials and are frequently sold as bedding plants for early spring color.

Perennials

Perennials are plants that persist for more than 2 years (Fig. 13-8). Many of our favored ornamentals, such as shrubs, trees, and bulbs, are perennials. However, horticultural usage of the word perennial has come to imply herbaceous, nonbulbous flowering perennials used in beds or borders. This approach will be used here, and the woody or bulbous perennials will be treated in separate sections.

Many perennials can be started from seed, and it is economical if extensive numbers are required. Drawbacks include the care of plants until they reach blooming size, usually about two years, and that many of the more modern varieties of perennials do not come true from seed. The latter are propagated asexually. The old-fashioned perennials can be profitably raised from seeds, but plants of the newer ones should be obtained from a nursery. Once these plants are established, it is always possible to increase their numbers through such asexual means as division and cuttings of stems and roots.

Figure 13-8. A few horticultural perennials. A. Some species and cultivars of *Rudbeckia* are short-lived perennials. Courtesy of National Garden Bureau, Inc. B. *Aquilegia*. These columbines are 'McKana Giants'. C. Evergreen candytuft (*Iberis sempervirens*). B and C are Herbst Brothers Seedsmen, Inc. D. Chrysanthemums are well-known perennials. U.S. Department of Agriculture photo now in National Archives (16-N-22832).

The perennial border must be carefully planned. It will last for years, and a poorly planned one will give years of dissatisfaction. Since there is a reasonable amount of work involved in establishing a perennial planting, the arrangement should certainly be right the first time. Things to consider in the plan are height, spacing, color, hardiness, time of bloom, and required cultural conditions. These factors are considered for a number of perennials in Table 13-2. The ideal perennial border or bed provides a succession of bloom throughout the growing season, but such an effect requires moderately high maintenance.

Table 13-2 does not include all the perennials in cultivation, but does include most of those more popularly cultivated in the various regions of the United States. Care should be observed in considering the hardiness zones. A perennial in California may very well be an annual or pot plant in the North. In fact, reference should be made to Table 13-1, which includes a large number of additional plants that are perennials in the warmer areas of the South. In addition, zones are not absolute. Factors such as microclimates must be considered, as well as additional environmental conditions that may eliminate a plant for your area, even though it would be of sufficient hardiness. An example of such a condition would be the arid-alkaline region found in the Southwest. One final point is that a listing such as in Table 13-2 cannot go beyond a species level because of space problems. The numbers of cultivars for chrysanthemum and iris alone would fill a table. The references cited at the chapter's conclusion will provide information on cultivars.

Planting time (and hence finished production time in a nursery) for perennials varies according to region. Many perennials can be planted in the North as soon as the ground can be worked in early spring. This appears to be better than early fall planting in the North; however, early fall planting is equally possible in warmer sections. This gives the perennial root systems time enough to become established. Late fall planting is to be avoided in the North, for the root systems fail to become established, and the plant is likely to be heaved from the soil during alternate freezing and thawing.

Perennial plantings require some additional efforts beyond those of watering, weeding, fertilizing, and spraying for insects. Some of the taller perennials will require staking. Perennials become crowded through natural propagation and will require some attention to division of the clump at intervals, which vary according to the species and existing cultivation. If this is not done, flowering will become sparser and even nonexistent. Division can usually be done most favorably in the early fall. Certain perennials can also be divided early in the spring. When this is done, care must be taken not to damage the young shoots. Other perennials may not last for more than 3 or 4 years and need propagation every few years to keep them at peak blooming. Still others may need little attention over several years.

Bulbs

The term bulbs, when used in a horticultural sense, does not have the narrow connotation attributed to it by botanists. To the horticulturist the word bulb suggests such propagules as corms, tubers, and thickened rhizomes, as well as actual bulbs. Most of the plants arising from these structures are simple-stemmed perennials. Generally, they are divided into two groups: spring and summer bloomers (Fig. 13-9). Most spring-flowering bulbs require a cold period and are limited to areas of minimal to extensive frost. Summer-blooming bulbs in some cases may be limited in cold hardiness and are treated as annuals or pot plants in the colder areas of the United States.

Spring-flowering bulbs are quite popular, since their blooming period precedes that of the annuals and bedding perennials covered previously. Summer-blooming bulbs offer color and diversity in formal plantings beyond the annuals and bedding perennials. Culture of either bulb group is relatively straightforward. The versatility of bulbs leads to their use in naturalized or formal plantings.

Bulbs are planted at various depths, depending on species and varieties. The depth is meant as the distance from the top of the bulb to the soil level. Planting depths also vary according to soil texture. Soils containing much sand (sandy loam) require a planting depth about one third greater than a sandy clay loam, and a clay loam would require a depth about one third less than that in the sandy clay loam. A general rule of thumb is that bulbs 2 inches or larger in diameter are planted at a depth of twice or thrice their diameter, and smaller bulbs are planted at depths three or four times their diameter. Approximate spacing is about twice the planting depth.

A well-drained soil with fertility maintained at a moderate level is essential, since many bulbs will be left undisturbed for many years. Others may peter out and need replacement more frequently. Still others will become crowded with decreasing blooms, and lifting, dividing, and replanting will be required. Those treated as annuals have similar soil fertility requirements.

Spring-blooming or hardy bulbs are planted in midfall up to when the ground freezes too hard to be dug. A mulch is often used on beds where the ground is subject to alternate freezing and thawing to prevent heaving of bulbs and to increase survival of half-hardy bulbs.

After flowering is over, the foliage should be left until it yellows and dies or is frost damaged. The foliage is involved in manufacturing food, which will be translocated to the bulb and stored for the following year's growth. Good soil fertility is essential at this time, and supplemental fertilization may be required. Quite often annuals or other perennials may be used to hide the unsightly or uninteresting foliage of spring bulbs at this stage.

Summer-flowering or tender bulbs are usually planted when the danger of spring-killing frosts is past. Frost-tender bulbs after a light frost

(A)

(B)

(C)

Figure 13–9. Some examples of horticultural bulbs. A. Snowdrops (*Galanthus* cv.). B. *Crocus*. C. Daffodil (*Narcissus* cv.). D. *Hyacinthus* sp. A through D are spring flowering. U.S. Department of Agriculture photos. E. *Lilium* 'Enchantment', summer flowering. Courtesy of Oregon Bulb Farms. F. *Dahlia*, single bedding type used for summer flowers. William P. Raffone, Jr. photo. G. *Caladium*, noted for summer foliage outdoors and as a pot plant. Author photo.

(D)

(E)

(F)

(G)

(30° to 32°F, –1° to 0°C) should be trimmed of dead and damaged foliage and lifted from the soil. They are next cured of wounds or bruises under conditions that minimize shriveling and disease problems. These curing conditions are 2 to 4 weeks at 60° to 70°F (16° to 21°C) and 40 to 50 percent relative humidity. After curing, a dusting with a fungicide or insecticide is often used to eliminate disease problems, and waxing is sometimes used to prevent shriveling. Storage conditions for the various species vary, but the relative humidity must be high enough to prevent shriveling, but not high enough to encourage disease, and cool temperatures below 50°F or 10°C are best. Tender bulbs are often packed in sand, sawdust, or peat moss and inspected periodically for disease and/or shriveling. Moisture is added as required. Commercial horticulturists use temperature and relative humidity-regulated storage facilities.

Spring-flowering bulbs may be stored in cool, well-ventilated areas out of direct sunlight. Temperature will vary according to the type of bulb, but the homeowner may use a range between 40° and 50°F (4.5° and 10°C). This is needed for fall bulbs upon receipt by the homeowner if they are not to be planted soon. Long-term storage by commercial horticulturists after harvesting bulbs, of course, requires temperature and relative humidity regulation as determined by bulb type.

Spring-flowering bulbs may be forced indoors for early winter color or for holidays such as Easter. Bulbs that are usually forced and sold potted include *Hyacinthus orientalis* (hyacinth), *Narcissus* (includes daffodil, narcissus, jonquil), *Tulipa* (tulip), *Crocus, Muscari* (grape hyacinth), *Iris danfordiae,* and *I. reticulata.* Daffodils and tulips are also sold as cut flowers after being forced. Forcing is the only way some of the warmer areas can use these bulbs, because of their cold requirement. The procedure is relatively easy, but strict attention to a time and temperature program is essential if the grower desires blooming bulbs for a specific day. This type of treatment is beyond the scope of this text, but information can be obtained from the references at the end of the chapter.

The art of forcing, for those who are not concerned with a specific day, is as follows (Fig. 13–10). First, the cold requirement of the bulbs must be satisfied. Cold temperatures are needed to promote flower stem elongation. This can be done prior to (precooling) or after potting them, and the cold treatment can be naturally or artificially supplied. A temperature between 41° to 48°F (5° to 9°C) supplied through refrigeration can be used with potted or unpotted bulbs. Hyacinths may also be forced in special hyacinth bulb glasses filled with water at similar temperatures. The treatment lasts for about 8 to 12 weeks, and no natural light should be provided at this time. Rooting also occurs with potted bulbs during the cold storage period, but precooled bulbs can be rooted at warmer temperatures. Potted bulbs are generally ready for flower forcing when roots are seen to come from the drainage hole of the pot. This type of treatment can be done whenever bulbs are available. If it is done in November or December, bulbs can be flowered during the winter.

(A)

(C)

(B)

(D)

(E) 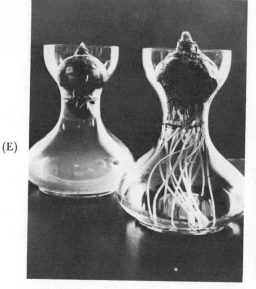  (F)

Figure 13-10. Steps in bulb forcing. A. Place crocking over drainage hole in clean pot to prevent soil leakage. B. Place growing media in pot to height such that top of bulb will be about ½ inch below top of pot. C. Fill with growing media around bulbs leaving bulb tip slightly exposed or barely covered. Water and place in cold storage (see text). D. When shoots appear and root growth is good, put pots in cool area with indirect light. E. If growing hyacinths in hyacinth glasses, move to cool, indirectly lit area when appearance is similar to glass on right. F. After a few days the shoots become green and pots are moved to a bright sunny area with cool temperatures. U.S. Department of Agriculture photos.

Potted bulbs can also be placed outdoors in the late fall in a cold frame and covered over with several inches of hay, dry leaves, or peat moss. An unheated garage may also be used, with the pots placed in boxes and covered over in the same manner.

Whichever form of cold treatment is used, one thing must be remembered. The soil must not dry out. This should be checked periodically. A growth of grayish mold may be observed on the soil. It can be ignored, as it will disappear when the pots are brought into a sunny, warm environment.

Once past the cold treatment, bulbs can be moved into a warmer but not sunny area to allow acclimation of the new sprouts. Temperatures may range from 50° to 68°F (10° to 20°C). The higher the temperature, the more rapid the flowering and the longer the flower stalk. Pots should be moved into the direct sun after a few days in dim light or after the whitish sprouts become green.

Examples of spring- and summer-flowering blooms and their cultural information are shown in Table 13-3. In some cases only the genus is indicated because of numerous species and cultivars, which may or may not all be readily available. Examples of cultivars recommended for forcing are shown in Table 13-4.

(B)

(A)

Figure 13–11. Ferns. A. *Adiantum pedatum* var. *aleuticum* (Aleutian maidenhair fern), a hardy fern of native origin. B. *Platycerium stemaria* (Triangular staghorn fern), a tender fern of tropical origin. American Society for Horticultural Science photos.

Ferns

Ferns, although not a substantial part of ornamental horticulture, are still of value for three reasons (Fig. 13–11). First, they are an important group of nonflowering, foliage plants that can be useful in perennial borders, rock gardens, or naturalized plantings of a woodsy nature. Second, their foliage is valued by the florist for floral arrangements. Finally, they are attractive pot plants found in greenhouses and homes.

Two groups of ferns are generally recognized by the horticulturist. The first consists of hardy ferns of native and exotic origin, which can be cultivated outdoors. The second group includes ferns originating in tropical areas, which restricts their use in most parts of the United States to the greenhouse and home.

Transplanting of ferns requires about the same effort as for flowering perennials. Soil and light conditions should be similar to the natural environment of the ferns' natural habitat. Many ferns are partial to full or partial shade when grown outdoors and require filtered sunlight or lower intensities of light indoors. Ferns often grow in soils having a high humus level and a covering of organic litter, such as leaves. Since the rhizomes are often near the soil-litter interface, it is best not to cultivate in areas planted to ferns. Most ferns require added soil moisture during dry periods and usually high humidity levels in the greenhouse or home. Fertilizer requirements are minimal for ferns, particularly since the organic matter and litter provide small amounts of nutrients.

Many ferns of interest to the horticulturist are shown in Table 13–5.

Introduction to Horticultural Production / Part III

Lawns

Lawns are generally regularly mowed plantings of grass (Fig. 13-12). Their purpose is to enhance the visual appeal of open spaces around buildings and parks, to provide useful outdoor living space, and to improve environmental quality. The latter consists of soil improvement and conservation, as well as a modifying influence on drastic changes in soil-air temperatures.

Grasses are generally used for lawns, since their basal meristem allows mowing without plant loss. One exception is *Dichondra micrantha,* which is used as a grass substitute in the southwest United States and especially California. Ground covers other than these are utilized, but not in areas requiring mowing. Their uses will be covered in the next section.

The use of grass species is dictated by climate. The main two climatic areas for grasses are the North and South. The first climatic region consists of the northern two thirds of the United States. The dividing line between the northerly and southerly regions may be considered to extend from Washington, D.C., through northern Tennessee and Arizona to San Francisco, California.

Figure 13-12. Lawns are an accepted part of the home landscape. U.S. Department of Agriculture photo.

The species of choice for the North include *Poa pratensis* (Kentucky bluegrass), *Lolium perenne* (perennial ryegrass), *Festuca rubra* (fine fescue), and *Agrostis* sp. (bent grasses). Grasses of choice for the south include *Cynodon dactylon* (Bermuda grass), *Zoysia* sp. (zoysia), *Stenotaphrum secundatum* (St. Augustine grass), *Paspalum notatum* (Bahia grass), and *Eremochloa ophiuroides* (centipede grass). Characteristics of each and some available cultivars are shown in Table 13–6. In addition, some grasses utilized for difficult or specialized situations are indicated.

Lawns can be planted either from seed or by vegetative means, such as sod, sprigs, or plugs. Seeding is more economical, and, unlike with sod, there is minimal danger of introducing diseases, weeds, or insects. Sod offers the advantage of quick, but costly cover. Sprigs, individual stems, and plugs (biscuits of sod) are economical means of establishing grasses that do not come true from seed.

In the North, seeding is done either in early Spring or, better still, from mid-August in the northernmost states to mid-September for the lower reaches of the North. Vegetative establishment is the method of choice in the South.

Seedbed preparation is the same for either seed or vegetative means. Basically, the bed must be cultivated to at least 3 inches, such that soil lumps larger than a golf ball and vegetative and inorganic debris are not evident. Excessive cultivation is avoided to minimize breakdown of soil structure. Fertilizer and lime are added according to soil tests (see Chapter 7). Organic matter need not be added, unless the soil structure is poor, since the established grass will build up organic matter. The soil is settled by watering, not rolling, which compacts the soil. If light rolling is deemed necessary, it should be done when the soil is not wet.

Seed is either hand or mechanically sown. Unless the surface is not crumbly, no light dragging or raking is required after sowing. The seed may be covered with a light straw or excelsior mulch or woven net. This prevents washing of seed by rain, conserves moisture, and reduces seed losses to birds. Watering is especially important during germination and subsequent establishment of seedlings or sod.

Sections of established lawns can be rejuvenated as follows. Vegetation is removed by either mechanical or chemical means. The surface is roughed, such as by hand raking or scarifying machines. Seeding and subsequent care are the same as outlined previously.

Established lawns require regular mowing, periodic fertilization and adjustment of soil pH, irrigation when needed, and pest control. Mowing height varies according to grass type as follows: northern bent and southern Bermuda grasses, 1 inch; Kentucky bluegrass, perennial ryegrass, fescue, Bahia, and St. Augustine grasses, 2 or 3 inches; improved varieties of Kentucky bluegrass, most zoysias, centipede, and common Bermuda grasses, 1½ to 2 inches. Mowing is required when the grass exceeds 50 percent of its cut height.

Fertilization and liming are best done according to soil test. Fertilization is commonly done in the spring and fall with a complete fertilizer (N-P-K) at a rate of 1 to 2 pounds of nitrogen per 1000 square feet. Slow-release fertilizers in addition to quick-release forms are used to extend nutrient availability. A pH of 6 is satisfactory and obtained through adding lime to acid soils or sulfur to alkaline soils.

Lawns require about 1 inch of water per week. If natural rainfall is less, some form of irrigation must be provided to maintain the lawn in optimal condition. Weeds, diseases, and insects can be problems, and preventative or control measures will be required. A more specialized text should be consulted for details.

Ground Covers

Plants selected for ground covers are herbaceous perennials or low woody plants requiring minimal care (Fig. 13–13). They are usually used in place of lawns on areas where lawns fare poorly or not at all or on sites where lawn care is difficult, such as slopes. As a rule, ground covers are not suitable for walking or recreation.

Beds for ground covers should be prepared the same as for lawns, except that they should be prepared to 10 inches deep if possible. Spacing is determined by several factors: cost, size of area to be covered, rate of growth, and the importance attached to the time for achieving full cover. A mulch is important for minimizing weed competition and maximizing water retention during the establishment of the ground cover. Once a ground cover is established, weeding is minimal. Fertilization and watering are done as needed. A number of plants suggested for ground covers are indicated in Table 13–7.

Woody Perennials

Woody perennials include both deciduous and evergreen trees, shrubs, and vines (Figs. 13–14 through 13–17). Woody plants have great value in the overall landscape because of their enduring, year-around effect. As such, they are the mainstay in landscaping planning, whether utilized for background, screening, foundation plantings, visual impact, shade, or windbreaks.

Obviously, the woody plant group is very diverse, but there are some factors common to all that should be considered when choosing plants. Foremost is the habit of growth, which determines the ultimate height and width of the mature plant. The placement of a plant in a landscape will ultimately be determined by this fact. Hardiness is also important for determining the ultimate success of a plant in a given area. As before, it must be stressed that hardiness zones (Fig. 13–18) are at best a guide to relative hardiness. Too often, local environmental conditions will increase or decrease the hardiness of a particular plant in a given area, or the hardiness limitations of a plant may not be truly known, but only guessed at.

(A)

Figure 13–13. Ground covers. A. *Euonymus fortunei* (Winter-creeper). B. *Sedum acre* (Gold moss stonecrop). C. Vinca minor (Periwinkle). U.S. Department of Agriculture photos.

(B)

(C)

(A)

(B)

Figure 13-14. Deciduous trees. A *Ulmus procera* (English elm). B. Cultivar of *Malus floribunda* (Japanese flowering crabapple). U.S. Department of Agriculture photos. C. *Salix babylonica* (Weeping willow). William P. Raffone, Jr. photo.

(C)

(A)

(B)

(C)

(D)

Figure 13–15. Evergreen trees. A. *Cedrus atlantica* (Atlas cedar), a narrow-leaved (needled) evergreen. B. Close-up of Atlas cedar showing needles and cones. American Society for Horticultural Science photos. C. *Washingtonia filifera* (Desert fan palm), a tropical broad-leaved evergreen. U.S. Department of Agriculture – Soil Conservation Service photo by L.N. Gooding. D. *Magnolia grandiflora* (Southern magnolia), a broadleaved evergreen. U.S. Department of Agriculture photo.

(A)

(B)

(C)

Figure 13-16. Evergreen and deciduous shrubs. A. *Thuja occidentalis* 'Minima' (cultivar of American arborivitae), a narrow-leaved evergreen. American Society for Horticultural Science photo. B. *Pieris* 'Brouwer's Beauty', a broad-leaved evergreen. C. An unknown deciduous cultivar of Azalea (technically included with *Rhododendron*). William P. Raffone, Jr. photos. D. An unnamed red-budded selection of *Kalmia latifolia* (Mountain laurel), a broad-leaved evergreen. Author photo.

(D)

445

(A)   (B)

Figure 13–17. Vines. A. *Passiflora* x *alatocaerulea* (hybrid passion flower). B. Juvenile stage of *Monstera deliciosa* (Ceriman or Mexican breadfruit). Leaves develop more leaf splittings which produce a more symmetrical pattern when the plant is mature. These two vines are greenhouse plants in the North, but are grown outdoors in the deep South. Author photos.

Figure 13–18. Zones of plant hardiness, USDA version. All references to zones of hardiness in this text are based on this map. U.S. Department of Agriculture photo.

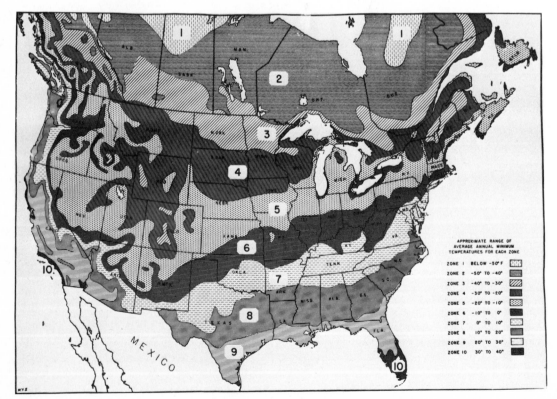

There is no substitute for localized knowledge of the plants that are hardy in your area.

Two other factors are important: permanence and foliage effect. The plant should be long-lived in your locality, considering the investment of time and money put into woody plants. If you know a plant is generally not long-lived, avoid it. For example, a plant may be hardy in your area, but require cool, moist summers, and your summers are hot and dry. Some plants may be highly susceptible to certain pests and/or diseases that prevail in your area.

Foliage effect can be quite important, since you may view the plant throughout the change of seasons year after year. Several factors are involved in foliage effect: size, texture, color, arrangement on the stem, and appearance and condition throughout the seasons. The bark color and texture, flowers and fruit, and the branching pattern are also factors that contribute to the overall effect of appearance and condition throughout the year.

The major means of producing woody perennials in the production nursery is in containers. Balled and burlapped stock is still produced, especially for larger evergreen or deciduous trees and shrubs. Deciduous trees, to a lesser degree, are also sold in bare-root form. Very small needled evergreens, such as would be used to establish a wood lot or forest, are sometimes sold bare root. Most evergreens cannot be handled successfully in bare-root form.

Container stock can be transplanted successfully anytime from early spring through fall. Balled and burlapped stock can also be planted during this time period. The best planting times for both in the North are early spring while in the dormant stage prior to growth, and in the fall and winter in the South after cessation of growth just as dormancy is about to start. Bare-root stock should only be planted at these times while dormant. These preferred planting times allow the plant to recover from transplanting shock without the added stress produced by high rates of transpiration during warmer weather. Container-grown stock is most likely to recover sooner than other types, since their roots experience less transplanting shock.

If the nursery stock cannot be planted within 2 days, some precautions are necessary. Bare-root stock should be heeled in; that is, a trench is dug and the plants laid in on a 45° angle, watered, and then the roots covered with soil (Fig. 13-19). Balled and burlapped or container stock can be placed temporarily in a shady location and the root ball kept moist.

When ready to plant the shrub, tree, or vine, dig a hole adequate to accommodate the root ball or roots without crowding (Fig. 13-20). If necessary, organic matter can be incorporated into the backfill soil to improve it or sand or gravel may be incorporated to improve drainage. If the root ball is in natural burlap, slit it with a knife to aid root penetration. Plastic burlap should be removed after the ball is in place. Container-grown plants should have the root mass quarter-scored with a knife or

Figure 13-19. Bare root stock after being heeled in. U.S. Department of Agriculture photo.

Figure 13-20. Planting a bare root rose. The hole should be big enough to accomodate the spread out root mass and deep enough such that the plant will be at the same level as it was in its previous growing site. Fill the hole gradually with soil that has been amended with fertilizer and soil conditioners as indicated by soil tests. Leave no air spaces and tamp at intervals with foot. Add water to partially filled hole. Complete filling the hole and firm the soil around the rose. If planted in the fall, use a mound of soil to provide winter protection. Some minor variations are used with balled and burlapped or container nursery stock as indicated in the text. Courtesy of All-America Rose Selections.

All-America Rose Selections

spading shovel. Otherwise, the roots will continue to grow in a constricted pattern with detrimental effects on the long-range adjustment of the plant. The plant should be placed at the same soil depth it grew in previously. New plants should be watered and mulched to carry them through the critical establishment period. Once they are established, fertilization, pruning, watering, and so on, should be as indicated for normal culture.

Because of dryness, failure is more apt to occur with fall-planted rather than spring-planted material. Often the growing media around the root ball drains faster and dries quicker than the site soil; this differential drying bears close watch during the establishment period. Adequate moisture should be supplied if needed right up to a hard freezing. An antidessicant for broad-leaved evergreens may help to slow transpiration on certain bright warm days in winter when water losses occur. However, frozen soil prevents compensating water uptake, so the antidessicant may not be sufficient. A tree wrap can prevent drying, splitting, and scalding of bark on trees until they are established. Staking, based on recent evidence, would appear to be unnecessary, as stronger trees are produced without it. Staking would be in order if a specimen were susceptible to wind damage or shifting of the root ball.

Information for various deciduous trees (Table 13-8) and deciduous shrubs (Table 13-9) follows. Coniferous evergreens are indicated in Table 13-10 and broad-leaved evergreens in Table 13-11. Vines are found in Table 13-12.

House and Greenhouse Plants

Success with house or greenhouse plants (Fig. 13-21) depends upon the ability of the horticulturist to provide conditions similar to those found in the native habitat of the plant. Obviously, this is more easily accomplished in a greenhouse than in the home environment where only minimal adjustments of light levels, temperature, and humidity are feasible in view of the comfort level of human occupants. Plants that succeed in the home usually have a wider tolerance for deviations from the environmental norm than do those that are only successfully grown in the greenhouse. Therefore, the key to growing house or greenhouse plants lies first with the ability of the horticulturist to associate normal environmental conditions with a given plant, and then to provide these conditions as closely as possible within the limits of accepted cultural practices.

The regulation of conditions within a greenhouse or home was discussed in Chapter 8. The use of artificial lighting in these environments was also covered there. Potting soils and plant protection were discussed in Chapters 6 and 10, respectively. Great caution must be observed with the selection and application of pesticides in enclosed environments, such as the home or greenhouse. Cultural requirements of plants used in the home and greenhouse are given in Table 13-13. These are the most com-

(A)

(B)

(C)

(D)

Figure 13-21. A few house and greenhouse plants. A. Ornamental pepper 'Holiday Cheer'. Courtesy of All-America Selections. B. *Schlumbergera bridgesii* (Christmas cactus). C. Greenhouse plants as viewed from the author's dining room. D. *Citrus limon* 'Ponderosa' (Ponderosa lemon) showing fruit, flowers and foliage. B, C, D: Author photos.

mon plants found in houses and greenhouses. It should be realized that many other plants found in the other tables, such as bulbs, ferns, and shrubs, can be cultured under greenhouse conditions if the prevailing climate rules them out for outdoor consideration.

The three most likely areas for failure with these plants are overwatering, failure to observe dormancy, and lack of insect control. Frequent overwatering results in an increasing loss of soil aeration, thus reducing root respiration to a harmful level. It also enhances environmental conditions favorable to the growth of pathogens involved in rotting of roots. Failure to recognize dormancy results in overwatering and excessive fertilization during a portion of the plant's cycle when these requirements are minimal. Over the long run this is detrimental to the plant. Constant debilitation of plants by insect attack often increases susceptibility to secondary infections by pathogens. If left unchecked, the plant is eventually lost.

FURTHER READING

American Horticultural Society, *Environmentally Tolerant Trees, Shrubs, and Ground Covers* (rev. ed.). Mount Vernon, Va.: American Horticultural Society, 1977.

Ball, V., *Ball Bedding Book*. West Chicago, Ill.: George J. Ball, Inc., 1977.

———, *Ball Red Book* (13th ed.). West Chicago, Ill.: George J. Ball, Inc., 1975.

Beard, J. B., *Turfgrass: Science and Culture*. Englewood Cliffs, N.J.: Prentice-Hall, Inc., 1973.

Bloom, A., *Perennials for Your Garden*. Nottingham, Eng.: Floraprint Ltd., 1971.

Carpenter, P. L., T. D. Walker, and F. O. Lanphear, *Plants in the Landscape*. San Francisco: W. H. Freeman & Company Publishers, 1975.

DeHertogh, A., *Holland Bulb Forcer's Guide* (2nd ed.). New York: Netherlands Flower-Bulb Institute, 1977.

Dirr, M. A., *Manual of Woody Landscape Plants* (rev. ed.). Champaign, Ill.: Stipes Publishing Co., 1977.

Faust, J. L., *The New York Times Book of Home Landscaping*. New York: Alfred A. Knopf, Inc., 1972.

———, *The New York Times Book of House Plants*. New York: Quadrangle/The New York Times Book Co., 1973.

Foster, C. O., *Organic Flower Gardening*. Emmaus, Pa.: Rodale Press, Inc., 1975.

Foster, H. L., *Rock Gardening: A Guide to Growing Alpines, and Other Wildflowers in the American Garden*. New York: Crown Publishers, Inc., 1968.

Graf, A. B., *Exotica 3: Pictorial Encyclopedia of Exotic Plants from Tropical and Near-tropic Regions* (7th ed.). E. Rutherford, N.J.: Roehrs Co., Inc., 1974.

Harrison, C. R., *Ornamental Conifers*. New York: Macmillan, Inc., 1975.

Hoshizaki, B. J., *Fern Growers Manual*. New York: Alfred A. Knopf, Inc., 1975.

Jacobsen, H., *A Handbook of Succulent Plants* (3 volumes, English ed.). Poole, Eng.: Blandford Press, 1960.

Kingsbury, J. M., *Poisonous Plants of the United States and Canada*. Englewood Cliffs, N.J.: Prentice-Hall, Inc., 1964.

Laurie, A., D. C. Kiplinger, and K. S. Nelson, *Commercial Flower Forcing*. (8th ed.). New York: McGraw-Hill Book Co., 1979.

——, and V. C. Ries, *Floriculture*. New York: McGraw-Hill Book Co., 1950.

Long, R. W., and O. Lakela, *A Flora of Tropical Florida*. Coral Gables, Fla.: University of Miami Press, 1971.

Mastalerz, J. W., ed., *Bedding Plants: A Manual on the Culture of Bedding Plants as a Greenhouse Crop*. University Park, Pa.: Pennsylvania Flower Growers, 1976.

Mathias, M. E., ed., *Color for the Landscape: Flowering Plants for Subtropical Climates*. Arcadia, Calif.: California Arboretum Foundation, Inc., 1976.

Nehrling, A., and I. Nehrling, *The Picture Book of Annuals*. Great Neck, N.Y.: Hearthside Press, Inc., 1966.

——, *Propagating House Plants*. Great Neck, N.Y.: Hearthside Press, Inc., 1971.

Poincelot, R. P., *Gardening Indoors with House Plants*. Emmaus, Pa.: Rodale Press, Inc., 1974.

Rehder, A., *Manual of Cultivated Trees and Shrubs*. New York: Macmillan, Inc., 1940.

Rickett, H. W., *Wildflowers of the United States* (7 volumes). New York: McGraw-Hill Book Co., 1966.

Rockwell, F. F., and E. C. Grayson, *The Complete Book of Bulbs* (revised by M. J. Dietz). New York: J. B. Lippincott Co., 1977.

Schery, R. W., *Lawn Keeping*. Englewood Cliffs, N.J.: Prentice-Hall, Inc., 1976.

Simonds, J. O., *Landscape Architecture: The Shaping of Man's Environment.* New York: McGraw-Hill Book Co., 1961.

Staff of the L. H. Bailey Hortorium, *Hortus Third.* New York: Macmillan, Inc., 1976.

Taylor, N., *The Guide to Garden Shrubs and Trees.* New York: Crown Publishers, Inc., 1965.

Thomas, G. S., *Perennial Garden Plants.* New York: David McKay Co., Inc., 1976.

U.S. Department of Agriculture, *Landscape for Living.* Washington, D.C.: U.S. Government Printing Office, 1972.

Weber, N. M., *How to Plan Your Own Home Landscape.* Indianapolis, Ind.: Bobbs-Merrill Co., Inc., 1975.

Wilson, H. V. P., *Successful Gardening with Perennials.* Garden City, N.Y.: Doubleday & Co., Inc., 1976.

Wyman, D., *Dwarf Shrubs.* New York: Macmillan, Inc., 1975.

, *Ground Cover Plants.* New York: Macmillan Inc., 1956.

, *Shrubs and Vines for American Gardens.* New York: Macmillan, Inc., 1969.

, *Trees for American Gardens.* New York: Macmillan, Inc., 1965.

TABLE 13-1. ANNUALS

Common Name	Scientific Name	Germination Data					Cultivation Data			
		Sowing Time	Light Needs	Germination Time Temperature (days)	Spacing (in.)	Height (in.)	Color	Sun Water Temperature	Bloom Time	Best Use
African daisy	*Arctotis*	HA	—	10	18–24	12–24	MC	S	SuAA	BC
Ageratum[a]	*Ageratum*	TA, SAB6	L	8W	8	4–9	WBP	S	SuEA	ER
Alyssum, sweet	*Lobularia maritima*	HA	—	8W	6–12	3–12	WVRo	SPS	LSEA	BEF
Baby-blue-eyes[a]	*Nemophila menziesii*	HA	—	10	9	20	WB	S	ES	E
Baby's breath	*Gypsophila*	HA	—	10W	10–12	30	WP	S	LSEA	CDHb
Balsam[a]	*Impatiens balsamina*	TA, SAB6	—	8W	12–14	30	MC	S	Eb	BW
Bachelor's button[a]	*Centaurea cyanus*	HA	D	10M	10–12	12–48	MC	SD	LSSu	BC
Begonia	*Begonia*	SAB6	L	15W	6–10	9–15	RWP	SSh	Su	EBW
Bellflower	*Campanula*	HA	—	20W	15–20	24–48	BWP LV	SPS	LSSu	CHb
Bells-of-Ireland[a]	*Molucella laevis*	HA	L	25	12	24–36	W	SPS	SuLSu	CBD
Browallia	*Browallia*	TA, SAB6	L	15W	8–10	12–15	BW	SPS	Eb	BW
Bugloss	*Anchusa*	HA	—	10	12	18	BW	S	LSEA	CHb
Butterfly flower[a]	*Schizanthus pinnatus*	TA, SAB6	D	20C	24–36	18–48	MC	SPS	Su	BCW
California poppy	*Eschscholzia californica*	HA	—	10	8	12–15	YO	S	Eb	B
Candytuft	*Iberis*	HA, SAB6	—	20W	8	12–15	WPRoPu	S	LSESu	ER
Cape marigold	*Dimorphotheca sinuata*	HA, SAB6	—	10	8–10	12–18	MC	SHD	Su	Hb
Castor bean	*Ricinus communis*	TA, SAB6	—	15	—	60+	—	S	F	Sp
Chamomile	*Matricaria*	HA, SAB6	L	10	12–15	12–24	YW	S	Eb	BCE
China aster[a]	*Callistephus*	TA, SAB6	—	8W	12–15	12–36	MC	S	Su	BC
Chinese-forget-me-not	*Cynoglossum amabile*	HA	D	5M	12–18	18–30	B	S	Eb	BC
Cigar flower	*Cuphea*	TA	L	8W	12	24+	RW	SPS	Eb	B
Clarkia, rose[a]	*Clarkia unguiculata*	HA	—	5W	8–12	24–36	MC	S	SuLSu	BC
Cockscomb	*Celosia*	TA, SAB6	—	10W	12–24	12–48	MC	S	SuLSu	BCD
Coleus	*Coleus blumei*	TA, SAB6	L	10M	12–18	12–36	MC	SPS	F	BW
Coneflower	*Rudbeckia*	HA	—	20W	24	12–36	Y	SH	LSSu	BCHb
Cornflower	*Centaurea*	HA	—	10	10–12	24–36	MC	S	LSSu	BC
Coreopsis	*Coreopsis*	HA	L	5	6–10	12–36	Y	S	SuEA	CHb
Cosmos	*Cosmos*	TA, SAB6	—	5W	18–20	30–72	MC	S	SuEA	BCS

Common name	Scientific name									
Creeping zinnia[a]	*Sanvitalia procumbens*	HA	L	10	6–12	6	Y	S	SuEA	ER
Cupid's dart	*Catananche*	HA	—	20	9–12	24	BW	S	LSSu	CHb
Cup flower	*Nierembergia caerula*	TA, SAB6	—	15W	6–12	6–9	B	S	Su	BE
Cup and saucer	*Cobaea scandens*	TA, SAB6	—	15W	24–36	Vine	PW	S	Su	SW
Dahlia	*Dahlia* (Unwin hybrids)	TA, SAB6	—	5W	12–15	12–24	MC	S	LSuAA	BC
English daisy	*Bellis perennis*	HA, SAB6	L	8	4–6	6	WPRo	SPS	ESESu	BEW
Everlasting	*Helipterum*	TA, SAB6	—	15	6	12	WRo	SH	Su	C
Flowering flax[a]	*Linum grandiflorum*	HA, SAB6	L	25	6–10	12–15	PR	S	LSSu	BR
Flowering tobacco	*Nicotiana alata*	TA, SAB6	L	20W	24–36	24–48	MC	S	SuEA	BF
Four-o'clock	*Mirabilis jalapa*	HA, SAB6	—	5	12–20	15–30	MC	S	LSAA	B
Gaillardia	*Gaillardia*	HA, SAB6	—	20W	12	18–24	YOR	S	SuEA	CHb
Geranium	*Pelargonium*	HA, SAB10	—	20H	12–16	18–24	RPW	S	LSAA	BRW
Gilia	*Gilia*	HA	—	10	10	18–24	MC	S	SuEA	Hb
Globe amaranth[a]	*Gomphrena globosa*	HA, SAB6	D	15M	12	12–24	MC	SH	SuEA	BCD
Golden cup	*Hunnemannia fumariaefolia*	TA	—	15W	9–12	18–24	Y	S	Su	BC
Heliotrope	*Heliotropium*	TA, SAB6	—	25W	12–24	12–20	VL	SPS	Eb	BWF
Immortelle[a]	*Xeranthemum annum*	HA	—	10	8	24–36	WRoP	S	Su	CD
Kingfisher daisy[a]	*Felicia bergerana*	HA, SAB6	—	15	6	12	B	SPSM	LSuEA	CHb
Larkspur, annual	*Delphinium*	HA	D	20C	10–12	24–48	MC	SPS	LSESu	CHb
Lobelia	*Lobelia*	HA, SAB6	—	20W	6	4–6	BWP	SPS	Eb	EHbW
Love-in-a-mist[a]	*Nigella damascena*	HA	—	8	5–9	18–24	BW	S	Su	BD
Love-lies-bleeding[a]	*Amaranthus*	TA, SAB6	—	10W	12–18	48–72	R	SH	Su	B
Lupine	*Lupinus*	HA	—	20SC	8–10	12–36	MC	S	Su	CHb
Malope[a]	*Malope*	HA	—	20	9	36	RoPu	S	Su	BC
Marigold	*Tagetes*	HA, SAB6	—	5W	12–24	6–48	YGOR	S	Eb	BCEW
Mignonette[a]	*Reseda odorata*	HA	—	5W	6–10	12–18	Y	SPS	Eb	FBW
Monkey flower	*Mimulus*	HA	—	10S	10–12	18	YR	SPSM	Eb	BRW
Morning glory	*Ipomoea*	TA	—	55M	10–12	Vine	MC	S	SuEA	S
Nasturtium	*Tropaeolum*	TA	D	10M	12–36	7–48	MC	S	LSAA	BE
Painted tongue	*Salpiglossis sinuata*	TA, SAB6	D	15W	8–18	24	MC	S	SuEA	BC
Patience plant	*Impatiens wallerana*	TA	L	15W	12–18	6–18	PRW	SSH	Eb	BEW
Periwinkle	*Catharanthus roseus*	TA, SAB6	L	15W	6–12	12–18	PRoW	S	Eb	BW
Petunia	*Petunia*	TA, SAB6	L	10W	10–12	6–24	MC	S	Eb	BW
Phacelia	*Phacelia*	HA	—	25	6	12	RW	S	LSAA	BE
Phlox, annual[a]	*Phlox drummondii*	HA, SAB6	D	10M	6–8	8–18	MC	SH	LSSu	B
Pimpernel	*Anagallis*	TA	—	21	6–10	2–18	RWB	SC	Su	R
Pinks, annual	*Dianthus*	HA, SAB6	—	5W	8–10	6–24	MC	S	Eb	CEHbF
Poppy	*Papaver*	HA	D	10W	10–12	12–20	MC	S	Su	CHbR
Portulaca[a]	*Portulaca*	HA	D	10W	4	6	MC	SD	SuEA	BER

TABLE 13-1. CONTINUED

Common Name	Scientific Name	Germination Data			Spacing (in.)	Height (in.)	Color	Cultivation Data	Bloom Time	Best Use
		Sowing Time	Light Needs	Germination Time (days) Temperature				Sun Water Temperature		
Pot-marigold[a]	Calendula officinalis	HA, SAB6	D	10W	8-12	18-24	GOY	S	Eb	BCW
Prince's feather[a]	Polygonum orientale	HA	—	20	12-18	60	P Ro	SPS	LSuEA	S
Salvia	Salvia	TA	L	15W	6-12	24-36	RB	S	Eb	CBHb
Scabious, sweet[a]	Scabiosa atropurpurea	HA, SAB6	—	10W	6-9	24-36	MC	S	SuEA	CHb
Scarlet-runner bean	Phaseolus coccineus	TA	—	5	12	Vine	RW	S	Su	S
Snapdragon	Antirrhinum	HA, SAB6	L	10M	6-12	12-36	MC	S	ESuEA	BCW
Soapwort	Saponaria	HA	D	10	6	12	WR	S	LSESu	ER
Spider flower[a]	Cleome hasslerana	HA	L	10	20	36-60	PPuW	S	SuEA	BC
Star of Texas[a]	Xanthisma texana	HA, SAB6	—	25	12	24-36	Y	S	Eb	BC
Statice	Limonium	TA, SAB6	—	15W	10-12	24	BRoLYPu	S	LSSu	BCD
Stock	Matthiola	HA, SAB6	—	5W	12	12-30	MC	S	LSSu	BCF
Straw-flower	Helichrysum bracteatum	HA, SAB6	L	5W	8-10	24-36	MC	S	Su	BCD
Summer cypress[a]	Kochia scoparia	HA, SAB6	—	15W	15-20	36-60	R	S	F	S
Sunflower	Helianthus	TA	—	5	12-36	18-120	Y	SDH	LSu	CS
Swan River daisy[a]	Brachycome iberidifolia	TA, SAB6	—	15	6-9	18-24	MC	S	LSESu	E
Sweet pea[a]	Lathyrus odoratus	HA	D	15SC	6	Vine	MC	SC	ESESu	C
Tahoka daisy[a]	Machaeranthera tanacetifolia	HA	—	30	6	12-24	BV	SH	LSLSu	BC
Tassel flower[a]	Emilia javanica	HA, SAB6	—	8	4-5	18-24	RO	SD	LSLSu	SD
Tidy tips[a]	Layia platyglossa	HA	L	8	9	12-24	YW	S	ESu	BC
Tithonia[a]	Tithonia rotundifolia	HA	D	10W	24-48	48-72	YO	SD	LSuAA	CS
Toadflax	Linaria	HA	—	10	6-12	12-18	MC	S	LSSu	ER
Torenia[a]	Torenia fournieri	TA, SAB6	—	15W	6-8	10-14	BY	SH	Eb	BW
Tree mallow	Lavatera	TA, SAB6	—	20	24	60-96	RORW	S	LSSu	HbS
Verbena	Verbena	TA, SAB6	D	20M	10-12	8-15	MC	S	SuAA	BC
Viper's bugloss	Echium	TA	—	8	18	36	MC	SD	Eb	R
Winged everlasting	Ammobium alatum	HA, SAB6	—	5	12	24-36	W	S	Su	CD
Zinnia	Zinnia	TA, SAB6	—	5W	8-15	10-36	MC	SH	Eb	BC

aTrue annuals. Remaining genera or species consist in part or in total of perennials or biennials or a mixture of both. However, they are all treated as annuals and would be perennials only in zones where they are winter hardy.

Explanation of Table Symbols

Sowing time:

TA, tender annual, sow after all danger of frost is past.

HA, hardy annual; North, sow after danger of heavy spring frost is past or with some risk in late fall just before ground freezes. South and Pacific Coast areas, sow in fall and winter.

SAB 6, annual, slow to bloom or longer blooming period desired: start indoors 6 to 10 weeks prior to setting out plants.

SAB 10, same, but start 10 to 12 weeks ahead.

Light needs:

L, seeds need light to germinate; do not cover with soil or propagation media, but press seeds lightly into media.

D, seeds require darkness to germinate; cover with media; fine seeds only pressed into soil should be kept dark.

—, light and darkness do not affect germination or this seed has not been characterized for light needs.

Germination time and temperature:

Time given in days and assumes reasonably optimal germination conditions.

Temperature: C, 55–60°F, 12.8–15.6°C
M, 65–70°F, 18.3–21.1°C
W, 70–75°F, 21.1–23.9°C
H, 75–80°F, 23.9–26.7°C
No entry, unavailable or unknown (assume about 70°F, 21.1°C).

Color:

W, white   Y, yellow   Ro, rose
B, blue   O, orange   L, lavender
P, pink   Pu, purple   G, gold
V, violet   R, red   MC, many colors

Sun, water, and temperature:

S, sun   M, requires extra moisture
SPS, sun or partial shade   D, tolerates dryness
SSH, semishade   C, prefers cool weather
                                 H, tolerates hot weather

Bloom time:

ES, early spring   LSu, late summer   Eb, tends to be
LS, late spring   EA, early autumn   everblooming
ESu, early summer   LA, late autumn
Su, summer   AA, all autumn
F, foliage plant

Best use:

B, bedding   S, screening   Hb, hardy border
E, edging   H, hedging   Sp, specimen plant
C, cut flower   R, rock garden   D, suitable for drying
G, ground cover   W, window box,
                       porch box, container
                    F, fragrance

TABLE 13–2. PERENNIALS.

Scientific Name	Common Name	Height (in.)	Color	Hardiness Zones	Spacing (in.)	Time of Bloom	Soil	Light	Remarks
Achillea filipendulina	Fern-leaf yarrow	48–60	Y		18			FS	Di
A. millefolium	Common yarrow	36	WP	3	24	J–S	D	FS	
A. ptarmica	Sneezewort	24	W	3	24	J–S	D	FS	
A. tomentosa	Woolly yarrow	12	Y	3	12	J–S	D	FS	
Aconitum carmichaelii	Aconite	72	B-Pu	2	18	Au–S	N	Sh	C, Di, Se
A. napellus	Helmet flower	36	Pu	2	18	Ju–Au	N	FS	Ci, RG
Adonis vernalis	Pheasant's eye	18	YW	3	12	Ma–A	N	FS	
Aeonium canariense	Canary Island aeonium	18	Y	9	18	Ma–A	D	FS	Cu
Agapanthus africanus	African lily	20	Vi-B	9	18	Ju–Au	N	FS	C, Di
Ajuga genevensis	Bugleweed	4–16	BPW	2	6–10	M–J	N	Sh	Di, GC
A. pyramidalis		2–12	B	2	6–10	M–J	N	Sh	
A. reptans		10	BPW	2	6–10	M–J	N	Sh	
Aloe brevifolia	Aloe	16	R	9	12	F–Ma	D	FS	Cu, Se, GC
A. variegata	Patridge breast aloe	9	R	9	10	F–Ma	N	FS	Cu, Se
Amsonia tabernaemontana	Bluestar	42	B	5	18	A–Ju	N	Sh	C, Di, Se
Anchusa officinalis	Bugloss	24	B	4	18	Ju–S	N	FS	C, Di, Se
Anemone hupehensis	Japanese anemone	12–36	Ro	5	18	S–O	N	Sh	Di, Se
A. pulsatilla	Pasque-flower	12	PuW	5	12	A–M	N	Sh	Di, RG, Se
Antennaria dioica	Pussy-toes	10	RoW	3	8	M–J	N	Sh	Di, RG, Se
Anthemis tinctoria	Golden marguerite	36	Y	3	36	J–O	D	FS	C, Di, Se
Aquilegia alpina		12–36	MC	3	12–18	M–J	N	Sh	C, Di, RG, Se
A. caerula		12–36	MC	3	12–18	M–J	N	Sh	C, Di, RG, Se
A. chrysantha	Columbine	12–36	MC	3	12–18	M–J	N	Sh	C, Di, RG, Se
A. flabellata		12–36	MC	3	12–18	M–J	N	Sh	C, Di, RG, Se
A. vulgaris		12–36	MC	3	12–18	M–J	N	Sh	C, Di, RG, Se
Arabis caucasica	Wall rock cress	12	W	3	8	A–M	N	Sh	C, Cu, Di, RG, Se
Armeria maritima	Thrift	12	PW	4	10	A–M	D	FS	C, Cu, RG, Se
Artemisia ludoviciana	Western mugwort	36	Gray-white leaves	4	24	—	N	FS	Di
Aruncus dioicus	Goatsbeard	48–84	W	4	48	J–Ju	M	Sh	Se
Asclepias tuberosa	Butterfly weed	36	O	3	12	Au–S	D	FS	C, Di, Se
Aster alpinus	Alpine aster	9–12	BViRoW	3	12	M–J	N	FS	C, Cu, Di, RG
A. × frikartii	—	24	Vi-B	5	15	J–O	M	FS	C, Cu, Di
A. novae-angliae	New England aster	72	RoViW	3	24	Au–S	M	FS	C, Di
Astilbe × arendsii	Spirea	24–48	PPuR	6	24–36	J–Au	M	FS, Sh	C, Di, Se

Astrantia major	Masterwort	36	RoW	7	18	Ju-S	M	FS, Sh	Di
Aubrieta deltoidea	Purple rockcress	6	R-Pu	4	6	A-J	N	FS	Di, GC, RG, Se
Aurinia saxatilis	Basket of gold	6	Y	3	6	A-M	N	FS, Sh	Di, RG
Baptisia australis	Blue false indigo	72	B	3	24	M-J	N	FS, Sh	C, Di, Se
Bergenia × schmidtii		12–20	Ro	4	18	M	M	Sh	Di, GC, Se
Brunnera macrophylla	Siberian bugloss	18	B	3	24	M-J	M	FS, Sh	Di, GC
Calamintha grandiflora	Calamint	18	P	6	18	Ju-Au	N	FS	Cu, Di
Caltha palustris	Marsh-marigold	12–24	Y	3	18	A	M	Sh	Di, Se
Campanula carpatica	Tussock bellflower	18	B	4	12	Ju	N	FS	C, Cu, Di, RG, Se
C. elatines	Adriatic bellflower	8	BW	7	12	J-Ju	N	FS, Sh	C, Cu, Di, RG, Se
C. persicifolia	Willow bellflower	36	ViW	5	12	J-Ju	N	FS	C, Cu, Di, RG, Se
C. rotundifolia	Bluebell	18	B	4	12	J-O	N	Sh	C, Cu, Di, RG, Se
Centaurea montana	Mountain bluet	18	BW	3	24	M-Ju	N	FS	Di
Centranthus ruber	Red valerian	36	RRoW	5	18	J-Au	D	FS, Sh	Di, Se
Cerastium tomentosum	Snow-in-summer	6–10	W	3	6	M-J	N	FS	Cu, Di, RG, Se
Chelone obliqua	Rose turtle-head	36	P-Pu	6	18	Au-O	M	FS	Di, Se
Chrysanthemum coccineum	Pyrethrum	24	PRW	4	18	J-Ju	N	FS	C, Di, Se
C. frutescens	Marguerite	36	WY	10	36	Ju-Au	N	FS	C, Cu
C. morifolium	Florists' chrysanthemum	24–60	MC	5	18–24	S-N	N	FS	C, Cu, Di, Se
C. × superbum	Shasta daisy	36	W	4	18	J-O	N	FS	C, Di
Chrysogonum virginianum	—	12	Y	7	12	A-J	M	Sh	Di, Se
Cimicifuga racemosa	Black cohosh	36–72	W	3	24	Ju-S	M	FS, Sh	C, Di, Se
Convallaria majalis	Lily-of-the-valley	8	PW	2	10	M-J	N	FS, Sh	C, Di, GC
Coreopsis auriculata	Tickseed	12–24	Y	5	12–18	J-Au	N	FS	C, Cu, Di, GC, RG, Se
Delphinium elatum	Larkspur	72	B	3	24	J-S	N	FS	C, Cu, Di, Se
D. grandiflorum		12–42	BW	3	12–18	J-S	N	FS	C, Cu, Di, Se
Dianthus deltoides	Maiden pink	6–18	PRW	3	12	M-J	N	FS	C, Cu, Di, RG, Se
D. plumaris	Cottage pink	16	PRoW	4	18	M-J	D	FS	C, Cu, Di, RG, Se
Dicentra eximia	Turkey corn	24	P	4	18	M-Au	D	FS	C, Di, Se
D. formosa	Western bleeding-heart	18	RoW	4	18	M-S	N	FS, Sh	C, Se
D. spectabilis	Bleeding-heart	24	PW	4	18	M-J	N	FS	C, Se
Dictamnus albus	Dittany, gas plant	36	PuRoW	3	24	J	N	FS	C, Di, Se
Digitalis purpurea	Common foxglove	36	PPuW	4	18	J-Ju	N	FS, Sh	C, Di, Se
Doronicum cordatum	Leopard's-bane	10–30	Y	4	24	A-M	N	FS, Sh	C, Di, Se
Echinacea purpurea	Purple coneflower	36	Ro-PuW	4	18	Ju-Au	D	FS	C, Di, Se
Echinocactus grusonii	Barrel cactus	48	Y	10	60	A	D	FS	Se
E. ingens	Large barrel cactus	60	Y	9	72	M-J	D	FS, Sh	C, Di, Se
Echinops ritro	Small globe thistle	24	B	4	24	Ju-Au	D	Sh	C, Di, Se
Epimedium alpinum	—	12	Y	7	12	M-J	N	FS	C, Di
E. grandiflorum	—	12	WVi	7	12	M-J	N	Sh	C, Di

TABLE 13–2. CONTINUED

Scientific Name	Common Name	Height (in.)	Color	Hardiness Zones	Spacing (in.)	Time of Bloom	Soil	Light	Remarks
Eryngium alpinum	—	24	BW	5	18	J-Ju	D	FS	Di, Se
Eupatorium coelestinum	Mist flower	36	B	3	18	Au-O	M	FS, Sh	Di, Se
E. maculatum	Joe-Pye weed	72	Pu		36	Au-S	M	FS, Sh	Cu, Se
Euphorbia epithymoides	Cushion euphorbia	12	Y	4	12	Ju-Au	D	FS, Sh	Cu, Se
E. myrsinites	—	12	Y	5	12	A-M	D	FS, Sh	
Filipendula purpurea	Meadowsweet	48	RW	7	24	J-Ju	N	FS, Sh	Di, Se
F. vulgaris	Dropwort	36	W	4	24	Ju-Au	N	FS, Sh	
Gaillardia × *grandiflora*	Blanket flower	30	ORY	3	18	Ju-Au	D	FS	C, Cu, Di, Se
Galium odoratum	Woodruff	12	W	5	12	M-J	M	Sh	C, Di, RG, Se
Gazania ringens	Treasure flower	16	OY	9	18	Ju-Au	N	FS	C, Cu, Di, GC, Se
Geranium maculatum	Wild cranesbill	24	Ro-PuW	5	18	M-J	N	FS	Di, RG, Se
G. sanguineum	Cranesbill	18	R-PuW	4	18	M-Au	N	FS	
Gypsophila paniculata	Baby's-breath	36	W	3	48	Ju-Au	D	FS	C, Cu, Di, RG, Se
G. repens	—	10	WPRo	4	12	J-Au	D	FS	
Helenium autumnale	Sneezeweed	60	RY	4	18	Ju-Au	M	FS	C, Cu, Di, Se
Helianthus decapetalus	Thin-leaf sunflower	60	Y	3	24	Ju-S	N	FS	C, Di, Se
Heliopsis helianthoides	Oxeye	60	OY	3	24	Ju-O	N	FS	C, Cu, Di, Se
Helleborus niger	Christmas rose	12	W	4	18	F-M	M	Sh	Di, Se
Hemerocallis fulva	Orange daylily	60	OR-O	3	36	Ju-Au	N	FS	Di
H. lilioasphodelus	Yellow daylily	36	Y	3	18	J	N	FS	
Hepatica acutiloba	Liverleaf	9	B-W	5	18	Ma-A	N	Sh	Di, Se
Heuchera sanguinea	Coral bells	24	RW	4	12	J-Au	N	FS	C, Di, Se
Hibiscus moscheutos	Common rose mallow	96	PRoW	6	36	Au-S	M	FS	Di, Se
Hosta fortunei	Plantain lily	24-36	L	3	24	M-Ju	M	Sh	Di
H. lancifolia						Ju-Au			
H. sieboldiana						M-Ju			
H. undulata						M-J			
Iberis sempervirens	Edging candytuft	12	W	3	12	A-J	N	FS, Sh	C, Cu, Di, Se
Iris brevicaulis	Lamance iris	18	BB-Pu	5	12	M-J	M	FS	
I. cristata	Dwarf crested iris	4	WY	3	12	M-J	N	Sh	
I. × *germanica*	Flag	30	Vi	3	12	M-J	N	FS	
I. kaempferi	Japanese iris	24	R-PuPW	5	18	J-Ju	M	FS, Sh	C, Di
I. sibirica	Siberian iris	45	B-P	3	18	J	M	FS, Sh	
I. spuria	Butterfly iris	24	B-PWY	5	18	J-Ju	M	FS	
Kalanchoe blossfeldiana		12	R	10	18	S-N	D	FS	Cu, Se
Kniphofia uvaria	Poker plant	48	R	6	24	S-O	N	FS	Di

Lavandula angustifolia	English lavender	36	BPuRoW	6	48	J	N	FS	Cu, Di
Liatris scariosa	Blazing star	36	PW	3	12	S	N	FS, Sh	C, Di, Se
L. spicata		60	P	3	12	S	M	FS, Sh	C, Di, Se
Linum perenne	Perennial flax	24	BW	4	24	M-Au	N	FS	Cu, Di, Se
Liriope muscari	Big blue lilyturf	18	ViW	6	18	Ju-S	N	FS, Sh	Di, GC
Lupinus polyphyllus	Lupine	60	BRRoW	3	24	J-Ju	N	FS	C, Di
Lychnis chalcedonica	Maltese cross	24	RRoW		12	J-Ju	N	FS	Di, Se
L. viscaria	German catchfly	36	PPuW	3	18	M-J	N	FS	Di, Se
Lythrum salicaria	Purple loosestrife	72	PuRo	3	18	J-S	M	FS	C, Di
Mertensia virginica	Bluebells	24	BPW	3	18	A-M	N	Sh	RG, Se
Monarda didyma	Bee balm	48	PRRoViW	4	12	J-Au	N	FS	C, Di
Myosotis scorpioides	Forget-me-not	18	B	4	18	A-S	M	Sh	Di, Se
Nepeta grandiflora	Catmint	36	B	4	36	J-S	N	FS	Di, RG, Se
Oenothera fruticosa	Sundrops	24	Y	4	18	Ju-Au	D	FS	Di, Se
Opuntia basilaris	Beaver tail	144	RoY	6	—	A-M	D	FS	Cu, Se
O. erinacea	Grizzly bear cactus	18	RYW	3	60	A-M	D	FS	Cu, Se
Paeonia lactiflora	Chinese peony	36	PRW	3	24	M-J	N	FS	C, Cu, Di, Se
P. officinalis	Common peony	24	PRW	3	24	M-J	N	FS	C, Cu, Di, Se
P. suffruticosa	Tree peony	84	PRWY	6	72	M-J	N	FS	C, Cu, Di, Se
Papaver orientale	Oriental poppy	48	OPRW	3	36	M-J	N	FS	RG, Se
Penstemon barbatus	Beard-tongue	72	RRo	3	24	J-Au	N	FS	Di, Se
Phlox carolina	Thick-leaf phlox	48	PPuW		24	J-Au	N		Di
P. divaricata	Wild sweet william	18	LVi-BW	3	18	A-M	N	FS	Di, RG
P. paniculata	Perennial phlox	72	LP-PuRW		36	Ju-O	N	FS	Di
P. subulata	Moss pink	6	PR-PuVi-PuW		12	A-M	N		Di, GC
Physalis alkekengi	Chinese lantern	24	W, red fruit	3	24	Au-S	N	FS	Cu, Di
Physostegia virginiana	Obedience	48	PRRo-PuW	3	24	Ju-S	N	FS	C, Di, Se
Platycodon grandiflorus	Balloon flower	30	BLW	3	18	J-S	N	FS	Di, Se
Polemonium caeruleum	Jacob's ladder	36	BW	2	24	M-J	N	FS	Di, Se
Polygonatum biflorum	Small Solomon's seal	36	W	3	12	M-J	N	Sh	Di
Polygonum capitatum	Knotweed	3	P	7	12	A-O	N	FS	Di, GC, Se
Primula auricula	Auricula	8	MC	3	12	A-M	N	FS	Cu, Di
P. denticulata		12	P-PuPu	4	12	A-M	N	FS	C, Di, Se
P. japonica	Primrose	30	PuRW	5	18	M-Ju	M	Sh	Di, Se
P. juliae		3	RRoW	5	10	J-S	N	FS	Di, Se
P. polyantha	Polyanthus	12	MC	3	12	A-M	N	FS	Di
P. vulgaris	English primrose	6	BPuY	6	12	A-M	N	FS	C, Di, RG, Se
Ranunculus repens	Butter daisy	24	Y	4	12	M-Au	N	FS	C, Di, GC
Romneya coulteri	California tree poppy	84	W	9	36	M-Au	N	FS	Di, Se
Rudbeckia fulgida	Coneflower	36	O-Y	4	18	Ju-Au	N	FS	C, Cu, Di, Se
Salvia azurea	Blue sage	72	B	3	24	Au-O	N	FS	Cu, Di, Se
Scabiosa caucasica	Scabious	24	BLW	3	24	Ju-Au	N	FS	C, Di, Se

461

TABLE 13–2. CONTINUED

Scientific Name	Common Name	Height (in.)	Color	Hardiness Zones	Spacing (in.)	Time of Bloom	Soil	Light	Remarks
Sedum acre	Golden-carpet	5	Y	4	6	M-J	D	FS	Cu, Di, GC, Se
S. spectabile	Stonecrop	22	PPuW	3	18	Au-S			Cu, Di, RG, Se
Stokesia laevis	Stoke's aster	24	BLPuRoW	5	18	Ju-Au	N	FS	Cu, Di, Se
Tanacetum vulgare	Common tansy	36	Y	4	18	Ju-Au	D	FS	Di, Se
Teucrium hyrcanicum	Germander	30	PuR	4	24	Au-S	N	FS	Cu, Di, RG, Se
Thalictrum aquilegifolium	Meadow rue	36	Li-RoOPuW	4	12	M-J			C, Di, Se
T. polygamum	Tall meadow rue	84	W	4	24	Ju-S	N	FS	
T. speciosissimum	—	60	Y	5	24	Ju-S			
Thermopsis caroliniana	Carolina lupine	60	Y	3	24	J-Ju	N	FS	C, Di, Se
Tiarella cordifolia	Foamflower	12	RW	4	12	A-Ju	N	Sh	Di, RG, Se
Tradescantia virginiana	Common spiderwort	36	Vi-PuW	4	24	J-Ju	N	Sh	Cu, Di, Se
Trollius europaeus	Globeflower	24	Y	4	18	M-Ju	M	FS, Sh	C, Di, RG, Se
Valeriana officinalis	Common valerian	42	LPRW	3	36	Ju-Au	N	FS	C, Di, Se
Veronica incana		24	BP	3	12	J-Ju			
V. latifolia	Speedwell	24	BRoW	4	12	M-J	N	FS	C, Di, Se
V. longifolia		48	BL	4	18	Ju-S			
V. spicata		18	BP	3	18	J-Ju			
Vinca minor	Common periwinkle	5	BL-BW	4	24	A-M	N	Sh	Cu, Di, GC
Viola cornuta	Horned violet	12	PuW	6	24	Ju-Au	N	FS, Sh	Cu, Di, RG, Se
V. odorata	Sweet violet	6	ViW	3	12	A-M			
V. tricolor	European wild pansy	12	PuWy	3	12	A-M			
Yucca filamentosa	Adam's needle	60	W	4	24	Au	D	FS	Cu, Se

Explanation of Table Symbols

Color:

W, white	Y, yellow	MC, many colors	
B, blue	R, red		
P, pink	L, lavender		
Pu, purple	Ro, rose		
O, orange	Vi, violet		

Hardiness zones:
See Fig. 13–18. Zone of hardiness limitation given.

Time of Bloom:

F, February   Ju, July
Ma, March   Au, August
A, April   S, September
M, May   O, October
J, June   N, November

Soil:
M, suitable for moist to wet soil
D, suitable for dry soil
N, suitable for normal soil neither too moist or dry

Light:
Sh, can tolerate shade
FS, full sun

Remarks:
C,   good for cut flowers
Di, propagate by division
Cu, propagate by cuttings
Se, propagate by seeds
GC, good for ground cover
RG, good for rock garden

462

TABLE 13–3. SPRING- AND SUMMER-FLOWERING BULBS

Scientific Name	Common Name	Height (in.)	Depth of Planting (in.)	Spacing (in.)	Color	Zone of Hardiness	Time of Bloom	Propagule	Remarks
Achimenes sp.	Monkey-faced pansy	12–24	0.5–1	6–12	PPuRW	10	SuF	Rh, Cu	Sh, Sl, T/P
Acidanthera bicolor	Peacock orchid	12–24	4	6	W	10	SuF	Co	—
Agapanthus africanus	African lily	20	1–2	24	V-B	9	SuF	Rh, S	T/P
A. orientalis	—	36			BW				
Albuca canadensis	—	18–36	6	12	Y	10	SuF	Bu	T/P
A. nelsonii	—	60			W				
Allium sp.	—	1–36	1–3	12–30	BLPRoRW	4–6	SpF, SuF	Bu, Rh, S	—
Alstroemeria	Lily-of-the-Incas	12–36	5–8	18–36	PuRY	7–10	SuF	TR, S	Sh, Sl, T/P
Amaryllis belladonna	Belladonna lily	18	5	12–18	Ro-RW	9	SuF	Bu	Sl, T/P
Anemone sp.	Windflower	3–36	2–3	6–18	BPPuRRoW	3–7	SpF, SuF	Th, Tu, S	Sh, T/P
Anthericum liliago	St. Bernard's lily	36	3–4	18	W	7	SuF	D, S, TuR	—
Arisaema dracontium	Green dragon		2–4	12–24	G-Pu	5	SpF	S, Tu	Sh
A. speciosum	Cobra lily	12–18							
A. triphyllum	Jack-in-the-pulpit								
Babiana stricta	Baboon flower	8	3	9	PuY	7	SpF	Co, S	T/P
Begonia × *tuberhybrida*	Hybrid tuberous begonia	6–24	1–3	6–18	MC	10	SuF	S, Tu	Sh, Sl, T/P
Belamcanda chinensis	Blackberry lily	48	1–2	12–18	O	6	SuF	Rh, S	—
Bessera elegans	Coral-drops	36	4	12	PuR	9	SuF	Co	T/P
Bloomeria crocea	Golden stars	18	3	3	O-Y	7	SuF	Co, S	—
Brodiaea sp.	—	12–18	4	4–8	LPV	7	SuF	Co, S	T/P
Bulbocodium vernum	Spring meadow saffron	4	3–4	6	V-Pu	6	SpF	Co	T/P
Caladium sp.	Caladium	6–24	1–2	12–24	Foliage	10	—	S, Tu	Sh, Sl, T/P
Calochortus sp.	Mariposa	6–30	2–3	6–24	MC	4	SuF	Bu, S	—
Camassia	Camass	30–48	3–4	4–6	BB-VW	6	SpF	Bu, S	—
Canna × *generalis*	Canna	24–96	3–4	12–24	OPRRoWY	9	SuF	Rh, S	Sl
Chionodoxa luciliae	Glory-of-the-snow	6	3	4–6	BPW	5	SpF	Bu	—
Chlidanthus fragrans	Perfumed fairy lily	10	2	4–6	Y	9	SuF	Bu	—
Claytonia sp.	Spring beauty	12	2	4–6	RoW	7	SpF	Co	Sh
Clivia sp.	Kaffir lily	18–24	1	18	OR-Y	10	SpF	TR	T/P
Colchicum sp.	Autumn crocus	4–6	3	9–12	PuWY	5–7	SuF	Co, S	—
Colocasia esculenta	Taro	36–84	2–3	36–72	Foliage	9	—	Co	T/P
Crinum sp.	Crinum lily	18–48	6	24–36	PRW	7–9	SuF	Bu	—
Crocosmia × *crocosmiiflora*	Montebretia	36–48	3–4	4	O-R R	8	SuF	Co, S	T/P
Crocus sp.	Crocus	3–6	3–4	3	LPuWY	6	SpF, SuF	Co	T/P
Curtonus paniculatus	—	48	4	8–12	RY	7	SuF	Co	—

TABLE 13-3. CONTINUED

Scientific Name	Common Name	Height (in.)	Depth of Planting (in.)	Spacing (in.)	Color	Zone of Hardiness	Time of Bloom	Propagule	Remarks
Cyclamen sp.	Persian violet	3-12	2	6-12	PPuRRoW	6-9	SpF, SuF	S, Tu	Sh, T/P
Cypella herbertii	—	20	4	8-12	L-Pu O-Y	9	SuF	Co, Se	—
Cytanthus sp.	Fire lily	12-18	4	8-12	RW	7	SuF	Bu	—
Dahlia coccinia × *pinnata*	Dahlia	2	6	36-48	MC	10	SuF	Cu, S, TuR	SI
Endymion sp.	Wood hyacinth	12-20	2	4-6	BPRo-PuW	5-6	SpF	Bu	Sh
Eranthis sp.	Winter aconite	2-8	2	4-6	WY	5	SpF	Tu	—
Eremurus sp.	Desert-candle	24-120	4-6	18-36	OPWY	4-7	SuF	S, TR	—
Erythronium sp.	Adder's tongue	8-24	3-5	6-18	PPuRoWY	3-6	SpF	Co, S	T/P
Eucharis grandiflora	Amazon lily	24	0-1	6-12	W	9	SpF	Bu, S	T/P
Eucomis sp.	Pineapple lilly	12-72	6	12	GW	7	SuF	Bu, S	T/P
Freesia sp.	—	8-18	3	6-9	BPPuRoWY	9	SpF	Co	T/P
Fritillaria sp.	Fritillary	4-48	3-4	6-24	MC	3-8	SpF	Bu	—
Galanthus sp.	Snowdrop	8-12	3-4	6-12	W	4	SpF	Bu, S	—
Galtonia candicans	Summer hyacinth	48	6	12-18	W	6	SuF	Bu, S	—
Gladiolus sp.	Corn flag	6-48	4-6	5-8	MC	7-9	SuF	Co, S	SI
Gloriosa sp.	Gloriosa lily	36-96	4	18-36	PuRY	10	SuF	S, Tu	—
Habranthus brachy-andrus	—	12	0-1	2-3	R-Pu	10	SuF	Bu	—
Haemanthus sp.	Blood lily	10-18	0-1	8-18	RW	9	SuF	Bu	T/P
Hippeastrum hybrids	Amaryllis	36	1-2	12	OPRW	9	SuF	Bu, S	T/P
Hyacinthus orientalis	Hyacinth	18	5-6	9	BPPuRRoWY	6	SpF	Bu, S	T/P
Hymenocallis narcis-siflora	Basket flower	24	4	8	WY	7	SuF	Bu, S	T/P
Ipheion uniflorum	Spring starflower	8	3	4-6	W	6	SpF	Bu	—
Iris sp.	Iris	3-40	1-4	3-24	MC	3-8	SpF	Bu, Rh, S	—
Ixia sp.	Corn lily	6-36	2-3	4-6	BLPPuRWY	7	SpF, SuF	Co	T/P
Ixiolirion tataricum	Siberian lily	16	3	9	B	7	SpF	Bu	—
Lachenalia aloides	Cape cowslip	12	2-3	4-6	OO-YY	9	SpF	Bu	T/P
Lapeirousia sp.	—	6-18	5	6	MC	7	SuF	Co	T/P
Leucocoryne ixioides	Glory-of-the-sun	12	2-3	4-6	BLW	9	SpF	Bu	—
Leucocrinum mon-tanum	Sand lily	7	4	8	W	5	SpF	Rh, S	—
Leucojum sp.	Snowflake	9-12	3	6-9	W	5	SpF	Bu	—
Lilium sp.	Lily	10-96	4-10	8-36	OPPuRWY	3-10	SuF	B, Bu	T/P
Liriope sp.	Lilyturf	8-18	1-2	12-24	LVW	5-6	SuF	Di, Rh, Tu	Sh
Lycoris sp.		12-24	3-6	6-18	PRWY	6-7	SuF	Bu	—
Milla biflora	Mexican-star	12	2	4-6	BPW	9	SuF	Co	T/P

Genus	Common name	Height	Depth	Color	Zone	Time of bloom	Propagule	Remarks
Moraea sp.	Butterfly iris	12–48	1–4	BLO-RPuWY	9	SuF	Co	T/P
Muscari sp.	Grape hyacinth	4–18	2	BVWY	3	SpF	Bu, Se	–
Narcissus sp.	Narcissus, daffodil and jonquil	4–18	4–5	OWY	5–7	SpF	Bu	T/P
Nerine sp.	—	8–36	3	RW	9	SuF	Bu	T/P
Nomocharis sp.	—	24–36	6	PRoWY	5	SuF	Bu	T/P
Ophiopogon sp.	Lilyturf	6–36	1–2	PuW	7	SuF	Rh, TR, TuR	Sh
Ornithogalum sp.	—	2–36	3–6	G-WO-RWY	5–8	SpF, SuF	Bu	T/P
Oxalis sp.	Wood sorrel	3–16	0.5–2	MC	4–9	SpF, SuF	Bu, Rh, S, Tu	—
Pancratium sp.	—	12–24	3	W	8	SuF	Bu, S	—
Polianthes tuberosa	Tuberose	42	1–2	W	9	SuF	Rh	T/P
Polygonatum sp.	Solomon's seal	18–72	1–2	GY	3–4	SuF	Rh, S	Sh
Puschkinia scilloides	Striped squill	6	2	BW	5	SpF	Bu	—
Ranunculus asiaticus	Persian buttercup	18	2	OPRWY	8	SuF	TuR	T/P
Scilla sp.	Squill	6–36	2	BPuRoW	3–5	SpF	Bu	T/P
Sinningia speciosa	Gloxinia	12	0.5–1	MC	10	SuF	Cu, S, Tu	T/P
Smilacina sp.	False Solomon's seal	12–18	1–2	PPuW	4	SpF	Di, Rh, S	Sh
Sparaxis sp.	Wandflower	15–36	3	PuRoWY	10	SpF, SuF	Co	—
Sprekelia formosissima	Jacobean lily	18–24	1	R	9	SpF, SuF	Bu	T/P
Sternbergia lutea	Winter daffodil	12	6	Y	7	SuF	Bu, S	—
Tigridia pavonia	Tiger flower	24	2–3	O-RRRoWY	7	SuF	Rh, S	Sh, T/P
Tricyrtis sp.	Toad lily	20–36	1–2	WY	7	SuF	S, TR	Sh
Trillium sp.	Wake-robin	6–30	1–2	GPPuWY	4–8	SuF	Rh, S	T/P
Tritonia	—	12–18	3	PRRoW	6	SuF	Bu	T/P
Tulbaghia violacea	Society garlic	9–12	2–3	L	9	SuF	Bu, S	Sh
Tulipa sp.	Tulip	6–24	4	MC	5	SpF	Co	T/P
Veltheimia viridifolia	—	24	2–3	P-Pu	9	SpF, SuF	Bu	T/P
Watsonia sp.	Bugle lily	24–72	4	OPPuRRoW	8	SuF	Rh, S	Sh
Zantedeschia sp.	Calla lily	24–36	2–3	OPRWY	8–10	SuF	Co	T/P
Zephyranthes sp.	Zephyr lily	6–12	4	OPRWY	8	SuF	Bu	T/P

Explanation of Table Symbols

Height: Both individual and range values are expressed as maximal potential.
Height can be less if environmental conditions are less than optimal or if cultivars were bred for height reduction.

Color:
B, blue
G, green
L, lavender
MC, many colors
O, orange
P, pink
Pu, purple
R, red
Ro, rose
V, violet
W, white
Y, yellow

Time of bloom:
SpF, spring flowering
SuF, summer flowering (can include early autumn)

Propagule:
B, bulbil
Bu, bulb or offset
Co, corm or cormel
Cu, cutting
D, division
Rh, intact or divided rhizome
S, seed
SD, simple division of plant
TR, thickened roots and divisions thereof
Tu, Intact or divided tuber
TuR, tuberous roots and divisions thereof

Remarks:
Sh, some or all good for partial shade
SI, start indoors of in greenhouse for early bloom
T/P, good tub or pot plant

TABLE 13-4. BULB CULTIVARS USEFUL FOR FORCING[a]

Tulipa (Tulip)

Abra	Couleur Cardinal	Madame Spoor	Preludium
Albury	Danton	Makassar	Prince Charles
Angélique	Diplomate	Merry Widow	Princess Irene
Apricot Beauty	Edith Eddy	Mirjoran	Prominence
Arma	Golden Eddy	Monte Carlo	Red Riding Hood
Attila	Henry Dunant	Murillo Tulips	Robinea
Bellona	Hibernia	Olaf	Ruby Red
Bing Crosby	Hugo Schlösser	Orange Monarch	Snowstar
Blenda	Invasion	Orange Sun	Stockholm
Cassini	Jimmy	Palestrina	Thule
Charles	Karel Doorman	Paris	Topscore
Christmas Gold	Kareol	Paul Richter	Trance
Christmas Marvel	Kees Nelis	Pax	Yellow Present
Comet	La Suisse	Peerless Pink	Yokohama
Coriolan	Los Angeles	Plaisir	

Hyacinthus (Hyacinth)

Amethyst	Blue Jacket	Lady Derby	Ostara
Amsterdam	Carnegie	L'Innocence	Pink Pearl
Anna Marie	Delft Blue	Madame Kruger	Queen of the Pinks
Bismarck	Eros	Marconi	White Pearl
Blue Giant	Jan Bos	Marie	

Narcissus (Narcissus, Daffodil)

Barrett Browning	February Gold	Ice Follies	Peeping Tom
Bridal Crown	Flower Drift	Joseph MacLeod	Standard Value
Carlton	Flower Record	King Alfred	Tête-a-Tête
Dutch Master	Geranium	Magnet	Van Sion
Explorer	Gold Medal	Mount Hood	Yellow Sun

Crocus

Flower Record	Peter Pan	Purpureus Grandiflorus	Victor Hugo
Joan of Arc	Pickwick	Remembrance	Yellow
King of the Striped			

Iris
  Iris danfordiae
  Iris reticulata cultivars

Cantab	Joyce
Harmony	J. S. Dijt
Hercules	

Muscari (Grape Hyacinth)

Blue Spike	Early Giant

[a]For detailed forcing schedules (time, temperature, and so on) consult A. DeHertogh, *Holland Bulb Forcer's Guide* (see Further Reading).

466

TABLE 13-5. FERNS

Scientific Name	Common Name	Frond Length (in.)	Zone of Hardiness	Propagule	Culture	Remarks
Adiantum capillus-veneris	Southern maidenhair fern	18	9	Di, Sp	MS, Sh	O, GH, H
A. hispidulum	Australian maidenhair fern	6–12	10	Di, Sp	MS, Sh	O, GH
A. pedatum	American maidenhair fern	18–24	4	Di, Sp	MS, Sh	O, GH
A. raddianum	Delta maidenhair fern	16	10	Di, Sp	MS, Sh	GH
Aglaomorpha meyeniana	Bear's paw fern	12–36	10	Di, Sp	PS, HH	GH
Asplenium bulbiferum	Mother spleenwort	24	10	Di, Sp, VB	MS, Sh	GH, H
A. nidus	Bird's nest fern	12	10	Di, Sp	MS, Sh	GH
A. trichomanes	Lobed spleenwort	6–12	6	Di, Sp	MS, Sh	O
A. pinnatifidum	Maidenhair spleenwort	6	4	Di, Sp	AS, MS, Sh	O
Athyrium filix-femina	Lady fern	1	4	Di	LS, MS, Sh	O
Botrychium virginianum	Rattlesnake fern	8–30	4	Di, Sp	MS, Sh	O
Camptosorus rhizophyllus	Walking fern	4–9	4	Sp, TR	MS, Sh	O
Cyrtomium falcatum	Holly fern	30	10	Sp	LS, MS, Sh	GH, H
Cystopteris bulbifera	Berry bladder fern	30		Di, FB, Sp	LS, MS, Sh	
C. fragilis	Brittle fern	8–12	4	Di, Sp	MS, Sh	O
Davallia canariensis	Deer's foot fern	18		Di	MS, Sh	GH, H
D. fejeensis	Rabbit's foot fern	12–30	10	Di	MS, Sh	GH
D. trichomanoides	Squirrel-foot fern	18		Di	MS, Sh	O, GH
Dennstaedtia punctilobula	Hay-scented fern	30	4	Di, Sp	MS, Sh	O
Diplazium pycnocarpon	Silvery spleenwort	30	4	Di	LS, MS, Sh	O
Dryopteris austriaca intermedia	Intermediate shield fern	36	4	Di, Sp	MS, Sh	O
D. austriaca spinulosa	Spinulose wood fern		4	Di, Sp	MS, Sh	
Humata tyermannii	Bear-foot fern	12	10	Di, Sp	MS, Sh	O
Lygodium japonicum	Japanese climbing fern	96–120	8	Di	MS, Sh	O, GH
Matteuccia pensylvanica	Ostrich fern	72–108	3	Sp	MS, Sh	O, GH
Nephrolepis exalta 'Bostoniensis'	Boston fern	72–96		Di, Sp	MS, Sh	O
N.e. 'Compacta'	—	24–48	10	Di, Sp	MS, Sh	
N.e. 'Elegantissima'	Dwarf Boston fern	60–72		Di, RR	MS, Sh	GH, H
Onoclea sensibilis	Sensitive fern	54	4	Di, Sp	MS, S-Sh	O
Osmunda cinnamomea	Cinnamon fern	60	4	Sp	MS, Sh	O
O. claytoniana	Interrupted fern	48	4	Sp	AS, MS, Sh	
Pellaea rotundifolia	Button fern	12	10	Di, Sp	MS, Sh	O, GH
Platycerium bifurcatum	Common staghorn fern	36	10	Di, Sp	HH, MS, Sh	GH
Polypodium aureum	Rabbit's foot fern	48	10	Di, Su	MS, Sh	GH
P. vulgare	European polypody	36	4	Di, Sp	MS, Sh	O

TABLE 13-5. CONTINUED

Scientific Name	Common Name	Frond Length (in.)	Zone of Hardiness	Propagule	Culture	Remarks
Polystichum acrostichoides	Christmas fern	24	4	Di, Sp	MS, Sh	O
P. braunii	Shield fern	18		Di, Sp	MS, Sh	GH, H
Pteris cretica	Cretan brake	36	10	Di	MS, Sh	GH
P. quadriaurita	Silver fern	72		Di, Sp		GH, H
P. tremula	Australian brake	24	3	Di	AS, MS, S-Sh	O
Thelypteris noveboracensis	New York fern	24	4	Di, Sp	MS, Sh	O
Woodsia ilvensis	Rusty woodsia	10	5			O
Woodwardia areolata	Netted chain fern	15		Di, Sp	AS, MS, Sh	O
W. virginica	Virginia chain fern	24				

Explanation of Table Symbols
Frond length is maximal length for optimal environmental conditions exclusive of dwarf cultivars.

Propagule:
Di, division of clump and rhizomes
FB, frond bulblike bodies
RR, rooting runners
Sp, spores
Su, suckers
TR, tip rooting
VB, vegetative buds on frond

Culture:
AS, acid soil
HH, high humidity over 70%
LS, limestone soil
MS, moist soil
Sh, partial to full shade
S-Sh, sunny to partial shade

Remarks:
GH, suitable for greenhouse
H, suitable for home
O, outdoors in garden or naturalized setting

468

TABLE 13–6. LAWN GRASSES

Scientific Name	Common Name	Advantages	Disadvantages	Cultivars
		Northern Grasses		
Poa pratensis	Kentucky bluegrass	Attractive northern grasses. Self-reliant. Easy care. Strong sod. Resistant to wear. Good recuperative powers. Tolerates light shade. Best growth in cool weather. Perennial. Good texture.	Weakest in hot weather. Some slowness for seeding to become established.	Adelphi. Arboretum. Arista. Baron. Bonnieblue. Fylking. Galaxy. Glade. Majestic. Merion. Nugget. Pennstar. Prato. Plush. Sodco. Sydsport. Touchdown.
Festuca rubra	Fine fescue	Good companion for bluegrass. Adapted to shade, poor soil, drought, minimal fertilization. Perennial.	Weak in warm, humid weather.	Chewings. Creeping Red. Illahee. Pennlawn. Rainer. Wintergreen.
Agrostis sp.	Bent grasses	Finest texture. Very elegant. Good in moist climates. Perennial.	Require extensive fertilization, watering, mowing to keep appearance. Weak in hot, muggy weather. Sensitive to snowmold.	Astoria. Emerald. Highland. Kingstown. Penncross. Seaside.
Lolium perenne	Perennial rye grass	Quick sprouting. Attractive. Perennial.	Does not mow cleanly. Weak during extremes of weather.	Citation. Derby. Diplomat. Manhattan. NK-200. Pennfine. Yorktown.
		Southern Grasses		
Cynodon dactylon	Bermuda grass	Fast growing. Aggressive. Attractive texture, color. Good recuperation. Good under warm, humid conditions. Perennial.	Thatches. Requires frequent mowing, fertilization, watering for best performance. Goes dormant near freezing. Poor in shade or below 50° F.	Midway. Santa Ana. Sunturf. Tifdwarf. Tifgreen. Tiflawn. Tifway. Tufcote. U-3.
Zoysia sp.	Zoysia	Minimal mowing. Dense. Attractive. Minimal attention. Tough. Durable. Perennial.	Slow to establish. Requires heavy-duty mower. Thatches heavily. Slow recuperation. Long winter dormancy. Billbug susceptible.	Emerald. Meyer. Midwest.

TABLE 13-6. CONTINUED

Scientific Name	Common Name	Advantages	Disadvantages	Cultivars
Stenotaphrum secundatum	St. Augustine grass	Reasonably attractive. Shade tolerant. Salt-spray tolerant. Modest attention. Good for extreme South. Perennial.	Coarse. Thatches. Susceptible to cinchbug.	Bitter Blue. Floratine. Floratam Roselawn.
Eremochloa ophiuroides	Centipede grass	Modest attention. Requires minimal care. Tolerates shade and heat.	Sensitive to high fertility or alkalinity. Slow to establish. Cold intolerant. Sensitive to chlorosis.	Oaklawn.
Paspalum notatum	Bahia grass	Highly adaptable. Modest attention. Tolerates shade and heat.	Coarse. Cold intolerant.	Argentine. Paraguay. Pensacola. Wilmington.
		Specialty Grasses		
North				
Agropyron sp.	Wheatgrass	Poor turf, but tolerates dryness in north.		
Agrostis alba	Redtop	Tolerates soggy areas.		
Festuca ovina	Sheep fescue	Meadow and forage grass.		
Lolium multiflorum	Italian rye grass	Good for rapid, but only temporary, cover.		
Poa trivialis	Rough bluegrass	Good for shady, moist areas. Poor for traffic.		
Puccinellia distans	Alkaligrass	Adapted for saline, arid areas.		
South				
Axonopus sp.	Carpet grass	Tolerates soggy areas.		
Buchloë dactyloides	Buffalo grass	Used in dry areas that cannot be irrigated.		

TABLE 13-7. GROUND COVER PLANTS

Scientific Name	Common Name	Height (in.)	Zone of Hardiness	Remarks	Propagule
Acaena microphylla	Redspine sheepburr	1	7	CFr, E, NS	Cu, Di, Se
Achillea millefolium	Common yarrow	6–30	3 ⎫	D, F, Su, NS, ST	⎫ Di, Se
A. tomentosa	Wooly yarrow	12	⎭		⎭
Aegopodium podagraria	Goutweed	14	4	CFo, D, DS, IR, M, NS, Sh, Su	Di
Ajuga repens	Carpet bugleweed	10	5	CFo, D, F, IR, NS, Sh, Su, W	Di, Se
Akebia quinata	Five-leaf akebia	Vine	5	B, D, IR, NS, Sh, Su, P	Di, L, RCu, Se
Aloe sp.	Aloe	3–12	9	E, DS, F, Su	Cu, Se, Su
Andromeda polifolia	Bog rosemary	12	3	AS, E, Su, WS	Cu, L, Se
Arabis alpina	Mountain rock cress	10	4	D, F, Su, W	Cu, Di
Arctostaphylos uva-ursi	Common bearberry	12	2	B, CFo, CFr, DS, E, Su	Cu
Armeria maritima	Common thrift	12	3	D, Sh, ST, W, WS	Di
Asarum caudatum	—	7	5	Sh, NS, WS	Di
Cerastium tomentosum	Snow-in-summer	10	3	CFo, D, IR, NS, ST, W	Cu, Di, Se
Comptonia peregrina	Sweet fern	60	3	AS, B, D, DS, NS, ST, WS	Di, L, Se
Convallaria majalis	Lily-of-the-valley	8	3	CFr, D, F, IR, NS, Sh	RD
Cornus canadensis	Bunchberry	9	3	AS, CFr, E, F, NS, Su, WS	L
Coronilla varia	Crown vetch	24	4	B, D, DS, F, IR, Sh, Su	Di, Se
Cotoneaster adpressus	—	10	5	CFr, D, Su, ST	Cu
C. dammeri	—	12	6	CFr, E, Su, ST, WS	Cu, Di
C. horizontalis	Rock cotoneaster	36	6	B, CFr, D, Su, ST	Cu, L
Cytisus decumbens	—	8	6	DS, E, F, Su, ST	Cu, Se
Dichondra micrantha	Dichondra	3	10	E, IR, M, NS, Sh, Su, W	Se
Diervilla lonicera	Bush honeysuckle	48	4	B, D, DS, NS, Su	Cu, Di
Duchesnea indica	Indian strawberry	2	6	CFr, D, M, Sh, Su, W	Di
Epimedium grandiflorum	—	12	4	D, F, Sh	Di
Erica carnea	Spring heath	12	6	AS, DS, E, F, IR, Su, ST, W	Cu, Di, L
Euonymus fortunei	Wintercreeper	Vine	5	B, CFo, CFr, D, E, IR, NS, Su, W	Cu, Di
Fragaria chiloensis	Beach strawberry	6	5	CFr, D, F, Su	RuD
Galax urceolata	Wandflower	6	4	AS, E, F, Sh, WS	Di
Gaultheria procumbens	Wintergreen	2	4	AS, CFr, E, Sh	Di
Glechoma hederacea	Ground ivy	3	4	D, NS, Su	Cu
Hedera helix	English ivy	Vine	6	B, E, IR, NS, Sh, Su	Cu, Di, L
Hypericum calycinum	Rose-of-Sharon	12	7	B, DS, E, F, IR, NS, Sh, ST	Cu, Di, Se
Juniperus chinensis cv.	Juniper	12	5 ⎫	⎫	⎫
J. conferta	Shore juniper	12	6 ⎬	⎬ B, CFo, CFr, DS, E, NS, P, Su, ST	⎬ Cu, L
J. horizontalis cv.	Creeping juniper	18	3 ⎭	⎭	⎭

TABLE 13-7. CONTINUED

Scientific Name	Common Name	Height (in.)	Zone of Hardiness	Remarks	Propagule
Lantana montevidensis	Weeping lantana	24	10	E, F, NS, Su	Cu, Se
Liriope spicata	Creeping lilyturf	10	6	E, F, NS, Sh, Su	Di
Lonicera japonica	Japanese honeysuckle	Vine	6	B, CFr, D-E, F, IR, NS, P, Sh, ST, Su	Cu, Di
Mahonia repens	—	12	6	CFr, E, F, NS, Sh, Su	Di, RCu
Mitchella repens	Partridgeberry	2	4	AS, CFr, E, M, Sh, W, WS	Di
Pachysandra procumbens	Alleghany pachysandra	12 ⎫	5	D, F, NS, Sh	Cu, Di
P. terminalis	Japanese pachysandra	⎭		CFo, CFr, E, IR, NS, Sh	
Phlox subulata	Moss pink	6	3	D-E, DS, F, Su, W	Cu, Di, Se
Potentilla crantzii	—	12	7	DS, E, F, M, NS, Su, ST, W	Di, Se
Rosa wichuraiana	Memorial rose	12	6	B, CFr, D-E, F, IR, NS, P, Su, ST	Cu, Di, L, Se
Rubus laciniatus	Cut-leaf blackberry	24	6	B, CFr, D, DS, IR, NS, P, Su	Cu, Di, L
Sasa sp.	Bamboo	12-42	6	B, D-E, IR, NS, Sh	Di
Sedum sp.	Stonecrop	2-8	4	DS, E, F, IR, NS, Sh, Su, ST	Cu, Di
Teucrium chamaedrys	Germander	12	6	D, Sh, Su	Cu, Di
Thymus sp.	Thyme	1-8	4	D-E, DS, IR, M, NS, Su, ST, W	Di, Se
Vaccinium angustifolium	Lowbush blueberry	12	2	AS, B, CFr, D, DS, Sh, Su	Cu, Di, L
Veronica sp.	Speedwell	4-24	4	D-E, DS, IR, M, NS, Su, W, WS	Cu, Di, Se
Vinca minor	Common periwinkle	6	5	E, F, IR, NS, Sh, Su, W	Cu, Di, RCu
Viola sp.	Violet	4-16	3-7	D, F, M, NS, Sh, Su, W	Di, Se
Xanthorhiza simplicissima	Shrub yellowroot	24	5	B, D, NS, P, Sh, Su, WS	Di, RCu, Se

Explanation of Table Symbols

Height: maximal potential under optimal environmental conditions exclusive of dwarf cultivars.

Remarks:

AS, for acid soils
B, useful for banks to prevent erosion
CFo, colorful foliage
CFr, colorful fruit
D, loses some or all leaf cover during winter (deciduous)
DS, for dry soil
E, evergreen
F, colorful or interesting flowers
IR, increases rapidly, may require control for more vigorous ones
M, will withstand occasional mowing
NS, for normal soil
P, provides protective barrier

Sh, tolerates shade
ST, tolerates seashore conditions
Su, full sun
W, will withstand occasional walking
WS, for wet soil

Propagule:
Cu, cutting
Di, division
L, layering
RCu, root cutting
RuD, runner division
Se, seed

472

TABLE 13–8. DECIDUOUS TREES

Scientific Name	Common Name	Height[a] (ft)	Zone of Hardiness	Remarks
Acer argutum	—	25	6	Maples are used extensively for lawns, parks, and
A. barbatum	Southern sugar maple	50	7	streets. Strong and free growing, but shallow
A. buergeranum	Trident maple	50	6	rooted. Noted for shade and autumn color.
A. campestre	Hedge maple	35	5	
A. carpinifolium	Hornbeam maple	30	6	
A. ginnala	Amur maple	20	5	
A. glabrum	Rocky mountain maple	30	5	
A. griseum	Paperbark maple	40	6	Interesting bark.
A. japonicum	Japanese maple	20–30	5	Many cultivars. Attractive foliage.
A. macrophyllum	Oregon maple	100	7	Environmentally tolerant in cities.
A. negundo	Box elder	50–70	3	Environmentally tolerant in cities. Many cultivars.
A. palmatum	Japanese maple	20–50	5	Many cultivars. Attractive, colorful foliage.
A. platanoides	Norway maple	70	4	Environmentally tolerant in cities. Many cultivars.
A. pseudoplatanus	Sycamore	100	5	Many cultivars.
A. rubrum	Red maple	120	3	Environmentally tolerant in cities. Many cultivars.
A. saccharinum	Silver maple	130	3	Many cultivars. Poor though, because weak-wooded.
A. saccharum	Sugar maple	130	3	Environmentally tolerant in cities. Many cultivars.
A. truncatum	Shantung maple	75	5	—
Amelanchier laevis	Serviceberry	40	5	White flowers followed by purple-black fruit.
Betula papyrifera	Canoe birch	100	3	Attractive whitish bark. Environmentally tolerant in cities.
B. pendula	European white birch	60		Short-lived. Borer susceptible.
Carpinus betulus	European hornbeam	70	4	Environmentally tolerant in cities. Somewhat difficult to transplant.
C. caroliniana	American hornbeam	40		—
Celtis australis	Mediterranean hackberry	80	7	Environmentally tolerant in city.
C. laevigata	Sugarberry	100	6	Environmentally tolerant in city.
C. occidentalis	Nettle tree	120	4	Susceptible to witches-broom growth.
Cercidiphyllum japonicum	Katsura tree	100	5	Environmentally tolerant in city. Excellent, but not well-known.
Cercis canadensis	Redbud	40	5	Attractive flowers. Environmentally tolerant in city.
Chionanthus virginicus	White fringetree	30	5	Attractive flowers.
Cladrastis lutea	Yellowwood	50	4	Fragrant white flowers. Environmentally tolerant in city.
Cornus florida	Flowering dogwood	40	5	Attractive flowers. Colorful fruit. Susceptible to borers. Environmentally tolerant in city.

473

TABLE 13-8. CONTINUED

Scientific Name	Common Name	Height[a] (ft)	Zone of Hardiness	Remarks
Crataegus crus-galli	Cockspur	30		Thorns can be hazardous to small children. Attractive flowers and fruits. Few cultivars found without thorns. Environmentally tolerant in city.
C. laevigata	English hawthorn	25	5	
C. monogyna	English hawthorn	30		
C. phaenopyrum	Washington thorn tree	25		
Elaeagnus angustifolia	Russian olive	20	3	Attractive foliage, flowers, fruits. Environmentally tolerant in city.
Evodia danielii	Korean evodia	50	5	Attractive foliage, flowers, fruits. Not well known.
Fagus sylvatica	European beech	80	5	Majestic. Many excellent cultivars. Environmentally tolerant in cities.
Franklinia alatamaha	Franklin tree	30	6	Attractive flowers. Good fall color.
Fraxinus excelsior	European ash	140	4	
F. mariesii	—	25	7	Attractive purple fruit.
F. pennsylvanica	Red ash	60	3	Adaptable. Get seedless cultivars.
F. velutina	Velvet ash	50	6	
Ginkgo biloba	Maidenhair tree	120	5	Environmentally tolerant in city. Avoid female which has foul-smelling fruit.
Gleditsia triacanthos	Honey locust	140	5	Overused. Thorns hazardess, but thornless cultivars exist. Fruit attractive, but litters. Environmentally tolerant in city. Watch for insects and disease.
Gymnocladus dioicus	Kentucky coffee tree	100	5	Environmentally tolerant in city. Interesting bark.
Halesia carolina	Wild olive	40	5	Attractive flowers. Environmentally tolerant in city.
Hamamelis × intermedia	—			
H. japonica	Japanese witch hazel	30	6	Attractive, very early flowers.
H. mollis	Chinese witch hazel			
Koelreuteria paniculata	Varnish tree	45	6	Attractive flowers, fruit. Environmentally tolerant in city.
Laburnum × watereri	Goldenchain tree	15	6	Attractive flowers.
Larix decidua	European larch	100	3	
L. kaempferi	Japanese larch	90	5	Good fall color. Watch for larch casebearer. Cones.
L. laricina	American larch	80	2	
Liquidambar styraciflua	Sweet gum	120	6	Good fall color.
Liriodendron tulipifera	Tulip tree	150	5	Weak-wooded. Environmentally tolerant in city.
Magnolia acuminata	Cucumber tree	100	5	
M. campbelli	Chinese magnolia	80	9	Attractive flowers. Environmentally tolerant in city.
M. × soulangiana		30	6	
M. stellata	Star magnolia	25	5	

Scientific name	Common name	Height	Rating	Remarks
Malus arnoldiana	Arnold crab apple	20	5	Attractive flowers and fruit. Many excellent cultivars. Environmentally tolerant in city.
M. baccata	Siberian crab apple	50	3	
M. floribunda	Japanese flowering crab apple	30	5	
Nyssa sylvatica	Pepperidge	100	5	Good fall color.
Oxydendrum arboreum	Sourwood	80	5	Attractive flowers. Good fall color. Environmentally tolerant in city.
Phellodendron amurense	Corktree	50	4	Good bark pattern when mature. Environmentally tolerant in city.
Platanus acerifolia	London plane tree	120	5	Interesting bark. Drops litter. Overplanted. Environmentally tolerant in city.
P. occidentalis	Eastern sycamore	150	4	
P. orientalis	Oriental plane tree	100	6	
Populus alba	White poplar	90	4	Fast growing. Weak-wooded. Short-lived. Clog drains. Survives in areas where other trees cannot.
P. × canadensis	Carolina poplar	150	4	
P. tremuloides	Quacking aspen	90	1	
Prunus padus	Bird cherry	40	4	Environmentally tolerant in cities. Attractive flowers and fruits. Many excellent cultivars.
P. sargentii	Sargent cherry	50	5	
P. serotina	Black cherry	80	4	
P. serrulata	Japanese flowering cherry	60	6	
Pyrus calleryana 'Bradford'	Bradford pear	50	5	Attractive flowers. Good fall color. Environmentally tolerant in city.
Quercus acutissima	Sawtooth oak	50	7	Environmentally tolerant in city. Good fall color.
Q. alba	White oak	100	4	Long-lived. Strong trees.
Q. imbricaria	Shingle oak	60	5	
Q. macrocarpa	Bur oak	80	4	
Q. palustris	Pin oak	80	5	
Q. prinus	Basket oak	100	6	
Q. rubra	Red oak	80	5	
Q. velutina	Black oak	100	5	
Robinia pseudoacacia	Black locust	80	4	Attractive fragrant flowers. A tree for difficult areas.
Salix alba	White willow	75	2	Somewhat weak-wooded and littery. Environmentally tolerant in city.
S. babylonica	Weeping willow	30	5	
S. caprea	Pussy willow	25	5	Noted for early spring catkins.
Sassafras albidum	Sassafras	60	5	Good fall color.
Sophora japonica	Japanese pagoda tree	80	5	Attractive flowers. Environmentally tolerant in city.
Sorbus alnifolia	Korean mountain ash	60	6	Good fall color. Attractive fruit. Watch out for disease.
S. aucuparia	European mountain ash	60	2	
Stewartia koreana	Korean stewartia	50	6	Attractive flowers, fall color, and bark. Environmentally tolerant in city.
S. pseudocamellia	Japanese stewartia			

TABLE 13-8. CONTINUED

Scientific Name	Common Name	Height[a] (ft)	Zone of Hardiness	Remarks
Styrax japonicus	Japanese snowbell	30	5	Attractive flowers and fruit.
S. obassia	Fragrant snowbell			
Syringa reticulata	Japanese tree lilac	30	4	Attractive flowers. Environmentally tolerant in city
Taxodium distichum	Bald cypress	150	5	Good for wet areas.
Tilia americana	American linden	130	3	
T. cordata	Small-leaved European linden	100	4	Environmentally tolerant in city. Fragrant flowers.
T. platyphyllos	Large-leaved linden	130	4	
T. tomentosa	Silver linden	90	4	
Ulmus americana	American elm	120	2	Susceptible to Dutch elm disease.
U. parvifolia	Chinese elm	50	6	Somewhat resistant to Dutch elm disease. Attractive bark.
Viburnum rufidulum	Southern black haw	30	6	Attractive flowers and fruit.
V. sieboldii	Siebold viburnum	20	5	

[a]Maximal potential under optimal environmental conditions, exclusive of dwarf cultivars.

TABLE 13–9. DECIDUOUS SHRUBS

Scientific Name	Common Name	Height[a] (ft)	Zones of Hardiness	Remarks
Abelia × *grandiflora*	Glossy abelia	6	6	Attractive flowers. Semievergreen. Good texture. Environmentally tolerant in city.
A. schumannii	—	5	7	Attractive flowers.
Acanthopanax sieboldianus	—	9	5	Environmentally tolerant in city.
Aesculus parviflora	Buckeye	15	5	Excellent form and texture. Attractive flowers.
Amelanchier florida	Pacific serviceberry	10	5	Attractive flowers and fruit.
Amorpha canescens	Lead plant	4	3	Attractive gray foliage. Environmentally tolerant in city.
Aronia arbutifolia	Chokeberry	12	6	Attractive fruit left alone by birds. Environmentally tolerant in city.
Berberis koreana	Korean barberry	4	5	Attractive fruit, flower, and foliage. Excellent thorny hedge. Good fall color. Environmentally tolerant in city.
B. thunbergii	Japanese barberry	5	5	
Buddleia alternifolia	Fountain buddleia	12	6	Attractive flowers.
Callicarpa bodinieri	—	10	5	Attractive fruit.
C. japonica	Japanese beautyberry	5	5	
Calycanthus floridus	Carolina allspice	10	5	Fragrant flowers. Environmentally tolerant in city.
Caragana arborescens	Siberian pea shrub	20	3	Attractive flowers.
C. maximowicziana	—	5	3	
Chaenomeles japonica	Lesser flowering quince	4	5	Attractive flowers. Somewhat thorny. Attractive fruit.
C. speciosa	Japanese quince	10	5	
C. × *superba*	—	5	5	
Clethra alnifolia	Sweet pepperbush	10	5	Attractive, fragrant flowers. Environmentally tolerant in city.
Comptonia peregrina	Sweet fern	5	3	Good bank cover. Naturalistic.
Cornus alba	Tartarian dogwood	10	3	Colorful winter bark. Attractive fruit. Environmentally tolerant in city.
C. mas	Cornealian cherry	20	5	Attractive flowers and fruit.
C. racemosa	Panicled dogwood	15	5	Attractive flowers and fruit.
C. sericea	Red-osier dogwood	10	3	Colorful winter bark. Environmentally tolerant in city.
Corylopsis glabrescens	Fragrant winter hazel	20	6	Early flowers.
C. spicata	Spike winter hazel	6	6	
Cotinus coggygria	Smoke tree	15	5	Attractive flowers.

TABLE 13–9. CONTINUED

Scientific Name	Common Name	Height[a] (ft)	Zones of Hardiness	Remarks
Cotoneaster apiculatus	Cranberry cotoneaster	3	5	Attractive fruit. Watch for fire blight.
C. bullatus	—	10	6	
C. divaricatus	—	6	5	
C. horizontalis	Rock cotoneaster	3	6	*C. horizontalis* is semievergreen.
C. multiflorus	—	6	6	
C. racemiflorus	—	8	5	
Cytisus scoparius	Scotch broom	10	6	Attractive flowers.
Daphne genkwa	—	3	5	
D. giraldi	—	2.5	6	Attractive flowers. Culture difficult.
D. mezereum	February daphne	5	5	
Deutzia × candelabrum	—	6	6	
D. discolor	—	6	6	
D. gracilis	Slender deutzia	6	5	Attractive flowers Watch for winterkill. Environmentally tolerant in city.
D. × kalmiiflora	Kalmia deutzia	6	6	
D. × lemoinei	Lemoin deutzia	7	4	
D. × magnifica	Showy deutzia	6	6	
D. scabra		7	6	
Dirca palustris	Leatherwood	6	5	Good for wet, shady areas. Attractive flowers.
Elaeagnus multiflora	Cherry elaeagnus	6	5	Attractive fruit.
Enkianthus campanulatus		30	5	Attractive flowers. Good fall color. Environmentally tolerant in city.
Euonymus alata	Winged euonymus	8	4	Excellent fall color. Interesting stems. Environmentally tolerant in city.
Exochorda giraldii	Pearlbush	15	5	Attractive flowers.
E. racemosa		12		
Forsythia × intermedia	Golden bells	10	5	Attractive flowers. Environmentally tolerant in city.
F. suspensa	Weeping forsythia			
Fothergilla gardenii	Witch alder	3	6	Attractive flowers and good fall color.
F. major		10		
Genista hispanica	Spanish broom	2	7	Attractive flowers. Spiny.
Hamamelis vernalis	Vernal witch hazel	6	5	Early flowers and good fall color.
Hibiscus rosa-sinensis	Chinese hibiscus	8	9	Attractive flowers. Environmentally tolerant in city.
H. syriacus	Rose-of-Sharon	10	6	
Hydrangea macrophylla	French hydrangea	8	6	Attractive flowers.
H. quercifolia	Oakleaf hydrangea	6	5	Attractive flowers.
Hypericum prolificum	Shrubby St. John's wort	5	5	Attractive flowers.

Ilex decidua	Possum haw	30	6	Attractive foliage and fruit.
I. laevigata	Smooth winterberry	12	5	
I. serrata	Japanese winterberry	10	5	
I. verticillata	Winterberry	15	4	
Indigofera kirilowii	Kirilow indigo	4	5	Attractive flowers.
Kerria japonica	Japanese rose	8	5	Attractive flowers. Good winter stem color. Environmentally tolerant in city.
Kolkwitzia amabilis	Beauty bush	15	5	Attractive flowers. Environmentally tolerant in city.
Lespedeza bicolor	Shrub bushclover	10	5	Attractive flowers.
Ligustrum amurense	Amur privet	15	4	Used as hedges. Environmentally tolerant in city.
L. obtusifolium	Border privet	10	4	
L. vulgare	Common privet	15	5	
Lonicera alpigena	Alps honeysuckle	10	6	
L. × bella	—	6	5	
L. fragrantissima	Winter honeysuckle	10	6	Attractive flowers and fruit. Environmentally tolerant in city.
L. maackii	Amur honeysuckle	15	3	
L. tatarica	Tatarian honeysuckle	10	4	
Malus brevipes	Nippon crab apple	15	6	Attractive flowers fruit. Environmentally tolerant in city.
M. sargentii	Sargent crab apple	8	5	
Myrica pensylvanica	Bayberry	9	3	Somewhat salt tolerant. Good for poor soil sites.
Niviusia alabamensis	Snow wreath	6	6	Attractive flowers.
Philadelphus coronarius	Sweet mock-orange	9	5	Attractive flowers. Environmentally tolerant in city.
P. × lemoinei	Lemoine mock-orange	8	5	
P. × virginalis	—	9		
Photinia villosa	Oriental photinia	15	5	Attractive flowers and fruit. Good fall color. Susceptible to fire blight.
Physocarpus opulifolius	Eastern ninebark	10	2	Attractive flowers and fruit. Bark somewhat interesting. Environmentally tolerant in city.
Poncirus trifoliata	Hardy-orange	20	7	Thorny. Good hedge. Attractive flowers and fruit.
Potentilla fruticosa	Shrubby cinquefoil	4	2	Attractive flowers. Environmentally tolerant in city.
Prunus fruticosa	European dwarf cherry	4	4	Attractive flowers and fruits.
P. glandulosa	Flowering almond	5	4	
P. japonica	Japanese flowering almond	4	4	
P. tomentosa	Nanking cherry	5	3	
P. triloba	Flowering almond	10	3	
Rhamnus cathartica	Common buckthorn	12	3	Hedge in difficult situations.
Rhododendron sp.	Azalea	1.5–9	3–7	Attractive flowers. Numerous cultivars and hybrids of deciduous azaleas.
Rhodotypos scandens	Jetbead	6	5	Fruit interesting. Environmentally tolerant in city.
Rhus aromatica	Fragrant sumac	8	4	Attractive flowers and fruit. Good fall color. Environmentally tolerant in city.
Ribes alpinum	Mountain currant	8	2	Good hedge.

TABLE 13-9. CONTINUED

Scientific Name	Common Name	Height[a] (ft)	Zones of Hardiness	Remarks
Rosa sp. and cv.	Rose	1.5–15	2–7	Attractive flowers and fruits. Many fragrant. Choices extremely numerous.
Shepherdia canadensis	Buffalo berry	8	3	Attractive foliage and fruit. Environmentally tolerant in city.
Sorbaria sorbifolia	Ural false spirea	6	2	Attractive flowers.
Spiraea albiflora	Japanese white spirea	2	5	Attractive flowers. *S. prunifolia* and *S. thunbergii* have some fall color. Environmentally tolerant in city.
S. × *arguta*	Garland spirea	6	5	
S. × *billiardii*	Billiard spirea	6	5	
S. × *bumalda*	—	2	5	
S. japonica	Japanese spirea	6	6	
S. prunifolia	Bridalwreath	6	5	
S. thunbergii	Thunberg spirea	5	5	
S. × *vanhouttei*	Vanhoutte spirea	6	5	
Stephanandra incisa	Lace shrub	8	5	Attractive foliage with good fall color.
Symphoricarpos albus	Snowberry	3	3	Attractive fruit. Environmentally tolerant in city.
Syringa × *chinensis*	Chinese lilac	15	3	Attractive, often fragrant flowers. Only *S. oblata* has good fall color. Environmentally tolerant in city.
S. laciniata	Cut-leaf lilac	6	5	
S. meyeri	—	6	6	
S. microphylla	Little-leaf lilac	6	4	
S. oblata	Korean early lilac	12	4	
S. × *persica*	Persian lilac	6	5	
S. villosa	Late lilac	10	2	
S. vulgaris	Common lilac	20	4	
Tamarix parvifolia	Small-flowered tamarisk	9	5	Attractive flowers. Salt tolerant.
T. ramosissima	Five-stamen tamarisk	18	3	
Viburnum alnifolium	Hobblebush	10	4	
V. × *burkwoodii*	Burkwood viburnum	6	5	
V. carlesii	Fragrant viburnum	5	5	
V. cassinoides	Withe-rod	12	4	Attractive flowers, foliage, and fruits. Good fall color. *V.* × *burkwoodii* and *V. carlesii* are fragrant. Some are known to be environmentally tolerant in cities.
V. dilatatum	Linden viburnum	10	5	
V. opulus	Cranberry bush	12	3	
V. plicatum	Japanese snowball	10	5	
V. prunifolium	Black haw	15	3	
V. sargentii	Sargent cranberry bush	12	6	
V. sieboldii	Siebold viburnum	10	5	
V. trilobum	American cranberry bush	12	2	
Vitex negundo	—	20	6	Attractive flowers.
Weigela floribunda	—	10	6	Attractive flowers. Environmentally tolerant in city.
W. florida	Old-fashioned weigela	10	5	

[a]Maximal potential under optimal environmental conditions, exclusive of dwarf cultivars.

TABLE 13-10. CONIFEROUS AND NEEDLED EVERGREENS

Scientific Name	Common Name	Height[a] (ft)	Zones of Hardiness	Remarks
Abies concolor	White fir	100	4	Pyramidal shape. Slow growers. Do poorly in polluted air or where growing season is hot.
A. homolepis	Nikko fir	100	5	
A. pinsapo	Spanish fir	75	7	*A. concolor* is most adaptable and has a dwarf cultivar.
A. procera	Noble fir	150	6	
A. spectabilis	Himalayan fir	100	6	
A. veitchii	Veitch fir	75	4	
Calocedrus decurrens	California incense cedar	100	6	Columnar. Should be used more. Needs moist atmosphere.
Cedrus atlantica	Atlas cedar	100	7	Pyramidal. Excellent for specimens. *C. atlantica* and *deodara* somewhat environmentally tolerant in city. Dwarf cultivar for *C. libani*.
C. deodara	Deodar cedar	150	7	
C. libani	Cedar-of-Lebanon	100	6	
Cephalotaxus harringtonia	Harrington plum yew	30	6	Rounded and bushy. Poor in hot, dry areas.
Chamaecyparis lawsoniana	Lawson cypress	100	6	Pyramidal to almost columnar. *C. obtusa* can tolerate drier atmosphere, otherwise moist atmosphere needed. Numerous cultivars. Several are dwarf cultivars.
C. nootkatensis	Nootka cypress	100	5	
C. obtusa	Hinoki cypress	120	5	
C. pisifera	Sawara cypress	120	5	
C. thyoides	White cedar	90	5	
Cryptomeria japonica	Japanese cedar	150	8	Pyramidal to conical. Fast growing. Numerous cultivars, some are dwarf.
Cunninghamia lanceolata	Common china-fir	120	7	Pyramidal.
× *Cupressocyparis leyandii*	Leyland cypress	50	6	Pyramidal. Fast growing. Environmentally tolerant in city.
Cupressus arizonica	Arizona cypress	30	8	Various: pyramidal, columnar, rounded. Mostly for mild climates.
C. bakeri	Modoc cypress	90	5	
C. lusitanica	Portuguese cypress	75	9	
C. macrocarpa	Monterey cypress	40	8	
C. sempervirens	Italian cypress	80	8	
Dacrydium cupressinum	Rimu	100	10	Confined to frost-free areas.
D. laxifolium	Mountain rimu	1		
Fitzroya cupressoides		40	8	Very slow growing. Straggly.
Glyptostrobus lineatus	Chinese water pine	—	9	Needles turn rusty red in fall. Deciduous.
Juniperus chinensis	Chinese juniper	60	4	Various: prostrate, conical, columnar, pyramidal, and rounded. Adaptable: tolerant toward hot and dry, city conditions. Numerous varieties and cultivars, many of which are dwarf.
J. communis	Common juniper	35	3	
J. horizontalis	Creeping juniper	1	3	
J. sabina	Savin juniper	10	3	
J. scopulorum	Rocky-mountain juniper	30	4	
J. squamata	Singleseed juniper	3	5	
J. virginiana	Red cedar	75	3	

481

TABLE 13–10. CONTINUED

Scientific Name	Common Name	Height[a] (ft)	Zones of Hardiness	Remarks
Keteleeria davidiana	—	100	7	Pyramidal. Tolerates some dryness.
Microcachrys tetragona	—	3	10	For frost-free areas.
Phyllocladus trichomanoides	Celery pine	70	10	Prostrate. Broad pyrimidal.
Picea abies	Norway spruce	150	3	Pyramidal to conical. Lower branch loss frequent.
P. engelmannii	Engelmann spruce	150	3	*P. abies, glauca,* and *pungens* environmentally
P. glauca	White spruce	100	3	tolerant in city. A few dwarf cultivars available.
P. omorika	Serbian spruce	100	4	*P. omorika, orientalis, pungens* are more tolerant
P. orientalis	Oriental spruce	120	5	of dry conditions.
P. pungens	Colorado spruce	100	3	
Pinus aristata	Bristle-cone pine	40	6	
P. bungeana	Lace-bark pine	75	5	
P. canariensis	Canary island pine	100	8	
P. cembra	Swiss stone pine	75	3	
P. contorta	Shore pine	30	7	
P. densiflora	Japanese red pine	100	5	
P. koraiensis	Korean pine	100	4	Pyramidal. Few shrubby. Most widely used ever-
P. monticola	Western white pine	200	6	green conifers. More tolerant of adverse soil and
P. mugo	Mountain pine	30	3	climatic conditions than *Abies* or *Picea.* En-
P. nigra	Austrian pine	100	4	vironmentally tolerant in city. Numerous culti-
P. parviflora	Japanese white pine	50	6	vars, many of which are dwarf.
P. pinea	Italian stone pine	80	8	
P. ponderosa	Western yellow pine	200	6	
P. radiata	Monterey pine	75	7	
P. resinosa	Norway pine	90	3	
P. strobus	White pine	120	3	
P. sylvestris	Scotch pine	100	3	
P. thunbergiana	Japanese black pine	130	5	
Podocarpus acutifolius	Plum fir	10	9	Shrubby to irregular upright. Warmer areas.
P. andinus	African yellowwood	45	8	
P. elongatus	Southern yew	70	9	
P. macrophyllus	Golden larch	45	8	
Pseudolarix kaempferi		130	6	Pyramidal. Gold fall color. Grows slowly. Decidu-
				ous.
Pseudotsuga menziesii	Douglas fir	200	5	Pyramidal. Numerous cultivars, some of dwarf
				type. Variety *glauca* is hardy to zone 4.
Saxegothaea conspicua	Prince Albert yew	40	8	Shrubby to erect.

Species	Common name			Description
Sciadopitys verticillata	Umbrella pine	100	6	Pyramidal. Excellent specimen.
Sequoia sempervirens	Redwood	300	7	Majestic. Restricted to West Coast.
Sequoiadendron giganteum	Giant sequoia	250	7	Conical.
Taiwania cryptomerioides	—	175	9	
Taxus baccata	English yew	60	7	Shrubby to treelike. Various: prostrate, rounded, columnar. Moderate growth rate. Require good drainage. Numerous cultivars; some are dwarf. Environmentally tolerant in city.
T. canadensis	Canadian yew	6	3	
T. cuspidata	Japanese yew	50	5	
T. × media	—	40	5	
Tetraclinis articulata	Arar tree	20	10	Conical. Tolerate dryness.
Thuja occidentalis	American arborvitae	60	3	Narrow to broad pyramidal. Best in moist atmosphere, poor in hot, dry areas. Numerous cultivars, some of which are dwarf. Environmentally tolerant in city.
T. orientalis (now *Platycladus orientalis*)	Oriental arborvitae	40	6	
T. plicata	Giant arborvitae	200	5	
Thujopsis dolabrata	Hiba arborvitae	100	7	Pyramidal.
Torreya californica	California nutmeg	70	7	Pyramidal to rounded. Slow growing.
T. nucifera	Kaya	75	7	
T. taxifolia	Stinking cedar	40	8	
Tsuga canadensis	Canada hemlock	80	3	Pyramidal. Poor in hot, dry areas. *T. canadensis* and *caroliniana* environmentally tolerant in city. Numerous cultivars of *T. canadensis*, some of which are dwarf.
T. caroliniana	Carolina hemlock	70	5	
T. diversifolia	Japanese hemlock	80	6	
T. sieboldii	Siebold hemlock	100	6	
Widdringtonia cupressoides	Berg cypress	12	10	Shrubby. Tolerant of hot dry conditions. Conical.
W. schwarzii	Willowmore cedar	100	10	

[a]Maximal potential under optimal environmental conditions, exclusive of dwarf cultivars.

TABLE 13-11. BROAD-LEAVED EVERGREEN SHRUBS AND TREES

Scientific Name	Common Name	Height[a] (ft)	Zones of Hardiness	Remarks
Abelia floribunda	Mexican abelia	6	9	Both shrubs have attractive flowers. Good in partial shade. A. × grandiflora has purplish fall foliage.
A. × grandiflora	Glossy abelia	6	6	
Acacia sp.	—	3–100	9–10	Attractive flowers. Fast growing, but short-lived shrubs and trees. Environmentally tolerant in city.
Acoelorraphe wrightii	Everglades palm	25	9	Attractive palm.
Arbutus menziesii	Madrona	100	7	Attractive flowers, fruit, and bark.
A. unedo	Strawberry tree	30	8	Good trees.
Arctostaphylos glandulosa	Eastern manzanita	8	7	
A. glauca	Big-berry manzanita	18	10	Shrubs with attractive flowers. Fruit and bark are interesting on some species.
A. stanfordiana	Stanford manzanita	6	7	
A. tomentosa	Wooly manzanita	6	7	
Aucuba japonica	Japanese aucuba	15	8	Attractive fruit. Some interesting varieties and cultivars, especially variegated variety. Shrubs are environmentally tolerant in city.
Bambusa glaucescens	Hedge bamboo	10	9	Attractive foliage. Interesting cultivars.
Berberis buxifolia	Magellan barberry	7	6	
B. candidula	Paleleaf barberry	2	6	
B. × chenaultii	Chenault barberry	4	6	Attractive foliage, flowers, and fruits. Environmentally tolerant in city. Often used for hedges.
B. darwinii	Darwin's barberry	8	8	
B. gagnepainii	Black barberry	6	6	
B. julianae	Wintergreen barberry	7	6	
Brahea brandegeei	San José hesper palm	40	9	Better palms exist.
B. edulis	Guadalupe palm	30	10	
Buxus microphylla	Little-leaf box	3	6	Useful for hedges. Numerous cultivars, including dwarf and variegated types.
B. sempervirens	Common box	20	6	
Callistemon citrinus	Crimson bottlebush	25	9	Attractive flowers. Useful in dry soils. Shrubby to treelike.
C. speciosus		20		
C. viminalis	Weeping bottlebush	20		
Camellia japonica	Common camellia	45	8	Attractive flowers. Useful in partial shade. Numerous varieties and cultivars. Environmentally tolerant in city.
Carpenteria californica	Tree anemone	6	7	Attractive flowers.

Scientific name	Common name	Height	Zone	Remarks
Ceanothus arboreus	Catalina ceanothus	20	9	Attractive flowers. Taller species approach tree form.
C. × delilianus	—	3	7	
C. griseus	—	8	8	
C. thyrsiflorus	Blueblossom	25	8	
Chamaerops humilis	European fan palm	20	9	Small palm. Few cultivars and varieties.
Choisya ternata	Mexican orange	10	8	Fragrant white flowers.
Cinnamomum camphora	Camphor tree	100	9	Useful for specimen.
C. cassia	Cassia-bark tree	40	10	
Cistus incanus	Purple rock-rose	3	8	Attractive flowers.
C. ladanifer	Laudanum	5	8	
Coccothrinax alta	Silver palm	30	10	Slow-growing palm
Cotoneaster congestus	Pyrenees cotoneaster	3	7	Attractive fruit. Environmentally tolerant in city.
C. conspicuus	Wintergreen cotoneaster	6	7	
C. glaucophyllus	Brightbead cotoneaster	6	7	
C. microphyllus	Small-leaved cotoneaster	6	7	
C. salicifolius	—	15	6	
Daphne cneorum	Garland flower	1	5	Attractive flowers, usually fragrant. Somewhat difficult to grow.
D. odora	Winter daphne	4	7	
D. retusa	—	3	7	
Dombeya × cayeuxii	—	30	10	Attractive flowers. Shrubs to small trees.
D. tiliacea	Scarlet dombeya	25		
D. wallichii	—	30		
Elaeagnus pungens	Thorny elaeagnus	15	7	Attractive flowers with fragrance. Numerous cultivars, some of which are dwarf or variegated.
Eriobotrya japonica	Loquat	25	7	Fragrant but inconspicuous flowers. Only fruits in warmer areas.
Escallonia × langleyensis	—	8	8	Attractive flowers. Numerous cultivars.
E. rubra	—	15		
Eucalyptus sp.	Eucalypt	15–150	9–10	Excellent trees. Fast growing.
Eucryphia glutinosa	—	30	9	Attractive flowers. Sometimes partly deciduous.
E. × intermedia	—	40		
Eugenia uniflora	Surinam cherry	25	10	Fragrant flowers. Fruit is attractive, used for jam. Useful for hedge.
Euonymus japonica	Spindle tree	15	7	*E. japonica* has numerous varieties and cultivars, which vary from shrubs to trees; many are variegated.
E. kiautschovica	Spreading euonymus	9		
Ficus benghalensis	Banyan tree	100	10	Trees limited to frost-free areas.
F. benjamina	Benjamin fig	50		
Fraxinus uhdei	Shamel ash	50	9	Good street tree.
Gardenia jasminoides	Common gardenia	6	8	Attractive, fragrant flowers.

TABLE 13-11. CONTINUED

Scientific Name	Common Name	Height[2] (ft)	Zones of Hardiness	Remarks
Garrya elliptica	Silk tassel	25	8	Attractive catkins. Shrubby, except *G. elliptica*
G. fremontii	—	15	8	
G. wrightii	—	10	7	
Hakea laurina	Sea-urchin tree	30	10	Attractive flowers.
Ilex X altaclarensis	—	45	7	
I. aquifolium	English holly	50	7	
I. cassine	Dahoon	40	7	
I. cornuta	Chinese holly	9	7	Attractive foliage and fruit. Numerous cultivars.
I. crenata	Japanese holly	15	7	Shrubs to trees. Environmentally tolerant in city.
I. glabra	Gallberry	10	5	
I. opaca	American holly	50	6	
I. vomitoria	Yaupon	24	7	
Kalmia angustifolia	Sheep laurel	3	2	Attractive flowering shrubs. Some very interesting
K. latifolia	Mountain laurel	10	5	cultivars.
K. microphylla	Western laurel	2	2	
K. poliifolia	Bog laurel	2	2	
Laurus nobilis	Laurel	40	8	Aromatic foliage.
Leucothoe fontanesiana	Dog-hobble	6	5	Semievergreen in more northerly regions. Attractive flowers. Good winter-colored foliage. Some variegated cultivars.
Ligustrum delavayanum	Delavay privet	10	8	
L. henryi	Henry privet	12	8	Excellent shrubs for hedges. Some useful for
L. japonicum	Wax-leaf privet	10	7	specimen. Variegated cultivars available.
L. lucidum	Glossy privet	30	8	
L. ovalifolium	California privet	15	6	
Lonicera nitida	Box honeysuckle	6	7	Attractive flowers and fruit. Somewhat salt tolerant.
Magnolia grandiflora	Southern magnolia	100	7	Attractive flowers and foliage. This tree is environmentally tolerant in city.
Mahonia aquifolium	Oregon grape	3	6	Attractive foliage and fruit. Environmentally
M. bealei	Leatherleaf mahonia	7	7	tolerant in city.
Malus angustifolia	Southern wild crab apple	25	7	Attractive flowers.
Melaleuca quinquenervia	Paperbark tree	25	10	Attractive flowers.
Michelia figo	Banana shrub	15	8	Fragrant, attractive flowers.

486

Myrtus communis	Myrtle	15	9	Attractive flowers and fruit. Dwarf and variegated cultivars.
Nandina domestica	Heavenly bambo	8	7	Attractive fruit. Good fall color.
Nerium oleander	Common oleander	20	8	Attractive flowers. Good in hot, dry areas.
Osmanthus delavayi	—	6	8	⎫
O. × fortunei	—	12	8	Attractive foliage, flowers, and fruit. Flowers are fragrant.
O. fragrans	Fragrant olive	30	8	
O. heterophyllus	Holly olive	20	7	⎭
Paxistima canbyi	Cliff green	1	6	⎫ Prostrate shrubs. *P. canbyi* has good fall color.
P. myrsinites	Oregon boxwood	1.5	6	⎭
Pernettya mucronata	Chilean pernettya	3	7	Attractive fruit.
Phoenix canariensis	Canary Island date	50	10	⎫
P. dactylifera	Date	100	10	Attractive ornamental palms for warm, dry areas. Fruit of *P. dactylifera* is true date.
P. reclinata	Senegal date palm	20	9	
P. roebelenii	Miniature date palm	6	9	
P. rupicola	Cliff date	20	9	⎭
Photinia glabra	Japanese photina	20	8	⎫ Attractive foliage and fruit.
P. serrulata	Chinese photina	40	7	⎭
Phyllostachys aurea	Golden bamboo	20	8	Attractive foliage.
Pieris floribunda	Mountain andromeda	6	5	⎫ Attractive flowers. Latter species has dwarf and variegated cultivars.
P. japonica	Japanese andromeda	10	6	⎭
Pittosporum crassifolium	Karo	35	10	⎫
P. eugenioides	Tarata	40	10	
P. phillyraeoides	Narrow-leaved pittosporum	30	9	Trees and shrubs. Attractive foliage, flowers, and fruit.
P. rhombifolium	Queensland pittosporum	80	10	
P. tobira	Japanese pittosporum	18	8	
P. undulatum	Victorian box	40	9	
P. viridiflorum	Cape pittosporum	25	10	⎭
Pritchardia martii	—	16	10	⎫ Striking palms seen only in few areas of United States.
P. pacifica	Fiji fan palm	30	10	⎭
Prunus caroliniana	Cherry laurel	40	7	Attractive foliage. Make good specimen.
P. ilicifolia	Holly-leaved cherry	25	10	⎫
P. laurocerasus	Cherry laurel	18	7	⎭
Pseudosasa japonica	Arrow bamboo	15	8	Attractive. Good screen or hedge.
Pyracantha angustifolia	Narrow-leaf fire thorn	12	7	⎫
P. atalantioides	Gibbs fire thorn	15	7	
P. coccinea	Scarlet fire thorn	15	7	Attractive fruit. Numerous cultivars.
P. crenulata	Nepal fire thorn	20	7	
P. koidzumii	Formosa fire thorn	12	8	⎭

TABLE 13–11. CONTINUED

Scientific Name	Common Name	Height[a] (ft)	Zones of Hardiness	Remarks
Quercus acuta	Japanese evergreen oak	40	8	
Q. agrifolia	California live oak	100	9	
Q. chrysolepsis	Canyon oak	90	8	
Q. ilex	Holly oak	60	8	Good specimen trees.
Q. suber	Cork oak	60	8	
Q. virginiana	Live oak	60	8	
Q. wislizenii	Interior live oak	75	8	
Raphiolepsis indica	Indian hawthorn	5	9	Attractive flowers. Environmentally tolerant in city.
R. umbellata	Yedda hawthorn	10	8	
Rhapidophyllum hystrix	Needle palm	6	7	Shrubby hardy palm.
Rhapis excelsa	Bamboo palm	10	9	Attractive clump palm.
Rhododendron sp.	Rhododendron	1.5–80	3–9	Extensive numbers of species, cultivars, and hybrids for evergreen rhododendrons. Few evergreen azaleas. Attractive flowers. Environmentally tolerant in city.
Roystonea regia	Cuban royal palm	75	10	Fast-growing, attractive palm.
Sabal minor	Dwarf palm	4	9	Attractive palms.
S. palmetto	Cabbage palm	90	8	
Sarcococca hookeriana	—	6	8	Attractive foliage and fruit.
S. ruscifolia	Fragrant sweet box	6	8	
Schinus molle	Pepper tree	50	9	Attractive fruits.
S. terebinthifolius	Brazilian pepper tree	20	10	
Skimmia japonica	Japanese skimmia	5	8	Fragrant flowers and attractive fruit. Environmentally tolerant in city.
S. reevesiana	Reeves skimmia	6	9	Fragrant flowers and attractive fruit.
Sophora secundiflora	Mescal bean	50	8	Attractive foliage, flowers, and fruit.
Stranvaesia davidiana	—	20	8	
Syzygium aqueum	Water-rose apple	30	10	
S. paniculatum	Brush cherry	40	10	Attractive flowers and fruit.
S. pycnanthum	Wild rose apple	20		
Thrinax morrisii	Key palm	30	10	Attractive palm.
T. parviflora	Florida thatch palm	30		
Trachycarpus fortunei	Windmill palm	40	8	Palm. Not good in extreme heat.
Umbellularia californica	California bay	80	8	Attractive fruit.

Vaccinium delavayi	Delavay blueberry	2.5	7	
V. ovatum	California huckleberry	10	7	Attractive foliage. Latter species used for florist's greens.
Viburnum davidii	David viburnum	3	7	
V. henryi	Henry viburnum	10	7	
V. japonicum	Japanese viburnum	6	7	
V. odoratissimum	Sweet viburnum	20	8	Attractive flowers and fruit.
V. suspensum	Sandankwa viburnum	6	8	
V. tinus	Laurustinus	10	7	
V. utile	—	6	7	
Washingtonia filifera	Desert fan palm	80	9	Extensively used palm.
W. robusta	Thread palm	90	10	
Yucca aloifolia	Spanish bayonet	25	8	
Y. baccata	Blue yucca	3	8	
Y. brevifolia	Joshua tree	40	8	
Y. elata	Soap tree	20	8	Interesting succulent foliage. Attractive flowers.
Y. filamentosa	Adam's needle	3	5	
Y. gloriosa	Palm lily	8	8	
Y. whipplei	Our Lord's candle	10	9	

[a]Maximal potential under optimal environmental conditions, exclusive of dwarf cultivars.

TABLE 13-12. DECIDUOUS AND EVERGREEN WOODY TO SEMIWOODY VINES

Scientific Name	Common Name	Zones of Hardiness	Remarks
Actinidia arguta	Bower actinidia	5	Deciduous. Twiners. Mostly foliage value. Latter species variegated.
A. kolomikta	—		
Akebia quinata	Five-leaf akebia	5	Deciduous. Twiner. Attractive flowers.
Allamanda cathartica	Common allamanda	10	Evergreen. Clambering. Attractive flowers. Some good cultivars.
Ampelopsis aconitifolia	Monks hood vine	5	
A. arborea	Pepper vine	7	Deciduous. Tendrils. Attractive fruit.
A. brevipedunculata	Porcelain ampelopsis	5	
Anredera cordifolia	Madeira vine	9	Evergreen. Twiner. Fragrant, attractive flowers.
Antigonon leptopus	Mexican creeper	9	Evergreen. Tendrils. Attractive flowers.
Aristolochia durior	Dutchman's pipe	5	Deciduous. Twiner. For screening. Interesting but hidden flower.
Asparagus falcatus	Sickle thorn	10	Evergreen. Climbing. Thorny. Fragrant flowers.
Bauhinia corymbosa	Phanera	10	Evergreen. Tendrils. Attractive flowers and fruit.
Beaumontia grandiflora	Herald's trumpet	10	Evergreen. Attractive, fragrant flowers.
Bignonia capreolata	Cross vine	7	Evergreen. Clinging. Attractive flowers.
Bougainvillea glabra	Paper flower	10	Evergreen. Twining. Attractive flowers.
B. spectabilis	Brazilian bougainvillea		
Campsis grandiflora	Chinese trumpet creeper	8	Deciduous. Holdfasts. Needs additional support. Attractive flowers.
C. radicans	Trumpet creeper	5	
Celastrus orbiculatus	Oriental bittersweet	5	Deciduous. Twining. Attractive fruit.
C. rosthornianus	Loesener bittersweet	7	
Cissus antarctica	Kangaroo vine	10	Evergreen, some only semievergreen. Tendrils. Interesting foliage.
C. discolor	Trailing begonia	10	
C. incisa	Marine ivy	6	
C. rhombifolia	Venezuela treebine	10	
Clematis sp.	Clematis	5–7	Deciduous, some evergreen. Tendrillike leaf stalks. Attractive flowers. Numerous cultivars.
Clerodendrum thomsoniae	Bleeding heart glorybower	9	Evergreen. Twining. Attractive flowers.
Cocculus carolinus	Carolina moonseed	7	Deciduous to evergreen. Twining. Attractive fruit.
Cryptostegia grandiflora	Rubbervine	10	Evergreen. Twining. Attractive flowers.
Distictis buccinatoria	Blood trumpet	9	Evergreen. Clinging. Attractive flowers.
D. laxiflora	—		
Eccremocarpus scaber	Glory flower	8	Evergreen. Tendrils. Attractive flowers. Can be used as an annual further north.
Euonymus fortunei	Wintercreeper	5	Evergreen. Clinging. Hardy. Attractive fruit, good fall color. Numerous cultivars.
Ficus pumila	Creeping ficus	9	Evergreen. Clinging. Attractive foliage.
Gelsemium sempervirens	Evening trumpet flower	8	Evergreen. Twining. Fragrant, attractive flowers.
Hedera canariensis	Algerian ivy	7	Evergreen. Clinging. Attractive foliage. Numerous cultivars, includes variegated ones.
H. helix	English ivy	6	

TABLE 13–12. CONTINUED

Scientific Name	Common Name	Zones of Hardiness	Remarks
Hibbertia scandens	Snake vine	10	Evergreen. Twining. Attractive flowers.
Hoya carnosa	Wax flower	10	Evergreen. Twining. Attractive foliage. Fragrant, attractive flowers.
Hydrangea anomala	Climbing hydrangea	5	Deciduous. Clinging. Attractive flowers.
Jasminum dichotomum	Goldcoast jasmine	10 }	First species evergreen, second semiever-
J. officinale	Poet's jasmine	8 }	green. Clambering climbers. Fragrant, attractive flowers.
Kadsura japonica	Scarlet kadsura	8	Evergreen. Twining. Attractive fruit.
Lonicera caprifolium	Italian woodbine	6	
L. etrusca	Cream honeysuckle	7	Evergreen through semievergreen and
L. flava	Yellow honeysuckle	6	some deciduous. Twining. Attractive,
L. hildebrandiana	Giant Burmese honeysuckle	10	often fragrant flowers and attractive
L. japonica	Japanese honeysuckle	5	fruit.
L. sempervirens	Trumpet honeysuckle	4	
Macfadyena unguis-cati	Cat's claw	9	Evergreen. Clinging. Attractive flowers.
Mandevilla laxa	Chilean jasmine	9	Evergreen. Twining. Attractive, fragrant flowers.
Menispermum canadense	Yellow parilla	4	Deciduous. Twining. Attractive fruit.
Muehlenbeckia complexa	Maidenhair vine	7	Evergreen. Twining.
Pandorea jasminoides	Bower plant	9	Evergreen. Twining. Attractive flowers.
Parthenocissus quinquefolia	Virginia creeper	4 }	Deciduous. Clinging. Good fall color.
P. tricuspidata	Boston ivy	5 }	Several cultivars.
Passiflora caerulea	Blue passionflower	8	Evergreen. Tendrils. Attractive flowers.
Philodendron sp.	Philodendron	10	Evergreen. Clinging. Attractive foliage.
Podranea brycei	Queen-of-Sheba vine	10	Evergreen. Climbing. Attractive flowers.
Polygonum aubertii	China fleece vine	5	Deciduous. Twining. Attractive flowers.
Pyrostegia venusta	Flame vine	10	Evergreen. Clinging. Attractive flowers.
Quisqualis indica	Rangoon creeper	10	Evergreen. Climbing. Attractive, fragrant flowers.
Rhoicissus capensis	Cape grape	10	Evergreen. Tendrils. Attractive fruit.
Saritaea magnifica	—	10	Evergreen. Tendrils. Attractive flowers.
Senecio confusus	Mexican flame vine	9 }	
S. macroglossus	Natal ivy	10 }	Evergreen. Twining. Attractive flowers.
S. mikanioides	German ivy	10 }	Variegated cultivar for *S. macroglossus*.
Solandra gutata	Goldcup	10	Evergreen. Climbing. Attractive fragrant flowers. Attractive fruit.
Solanum jasminoides	Potato vine	9	Evergreen. Twining. Attractive flowers.
Stauntonia hexaphylla	—	8	Evergreen. Twining. Attractive, fragrant flowers.
Stephanotis floribunda	Madagascar jasmine	10	Evergreen. Twining. Attractive, fragrant flowers.
Stigmaphyllon ciliatum	Brazilian golden vine	10	Evergreen. Twining. Attractive flowers.
Tecomaria capensis	Cape honeysuckle	9	Evergreen. Twining. Attractive flowers.
Thunbergia grandiflora	Blue trumpet vine	9	Evergreen. Twining. Attractive flowers.
Trachelospermum jasminoides	Star jasmine	9	Evergreen. Twining. Fragrant, attractive flowers.
Vitis coignetiae	Crimson glory vine	5	Deciduous. Tendrils. Good fall color.
Wisteria floribunda	Japanese wisteria }	5	Deciduous. Twining. Attractive flowers.
W. sinensis	Chinese wisteria }		Numerous cultivars.

TABLE 13–13. HOUSE AND GREENHOUSE PLANTS

Scientific Name	Common Name	Light	Humidity	Temperature	Soil	Water
Abutilon sp.	Flowering maple	B	CH	I	GP	E
Acalypha hispida	Chenille plant	} B	} CH	} I	} GP	} E }
A. wilkesiana	Jacob's coat					
Acanthocalycium sp.	Cacti	B	LH, CH	C	RD	D
Achimenes grandiflora	Magic flower	M	G	H	HR	E
Adiantum capillus-veneris	Southern maiden-hair fern		CH	H		
A. hispidulum	Australian maiden-hair fern		G	H		
A. pedatum	American maiden-hair fern	L	G	I	HR	S
A. raddianum	Delta maidenhair fern		G	H		
Adromischus sp.	—	B	LH, CH	I	RD	D
Aechmea sp.	Living vase	M	CH	H	OM	E
Aeonium sp.	—	M	LH, CH	I	RD	D
Aeschynanthus radicans	Lipstick plant	M	G	H	HR	E
Agapanthus africanus	African lily	} B	} CH	} I	} GP	} E }
A. orientalis	—					
Agave americana	Century plant	B	CH	I	RD	D
Aglaomorpha meyeniana	Bear's paw fern	M	G	H	HR	E
Aglaonema commutatum	—	} L	} LH, CH	} H	} GP	} E }
A. modestum	Chinese evergreen					
Allamanda cathartica	Common allamanda	B	G	H	GP	E
Alocasia sp.	Elephant's ear plant	M	G	H	HR	E
Aloe sp.	Aloe	B	LH, CH	I	RD	D
Alpinia sp.	Ginger lily	M	G	H	GP	E
Alstroemeria sp.	Lily-of-the-Incas	B	CH	C	GP	E
Amaryllis belladonna	Belladonna lily	B	CH	I	GP	E
Angraecum sp.	Orchid	M	G	H	OM	E
Anoectochilus setaceus	Jewel orchid	M	G	H	HR	E
Ansella sp.	Orchid	B	G	H	HR	D
Anthurium sp.	Tailflower	L	G	H	HR	S
Aphelandra squarrosa	Zebra plant	M	CH	H	HR	E
Araucaria araucana	Monkey puzzle tree	} M	LH	} I	} HR	} E }
A. heterophylla	Norfolk Island pine		CH			
Ardisia crenata	Coralberry	M	CH	I	GP	E
Ariocarpus fissuratus	Living rock cactus	B	LH, CH	I	RD	D
Ascocentrum sp.	Orchid	M	G	H	OM	E
Asparagus asparagoides						
A. densiflorus	Asparagus fern	} M	} CH	} I	} GP	} E }
A. setaceus						
Aspidistra elatior	Cast-iron plant	L	LH, CH	I	GP	E
Asplenium bulbiferum	Mother spleenwort	} L	CH	} I	} HR	} S }
A. nidus	Bird's nest fern		G			
Astrophytum sp.	Star cactus	B	LH, CH	H	RD	D
Aucuba japonica	Japanese aucuba	M	CH	C	GP	D
Beaucarnea recurvata	Ponytail	B	CH	I	GP	D

Habit	Height	Use	Attraction	Cultivar Availability	Propagule
Er, Tr	3	GH	Fl, Fo	VC	Cu, Se
Er	3	GH	Fl, Fo	VC, OC	Cu
Er	1	H/GH	Fl, Fo	–	Se
Er, Tr	1	GH	Fl	DC, OC	Rh
Er	1, 2	H/GH GH GH GH	Fo	OC	Di, Sp
Er	1	H/GH	Fl, Fo	–	Cu, RL
Er	1, 2, 3	H/GH	Fl, Fo	DC, OC, VC	Su
Er	1, 2, 3	GH	Fl, Fo	–	Cu
Tr	1	GH	Fl	–	Cu, Se
Er	2, 3	H/GH	Fl	DC, VC	Di, Se
Er	3	H/GH	Fo	DC, VC	Se, Su
Er	2, 3	GH	Fo	–	Di, Sp
Er	2, 3	H/GH	Fo	DC, VC	Cu, Di
Cl	3	GH	Fl	DC,OC	Cu
Er	3	GH	Fo	VC	Cu, Se, SP
Er	1, 2, 3	H/GH	Fo	–	Cu, Se, Su
Er	3	GH	Fl, Fo	–	Di
Er	2, 3	GH	Fl, Fo	OC	RD, Se
Er	3	H/GH	Fl	–	Bu
Pe	2, 3	GH	Fl, Fr	–	ASMC, RhD
Er	1	GH	Fl	–	ASMC, RhD
Er	1, 2	GH	Fl	–	ASMC, RhD
Er	1, 3	GH	Fl, Fo, Fr	OC	Se, SP, Su
Er	2	H/GH	Fl, Fo	DC	Cu, Se
Er	3	H/GH	Fo	–	Se
Er	2, 3	H/GH	Fl, Fo, Fr	–	Cu, Se
Er	1	H/GH	FL, St	–	Se
Er	1	GH	Fl	–	ASMC, RhD
Cl, Er, Tr	2, 3	H/GH	Fo	DC OC	Se, Di
Er	3	H/GH	Fo	VC	Di
Er	2, 3	H/GH GH	Fo	–	Di, Sp, VB / Di, Sp
Er	1, 2	H/GH	Fl, St	OC	Se
Er	3	H/GH	Fo, Fr	VC, DC, OC	Cu, Se
Er	3	H/GH	Fo	DC	Se

TABLE 13-13. CONTINUED

Scientific Name	Common Name	Light	Humidity	Temperature	Soil	Water
Begonia Sp.	Begonia	M	CH, G	H	HR	E
Billbergia sp.	Vase plant	B	LH, CH	H	HR	D
Blechnum sp.	–	L	G	H	HR	S
Bougainvillea X *buttiana*	Bougainvillea	B	CH	H	GP	D
Bouvardia longiflora	Bouvardia	B	CH	I	GP	E
Bowiea volubilis	Climbing onion	M	CH	I	GP	E
Brassaia actinophylla	Australian umbrella tree	B	CH	H	GP	D
Brassavola sp.	Orchid	L	CH, G	I	OM	D
Brassia sp.	Orchid	M	G	I	OM	E
X *Brassocattleya*	Hybrid orchids	M	G	I	OM	D
X *Brassolaelia*	Hybrid orchids	M	G	I	OM	D
Browallia sp.	Bush violet	M	CH	I	HR	E
Brunfelsia sp.	–	M	CH	I	GP	E
Bulbophyllum sp.	Orchid	M	G	H	OM	E
Caladium sp.	Caladium	M	CH, G	H	HR	E
Calanthe sp.	Orchid	M	G	H	GP	E
Calathea sp.	Calathea	M	G	H	HR	E
Calceolaria sp.	Pouch flower	L-M	CH	C	HR	D
Calliandra haematocephala	Red powderpuff	B	CH	H	GP	E
Callisia sp.	Inch plant	M	LH, CH	I	GP	E
Camellia japonica	Common camellia	M	CH	C	HR	S
Campanula isophylla	Star-of-Bethlehem	B	CH	C	HR	E
Capsicum annuum cv.	Ornamental pepper	B	CH	H	GP	E
Caralluma sp.	Caralluma	B	CH	I	RD	D
Carissa grandiflora	Natal plum	B	CH	I	GP	E
Caryota mitis	Burmese fishtail palm	M	LH, CH	H	GP	S
Catasetum sp.	Orchid	M	G	H	OM	E
Cattleya sp.	Orchid	M	CH, G	I	OM	D
Cephalocereus sp.	Cacti	B	LH, CH	H	RD	D
Ceratozamia mexicana	Mexican horncone	B	CH	H	RD	D
Cereus Sp.	Cacti	B	LH, CH	H	RD	D
Ceropegia sp.	Ceropegia	M	LH, CH	I	RD	D
Cestrum nocturum	Night jessamine	B	CH	H	GP	E
Chamaedorea sp.	Palm	L	LH, CH	H	GP	E
Chirita sp.	Chirita	M	G	H	HR	E
Chlorophytum comosum	Spider plant	M	LH, CH	I	GP	E
Chrysanthemum sp.	Chrysanthemum	B	CH	C	GP	E
Cibotium sp.	Tree fern	M	CH	H	HR	E
Cissus antarctica	Kangaroo vine	M	CH	H	GP	D
C. rhombifolia	Grape ivy	M	CH	H	GP	E
X *Citrofortunella mitis*	Calamondin orange	B	CH	I	GP	D
Citrus aurantiifolia	Lime	B	CH	I	GP	D
C. limon	Lemon	B	CH	I	GP	D
C. X *limonia*	Otaheite orange	B	CH	I	GP	D
Clerodendrum thomsoniae	Bleeding glory bower	M	G	H	GP	E
Clivia miniata	Kaffir lily	M	CH	I	GP	D
Codiaeum variegatum	Croton	B	CH, G	H	GP	E

Habit	Height	Use	Attrac-tion	Cultivar Availability	Propagule
Er, Tr	1, 2, 3	H/GH	Fo, Fl	DC, OC, VC	Cu, LC, RhD, Tu
Er	2, 3	H/GH	Fo, Fl	VC, OC	OS, Se
Er	2, 3	GH	Fo	—	Sp
Cl	3	GH	Fl	OC	Cu
Er	3	GH	Fl, Fra	—	Cu, RC
Cl	3	H/GH	Fo	—	BD, Se
Er	3	H/GH	Fo	—	AL, Cu, Se
Pe–Er	1, 2	H/GH	Fl, Fra	—	ASMC, RhD
Pe–Er	1, 2	GH	Fl, Fra	—	ASMC, RhD
Pe–Er	1, 2	GH	Fl	—	ASMC, RhD
Pe–Er	1, 2	GH	Fl	—	ASMC, RhD
Er–Pe	1, 2	GH	Fl	OC	Cu, Se
Er	3	GH	Fl, Fra	OC	Cu
Pe–Er	1, 2	GH	Fl	—	ASMC, RhD
Er	2, 3	GH	Fo	VC	Di, Se
Er	1, 2	GH	Fl	OC	ASMC, RhD
Er	1, 2, 3	GH	Fo	—	Cu, Di, Tu
Er	1, 2, 3	GH	Fl	—	Se
Er	3	GH	Fl	—	Cu, Se
Tr	1, 2	H/GH	Fo	—	Cu, Se
Er	3	GH	Fl	DC, OC, VC	Cu
Pe	1	GH	Fl	OC	Cu, Se
Er	3	H/GH	Fr	OC	Se
Er–Pe	1	GH	Fl, Fo	—	Cu, Se
Er	3	GH	Fl, Fr, Fra	DC	Cu, Se
Er	3	H/GH	Fo	—	Se
Er–Pe	2	GH	Fl	—	ASMC, RhD
Er–Pe	2	H/GH	Fl	OC	ASMC, RhD
Er	2, 3	H/GH	Fl, St	—	Cu, Se
Er	3	H/GH	Fo	—	Se, Su
Er	3	H/GH	Fl, St	—	Cu, Se
Tw	2, 3	H/GH	Fl, Fo	—	Cu, Se
Er	3	H/GH	Fl, Fra	—	Cu, Se
Er	3	H/GH	Fo	—	Se
Er	1, 2, 3	GH	Fl	—	Cu, Se
Pe	2, 3	H/GH	Fo	DC, VC	Di, Se
Er	2, 3	GH	Fl	DC, OC	Cu
Er	3	H/GH	Fo	—	SP
Cl }	2, 3 }	H/GH }	Fo }	DC }	Cu, Se
Er }	3 }	H/GH }	Fl Fo Fr Fra }	OC }	Cu, Se
Tw	3	GH	Fl	OC	Cu, Se
Er	2	H/GH	Fl	OC	Di, Se
Er	3	GH	Fo	VC	AL, Cu, Se

TABLE 13-13. CONTINUED

Scientific Name	Common Name	Light	Humidity	Tem-perature	Soil	Water
Coffea arabica	Coffee	M	CH	H	GP	E
Coleus × *hybridus*	Coleus	B	CH	H	GP	E
Colocasia esculenta	Taro	M	G	H	GP	S
Columnea sp.	Columnea	M	CH, G	H	HR	E
Commelina coelestis	Blue spiderwort	M	CH	I	HR	E
Conophytum sp.	Cone plant	B	CH	I	RD	D
Cordyline terminalis	Good luck plant	M	CH	H	GP	E
Coryphantha sp.	Cacti	B	CH	I	RD	D
Crassula sp.	Crassula	M–B	LH, CH	I	RD	D
Crocus sp.	Crocus	B	CH	C	GP	E
Crossandra infundi-buliformis	Firecracker flower	M	G	H	HR	E
Cryptanthus sp.	Earth-star	M	LH, CH	H	HR	D
Cuphea sp.	Cuphea	B	CH	I	GP	E
Cyanotis sp.	—	B	CH	I	GP	D
Cycas sp.	Bread palm	M	LH, CH	I	GP	D
Cyclamen sp.	Cyclamen	M	G	C	HR	E
Cychnoches sp.	Swan orchid	M	CH, G	H	OM	E
Cymbidium sp.	Orchid	M	CH	C–I	HR	E
Cyperus alternifolius	Umbrella plant	M	LH, CH	I	GP	S
Cyrtomium falcatum	Holly fern	L	LH, CH	I	GP	E
Darlingtonia californica	Pitcher plant	L	G	C	HR	S
Davallia canariensis	Deer's foot fern	L	LH, CH	I–H	HR	E
Delosperma sp.	—	B	CH	I	RD	D
Dendrobium sp.	Orchid	M	CH, G	I–H	OM	D
Dendrochilum sp.	Chain orchid	M	G	I–H	OM	E
Dianthus caryophyllus	Carnation	B	CH	C	GP	D
Diastema sp.	Diastema	M	G	H	HR	E
Dieffenbachia sp.	Dumb cane	M	LH, CH	H	GP	D
Dionaea muscipula	Venus's flytrap	B	G	C	SM	S
Dioon edule	Cycad	M	CH	H	GP	D
Dioscorea elephantipes	Elephant's foot	M	G	H	RD	E
Dizygotheca ele-gantissima	False aralia	M	CH	H	GP	E
Dracaena sp.	Dracaena	M	LH, CH	H	GP	S
Drosera sp.	Sundew	B	G	C	HR	S
Dryopteris sp.	Fern	L	G	I	HR	E
Echeveria sp.	Hen-and-chickens	B	LH, CH	I	RD	D
Echinocactus sp.	Cactus	B	LH, CH	I–H	RD	D
Echinocereus sp.	Hedgehog cactus	B	LH, CH	I–H	RD	D
Echinopsis sp.	Sea-urchin cactus	B	LH, CH	I–H	RD	D
Epidendrum sp.	Buttonhole orchid	M	CH, G	I–H	OM	D
Epiphyllum sp.	Orchid cactus	M	CH	I	RD	D–E
Episcia sp.	Flame violet	M	CH, G	H	HR	E
Espostoa sp.	Cactus	B	LH, CH	I	RD	D
Eucharis grandiflora	Amazon lily	M	G	H	GP	E
Eucomis sp.	Pineapple lily	B	CH	I	GP	E
Euonymus japonica	Spindle tree	M	CH	C	GP	E
Euphorbia sp.	Euphorbia	B	CH	I–H	RD	D
× *Fatshedera lizei*	Aralia ivy	B	CH	C–I	GP	E
Fatsia japonica	Japanese fatsia	M	LH, CH	C	GP	E

Habit	Height	Use	Attraction	Cultivar Availability	Propagule
Er	3	GH	Fl, Fr	OC	Cu, Se
Er	2	H/GH	Fo	VC	Cu, Se
Er	3	GH	Fo	OC	Tu
Tr	2	H/GH	Fl	OC	Cu, Se
Tr	2	H/GH	Fl, Fo	OC, VC	Cu, Se, Tu
Er	1	H/GH	St	—	Cu, Se
Er	3	H/GH	Fo	OC, VC	Cu, RLa, Se
Er	1	H/GH	St	—	Se
Er	1, 2, 3	H/GH	Fl, Fo	DC, OC, VC	Cu, Se
Er	1	H/GH	Fl	OC	Co
Er	1, 2	GH	Fl	—	Cu
Er	1	H/GH	Fl, Fo	OC, VC	OS
Er	1, 2, 3	GH	Fl	—	Cu, Se
Pe–Tr	1, 2	H/GH	Fo	—	Cu, Se
Er	3	H/GH	Fo	—	Se
Er	1	GH	Fl, Fo	OC	Se, Tu
Pe–Er	1, 2	H/GH	Fl	—	ASMC, RhD
Pe–Er	1, 2	GH	Fl	OC	ASMC, RhD
Er	2, 3	H/GH	Fo	OC	Di, Se
Er	3	H/GH	Fo	—	Sp
Er	3	GH	Fl, Fo	—	Di, Se
Er	2	H/GH	Fo	—	Di
Er	1, 2, 3	H/GH	Fl, Fo	—	Cu, Se
Pe–Er	1, 2	H/GH	Fl	OC	ASMC, RhD
Pe–Er	1, 2	GH	Fl	—	ASMC, RhD
Er	2, 3	GH	Fl, Fra	—	Cu
Er	1	GH	Fl	—	Rh
Er	1, 2, 3	H/GH	Fo	OC, VC	AL, Cu, SP
Er	1	GH	Fl, Fo	—	Se
Er	3	H/GH	Fo	—	Se
Tw	3	GH	Fo	—	Cu, Se, Tu
Er	3	H/GH	Fo	—	Cu
Er	3	H/GH	Fo	VC	Cu, Se
Er	1	GH	Fl, Fo	—	Di, RhD, Se
Er	2, 3	GH	Fo	—	Di, Sp
Er	1	H/GH	Fl, Fo	—	Cu, OS, Se
Er	3	H/GH	Fl, St	OC	Se
Er	1, 2	H/GH	Fl, St	—	Se
Er	1, 2	H/GH	Fl, St	OC	Se
Er–Pe	1	H/GH	Fl, Fra	—	ASMC, RhD
Pe	2, 3	H/GH	Fl	—	Cu, Se
Tr	2	H/GH	Fl, Fo	OC, VC	Cu, Se
Er	3	H/GH	Fl, St	—	Se
Er	2	GH	Fl, Fra	—	OS, Se
Er	2, 3	GH	Fl	—	OS, Se
Er	3	H/GH	Fo	DC, OC, VC	Cu, Se
Er	1, 2, 3	H/GH	Fl, Fo	DC, OC, VC	Cu, Se
Er	3	H/GH	Fo	VC	Cu, Se
Er	3	H/GH	Fo	DC, VC	Cu, RC, Se

TABLE 13-13. CONTINUED

Scientific Name	Common Name	Light	Humidity	Tem-perature	Soil	Water
Faucaria sp.	Tiger-jaws	B	LH, CH	I	GP	D
Felicia sp.	—	B	CH	I	GP	E
Fenestraria rho-palophylla	Baby-toes	B	CH	H	RD	D
Ferocactus sp.	Barrel cactus	B	CH	I	RD	D
Ficus sp.	Fig	M	LH, CH	H	GP	E
Fittonia verschaffeltii	Nerve plant	M	G	H	GP	E
Fouquieria sp.	—	B	CH	H	RD	D
Frailea sp.	Cacti	B	LH, CH	H	RD	D
Freesia sp.	Freesia	B	CH	C	GP	D
Fuchsia	Lady's-eardrops	L	CH	I–C	GP	E
Gardenia jasminoides	Gardenia	B	CH	H	HR	E
Gasteria sp.	Cow-tongue cactus	M	LH, CH	I	RD	D
Gelsemium sempervirens	Yellow jessamine	M	G	I	GP	E
Gerbera jamesonii	Transvaal daisy	B	CH	C	GP	D
Gesneria sp.	—	M	G	H	HR	E
Gibbaeum sp.	Flowering quartz	B	CH	H	RD	D
Gloriosa sp.	Gloriosa lily	B	CH, G	H	GP	E
Glottiphyllum sp.	Tongueleaf	B	CH	I	RD	D
Gloxinia perennis	Canterbury-bells gloxinia	M	G	H	HR	E
Gongora sp.	Orchid	L	G	H–I	OM	D
Graptopetalum	—	B	CH	I	RD	D
Greenovia	—	B	CH	I	RD	D
Gymnocalycium	Chin cactus	B	LH, GH	C	RD	D
Gynura aurantiaca	Velvet plant	B	CH	H	GP	E
Haageocereus sp.	Cactus	B	CH	I	RD	D
Haemanthus sp.	Blood lily	B	CH	I	GP	E
Haemaria discolor	Gold-lace orchid	L	G	H	HR	E
Harrisia sp.	Cactus	B	CH	I	RD	D
Hatiora salicornioides	Drunkard's dream	M	CH	H	HR	E
Haworthia sp.	Wart plant	M	LH, CH	I	RD	D
Hechtia sp.	Bromeliads	M	CH	I	HR	D
Hedera sp.	Ivy	B	CH	C–I	GP	E
Hedychium sp.	Ginger lily	M	CH	H	GP	S
Heliconia sp.	Lobster-claw	M	G	H	GP	E
Heliocereus sp.	Cactus	M	CH	H	RD	D
Heliotropium arbo-rescens	Heliotrope	M	CH	I	GP	E
Hesperantha sp.	—	B	CH	I	GP	D
Hippeastrum sp.	Amaryllis	B	CH	I–H	GP	E
Hoodia sp.	—	B	CH	I	RD	D
Howea sp.	Sentry palm	M	CH	I	GP	E
Hoya sp.	Wax vine	B	CH	I–H	GP	D
Huernia sp.	Dragon flower	M	CH	I–H	RD	D
Humata tyermannii	Bear-foot fern	M	G	I	GP	D
Hyacinthus orientalis	Hyacinth	B	CH	C	GP	E
Hylocereus sp.	Cactus	B	CH	H	HR	E
Hymenocallis sp.	Spider lily	B	CH	I–H	GP	E
Hypoestes phyllostachya	Polka-dot plant	M	CH	H	HR	E
Idria columnaris	Boojum tree	B	CH	H	RD	D

498

Habit	Height	Use	Attraction	Cultivar Availability	Propagule
Er	1	H/GH	Fl, Fo	—	Se
Er	1, 2, 3	GH	Fl	—	Cu, Se
Er	1	GH	Fl, Fo	—	Se
Er	1, 2, 3	GH	Fl, St	—	Se
Er, Cl	1, 2, 3	H/GH	Fo, Fr	DC, OC, VC	AL, Cu, RD
Tr	1	GH	Fo	—	Cu
Er	3	GH	Fl, Fo	—	Cu, Se
Er	3	H/GH	Fl, St	—	Se
Er	1, 2	GH	Fl, Fra	—	Co
Tr	2, 3	GH	Fl	DC	Cu
Er	2, 3	H/GH	Fl, Fra	DC	Cu
Er	1	H/GH	Fl, Fo	—	LC, Se, Su
Cl	3	GH	Fl, Fra	—	Cu
Er	1, 2	GH	Fl	OC	Cu, Se
Pe–Er	1, 2, 3	GH	Fl	—	Cu, Se
Er	1	H/GH	Fl, Fo	—	Cu, Se
Cl	3	GH	Fl	—	DS, Se, Tu
Er	1	H/GH	Fl, Fo	—	Cu, Se
Er	2	GH	Fl	—	Rh
Er	1, 2	GH	Fl	—	ASMC, RhD
De	1	H/GH	Fl, Fo	—	RL, Se
Er	1	H/GH	Fo	—	Cu, Se
Er	1	H/GH	Fl, St	—	Se
Tr	2	H/GH	Fo	OC	Cu
Er, Pe	3	H/GH	Fl, St	—	Se
Er	1, 2	GH	Fl	—	OS
Er	1	GH	Fl, Fo	—	ASMC
Cl, Tr, De	3	GH	Fl, St	—	Se
Pe	3	GH	St	—	Cu, Se
Er	1	H/GH	Fo	OC, VC	Se, Su
Er	2, 3	H/GH	Fl, Fo	—	Se, Su
Cl	1, 2, 3	H/GH	Fo	DC, OC, VC	Cu, L, Se
Er	3	GH	Fl, Fr	—	RhD
Er	3	GH	Fl, Fo	—	RhD, Se
De, Er	2, 3	GH	Fl, St	—	Se
Er	3	GH	Fl, Fra	—	Cu, Se
Er	2, 3	GH	Fl	—	Co, Se
Er	2, 3	H/GH	Fl	OC	Bu, Se
Er	2	H/GH	Fl, St	—	Se
Er	3	H/GH	Fo	—	Se
Cl, Tw	3	H/GH	Fl, Fo, Fra	DC, OC, VC	Cu, L
Er, De	1	GH	Fl, St	—	Cu, Se
Er	1	GH	Fo	—	Di
Er	1, 2	GH	Fl, Fra	OC	OS
Cl	3	GH	Fl, Fra	—	Cu
Er	1, 2, 3	GH	Fl	DC, OC	BS, Se
Er	2, 3	H/GH	Fo	—	Cu, Se
Er	3	GH	Fo	—	Cu, Se

TABLE 13-13. CONTINUED

Scientific Name	Common Name	Light	Humidity	Temperature	Soil	Water
Impatiens sp.	Balsam	B	LH, CH	I	HR	E
Incarvillea sp.	—	B	CH	C	GP	D
Iresine sp.	Bloodleaf	B	CH	I	GP	E
Ixia sp.	Corn lily	B	CH	I	GP	D
Ixora sp.	—	B	CH–G	H	GP	E
Jasminum sp.	Jasmine	B	CH	H-I-C	GP	E
Jatropha sp.	—	B	G	H	RD	E
Justica sp.	Water willow	M	G	H	HR	E
Kaempferia sp.	—	M	G	H	GP	E
Kalanchoe sp.	Palm-beach bells	B	LH, CH	I	RD	D
Koellikeria erinoides	—	M	G	S	HR	E
X *Koellikohleria rosea*	—	M	G	S	HR	E
Kohleria sp.	Tree gloxinia	M	G	S	HR	E
Lachenalia sp.	Cape cowslip	B	CH	C	GP	E
Laelia sp.	Orchid	M	CH, G	C-I	OM	D
X *Laeliocattleya* sp.	Orchid	M	G	I	OM	D
Lagerstroemia sp.	—	B	CH	C	GP	E
Lampranthus sp.	—	B	CH	I	RD	D
Lantana sp.	Shrub verbena	B	CH	C-I	GP	D
Lapageria rosea	Chilean bellflower	L	G	C	HR	E
Lapeirousia sp.	—	B	CH	C	GP	D
Lemaireocereus sp.	Cactus	B	LH, CH	I	RD	D
Licuala sp.	Palm	M	G	H	GP	E
Lilium Sp.	Lily	M–B	CH	C-I	GP	D
Lithops sp.	Living-stones	B	CH	I	RD	D
Livistona sp.	Fan palm	M	G	I-H	GP	S
Lobivia sp.	Cob cactus	B	LH, CH	I	RD	D
Lockhartia sp.	Orchid	M	CH, G	H	OM	E
Lophocereus sp.	Cactus	B	CH	I	RD	D
Lycaste sp.	Orchid	M	G	C-I	OM	D
Lycoris sp.	—	B	CH	I	GP	D
Lygodium sp.	Climbing fern	M	G	I	GP	E
Macradenia sp.	Orchid	M	G	H	OM	D
Macrozamia sp.	Cycad	M	CH	I	GP	E
Malphigia coccigera	Miniature holly	B	CH	I	GP	E
Mammillaria sp.	Pincushion cactus	B	LH, CH	C-I	RD	D
Manettia inflata	Firecracker vine	M	CH–G	I	GP	E
Maranta leuconeura	Prayer plant	M	G	S	GP	E
Masdevallia sp.	Orchid	M	G	C	OM	E
Maxillaria sp.	Orchid	M	CH, G	C-I	OM	E
Melocactus sp.	Cactus	B	CH	H	RD	D
Microlepia sp.	Fern	M	G	I	HR	E
Miltonia sp.	Pansy orchid	L	CH, G	C-I	OM	E
Mimulus sp.	Monkey flower	B	CH	I	GP	D
Monadenium lugardiae	—	B	CH	I	RD	D
Monanthes sp.	—	B	CH	I	RD	D
Monstera deliciosa	Ceriman	M	LH, CH	H	GP	E
Mormodes sp.	Orchid	M	G	I	OM	D
Muehlenbeckia sp.	Wire plant	M	CH, G	I	GP	E
Muscari armeniacum	Grape hyacinth	B	CH	C	GP	E
Myrtus communis	Myrtle	B	CH	I-C	GP	D

Habit	Height	Use	Attraction	Cultivar Availability	Propagule
Er	1, 2, 3	H/GH	Fl	DC, OC, VC	Cu, Se
Er	1, 2, 3	GH	Fl	—	Di, Se
Er	3	GH	Fo	VC	Cu
Er	1, 2	GH	Fl	—	Co
Er	3	GH	Fl, Fo	OC	Cu, Se
Cl, Er	3	GH	Fl, Fra	—	Cu, L, Se
Er	2, 3	GH	Fl, Fo	—	Cu, Se
Cl, Er	3	GH	Fl	—	Cu, Se
Er	1, 2	GH	Fl, Fo	—	Rh, D
Er	1, 2, 3	H/GH	Fl, Fo	OC, VC	Cu, LC, Se
Er	1	GH	Fl, Fo	—	Rh
Er	1	GH	Fl, Fo	—	Rh
Er	1, 2, 3	GH	Fl, Fo	—	Rh
Er	1, 2	GH	Fl	OC	OS, Se
Er	1, 2	H/GH	Fl	OC	ASMC, RhD
Er	1, 2	GH	Fl	—	ASMC, RhD
Er	3	GH	Fl	DC, OC	Cu, Se
De, Er	1, 2	GH	Fl	—	Cu, Se
Er, Tr	3	GH	Fl	OC	Cu, Se
Cl	3	GH	Fl	—	Cu, L, Se
Er	1, 2	GH	Fl	—	Co
Er	3	H/GH	Fl, St	—	Se
Er	3	GH	Fo	—	Se
Er	3	GH	Fl	DC, OC	OS, Se
Er	1	GH	Fl, St	—	Cu, Se
Er	3	GH	Fo	—	Se
Er	1	H/GH	Fl, St	—	Se
Er	1, 2	H/GH	Fl	—	ASMC
Er	3	GH	Fl, St	—	Se
Er	1, 2	GH	Fl	—	ASMC, RhD
Er	2	GH	Fl	—	Bu
Cl	3	GH	Fo	—	Sp
Er	1	GH	Fl	—	ASMC, RhD
Er	2, 3	GH	Fo	—	Se
Er	3	H/GH	Fl, Fo, Fr	—	Cu, Se
Er	1	H/GH	Fl, St	DC, OC	Se
Cl	2, 3	GH	Fl	—	Cu, Se
Er–Pe	2	GH	Fo	OC	Cu, Di
Er	1	GH	Fl	OC	ASMC
Er	1, 2	H/GH	Fl	—	ASMC, RhD
Er	1, 2, 3	GH	Fl, St	—	Se
Er	3	GH	Fo	—	Sp
Er	1, 2	H/GH	Fl	OC	ASMC, RhD
Er	1, 2, 3	GH	Fl	—	Cu, Di, Se
Er	1, 2	GH	Fo	—	Cu
Er	1	GH	Fl, Fo	—	Cu
Cl	3	H/GH	Fo	—	Cu, Se
Er	1, 2	GH	Fl	OC	ASMC, RhD
Cl	3	GH	Fl	—	Cu, Se
Er	1	H/GH	Fl	—	OS, Se
Er	3	GH	Fl, Fo, Fr, Fra	DC, VC	Cu, Se

TABLE 13-13. CONTINUED

Scientific Name	Common Name	Light	Humidity	Temperature	Soil	Water
Narcissus sp.	Daffodil, narcissus jonquil	B	CH	C	GP	E
Nautilocalyx sp.	Gesneriads	M	G	H	HR	E
Nematanthus sp.	Gesneriads	M	G	H	HR	E
Neomarica sp.	Fan iris	M	CH	I	GP	S
Neoporteria sp.	Cactus	B	CH	I	RD	D
Neoreglia sp.	Bromeliad	M	LH, CH	H	HR	E
Nepenthes sp.	Pitcher plant	M	G	H	OM	S
Nephrolepsis sp.	Sword fern	M	CH, G	I	GP	E
Nerine sp.	—	B	CH	C	GP	D
Nerium oleander	Oleander	B	CH	I	GP	E
Nidularium sp.	Bromeliad	M	LH, CH	H	HR	E
Nierembergia sp.	Cupflower	M	CH	C	GP	E
Notocactus	Cactus	B	LH, CH	C	RD	D
Odontoglossum sp.	Orchid	L	G	C-I	OM	E
Oncidium sp.	Dancing lady orchid	M	G	C-I-H	OM	D
Onychium japonicum	Claw fern	L	G	C-I	HR	E
Ophrys sp.	Orchid	M	CH	C	HR	E
Opuntia sp.	Prickly pear	B	LH, CH	I-H	RD	D
Ornithogalum sp.	—	B	CH	C	GP	D
Orthophytum saxicola	Bromeliad	M	CH	H	HR	D
Oscularia sp.	—	B	CH	I	RD	D
Osmanthus sp.	Devilweed	B	LH, CH	C-I	GP	E
Othonna capensis	Little pickles	B	CH	I	RD	D
Oxalis sp.	Wood sorrel	B	CH	C-I	GP	D
Pachyphytum sp.	—	B	LH, CH	I	RD	D
Pachystachys sp.	—	M	G	H	GP	E
× *Pachyveria*	—	B	LH, CH	I	RD	D
Pandanus sp.	Screw pine	M	LH, CH	H	GP	D
Paphiopedilum sp.	Cypripedium	L	G	I-H	OM	E
Parodia sp.	Cactus	B	CH	I	RD	D
Passiflora sp.	Passionflower	B	CH, G	I-H	GP	E
Pedilanthus tithymaloides	Devil's backbone	M	CH, G	H	RD	E
Pelargonium sp.	Geranium	B	LH, CH	C-I	GP	D
Pellaea sp.	Cliff brake	L	G	I	HR	E
Pellionia sp.	—	M	G	H	GP	E
Pentas lanceolata	Star cluster	B	CH	I-H	GP	E
Peperomia sp.	Peperomia	M	LH, CH	H	GP	D
Petunia sp.	Petunia	B	CH	C-I	GP	D
Phalaenopsis sp.	Moth orchid	L	G	H	OM	E
Philodendron sp.	Philodendron	M	LH, CH	H	GP	E
Phoenix sp.	Date palm	M	CH, G	I-H	GP	S
Phragmipedium sp.	Lady slipper	M	G	I	OM	E
Piaranthus sp.	—	M	CH	C	RD	D
Pilea sp.	Pilea	M	LH, CH, G	H	GP	E
Pinanga sp.	Palm	M	G	H	GP	E
Pitcairnia sp.	Bromeliad	M	CH, G	H	HR	E
Pittosporum tobira	Japanese pittosporum	B	CH	I	GP	D

Introduction to Horticultural Production / Part III

Habit	Height	Use	Attrac-tion	Cultivar Availability	Propagule
Er	1, 2	H/GH	Fl	DC, OC	OS
Er, De	1, 2	GH	Fl	—	Cu, Se
Tr	1, 2	GH	Fl	—	Cu, Se
Er	2, 3	H/GH	Fl	—	Di
Er	1, 2, 3	H/GH	Fl, St	—	Se
Er	1, 2	H/GH	Fl, Fo	VC	OS
Cl	3	GH	Fo	—	Cu, Se
Er	2, 3	H/GH	Fo	DC, OC	Di, Ru
Er	1, 2, 3	GH	Fl	—	OS
Er	3	GH	Fl	OC, VC	Cu
Er	1, 2	H/GH	Fl, Fo	—	OS
Er, Tr	1, 2, 3	GH	Fl	—	Di
Er	1	H/GH	Fl, St	—	Se
Er-Pe	1, 2	GH	Fl	—	ASMC, RhD
Er-Pe	1, 2	GH	Fl	—	ASMC, RhD
Er	2	GH	Fo	—	Di, Sp
Er	1, 2	GH	Fl	—	ASMC, Tu
Er	1, 2, 3	H/GH	Fl, St	—	Cu, Se
Er	1, 2, 3	H/GH	Fl, Fo	—	OS
Er	1	H/GH	Fl, Fo	—	OS
Er	1	H/GH	Fl, Fo	—	Cu, Se
Er	3	H/GH	Fl, Fo, Fra	DC, OC, VC	Cu
Tr	1, 2	H/GH	Fl, Fo	—	Cu
Er	1, 2, 3	H/GH	Fl, Fo	—	Bu, Di, Se
Er	1, 2	H/GH	Fl, Fo	—	LC
Er	3	GH	Fl	—	Cu, Se
Er	1, 2	H/GH	Fl, Fo	—	LC
Er	2, 3	H/GH	Fo	—	Cu, Se
Er	1, 2	GH	Fl	OC	ASMC, Di
Er	1	H/GH	Fl, St	—	Se
Cl	3	GH	Fl, Fra	OC	Cu, Se
Er	3	GH	Fl, Fo	DC, VC	Cu
Er, Tr	1, 2, 3	H/GH	Fl, Fo, Fra	DC, OC, VC	Cu, Se
Er	1, 2, 3	GH	Fo	—	Di, Sp
Tr	2	GH	Fo	—	Cu, Di
Er	3	GH	Fl, Fo	—	Cu
Er, Tr	1, 2, 3	H/GH	Fo	DC, OC, VC	Cu, Di, RL
Er	1, 2	GH	Fl, Fra	OC	Cu, Se
Er-Pe	1, 2	GH	Fl	—	ASMC
Er, Cl	2, 3	H/GH	Fo	DC, VC	Cu, Se
Er	3	H/GH	Fo, Fr	—	OS, Se
Er	1, 2	GH	Fl	—	ASMC, Di
Er	1	H/GH	Fl, St	—	Cu, Se
Er, Tr	1, 2	H/GH	Fl, Fo	DC, VC	Cu
Er	3	GH	Fo	—	Se
Er	2, 3	GH	Fl, Flo	—	OS
Er	3	H/GH	Fo	VC	Cu, Se

TABLE 13-13. CONTINUED

Scientific Name	Common Name	Light	Humidity	Temperature	Soil	Water
Pityrogramma sp.	Gold fern	M	G	I	GP	E
Platycerium sp.	Staghorn fern	M	CH, G	I-H	OM	E
Plectranthus sp.	Swedish ivy	M	LH, CH	I-H	GP	E
Pleiospilos sp.	Living rock	B	CH	I	RD	D
Pleurothallis sp.	Orchid	M	CH	I	OM	E
Plumeria rubra	Nosegay	M	CH, G	H	GP	E
Polypodium sp.	Polypody	M	CH, G	H-I	HR	E
Polystichum sp.	Shield fern	L	CH, G	H-C	GP	E
Primula malacoides } *P. sinensis*	Primrose	M	G	I	GP	E
Protea sp.	Protea	B	CH	I	GP	D
Pseudopanax sp.	—	M	CH	I	GP	E
Psilotum nudum	Whisk fern	L	G	H	HR	E
Pteris sp.	Dish fern	L	LH, CH, G	I	HR	E
Pyrosia sp.	Felt fern	L	G	I	HR	E
Quesnelia sp.	Bromeliad	M	LH, CH	H	HR	E
Ranunculus asiaticus	Persian buttercup	B	CH, G	C	GP	E
Rebutia sp.	Crown cactus	B	LH, CH	C-I	RD	D
Reinhardtia sp.	Palm	M	G	H	HR	E
Reinwardtia indica	Yellow flax	B	CH, G	I	GP	E
Rhapis excelsa *R. humilis*	Bamboo palm Reed rhapsis	M	CH	I	HR	S
Rhipsalidopsis gaertneri	Easter cactus	M	CH	H	HR-RD	E
Rhipsalis sp.	Wickerware cactus	M	CH	H	HR-RD	E
Rhododendron sp.	Rhododendron and azalea	M	CH	C-I	HR	E
Rhoeo spathacea	Moses-in-the-cradle	M	LH, CH	I	GP	E
Rodriguezia sp.	Orchid	M	G	H	OM	E
Rohdea japonica	Lily of China	M	LH, CH	I-C	GP	E
Ronnbergia columbiana	Bromeliad	M	G	H	HR	E
Rosa sp.	Rose	B	CH, G	I-H	GP	E
Ruellia sp.	—	M	G	I-H	HR	E
Rumohra adiantiformus	Leather fern	M	G	I	GP	E
Russelia equisetiformis	Coral plant	B	CH	I	GP	D
Saintpaulia sp.	African violet	M	CH, G	H-I	HR	E
Salpiglossis sinuata	Painted-tongue	B	CH	C	GP	E
Sansevieria sp.	Devil's tongue	M	LH, CH	H	GP	D
Sarracennia sp.	Pitcher plant	B	CH, G	C	OM	S
Saxifraga stolonifera	Strawberry geranium	B	CH, G	C	GP	D
Schefflera sp.	Umbrella tree	B	CH	H	GP	D
Schizanthus sp.	Butterfly flower	B	CH	I	GP	E
Schlumbergera bridgessi *S. truncata*	Christmas cactus Thanksgiving cactus	M	CH	H-I	HR	E-D
Schomburgkia sp.	Cow-horn orchid	M	G	I	OM	D
Schwantesia sp.	—	B	CH	I	RD	D
Scilla sp.	Squill	B	CH	C	GP	E
Scindapsus pictus	Scindapsus	M	LH, CH	H	HR	D-E
Sedum sp.	Stonecrop	B	CH	I-C	GP	D

Habit	Height	Use	Attraction	Cultivar Availability	Propagule
Er	2, 3	GH	Fo	—	Di, Sp
De	3	GH	Fo	OC	Sp, Su
Er, De	2, 3	H/GH	Fo	—	Cu, Se
Er	1	H/GH	Fl, St	—	Se
Er	1, 2	GH	Fl	—	ASMC
Er	3	GH	Fl, Fra	—	Cu
Er	1, 2, 3	H/GH	Fo	OC	Di, Sp
Er	1, 2	H/GH	Fo	OC	Di, Sp
Er }	1, 2 }	GH }	Fl }	OC }	Se
Er	3	GH	Fl	—	Cu, Se
Er	3	GH	Fo	—	Cu, Se
Er	3	GH	Fo	—	Di, Sp
Er	2, 3	H/GH	Fo	OC	Di, Sp
Er-Tr	2	GH	Fo	—	Sp
Er	1, 2, 3	H/GH	Fl, Fo	—	OS, Se
Er	2	GH	Fl	OC	TuR
Er	1	H/GH	Fl, St	—	Se
Er	3	GH	Fo	—	Se
Er	3	GH	Fl	—	Cu
Er }	3 }	H/GH }	Fl }	— }	Di, Se
Er	2	H/GH	Fl, St	—	Cu, Se
Er-Pe	1, 2, 3	H/GH	St	—	Cu, Se
Er	3	GH	Fl	OC	Cu, Se
Er	1	H/GH	Fl, Fo	VC	Cu, Se
Er-Pe	1	GH	Fl	—	ASMC, RhD
Er	2	H/GH	Fl, Fo, Fr	VC	Rh
Er	3	GH	Fl, Fo	—	OS, Se
Er-Cl	2, 3	GH	Fl	OC	Cu
Er	1, 2, 3	GH	Fl	—	Cu, Di, Se
Er	2	GH	Fo	—	Di, Sp
Er	3	GH	Fl	—	Cu
Er	1	H/GH	Fl, Fo	OC, VC	Se, RL
Er	3	GH	Fl	—	Se
Er	2, 3	H/GH	Fo	OC, VC	Di, LC
Er	2, 3	GH	Fl, Fo	—	Se
Tr	2	H/GH	Fl, Fo	VC	Di, Ru, Se
Er	3	H/GH	Fo	—	AL, Se
Er	3	GH	Fl	OC	Se
Pe }	1, 2 }	H/GH }	Fl }	OC }	Cu, Se
Er	1, 2	GH	Fl	—	ASMC, RhD
Er	1	GH	Fl, St	—	Di, Se
Er	1, 2	GH	Fl	—	OS
Tr	3	H/GH	Fo	VC	Cu, Se
Er, De	1, 2, 3	H/GH	Fl, Fo	DC, OC	Cu, OS, Se

TABLE 13-13. CONTINUED

Scientific Name	Common Name	Light	Humidity	Temperature	Soil	Water
Selaginella sp.	Little club moss	L	G	H–I	HR	E
Selenicereus sp.	Night-blooming cereus	M	CH	I–H	HR	E
Senecio sp.	Senecio	M–B	LH, CH	I–H	RD	D
Setcreasea sp.	—	B	CH	H	GP	D
Sinningia sp.	Gesneriads	M	G	H	HR	E
Skimmia sp.	—	M	CH	C	GP	E
Smithiantha sp.	Temple bells	M	G	H	HR	E
Solandra sp.	Chalice vine	B	CH	H	GP	E
Solanum pseudo-capsicum	Jerusalem cherry	B	CH	I	GP	D
Sophronitis sp.	Orchid	M	G	C	OM	E
Sparmannia africana	African hemp	M	CH	I–H	GP	E
Spathiphyllum sp.	Spathe flower	L–M	LH, CH	H	HR	S
Sprekelia formosissima	Jacobean lily	M	CH	I	GP	D–E
Stanhopea sp.	Orchid	L	G	I	OM	D
Stapelia sp.	Carrion flower	B	LH, CH	I	RD	D
Stephanotis floribunda	Madagascar jasmine	B	G	H	GP	E
Strelitzia reginae	Bird-of-paradise	B	CH	I	GP	D
Streptocarpus sp.	Cape primrose	M	G	C–I	HR	D–E
Strobilanthes sp.	Mexican petunia	M	G	H	HR	E
Stultitia cooperi	—	B	LH, CH	I	RD	D
Tacca chantrieri *T. integrifolia*	Bat plant	M	G	H	GP	E
Tectaria cicutaria	Button fern	M	G	I–H	HR	E
Tetranema roseum	Mexican violet	M	G	I	GP	E
Thelocactus sp.	Cactus	B	CH	I	RD	D
Thunia alba	Orchid	M	G	H	OM	E
Tibouchina sp.	Glory bush	M	CH	I	GP	E
Tillandsia sp.	Bromeliad	M	CH, G	I–S	HR	D
Tolmiea menziesii	Piggyback plant	M	CH	I–C	GP	E
Trachelospermum jasminoides	Star jasmine	M	CH	I	GP	D
Tradescantia sp.	Spiderwort	M	CH	I	GP	D
Trevesia palmata	Snowflake plant	M	G	H	GP	E
Trichocereus sp.	Cactus	B	CH	I	RD	D
Trichodiadema barbatum	Pickle plant	B	CH	I	RD	D
Tripogandra multiflora	Tahitian bridal veil	M	CH, G	H	GP	D
Tulbaghia sp.	—	B	CH	C	GP	E
Tulipa sp.	Tulip	B	CH	C	GP	E
Urginea maritima	Sea onion	B	CH	I	GP	E
Vallota speciosa	Scarborough lily	B	CH	I	GP	E
Vanda sp.	Orchid	M	G	H–I	OM	E
Vriesea sp.	Bromeliad	M	CH	I–H	HR	D
Widdringtonia sp.	African cypress	B	CH	I	GP	E
Woodwardia sp.	Chain fern	M	CH	I	GP	E
Xylobium sp.	Orchid	M	G	I	OM	E
Yucca sp.	Yucca	B	CH	I	RD	D
Zamia sp.	Cycad	M	CH, G	I–H	HR	E
Zantedeschia sp.	Calla lily	M	CH	I–H	GP	E
Zebrina pendula	Wandering jew	M	CH, G	I–H	GP	E
Zingiber sp.	Ginger	M	G	H	GP	E
Zygopetalum sp.	Orchid	M	G	I	OM	D

Habit	Height	Use	Attraction	Cultivar Availability	Propagule
Er, Tr	1, 2, 3	GH	Fo	VC	Cu, Sp
Cl, Pe	3	GH	Fl	—	Cu, Se
Er, Tr	1, 2, 3	H/GH	Fl, Fo	OC	Cu, Di, Se
Er–Tr	2, 3	GH	Fl, Fo	—	Cu, Se
Er	1, 2, 3	GH	Fl	OC	Cu, LC, Se, Tu
Er	1, 3	GH	Fl, Fr, Fra	—	Cu, Se
Er	3	GH	Fl	—	Rh
Cl	3	GH	Fl, Fra		Cu
Er	3	H/GH	Fl, Fr	DC, OC	Cu, Se
Er	1	H/GH	Fl	—	ASMC, RhD
Er	3	H/GH	Fl, Fo	OC	Cu
Er	2, 3	H/GH	Fl, Fo	—	Di
Er	2	H/GH	Fl	—	Bu
Pe–Tr	2, 3	GH	Fl	—	ASMC, RhD
Er–Pe–De	1	H/GH	Fl, St	—	Cu, Se
Tw	3	GH	Fl, Fra	—	Cu
Er	3	GH	Fl	—	Di, Se, Su
Er	1, 2, 3	GH	Fl	—	Di, LC, Se
Er	2, 3	GH	Fl, Fo	—	Cu
Er–De	1	H/GH	Fl, St	—	Cu, Se
Er	2	GH	Fl, Fr	—	Di, Se
Er	2, 3	GH	Fo	—	Di, Sp
Er	1	GH	Fl	—	Se
Er	1	H/GH	Fl, St	—	Se
Er	3	GH	Fl	—	ASMC
Er	3	GH	Fl	—	Cu
Er, Pe	1, 2, 3	GH	Fl, Fo	—	OS, Se
Er–Pe	2	H/GH	Fo	—	RL
Tw	3	GH	Fl, Fra	OC, VC	Cu
Er–Tr	2, 3	H/GH	Fl, Fo	OC, VC	Cu, Di, Se
Er	3	GH	Fo	OC	Cu
Er	2, 3	H/GH	Fl, St	—	Se
Er	1	H/GH	Fl, Fo	—	Cu, Se
Tr	3	H/GH	Fl, Fo	—	Cu, Se
Er	1, 2	GH	Fl	—	OS, Se
Er	1, 2, 3	GH	Fl, Fo	DC, OC, VC	OS
Er	3	H/GH	Fl, Fo	—	Di, Se
Er	2	GH	Fl	—	OS
Er	1, 2, 3	GH	Fl, Fra	OC	ASMC
Er	1, 2, 3	H/GH	Fl, Fo	—	OS, Se
Er	3	GH	Fo	—	Cu, Se
Er–Pe	2, 3	GH	Fo	—	Sp, VB
Er–Pe	2	GH	Fl	—	ASMC, RhD
Er	3	H/GH	Fo	—	Cu, OS, Se
Er	2, 3	GH	Fo	—	Se
Er	2, 3	GH	Fl	DC	Di, OS, Se
Tr	3	H/GH	Fl, Fo	VC	Cu
Er	3	GH	Fl, Fo	—	RhD
Er	2, 3	GH	Fl	—	ASMC, RhD

Explanation of Symbols for Table 13–13.

Light:
L, low, 50 to 500 footcandles (538 to 5382 lumens). Precludes direct sun. Can be found at un-shaded north exposure or lightly shaded eastern and western exposures.
M, moderate, 1000 to 3000 footcandles (10,764 to 32, 292 lumens). Diffused sunlight. Objects barely cast shadows at these light levels. Can be found at unshaded eastern or lightly shaded south-ern exposures. Maximal potential of most fluorescent lights would be at lower end of this range.
B, bright, full sun, 4000 to 8000 footcandles (43,056 to 86,112 lumens). Can be found at un-shaded southern window.

Humidity:
LH, low relative humidity, about 20 to 30 percent, as might be found in an unhumidified home.
CH, average relative humidity as would be comfortable for people, varies between 40 to 50 per-cent. Requires some degree of humidification in most homes.
G, greenhouse humidity levels, over 50 percent relative humidity.

Temperature:
C, cool. Day: 55° to 60° F (13° to 15°C).
    Night: 40° to 45° F (5° to 7°C).
I, intermediate. Day: 70° F (21°C)
    Night: 50° to 55° F (10° to 13°C).
H, hot. Day: 80° to 85° F (27° to 30°C).
    Night: 62° to 65° F (16° to 18°C).

Soil:
GP, general-purpose potting medium; Moderate levels of humus.
HR, humus-rich potting medium.
RD, potting medium with rapid drainage.
OM, organic matter (osmunda fern fiber or shredded fir bark with charcoal and drainage material).
SM, sphagnum moss.

Watering:
D, drench; allow top couple of inches of soil to dry before next application.
E, even moisture; damp constantly, but not to point of sogginess to restrict aeration.
S, soak; keep at constant wetness.

Habit:
De, decumbent
Tw, twining
Er, erect
Tr, trailing
Cl, climbing
Pe, pendulous

Height:
1, one foot or less
2, between one to two feet
3, over two feet

Use:
H/GH, house greenhouse plant; tolerates wide enough deviation from optimal cultural conditions to grow in house environment, and does even better in the controlled greenhouse environment.
GH, greenhouse; requirements are too exacting to make it a good house plant.

Attraction:
FO, foliage
Fl, flowers
Fr, fruit
Fra, fragrance
St, stem, as of cactus or certain other succulents

Cultivar availability (beyond botanical species):
DC, dwarf cultivars
VC, variegated cultivars
OC, other cultivars, e.g., double flowers, flower color differences, differences in foliage

Propagule:
   AL, air layering
   ASMC, aseptic seeds and mericlones
   BD, bulb division
   Bu, bulb
   Co, corm, cormel
   Cu, stem cutting
   Di, division
   L, layering
   LC, leaf cutting
   OS, offset
   RC, root cutting
   RD, root division
   Rh, rhizome
   RhD, rhizome division
   RL, rooting of leaves
   RLa, root layering
   Ru, runner
   Se, seed
   Sp, spore
   SP, stem piece
   Su, sucker
   Tu, tuber
   TuD, tuber division
   TuR, tuberous root
   VB, vegetative bud

14

Olericulture

Olericulture is the branch of horticulture that deals with the production, storage, processing, and marketing of vegetables. Most of these areas, with the exception of production, have been covered in Part Two. This chapter will be devoted to production.

Economics

Vegetable production on the family or noncommercial level is substantial. In 1979 about 33 million American households had a vegetable garden; this is about 42 percent. The average figure for the last seven years is also about 40 percent, as indicated by Gallup polls. The average vegetable garden was about 595 square feet in 1979, as opposed to 560 square feet in 1976. In 1979, expenditures reached $2 billion, and the retail value of the harvest was $13 billion.

On the commercial level, vegetables are big business. During 1975 about 1.5 million acres was involved in vegetable production for fresh marketing, and 1.9 million acres for processing. Similar figures were observed for 1976. The combined tonnage of vegetable production from this acreage was around 25 million. This easily places vegetable production in the $4 billion commercial range.

In terms of fresh market production, the four leading states in 1975 with percentages of total production acreage were California (33.0 percent), Florida (15.4 percent), Texas (9.8 percent), and Michigan (3.9 per-

510

cent). Similarly, for vegetables utilized in processing, the states were Wisconsin (19.9 percent), California (19.2 percent), Minnesota (10.6 percent), and Washington (6.9 percent). Figures were similar in 1976.

In the United States the most important vegetables on the commercial horticultural level, based on the total crop value in 1975, are as follows in decreasing order: potato (white), tomato, lettuce, onion, corn (sweet), snap bean, cucumber, carrot, sweet potato, pea, celery, cabbage, muskmelon (cantaloupe), watermelon, pepper, asparagus, broccoli, cauliflower, lima bean, winter melon ('Honey Dew'), garlic, escarole, Brussels sprouts, artichoke, beet, eggplant, and taro. Other vegetables have minimal commercial value, but along with the preceding vegetables can be found in home gardens and roadside stands.

Most vegetable seeds sold by firms are mass produced under contract to independent growers. Many of these are situated on the West Coast and in the Western states, where climatic conditions are favorable for seed production. In 1976 about 142,000 acres of seeds was harvested with a combined production of 197 million pounds.

Production of vegetable bedding plants had a wholesale value of about $32 million in 1978. The leading state was California (44 percent), followed by Georgia (17 percent), Michigan (6 percent), and Ohio (6 percent).

Vegetable Production

Generally, vegetables can be grouped in the following categories, which are not always absolute. **Salad crops** include leafy vegetables that are mainly eaten raw, as in a salad. **Celery, cress, endive,** and **lettuce** are often included in this group. Certainly, other vegetables are eaten raw in salads, but their main use is for other purposes. The **legumes** (family Leguminosae) or **pulse crops** can form a symbiotic relationship with *Rhizobium* sp., which are nitrogen-fixing bacteria. The edible legumes include **lima beans, peas, snap beans,** and **soybeans.** The **solanaceous crops** (family Solanaceae) include the **eggplant, pepper,** and **tomato. Cucurbits** are **vine crops** (family Cucurbitaceae), which include the **cucumber, melon, pumpkin, squash,** and **watermelon.** The **cole crops** (family Cruciferae) encompass many *Brassica* species, such as **broccoli, Brussels sprouts, cabbage, cauliflower, Chinese cabbage, kohlrabi, kale, collards,** and **mustard.** The last three are often included in the group of crops called **greens,** which are the immature leaves and stems of plants prepared by boiling. **Chard, New Zealand spinach, spinach,** and **turnip greens** are also in the group called **greens.** The **onion group** includes the common **onion** and its close relatives, **garlic, leek,** and **shallot.** The **starchy root** and **modified stem crops** are grown for their underground edible portions. They include the **beet, celeriac, carrot, parsnip, potato, radish, rutabaga, salsify, sweet potato, taro,** and **turnip. Asparagus** and **rhubarb** are included in the **perennial crop** group, whereas some vegetables, such as **sweet corn** and **okra,** are classed as **miscellaneous vegetables.**

TABLE 14-1. VEGETABLES

Common Name	Cultivars and Hybrids	Germination (days) Seed Viability (years)	Seed Planting Depth (in.)	Requirements for 100 Row-Feet		Distance between Rows		Distance between Plants in Row (in.)	Days to Maturity (from Seed)	Remarks[a]
				Seeds	Plants	Mechanical Cultivation (ft)	Hand Cultivation (ft)			
Asparagus	*Asparagus officinalis*	10–15, 3	1–1.5, 6–8b	1 oz	75	4–5	1.5–2	18	3 yr	H
Beans										
Bush lima	*Phaseolus limensis*	4–8, 3	1–1.5	8 oz	—	2.5–3	2	3–4	60–75	T-1
Pole lima	*Phaseolus limensis*			8 oz	—	3–4	3	36–48	75–100	T-1
Bush snap	*Phaseolus vulgaris*			8 oz	—	2.5–3	2	3–4	40–65	T-O
Pole snap	*Phaseolus vulgaris*			4 oz	—	3–4	2	36	60–75	T-O
Beet	*Beta vulgaris*	6–10, 4	1	2 oz	—	2–2.5	1–1.33	2–3	40–65	H
Broccoli	*Brassica oleracea* (Botrytis group)	4–8, 4	0.5	1 pkt	50–75	2.5–3	2–2.5	14–24	60–100	VH
Brussels sprouts	*Brassica oleracea* (Gemmifera group)	4–8, 4	0.5	1 pkt	50–75	2.5–3	2–2.5	14–24	60–90	VH
Cabbage	*Brassica oleracea* (Capitata group)	4–8, 4	0.5	1 pkt	50–75	2.5–3	2–2.5	14–24	60–90	VH
Carrot	*Daucus carota* var. *sativus*	10–15, 3	0.5	0.5 oz	—	2–2.5	1.2–1.33	2–3	60–100	H
Cauliflower	*Brassica oleracea* (Botrytis group)	4–8, 4	0.5	1 pkt	50–75	2.5–3	2–2.5	14–24	60–100	T-O
Celeriac	*Apium graveolens* var. *rapaceum*	12–20, 5	0.125	1 pkt	200–250	2.5–3	1.5–2	4–6	100–140	T-O
Celery	*Apium graveolens* var. *dulce*	12–20, 5	0.125	1 pkt	200–250	2.5–3	1.5–2	4–6	100–140	T-O
Chard	*Beta vulgaris* var. *cicla*	6–10, 4	1	2 oz	—	2–2.5	1.5–2	6	40–65	H
Chinese cabbage	*Brassica rapa* (Pekinensis group)	4–8, 4	0.5	0.25 oz	—	2–2.5	1.5–2	8–12	70–90	H
Collards	*Brassica oleracea* (Acephala group)	4–8, 4	0.5	1 pkt	—	3–3.5	1.5–2	18–24	60–90	H
Corn, sweet	*Zea mays* var. *rugosa*	6–8, 1	2	2 oz	—	3–3.5	2–3	14–16	65–100	T-O
Cress										
Upland	*Barbarea verna*	— —	0.125–0.25	1 pkt	—	2–2.5	1.2–1.33	2–3	50–60	H
Water	*Nasturtium officinale*	— —	0.125–0.25	1 pkt	—	2–2.5	1.5–2	4–6	50–60	H

Crop		Scientific name									
Cucumber	Cucumis sativus	6–10, 5	0.5	0.5 oz	—	6–7	6–7	36	55–70	T-1
Eggplant	Solanum melongena	10–15, 5	0.5	1 pkt	50	3	2–2.5	36	100–140	T-1
Endive	Cichorium endivia	10–12, 5	0.5	0.5 oz	—	2.5–3	1.5–2	12	60–90	H
Garlic	Allium sativum	—, —	1.2	1 lb[c]	—	2.5–3	1.2–1.33	2–3	150–180[d]	H
Kale	Brassica oleracea (Acephala group) B. napus (Pabularia group)	4–8, 4	0.5	0.25 oz	—	2.5–3	1.5–2	12–15	55–65	H
Kohlrabi	Brassica oleracea (Gongylodes group)	4–8, 4	0.5	0.25 oz	—	2.5–3	1.2–1.33	5–6	55–70	T-H
Leek	Allium ampeloprasum (Porrum group)	6–12, 1	0.5–1	1 pkt	—	2.5–3	1.2–1.33	2–3	180	T-H
Lettuce — Cos	Lactuca sativa	4–8, 5	0.5	0.25 oz	100	2.5–3	1.2–1.33	8–10	60–80	VH
Lettuce — Head								12–15	75–90	
Lettuce — Leaf								6	30–60	
Melon — Netted (muskmelon)	Cucumis melo (Reticulatus group)	6–10, 5	1	0.5 oz	—	6–7	6–7	48–60	75–110	T-1
Melon — Winter (honeydew, casaba)	C. melo (Inodorus group)								85–130	
Mustard	Brassica juncea var. crispifolia	4–8, 4	0.5	0.25 oz	—	2.5–3	1.2–1.33	12	35–45	H
Okra	Abelmoschus esculentus	—, 2	1–1.5	2 oz	—	3–3.5	3–3.5	24	50–60	T-1
Onion — Plants	Allium cepa (Cepa group)	—, —	1–2	—	400	2–2.5	1.2–1.33	2–3	100[e]	VH
Onion — Seeds		6–12, 1	0.5–1	1 oz	—				100–110	
Onion — Sets		—, —	1–2	1–1.5 lb	—				60–80[f]	
Parsnip	Pastinaca sativa	12–20, 1	0.5	0.5 oz	—	2–2.5	1.5–2	2–3	95–110	H
Pea — Garden	Pisum sativum var. sativum	8–10, 3	2–3	0.5 lb	—	2–4	1.5–3	1–2	55–75	VH
Pea — Edible podded	var. macrocarpon									
Pepper	Capsicum annuum var. annuum	6–10, 3	0.5	1 pkt	50–75	3–4	2–3	18–24	100–130	T-1
Potato	Solanum tuberosum	—, —	4	5–6 lb[g]	—	2.5–3	2–2.5	10–18	90–120[h]	T-H
Pumpkin	Cucurbita pepo; C. mixta, C. maxima, C. moschata	4–8, 4	1–2	1 oz	—	5–8	5–8	36–48	80–100	T-O
Radish	Raphanus sativus	4–8, 4	0.5	1 oz	—	2–2.5	1.2–1.33	1–2	25–30	T-H
Rhubarb	Rheum rhabarbarum	—, —	3–4	25–35[i]	—	3–4	3–4	3–4	80–100[j]	H

TABLE 14-1. CONTINUED

Common Name	Cultivars and Hybrids	Germination (days) Seed (years) Viability	Seed Planting Depth (in.)	Requirements for 100 Row-Feet		Distance between Rows		Distance between Plants in Row (in.)	Days to Maturity (from Seed)	Remarks[a]
				Seeds	Plants	Mechanical Cultivation (ft)	Hand Cultivation (ft)			
Rutabaga	*Brassica napus* (Napobrassica group)	4–8, 4	0.25–0.5	0.5 oz	—	2–2.5	1.2–1.33	6–8	80–100	VH
Salsify	*Tragopogon porrifolius*	6–10, —	0.5	1 oz	—	2–2.5	1.5–2	2–3	120–150	VH
Shallot	*Allium cepa* (Aggregatum group)	—, —	1–2	1 lb[c]	—	2–2.5	1–1.5	2–3	90–110[d]	H
Soybean	*Glycine max*	4–8, —	1–1.5	0.5–1.0 lb	—	2.5–3	2–2.5	3–4	80–100	T-O
Spinach	*Spinacia oleracea*	6–10, 4	0.5	1 oz	—	2–2.5	1.2–1.33	3–4	40–50	VH
Spinach, New Zealand	*Tetragonia tetragonioides*	12–20, —	1–1.5	1 oz	—	3–3.5	3–3.5	18	70	T-O
Squash	*Cucurbita pepo, C. mixta, C. maxima, C. moschata*									
Summer		4–8, 4	1–2	0.5–1 oz	—	4–5	4–5	36	50–60	T-O
Fall and winter						8–12	8–12	48	85–120[e]	T-O
Sweet potato	*Ipomoea batatas*	—, —	2–3	5 lb[k]	75	3–3.5	3–3.5	12–14	120–150[e]	T-1
Taro	*Colocasia esculenta*	—, —	2–3	5–6 lb[l]	50	3.5–4	3.5–4	24	180–360[m]	T-1
Tomato	*Lycopersicon lycopersicum*	6–10, 3	0.5	1 pkt	35–75	3–4	2–3	18–36	105–130	T-O
Turnip	*Brassica rapa* (Rapifera group)	4–8, 4	0.25–0.5	0.5 oz	—	2–2.5	1.2–1.33	3–4	60–80	H
Watermelon	*Citrullus lanatus*	6–10, 5	1–2	1 oz	—	8–10	8–10	36–60	70–100	T-1

[a] VH, very hardy. Plant 4 to 6 weeks before frost-free date.
H, hardy. Plant 2 to 4 weeks before frost-free date.
T-H, half-hardy. Plant 1 to 3 weeks before frost-free date.
T-O, tender. Plant on frost-free date.
T-1, tender. Plant 1 week or more after frost-free date.
[b] For crowns.
[c] Cloves.
[d] From cloves.
[e] From plants.
[f] From sets.
[g] Tubers.
[h] From tuber sections.
[i] Root divisions.
[j] From root divisions.
[k] Seed roots.
[l] Corms.
[m] From corms.

514

Success with these vegetables is dependent upon a number of factors covered in Part Two. The soil must be prepared properly for sowing of seeds and setting out transplants. Soil fertility, drainage, aeration, and pH must be properly maintained. Crops must be put in at favorable times based on a knowledge of local last killing spring frost and first killing fall frost (see Fig. 13-7), days needed for maturity, crop responses expected from the variations in weather known to occur during the growing season, and marketing and processing demand. Plant crowding is avoided by removal of excess seedlings (thinning) and proper spacing of transplants. Water must be supplied if rainfall is insufficient; weeds must be controlled through cultivation, mulches, or herbicides; insects and diseases must be controlled through judicious use of pesticides tempered with biological controls. Finally, crops must be harvested at optimal times.

Propagation of vegetables is mainly by seed, which should be obtained from a reliable seed firm and be clean, properly labeled, fresh, and insect and disease free. It should be viable, whether fresh or stored. The length of viability is indicated in Table 14-1. Some vegetables are propagated asexually. These include asparagus, garlic, potato, rhubarb, shallot, sweet potato, and taro.

Some vegetable crops are grown from transplants for reasons of early harvest (especially in the North) and for economical use of space. The transplants are usually produced in a greenhouse or outdoor seedbed at seeding rates three to six times those practiced in directly seeded fields. This leaves some field space free for early crops, which can be followed by transplants. Plants commonly grown from transplants include celery, cole crops, onion, and solanaceous crops.

Optimal efficiency can be realized in olericulture by the use of certain horticultural practices. Restricted areas, such as small city gardens, as well as commercial olericultural operations can benefit from some of the following practices. It must be kept in mind that increased productivity per unit area increases the loss of nutrients also. Therefore, soil fertility must be maintained if increased productivity is to be realized.

Intensive Gardening. This method increases the plant density in an area (Fig. 14-1). Some rows and all pathways are eliminated by planting vegetables in blocks. Remaining rows delineate the blocks and serve as pathways. Blocks are sized so that the middle is accessible from either side.

Plants are closely spaced such that the foliage of mature plants just touches, but plants must not crowd each other. This foliage cover (a living mulch) helps to retard weed growth, loss of soil moisture, and sudden changes of soil temperature. While there is no exact rule to achieve this foliage coverage for all vegetables, an approximation is to decrease the recommended minimal spacing by 20 to 25%. Plants should also be staggered for more efficient use of space. The miniature or dwarf hybrid vegetables are particularly suited to intensive gardening.

Fig. 14-1. Intensive gardening makes efficient use of land and gives good productivity in city gardens, where land is limited and expensive. Author photo.

A further refinement of intensive planting is **vertical gardening** (Fig. 14-2). In a 20 by 20-foot garden, the potential growing area is 400 square feet. However, the inclusion of a 5-foot-high fence around the perimeter adds another 400 square feet. This vertical space is ideal for climbing vegetables. Nylon netting is strong and lasts for several seasons, and is an inexpensive substitute for a fence. Such space consumers as pole snap, or pole lima beans, cucumbers, small vine squash, muskmelons, small pumpkins, and even tomatoes can be supported on fences or netting.

Succession planting. This approach is not based on increased plant density, but expands yields by continuous use of the garden area throughout the gardening season. Since crops differ in days to maturity, spaces become available that can be planted to another crop.

Many variations of succession planting are possible. Early spring sowings are possible for some crops, such as peas, radishes, onions, early potatoes, broccoli, lettuce, and spinach. After the harvest, radishes could be followed by snap beans, and in turn by rutabagas, which are picked at the end of the season. This particular succession utilizes one space through-

Introduction to Horticultural Production | Part III

out the garden season. Lettuces, onion sets, and radishes might be succeeded by celery, snap beans, or carrots; early potatoes by late corn; and early peas by late cabbages, carrots, or beets. Succession plantings should not slack off during midsummer. Early corn can be replaced by turnips or spinach. Late-summer crops might be followed by quick-maturing vegetables such as radishes or leaf lettuce. Harvest periods of certain popular vegetables are readily extended by succession plantings of one variety or by utilizing early, mid, and late season varieties; corn and tomatoes are good choices for this succession.

Possibilities for succession are endless, but certain points must be observed. Vegetables should not be replaced in the same area with another of their kind, but with unrelated vegetables. Vegetables subject to attack by the same insects or disease should not succeed each other. A knowledge of days to maturity for each vegetable is essential. If only 30 days is left in the season, snap beans that require at least 50 days to harvest would be a poor choice. The preference of some crops for cool or hot weather is important. In midsummer, radishes, peas, or lettuce could not be planted as a filler, but would be satisfactory when cool weather arrived. Frost tolerance is also important. The first killing frost will destroy tomatoes, but not turnips, rutabagas or other frost-tolerant vegetables.

Fig. 14–2. This part of a vertical garden shows pole snap beans about to be harvested. Growing climbing vegetables this way instead of allowing them to spread out and waste valuable space is more efficient. Author photo.

Interplanting. This technique increases the plant density in an area through the planting of a slow-maturing vegetable between the rows of a fast-maturing one. After the earlier vegetable is harvested, the second vegetable takes over the newly available space. Because of this cohabitation, two harvests are produced in less time than with succession planting. Another form of interplanting mixes two vegetables that differ in their growth habits, but not necessarily in their days to harvest. For example, one crop may grow tall and the other may hug the ground.

Some caution is advisable in choosing vegetable pairs. If rates of growth differ too greatly or growth forms are too rampant, the slower or less aggressive plant may lose to its competitor.

Fig. 14–3. (left) Lettuce and onions are interplanted in this garden. Courtesy of National Garden Bureau, Inc.

Fig. 14–4. (right) Bush beans and lettuce are interplanted in this garden. Courtesy of National Garden Bureau, Inc.

Introduction to Horticultural Production / Part III

Indians were the first Americans to practice interplanting, when they planted pumpkins around hills of corn. As the corn grew upward, the pumpkins shaded the ground and kept the soil moist by their ground-hugging growth. Cucumbers or squash could also be substituted for pumpkins. Onions, leaf lettuce, and carrots are interplanted (Fig. 14-3), or onions can be placed between rows of snap beans. Often radishes and carrots are mixed, since the radishes mark the rows of slow-emerging carrots, which fill in the space left after removal of the radishes. Lettuce can be interplanted with cabbage or early beans (Fig. 14-4), late cabbages between early tomatoes, or radishes and spinach between late tomatoes, peppers, or eggplants. Winter squash between rows of early potatoes is another possibility.

Information on the various vegetables can be found in Table 14-1. Additional information is found in the remaining part of this chapter, which is devoted to the individual vegetables. Scientific names for the vegetables will only be cited in Table 14-1.

Careers in Olericulture

Seed firms need people experienced in olericulture, as well as in ornamental horticulture. Positions include a breeder who develops improved cultivars of vegetables, a propagator in charge of producing seed and vegetative propagules, an independent grower who contracts with seed firms for seed production, and sales personnel such as managers, dealers, and salesmen.

Vegetable production takes place on farms, many of which are family owned and operated. Others are owned on the corporate level. Some of these require a person experienced in olericulture to manage all stages of vegetable production. Processing companies usually have a person in the field who acts as a liaison and consultant between independent growers and the processing company. A similar position is also found with companies that produce horticultural machinery and chemicals.

There are numerous positions within the vegetable industries, all of which require varying degrees of knowledge of olericulture. A number of them occur with sales agencies that handle sales of vegetables for growers. These agencies require an overall manager, a storage supervisor, a supervisor for grading and packing, a field person who is a liaison and consultant between the agency and growers, and a buyer. Brokers resell vegetable wholesale lots obtained from growers or sales agencies, and buyers for retail chains obtain such lots from growers, agencies, or brokers. There are also promotional agencies on the state or national level for specific vegetables that would need a director. See G. M. Kessler under Further Reading in Chapter 13 for more detail.

The various positions described under education and research in the previous chapter on ornamental horticulture would also have similar positions for people knowledgeable in olericulture.

ASPARAGUS

Asparagus is a hardy perennial grown for its edible spring shoots. Mature height is 3 to 5 feet. The shoots themselves are somewhat tender and easily frost damaged.

There is a southern limit for asparagus, since it does best where the ground freezes to at least a few inches. This would preclude the Deep South. Commercial production is found mainly in California and Washington and around the east and west coasts of the Great Lakes. Best spear quality occurs at a soil pH of 6.5 to 7.5 and when the temperature for 4 to 5 nights preceding harvest is 60° to 65°F (15.6° to 18.3°C), and daytime temperatures are moderate. Temperatures below 55°F (12.8°C) at this time produce a purple, tougher spear.

Direct seeding is usually limited to the production of 1–year–old crowns, which are the propagule of choice for commercial and home plantings of asparagus. Harvesting should not be started with new crowns until the third spring in order to establish strong crowns. The harvest period lasts from 5 to 12 weeks, depending on the prevailing climatic conditions. A row of 50 to 100 feet is sufficient for a family of six persons. Spears are harvested when 5 to 8 inches above the ground surface.

Fig. 14–5. Harvested and bunched asparagus. U.S. Department of Agriculture photo.

The two main diseases affecting asparagus are rust (*Puccinia asparagi*) and Fusarium root rot (*F. oxysporum f. asparagi*). The use of strains resistant to these is recommended. Most of the strains are derived from two cultivars, 'Mary Washington' and 'Waltham Washington.' Insect pests include the common and spotted asparagus beetles (*Crioceris asparagi* and *C. duodecimpunctata,* respectively), and the garden symphalen (*Scutigerella immaculata*).

BEANS

Beans are warm-season annual legume crops that are not frost tolerant. The lima bean is somewhat more tender than the snap bean. Bean seeds, since they decay quickly, should not be planted until the soil has warmed to 50°F (10°C). Lima beans need a 3- to 4-month warm growing season, which makes their culture somewhat difficult in the far northern parts of the United States. Commercial production is extensive in Wisconsin, New York, Florida, and Oregon for snap beans and California for lima beans.

Basically, three types of beans are grown: the green or yellow snap bean with edible pods, the full-sized but immature lima and green shell beans not eaten with pods, and the mature, dry shell beans not eaten with pods. All three types occur as twiners or bush (dwarf) cultivars.

Beans, being legumes, can form a symbiotic relationship with bacterial species of the genus *Rhizobium*. Since these are nitrogen-fixing bacteria, it is good horticultural practice to encourage their growth. Inoculants of *Rhizobium* sp. are available, and their use is desirable if the soil has no *Rhizobium* present. Once introduced into the soil, they should persist, especially if legumes are continued in the soil. Soil pH should be in the range of 5.5 to 7.0.

Pole beans require more space and longer times until harvest, but their bearing period is longer than the bush forms. However, succession plantings of the bush forms at 2-week intervals, until only the minimal time required for harvest remains until the first killing frost, can extend the harvest period. Bush forms are especially useful for following early-maturing crops, such as spinach, radish, peas, lettuce, beets, or early potatoes. Snap beans are harvested when the pods are almost full sized with the beans one quarter to one third developed. Lima beans are picked with full-sized green pods and nearly full sized beans.

A number of insects and diseases attack beans. Insects include the Mexican bean beetle (*Epilachna varivestis*), leafhopper (*Empoascus fabae*), bean leaf beetle (*Cerotoma trifurcata*), aphid (*Aphis fabae*), and Japanese beetle (*Popillia japonica*). Diseases include anthracnose (*Colletotrichum lindemuthianum*), bacterial blight (*Xanthomonas phaseoli*), bean mosaic (*Marmor phaseoli* and *M. laesiofaciens*), rust (*Uromyces phaseoli*), and downy mildew (*Phytophthora phaseoli*). Disease problems can be reduced by the use of resistant varieties.

(A)

(B)

Fig. 14–6. Beans. A. Lima beans, bush cultivar. B. Snap beans, bush cultivar. U.S. Department of Agriculture photos.

Suggested varieties of bush green snap beans include 'Astro,' 'Bush Blue Lake,' 'Contender,' 'Greensleeves,' 'Harvester,' 'Tendercrop,' and 'Tendergreen.' Pole forms of green snap beans include 'Blue Lake,' 'Kentucky Wonder,' and 'Romano.' Yellow bush snap beans of choice are 'Brittle Wax,' 'Cherokee,' 'Goldencrop,' 'Pencil Pod,' 'Rustproof Golden,' and 'Surecrop.' 'Kentucky Wonder Wax' is a pole form of yellow snap beans. Green shell beans include 'Dwarf Horticultural,' 'French Horticultural,' and 'Low's Champion.' 'Fordhook,' 'Henderson,' 'Kingston,' and 'Thorogreen' are bush lima beans, whereas pole varieties include 'Burpee's Best,' 'Carolina,' 'Florida Butter Speckled,' 'King of the Garden,' and 'Prizetaker.' 'Pinto,' 'Red Kidney,' and 'White Marrowfat' are used for dry shell beans.

BEET

The beet is a root crop that is reasonably heat and cold resistant and can be grown in all parts of the United States. Commercial production is highest in Wisconsin, New York, and Texas. Beet leaves are also used as greens, and beet is used as a vegetable and as a commercial source of sugar.

Beets grow poorly in strongly acid soils. A well-tilled soil at pH 6.0 to 6.8 and devoid of rocks is best; otherwise, beet roots will be misshapen. Since the beet matures quickly, successive sowings at 2–week intervals are possible until insufficient time remains for full maturity prior to a killing frost. The best-quality beets are produced during the cooler part of the growing season and are about 1.25 to 1.50 inches in diameter.

Introduction to Horticultural Production / Part III

Insects that attack beets include the beet leafhopper (*Circulifer tenellus,* vector for curly top virus), beet leaf miner (*Pegomya betae*), and beet webworm (*Loxostege sticticalis*). Fungal pathogens include both leaf spot (*Alternaria* sp.) and scab (*Streptomyces scabies*). Popular early beets are 'Crosby Egyptian' and 'Early Wonder.' Other popular cultivars are 'Detroit Dark Red,' 'Golden Beet,' 'Long Blood,' 'Perfected Detroit,' and 'Red Ball.'

Fig. 14–7. (above) Beets showing edible roots and foliage used as greens. U.S. Department of Agriculture photo.

Fig. 14–8. (right) Broccoli showing foliage plus edible terminal and lateral green buds. Upper stem portions are also edible. U.S. Department of Agriculture photo.

BROCCOLI

Broccoli, a cole crop, is a cool-season vegetable grown for its edible terminal and lateral green buds and stems. It is somewhat more heat resistant than cauliflower. Since it is a cool-season crop, it is best grown in the spring and late summer. Commercial production is concentrated in California. Seeds are usually started indoors for the spring crop, and transplants about 4 to 5 weeks old are set out in the soil, which should be at pH 6.5 to 8.0. Broccoli should be harvested prior to the opening of the green buds.

Insects and diseases associated with broccoli are those that attack cabbage (which see). Popular cultivars include 'Atlantic,' 'De Cicco,' 'Green Comet,' 'Italian Green Sprouting,' 'Neptune,' 'Premium Crop,' and 'Spartan Early.'

BRUSSELS SPROUTS

Brussels sprouts are noted for the axillary buds on the main stem that develop into miniature heads. They are a cole crop and are hardier than cabbage. In the North they can be grown as a spring or late-summer crop, and in addition to these as a winter crop in the South. However, the best-quality Brussels sprouts are obtained when days are sunny and light frosts occur during the night. Because of this, Brussels sprouts are favored as a fall crop. California is the major commercial source. The harvest period is about 6 to 8 weeks, since the harvesting of the mature lower sprouts (1.0 to 1.25 inches in diameter) and the removal of the associated leaves does not end the usefulness of the plant. Further sprouts higher up the main stem develop, since the terminal crown maintains the vigor of the plant. Treatment with butanedioic acid mono-(2,2'-dimethylhydrazide) produces a larger number of simultaneous sprouts on a more compressed stem for mechanical harvest.

Transplants are usually produced in an outdoor seedbed or coldframe. Soil pH should be between pH 6.5 to 8.0. Transplants are put in the permanent field in late spring through summer. Disease and insect problems are those of cabbage (which see).

Suggested cultivars include 'Half Dwarf,' 'Improved Long Island,' and 'Jade Cross.'

CABBAGE

Cabbage, a cole crop noted for its heads of tightly folded, fleshy leaves, is a cool-season crop. Heads vary in shape from conical to round and flat globe. It is generally treated as a spring, early summer, and fall crop in most of the United States. In the South it can be carried over as a winter crop. The main states for commercial production are Texas, Florida, New York, Wisconsin, California, and North Carolina.

The spring crops of cabbage are frequently grown from transplants started 4 to 6 weeks earlier. Transplants can be set out when hard frosts cease. Later cabbage is either seeded directly in place or grown from transplants produced in an outdoor seedbed. Late cabbage can be started between rows of potatoes prior to harvest, or it can follow early potatoes, spinach, beets, peas, radishes, and other early crops. Cabbage is harvested when the head is firm and white and weighs 2 pounds or more.

Several insects attack cabbage and other cole crops. These include the cabbage worm (*Pieris rapae*), cabbage looper (*Trichoplusia ni*), aphid (*Brevicoryne brassicae*), root maggot (*Hylemya brassicae*), and black or red harlequin bugs (*Murgantia histrionica*). Diseases include the fungal-caused clubroot (*Plasmodiophora brassicae*), which can be controlled by maintaining a soil pH of 6.5 to 8.0. If clubroot is not present, pH of 6.0

Fig. 14-9. (above) Brussels sprouts showing foliage and edible axillary buds on the main stem. U.S. Department of Agriculture photo.

Fig. 14-10. (right) Cabbage foliage heads ready to harvest. U.S. Department of Agriculture/University of Maryland Vegetable Research Farm photo.

to 6.5 is optimal. Other diseases include black rot (*Xanthomonas campestris*), blackleg (*Phoma lingam*), and cabbage yellows (*Fusarium oxysporum*). Some cultivars possess varying degrees of disease resistance.

Cabbage cultivars for early crops include 'Copenhagen Market,' 'Early Jersey Wakefield,' 'Emerald Cross,' 'Golden Acre,' 'Resistant Golden Acre,' 'Stonehead,' and 'Superette.' Midseason and later cabbages include 'Globe,' 'Green Parade,' 'King Cole,' 'Red Acre,' 'Red Head,' 'Roundup,' 'Savoy Ace,' and 'Savoy King.' The late fall or winter cabbages are 'Danish Roundhead,' 'Green Winter,' 'Penn State Ballhead,' 'Rio Verde,' and 'Storing Strain 4409.'

Fig. 14-11. Freshly picked carrots showing foliage and edible roots. U.S. Department of Agriculture photo.

CARROT

Carrots are root crops grown for their edible, fleshy taproots, which vary from long and pointed through blunt and cylindrical. They are hardy and make a good fall, winter, and spring crop in the South, and a good summer through fall crop in the North. Best growth occurs when the temperatures range between 60° and 70°F (15.5° and 21.2°C). California and Texas are the main commercial producers.

Soil should be in the pH range of 6.0 to 8.0 and of a deep, friable nature free of debris and rocks. Otherwise, the roots will be of poor shape. Carrots can be started in early spring and sown at 3-week intervals for successive harvests. Carrots can be harvested from finger size to maturity.

The carrot rust fly (*Psila rosae*) is a serious insect pest. Proper timing as to sowing can minimize damage by the larvae of this fly; otherwise, control by insecticides is necessary. Diseases include leaf blight (*Cercospora carotae, Alternaria dauci*) and carrot yellows. The latter is a virus transmitted by a leaf hopper.

Carrot cultivars of choice are 'Danvers Half Long,' 'Gold Pak,' 'Imperator,' 'Nantes Half Long,' 'Oxheart,' 'Red Cored Chantenay,' 'Short 'n Sweet,' and 'Spartan Bonus.'

CAULIFLOWER

Cauliflower, a cole crop, is not quite as hardy as cabbage. It is also not as heat tolerant as cabbage and will not form heads if the temperatures are too high. Cauliflower is noted for its large edible head, which consists of a condensed, thickened malformed flower cluster. These should be harvested before they discolor and become loose.

Introduction to Horticultural Production / Part III

Fig. 14–12. Cauliflower showing foliage and edible flower cluster. U.S. Department of Agriculture photo.

Temperature requirements make it a spring, fall, and winter crop in the South, and a spring or fall crop in the North. California is the leading commercial producer. Cauliflower is often started from 4- to 6-week-old transplants set out in the spring or by direct seeding in the late spring through early summer. Soil pH can range from 6.0 to 8.0.

Heads of early cultivars are protected from sun injury by tying the leaves over the heads or buttons as they begin to form. Later cultivars often have incurving inner leaves, which makes this practice unnecessary. Insects and disease are the same as described for cabbage (which see).

Recommended cultivars include 'Early Snowball,' 'Purple Head,' 'Snow Crown,' 'Snow King,' and 'Super Snowball.'

CELERIAC AND CELERY

Celeriac is a root crop noted for its thickened, edible root. Celery, a salad crop, is noted for its edible fleshy petioles. Both are cool-weather but tender crops and are usually grown as winter and spring crops in the South and as spring or fall crops in the North. California and Florida lead in commercial production of celery. Celeriac is mostly a home garden product.

Soil should be rich, moist, and well drained. Soil pH can be from 6.0 to 8.0. Celery or celeriac is usually started outdoors with transplants that are 8 to 10 weeks old. Germination of seeds can be speeded by overnight soaking in water. Early celery is usually blanched in the field by mounding soil. Late celery may be partially field blanched and the final blanching completed in storage. Blanching reduces the vitamin A content somewhat and green celery is now preferred, so blanched celery is less common today. Celery can be harvested before it reaches full size if desirable.

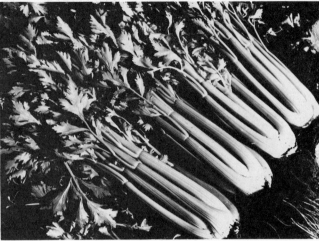

Fig. 14-13. A. Celeriac 'Alabaster' showing foliage and the edible root. Courtesy of Burpee Seeds. B. Celery showing foliage and edible petiole. U.S. Department of Agriculture photo.

Insects that attack celery include the carrot rust fly (*Psila rosae*) and the tarnished plant bug (*Lygus lineolaris*). Diseases include early and late blight (*Cercospora apii* and *Septoria petroselini*, respectively), bacterial leaf spot (*Pseudomonas apii*), root rot (*Phoma apiicola*), yellows (*Fusarium oxysporum*), and pinkrot (*Sclerotinia sclerotiorum*). Physiological disorders include black heart and cracked stem. The first results from calcium deficiency and nutrient imbalance, and the second from boron deficiency.

Good celery cultivars include 'Fordhook,' 'Giant Pascal,' 'Golden Plume,' 'Golden Self Blanching,' 'Summer Pascal,' 'Tendercrisp,' and 'Utah.' 'Alabaster' is a celeriac cultivar.

Fig. 14-14. Chard showing edible foliage. U.S. Department of Agriculture photo.

CHARD

Chard or Swiss chard is one of the vegetables termed greens. It is a type of beet grown for its edible leaves. If the outer leaves only are removed, the harvest period can be extended throughout the summer, since chard is well adapted to hot weather. Usually, only one sowing is necessary.

The culture and pests for chard are the same as for beet (which see). Cultivars include 'Fordhook Giant,' 'Lucullus,' and 'Vintage Green.'

Fig. 14-15. Chinese cabbage, heading culti-var. Courtesy of National Garden Bureau, Inc.

CHINESE CABBAGE

Chinese cabbage, a cole crop, is grown for its heads of leaves. The heading type is more commonly cultivated in the United States, and the nonheading cultivars are seen less frequently. It is a cool-weather crop primarily grown as a fall crop in the North and as a winter crop in the South.

It can be grown by seeding directly outdoors or from 5- to 6-week-old transplants. If planted too early, hot weather will force flowers before the head is fully developed. In addition, hot weather may adversely affect texture and flavor. Soil pH should be between 6.0 and 7.0.

Insects and diseases are similar to those of cabbage (which see). Suggested head-forming cultivars are 'Burpee Hybrid,' 'Michihli,' 'Nagaoka,' and 'Wong Bok.' 'Crispy Choy' is a nonheading cultivar.

COLLARD

Collard, a cole crop, is essentially a large form of kale, both of which are noted for their nonheading, loose leafy crowns used as greens. Collard is somewhat more tolerant of heat than cabbage. It can be grown as a spring and fall crop in the North, and as a fall, winter, and spring crop in the South. Its main use is as a winter crop in the South.

Soil pH should be 6.0 to 8.0. Seeds are usually sown directly in place, although 5- to 6-week-old transplants can be used. Leaves may be partially removed to extend the harvest. Leaves should be harvested before becoming tough and woody. Insects and diseases are similar to those of cabbage (which see). Suggested cultivars are 'Georgia,' 'Green Glaze,' 'Morris Heading,' and 'Vates.'

Fig. 14-16. (left) Collard, showing edible leafy crown. U.S. Department of Agriculture photo.

Fig. 14-17. (below) A well-filled ear of sweet corn. U.S. Department of Agriculture photo by T. O'Driscoll.

CORN (SWEET)

Sweet corn or sugar maize is a warm-season, frost-sensitive plant grown in most parts of the United States. It is noted for its edible, sweet, immature seed ears. Wisconsin, Minnesota, and Florida are the leading commercial states.

Corn is usually directly seeded for a summer and fall crop. Soil pH should be between 6.0 and 7.0. The harvest season is brief because of texture changes and enzymic conversion of sugar to starch. An extended harvest can be obtained through the use of early, midseason, and late cultivars or succession plantings at 2-week intervals. Some sowings can be made shortly before or after harvest of early crops such as peas, beets, radishes, lettuce, and others. Harvest is best when the silk first browns and the kernel juice is milky.

A number of insects attack corn: corn earworm (*Heliothis zea*), European corn borer (*Pyrausta nubilalis*), southern corn borer (*Diatraea crambidoides*), wireworm (*Agriotes mancus*), corn flea beetle (*Chaetochema ectypa*), army worm (*Pseudaletia unipuncta*), and chinch bug (*Blissus leucopterus leucopterus*). Birds, such as crows and starlings, frequently dig and eat freshly sown seed. Diseases include bacterial wilt (*Bacterium stewartii*), corn smut (*Ustilago maydis*), and corn blight (*Helminthosporum turcicum*). Resistant cultivars are available. Racoons are troublesome.

There are many hybrids of sweet corn. A few of the yellow hybrids are 'Earlibelle,' 'Early Golden Giant,' 'Early Sunglow,' 'F-M Cross,' 'Golden Beauty,' 'Golden Cross Bantam,' 'Illini Xtra-Sweet,' 'Ioana,' 'Iochief,' 'Mainliner E.H.,' and 'Spancross.' White hybrids are 'Country Gentlemen Hybrid,' 'Silver Queen,' 'Silver Sweet,' and 'Stowell's Evergreen Hybrid.' Mixed yellow-white hybrids are 'Bi-Queen,' 'Butter and Sugar,' and 'Honey and Cream.'

CRESS

Upland cress and watercress are salad crops noted for their edible leaves. Both are hardy. Upland cress is sown early in the spring and planted at 2-week intervals since it is a short-season crop. It will persist through the winter in the moderate regions of the United States.

Watercress is grown in wet surroundings, such as in shallow springs, brooks, or on river edges. It can be started from seeds or cuttings in the early spring, and once established will remain, since it is a perennial.

CUCUMBER

The cucumber is a warm-weather vine crop or cucurbit that does not withstand frost. Cucumbers can be grown over most of the United States. North Carolina, Michigan, and Florida are leading commercial producers. Cucumbers are grown for their edible fruits. There is also a parthenocarpic (seedless) cucumber with edible skin known as the European type. Cucumbers are the third most important vegetable in terms of commercial greenhouse production.

Cucumbers can be sown directly after all danger of frost is past, or transplants may be used if minimal root disturbance occurs. Germination is poor below 50°F (10°C). Soil pH can range from 5.5 to 8.0; 5.5 to 6.8 is best. A second sowing 4 to 5 weeks later can be used to extend the harvest season. If space is limited, cucumbers can be grown on fences, trellises, nylon net, and the like. Cucumbers should be picked when dark green and not overly large; otherwise, seeds become more noticeable.

A number of insects attack cucumbers. They are the striped cucumber beetle (*Acalymma vitlata*), which carries bacterial wilt and cucumber mosaic, spotted cucumber beetle (*Diabrotica undecimpunctata howardi*), which carries bacterial wilt, aphid (*Aphis gossypii*), and pickle worm (*Diaphania nitidalis*). Diseases include powdery mildew (*Erysiphe cichoracearum*), anthracnose (*Colletotrichum lagenarium*), scab (*Cladosporium cucumerinum*), angular leaf spot (*Pseudomonas lachrymans*), bacterial wilt (*Erwinia tracheiphila*), mosaic (a virus spread by aphids mostly), downy mildew (*Pseudoperonospora cubensis*), and leaf blight (*Alternaria cucumerina*). Some cultivars have multiple disease resistance.

Cultivars for pickling include 'Bounty,' 'Burpee Pickler,' 'Explorer,' 'Green Beauty,' 'Liberty Hybrid,' 'National Pickling,' 'Ohio MR17,' 'Pixie,' 'Saladin Hybrid,' and 'Wisconsin SMR 18.' Cultivars for slicing are 'Ashley,' 'Burpee Hybrid,' 'Challenger,' 'Cherokee 7,' 'Green Knight,' 'Marketmore,' 'Marketer,' 'Meriden T,' 'Poinsett,' 'Spacemaster' (a bush cultivar), 'Tablegreen,' and 'Victory.' 'Uniflora D' is a European type.

Fig. 14–18. (left) Cucumber showing foliage and edible fruit. Author photo.

Fig. 14–19. (below) Eggplant, showing foliage and edible fruit. U.S. Department of Agriculture photo.

Introduction to Horticultural Production / Part III

EGGPLANT

The eggplant is a warm-season solanaceous crop, which is frost sensitive. It is noted for its edible fruit. Florida and New Jersey are the leading states for commercial production.

The longer growing season and warmer temperatures of the South are ideally suited to the culture of eggplant. There the seeds may be sown directly in an outdoor seed bed and transplanted to the garden in about 6 to 8 weeks. In the North, eggplants must be started indoors about 8 weeks before the plants are set in place. All danger of frost must be past, and the daily mean temperature should be 60°F (15.5°C) or higher. Soil pH should be 5.0 to 6.0. Fruits should be harvested when shiny, as opposed to dull, to avoid hard seeds.

Insects that attack eggplant are the Colorado potato beetle (*Leptinotarsa decemlineata*), flea beetle (*Epitrix fuscula*), and aphid (*Aphis gossypii, Myzus persicae,* and *Macrosiphum euphorbiae*). Troublesome diseases include wilt (*Verticillium, Fusarium*) and blight (*Phomopsis vexans*). Some cultivars have disease resistance.

Earlier-maturing eggplants recommended for the North include 'Black Magic,' 'Early Beauty,' and 'New Hampshire.' Later cultivars are useful in the South: 'Black Beauty,' 'Burpee Hybrid,' 'Dusky,' 'Florida High Bush,' 'Ichiban,' and 'Jersey King.'

ENDIVE

Endive is a hardy, cool-season salad crop noted for its rosette of edible leaves. It is somewhat less sensitive to heat than lettuce. Both fringed or curled and broad-leaved cultivars exist. The latter is often referred to as escarole. Florida and New Jersey are the leading producers.

In the South, endive is mainly a winter crop, and in the North, a spring, summer, and fall crop. Endive is usually sown directly in place, and successive sowings at 2-week intervals are used to extend the harvest period. Endive can be blanched (if bitterness is objectionable) by loosely tying the leaves together. Insect and disease problems are minimal in endive.

The main cultivar for the fringed or curled leaves is 'Green Curled,' and for the broad leaves, 'Broad Leaved Batavian.'

GARLIC

Garlic is a hardy plant noted for its bulb, which can be broken apart into cloves. It can be grown successfully in both the North and South. Propagation is by cloves in the early spring. California is the main producer.

Culture and pests are similar to onion (which see). Early cultivars are of the white or Mexican type, and late ones the pink or Italian type. Harvest is after the leaves have wilted.

KALE

Kale, a cole crop, is a cool-season vegetable grown for its edible leaves, which do not form heads and are treated as greens. It is hardy and responds poorly to hot weather.

In the North, kale is grown as a spring and fall crop, and as a fall, winter, and spring crop in the South. Kale can follow early vegetables such as peas, radishes, early potatoes, and green beans. Soil pH should be 6.0 to 8.0. Seeds are usually sown in place and thinned, although 5- to 6-week-old transplants can be used. As with collard, kale may be partially or completely harvested. Generally harvest is best before the plants become large and tough.

Insect and disease pests are similar to those of cabbage (which see). Recommended cultivars include 'Blue Curled Scotch,' 'Curled Vates,' 'Dwarf Blue,' 'Dwarf Blue Scotch,' and 'Dwarf Siberian.'

Fig. 14-20. A. Kale, curly-type cultivar. Courtesy of National Garden Bureau, Inc. B. Kale, Scotch-type cultivar. U.S. Department of Agriculture photo.

(A)

KOHLRABI

Kohlrabi is a cole crop grown as a hardy cool-season vegetable. It is noted for the edible swollen portion of the stem near ground level, which should be harvested before it becomes tough and stringy (at most 3 inches in diameter).

In the North, kohlrabi is grown as a spring or fall crop, and as a spring, fall, and winter crop in the South. Seeds are usually sown in place, although 5- to 6-week-old transplants can be used. Soil pH should be in the 6.0 to 8.0 range.

Insect and disease problems are the same as for cabbage (which see). Cultivars include 'Earliest Erfurt,' 'Grand Duke,' 'Prima,' 'Primavera,' 'Purple Vienna,' and 'White Vienna.'

Fig. 14-21. (left) Kohlrabi 'Grand Duke' Hybrid. Courtesy of All-America Selections.

Fig. 14-22. (below) Leek, showing foliage and part of edible slight, soft bulb. U.S. Department of Agriculture photo.

LEEK

The leek is hardier than its relative, the onion. It is noted for its slight, soft bulb and sheaf of leaves. It is usually sown directly in place during the early spring, since it is a long-season crop. Soil pH should be 6.0 to 8.0. Leeks can be blanched by mounding soil at their base. They can be harvested when they become 1 to 1.5 inches in diameter.

Insects and diseases are similar as for onions (which see). Suggested cultivars are 'American Flag,' 'Broad London,' 'De Perpignan,' 'Monstrous Carentan,' and 'The Lyon.'

LETTUCE

Lettuce is a cool-season salad crop of good hardiness, but poor heat resistance. It is noted for its edible leaves. There are three basic types: head, which varies from firm and hard to loose and soft; leaf, clusters or bunches of leaves; and cos (romaine), long upright cylindrical heads. These are listed in order of increasing tolerance toward heat.

Fig. 14-23. Lettuce. A. Cultivar with firm, hard head. American Society for Horticultural Science photo. B. Cultivar with loose, soft head. C. 'Salad Bowl', a leaf lettuce. B and C courtesy of National Garden Bureau, Inc. D. 'Parris Island Cos', a cos lettuce. Courtesy of Herbst Brothers Seedsmen, Inc.

(A)

(B)

(C)

(D)

In the North, lettuce is a spring and fall crop, and a fall, winter, and spring crop in the South. California is the leading state in commercial production. Lettuce can be sown directly in place or 6-week-old transplants can be used. Soil pH can vary from 5.5 to 7.0. Successive plantings of the leaf type at 2-week intervals are often used to extend its harvest period. Early lettuce can be followed by a succession crop, such as cabbage, beet, celery, or green beans. Sometimes lettuce is grown as a companion crop with low-growing, longer-season crops, such as the cole crops. Lettuce is the second largest commercial vegetable crop grown in greenhouses.

There are few insects that defoliate lettuce, but some carry diseases. Troublesome diseases include lettuce drop (*Sclerotinia libertiana*), bottom rot (*Rhizoctonia solani*), downy mildew (*Bremia lactucae*), gray mold rot (*Botrytis cinerae*), and yellows (a virus carried by leaf hoppers). Tip burn is a physiological disorder resulting from high temperatures and low soil moisture.

Popular cultivars are as follows. Head-lettuce cultivars include 'Bibb,' 'Buttercrunch,' 'Crisphead,' 'Fordhook,' 'Fulton,' 'Great Lakes,' 'Imperial,' 'Pennlake,' 'Premier Great Lakes,' and 'White Boston.' Leaf-lettuce cultivars include 'Black-Seeded Simpson,' 'Grand Rapids,' 'Green Ice,' 'Oak Leaf,' 'Royal Oak,' 'Ruby,' 'Salad Bowl,' and 'Slobolt.' Cos cultivars include 'Dark Green,' 'Little Gem,' 'Parris White,' and 'Valmaine.'

MELON

Melons are warm-season vine crops (cucurbits) noted for their fruits. As treated here, the term melon includes the netted (muskmelon) and winter (honeydew, casaba) melons. The watermelon is treated separately. The term cantaloupe is often incorrectly used to indicate muskmelons. True cantaloupes are seen more in European horticulture. California leads in production, followed by Texas and Arizona.

Melons are seeded directly in place in the warmer areas. In the North, where the growing season is short, 4- to 5-week-old transplants can be used, if transplanting shock is minimal. Soil pH can be between 6.0 and 7.0. Clear or black (slower) plastic mulch can be used in the North to produce a soil temperature better suited for melon culture earlier than would occur naturally. The best muskmelons are harvested when the stem separates easily and cleanly from the melon.

Insect pests include those that attack cucumbers (which see). Diseases include fusarium wilt (*Fusarium oxysporium*), powdery mildew, downy mildew, and bacterial wilt (see cucumber).

Muskmelon cultivars include 'Ambrosia Hybrid,' 'Burpee Hybrid,' 'Delicious,' 'Fordhook Gem,' 'Granite State,' 'Hearts of Gold,' 'Luscious,' and 'Sampson Hybrid.' 'Golden Beauty' is a casaba cultivar; 'Honey Dew,' 'Honey Dew Green,' and 'Honey Drip' are honeydew cultivars.

(A)

(B)

Fig. 14-24. Melon. A. 'Earlisweet', a hybrid melon of the netted type. Courtesy of George J. Ball, Inc. B. 'Earli-Dew', a hybrid melon of the winter type. Courtesy of Petoseed Co., Inc. Wholesale Breeders • Growers.

MUSTARD

Mustard is a hardy plant grown for its edible leaves used as greens; some species are grown for their oil seeds used in the preparation of table mustard.

In the North it is treated as a spring or fall crop, and as a fall, winter, and spring crop in the South. Seeds are sown directly in place at 2-week intervals to provide successive harvests. Soil pH can be between 6.0 to 8.0.

Insect and disease problems are minimal. Suggested cultivars include 'Florida Broad Lead,' 'Fordhook,' 'Southern Giant Curled,' and 'Tendergreen.'

Fig. 14-25. Mustard showing edible leaves. Courtesy of National Garden Bureau, Inc.

OKRA

Okra, also known as gumbo, is a warm-season crop. It is grown for its edible, immature seed pod, especially in the South.

Seeds are planted directly in place after all danger of frost is past. Soil pH should be between 6.0 to 8.0. Pods should be picked to ensure a continuing supply. Insect and disease problems are minimal. Suggested cultivars include 'Clemson Spineless,' 'Dwarf Green,' 'Emerald,' 'Red Okra,' and 'Spineless Green Velvet.'

Fig. 14-26. Okra, showing foliage and edible immature seed pod. U.S. Department of Agriculture photo.

ONION

The onion is a hardy, cool-season crop grown for its edible bulb. Best conditions are good moisture and no extremes of heat or cold (55° to 75°F, 12.8° to 24°C). Texas, California, and New York lead in commercial production.

In the North, onions are a spring, summer, and fall crop, and a fall, winter, and spring crop in the South. Onions may be directly seeded in the garden or started from sets (immature onion bulbs less than ¾ inch in diameter) or grown from seedling plants. Soil pH should be from 6.0 to 6.5. Seeds or sets can be put out as early as the soil is workable, but seedlings should be delayed until all danger of frost is past. Onions may be harvested at pencil size or larger as green bunching onions. Mature onions are harvested when the tops fall over.

Several insects attack onions. The worst are the onion maggots (*Hylema antigua*) and onion thrips (*Thrips tabaci*). Troublesome diseases include onion smut (*Urocystis cepulae*), downy mildew (*Peronospora destructor*),

Fig. 14-27. Onion, showing foliage and edible bulb. Courtesy of National Garden Bureau, Inc.

bacterial soft spot (*Erwinia carotovora*), basal rot (*Fusarium oxysporum*), pink root rot (*Pyrenochaeta terrestis*), neck rot (*Botrytis* sp.), and yellow dwarf and yellows (viruses).

The time of bulb formation is a function of day length and is influenced somewhat by temperature. As such, cultivars tend to be regionally specific. This should be borne in mind when a cultivar is selected. A list of some cultivars, without regard to region or color (yellow, white, red), follows: 'Crystal White Wax,' 'Early Yellow Globe,' 'Ebenezer,' 'Empire,' 'Fiesta,' 'FM Hybrid,' 'Red Hamburger,' 'Southport Red Globe,' 'Southport Yellow Globe,' 'Sweet Spanish,' 'White Portugal,' 'White Sweet Spanish,' and 'Yellow Bermuda.'

PARSNIP

The parsnip is a hardy root crop that can be grown over much of the United States. However, extremes of hot weather cause poor seed germination and poor-quality roots. In the South the parsnip should be grown such that maturity occurs during the spring or early summer.

Soil can be between pH 6.0 and 8.0, and it should be free of debris or misshapen roots will be formed. Parsnip seed has limited viability, so only fresh seed should be used. Parsnips are sown directly in place. Parsnips can be harvested after frosts since cold improves their sweetness.

Insects and diseases are the same as those for carrot. Suggested cultivars are 'All American,' 'Avonresister,' 'Guernsey,' 'Hollow Crown,' 'Large Sugar,' 'Model,' and 'Offenham.'

Fig. 14-28. Parsnip, showing root shapes allowed in U.S. No. 1 grade. U.S. Department of Agriculture photo.

PEA

The pea is a hardy, cool-season legume noted for its edible green seeds. One variety is noted for both its edible pods and seeds. Dry mature pea seeds are also used in the production of pea soup. Wisconsin, Minnesota, and Washington lead in commercial production.

Peas are grown as a spring and fall crop in the North, and as a fall, winter, and spring crop in the South. Seeds are sown directly in place as early as the soil can be worked. A few succession plantings at 10-day intervals or planting both early- and late-maturing cultivars can be used to extend the harvest period. Soil pH can be from 5.5 to 6.5. The best yields

Fig. 14-29. 'Sugar Snap', an edible podded snap pea cultivar. Courtesy of All-America Selections.

are obtained from the spring crop in the North and the winter-spring crop in the South. The taller cultivars can be grown on brush, trellises, or netting to conserve space. An inoculum of *Rhizobium* sp. can be beneficial for yield, if they are not present in the soil. Pods should be picked when well developed and fully green, but before the peas harden.

Troublesome insects include the pea aphid (*Acyrthosipon pisi*), and pea weevil (*Bruchus pisorum*). Diseases include root rot (*Fusarium solani*), powdery mildew (*Erysiphe polygoni*), wilt (*Fusarium oxysporium*), and bacterial blight (*Pseudomonas pisi*). Some disease resistance can be found in certain cultivars. Garden pea cultivars include 'Alaska,' 'Blue Bantam,' 'Burpeeana Early,' 'Early Frosty,' 'Freezer 69,' 'Freezonian,' 'Frosty,' 'Gloriosa,' 'Green Arrow,' 'Hurst Beagle,' 'Little Marvel,' 'Onward,' 'Recette,' 'Sparkle,' 'Thomas Laxton,' 'Wando,' 'Waverex,' and 'Winfrida.' Edible-podded pea cultivars include 'Dwarf Gray Sugar,' 'Oregon Sugar Pod,' 'Sugar Snap,' and 'Sweetpod.'

The cowpeas or blackeyed peas grown in the South are not true peas, but are derived from *Vigna unguiculata*. They are tender and should not be planted until the danger of frost is past. Cultivars include 'Brown Crowder,' 'California Blackeye,' 'Dixilee,' 'Magnolia Blackeye,' 'Monarch Blackeye,' 'Queen Anne,' and 'White Crowder.' Nematodes can be a serious problem with these peas, although some cultivars have resistance to nematodes. The cowpea aphid (*Aphis crassivora*) is another pest.

PEPPER

Garden peppers are a frost-sensitive, warm-season solanaceous crop. They are noted for their green (immature) and yellow (mature) or red (mature) edible fruits, which vary in size and shape. The pungency also varies from mild to very hot. Florida, California, North Carolina, Texas, and New Jersey are leading commercial producers.

Peppers must be started indoors in the North, and 6- to 7-week-old transplants are set out when all danger of frost is past. In the South the seeds can be sown outdoors in a seed bed after all danger of frost is past, and the 6-week-old plants can be transplanted to the field. Blossoms fail to set fruit well if the temperatures drop below 55°F (13°C), or the temperatures are high for a prolonged period, or humidity or soil moisture is low, or if excess soil nitrogen is present. Soil pH can vary from 5.5 to 7.0. Green peppers are picked full sized, but before they become ripe (red or yellow); peppers are also utilized ripe.

Insect pests include the aphid (*Macrosiphum euphorbiae*), flea beetle, cutworms, common stalk borer (*Papaipema nebris*), pepper maggot (*Zonosemata electa*), pepper weevil (*Anthonomus eugenii*), and leaf miner (*Liriomyza brassicae*). Diseases include bacterial spot (*Xanthomonas vesicatoria*), blight (*Sclerotium rolfsii*), anthracnose (*Glocosporium piperatum*), tobacco mosaic virus, cucumber mosaic virus, and curly top virus.

Fig. 14-30. 'Bell Boy', a hybrid pepper. Courtesy of Petoseed Co., Inc. Wholesale Breeders • Growers.

The viruses are spread by aphids and leafhoppers. Physiological disorders that trouble tomatoes, blossom end rot and sun scald, also are found with peppers.

Cultivars of the mild or sweet bell-shaped form, usually eaten green or sometimes red, include 'Ace,' 'Bell Boy,' 'Bell Hybrid,' 'Better Belle,' 'Burpee Tasty Hybrid,' 'California Wonder,' 'Early Prolific,' 'Fordhook,' 'Golden Calwonder' (yellow), and 'Sunnybrook.' Long, pointed, mild cultivars include 'Aconcagua,' 'Cubanelle,' 'Long Yellow Sweet,' and 'Sweet Banana.' Small, hot cultivars are 'El Cid,' 'Mexican Chilhi,' and 'Tabasco.' Larger, long hot cultivars include 'Anaheim,' 'Jalapeno M,' and 'Long Red Cayenne.'

POTATO

The potato is a cool-season crop, but it only has moderate tolerance toward frost. It is noted for its edible, underground modified stem known as a tuber. Idaho, North Dakota, Maine, and Washington are the leading commercial sources.

Since tubers are formed in preference to vegetative growth only when days are on the shorter side with temperatures of 60° to 65° F (15.6° to 18.4°C), potatoes do not succeed as well in the South as in the North during midsummer. (Certain cultivars are recommended therefore for the South.) Potato cultivars of the early type can be planted 10 to 14 days before the last killing frost in the spring. Late cultivars can be planted a few or more weeks later. The tuber is cut into pieces, which have one or two buds ("eyes"). These seed pieces are cured by being placed in a cool

Fig. 14-31. Potato, showing edible tuber. U.S. Department of Agriculture photo.

(50° to 65°F, 10.0° to 18.4°C), dry area for 2 to 3 days until suberization of the cut surface occurs. Soil pH can vary from 5.0 to 6.0. Certified disease-free tubers should be utilized. Soil is hilled along the rows before the plant reaches a height of 10 inches. This prevents exposure of the tubers to light, which would cause greening of the tubers. Solanine, a toxin, is found in the green tissue. Potatoes are harvested after the vines mature and die.

A number of insects attack potatoes. These include Colorado potato beetle (*Leptinotarsa lineata*), leafhopper (*Empoasca fabae*), flea beetle (*Epitrix cucumeris*), aphid (*Macrosiphum euphorbiae*), and wireworm. Diseases associated with potatoes include scab (*Streptomycies scabies*), early blight (*Alternaria solani*), late blight (*Phytophthora infestans*), *Verticillium* wilt, black leg (*Erwinia astroseptica*), mosaic, and leaf roll. The last two are viruses spread by aphids. Cultivars are available with varying degrees of disease resistance.

Red-skinned cultivars include 'Chieftain,' 'Norland,' 'Red La Soda,' and 'Red Pontiac.' White-skinned cultivars include 'Chippewa,' 'Idaho,' 'Irish Cobbler,' 'Kennebec,' 'Norchip,' 'Russet Burbank,' 'Sebago,' 'Superior,' and 'Wauseon.' In the South, 'Early Gem,' 'Irish Cobbler,' 'Pungo,' 'Red La Soda,' and 'Red Pontiac,' are widely used.

PUMPKIN

The pumpkin is a cucurbit or vine crop that is sensitive to both frost and extreme heat. It is noted for its rounded, or at least symmetrical, fruit with orange to pale-orange flesh. The lines of distinction between winter squash and pumpkins are not exact, since both pumpkins and winter squashes can be found in all four cited species of *Cucurbita* in Table 14-1.

Since they are frost sensitive, pumpkins cannot be seeded outdoors until all danger of frost is past. They can also be started indoors and put out as 2-week-old transplants, if care is taken to minimize transplanting shock. Soil pH can range from 5.5 to 7.5. Pumpkins require much space and give minimal returns, so they are often grown as a companion crop with corn, since pumpkins can tolerate partial shade. Pumpkins can also be put in after the removal of early crops, such as radishes or lettuce. Pumpkins are picked when fully matured.

Insects that trouble pumpkins are those that affect squashes (which see). They include the striped cucumber beetle (carrier of bacterial wilt), squash bug, squash vine borer, and aphid. Diseases of concern are powdery mildew and sometimes wilt.

Pumpkin cultivars include the following: 'Big Max,' 'Cinderella' (bush form), 'Connecticut Field,' 'Dickinson,' 'Japanese Pie,' 'Kentucky Field,' 'Spirit Hybrid,' 'Small Sugar,' and 'Triple Treat.'

Fig. 14-32. (left) Pumpkin, showing edible fruit. Courtesy of National Garden Bureau, Inc.

Fig. 14-33 (below) Radishes, showing edible roots. U.S. Department of Agriculture photo.

RADISH

The radish is a root crop noted for its edible storage root. It is a cool-season crop that cannot tolerate heat. Radishes are grown during the spring and fall in the North, and in the fall, winter, and spring in the South.

Radishes can be directly sown in place as soon as soil is workable. The soil pH can range from 6.0 to 8.0. Radishes mature quickly and remain at optimal quality briefly, so succession sowings at 7–to 10–day intervals are possible. Radishes can be followed by other succession crops or interplanted with other crops requiring longer to reach maturity, such as carrots. The only pest of consequence is the radish maggot (*Hylemya brassicae*).

Suggested cultivars include 'Champion,' 'Cherry Belle,' 'Comet,' 'French Breakfast,' 'Scarlet Globe,' 'Sparkler,' and 'White Icicle.'

Fig. 14–34. Rhubarb 'Valentine'. William P. Raffone, Jr. photo.

RHUBARB

Rhubarb is a hardy perennial noted for its edible petiole or leaf stalk produced in the spring. The best rhubarb is grown where summers are cool and moist and winter freezes prevail. The mean summer and winter temperatures, respectively, should not exceed 75°F (24°C) and 40°F (4.5°C). As such, it is not suitable for cultivation in much of the South.

The fleshy roots can be divided in the spring. This is the method of choice for propagation, since seedlings do not come true. Leaf stalks should not be pulled the first year, only sparingly the second year, and more heavily thereafter. Flower stalks should be removed as they appear. Leaves are toxic as they contain oxalic acid. Rhubarb is relatively pest free. Soil pH can be 6.0 to 8.0. Leafstalks should be harvested in early spring when tender.

Suggested cultivars include 'MacDonald,' 'Valentine,' and 'Victoria.'

RUTABAGA

The rutabaga is a hardy, cool-season cole crop noted for its edible storage root. It is somewhat similar to turnip, but is a bit larger, more pungent in flavor, and requires a longer growing period. In the North,

Fig. 14-35. Rutabaga 'American Purple Top'.
U.S. Department of Agriculture photo.

rutabagas are generally grown as a fall crop. Turnips usually do better in the South than rutabaga. Harvest is usually when full size is reached.

Seeds are sown in place around very late spring to very early summer. Soil pH can be from 6.0 to 8.0. Because the rutabaga is from the cole group, it can be troubled by insects and diseases associated with cabbage (which see).

Suggested cultivars include 'American Purple Top,' 'Long Island Improved,' and 'Pandur.'

SALSIFY

Salsify or oyster plant is a hardy crop noted for its edible root. It can be grown in most parts of the country, except in the extremes of the North, as it requires a long growing season. Seed should be sown in place as soon as the soil is workable. Soil pH can be between 6.0 and 8.0. Insect and disease pests are minimal. The suggested cultivar is 'Sandwich Island.'

Fig. 14-36. Salsify 'Sandwich Island Mammoth'. Courtesy of Burpee Seeds.

Fig. 14-37. Shallot ready to use at immature (green onion) stage. Courtesy of Henry Field Seed & Nursery Co.

SHALLOT

The shallot is one of the onion crops. However, it is less noted for its bulbs, which are small and multiple, but more for its use in the immature stage when the young plants are used as green onions. Being hardy, shallots are sown as soon as the ground can be worked in the spring. They are planted in place by the small bulbs, as for onion sets. Soil pH can be between 6.0 and 7.0. Insect and disease pests are those of onions (which see). Cultivars include 'Dutch Yellow' and 'Giant Red.'

Fig. 14-38. Soybean showing foliage, seed pods, and edible seeds. U.S. Department of Agriculture photo.

SOYBEAN

The soybean is a tender, warm-season legume crop grown for its edible immature or mature seeds. It is planted after all danger of frost is past. It can be grown in most areas of the United States.

Soybeans are sown in place. An inoculant of *Rhizobium* sp. can be beneficial by allowing nitrogen fixation, if nitrogen-fixing bacteria are not present in the soil. Soil pH can be from 6.0 to 8.0. Insect and disease pests are minimal.

Suggested soybean cultivars include 'Bansei,' 'Fiskeby V,' 'Frostbeater,' 'Giant Green,' and 'Prize.'

SPINACH

Spinach is a hardy, cool-season crop grown for its edible leaves or greens. In the North it is a spring and fall crop, and a fall, winter, and spring crop in the South. High temperatures and long photoperiod induce seed formation and reduced leaf quality. California and Texas lead in production.

Seed can be sown in place as soon as the soil is workable. Soil pH can vary from 6.0 to 7.0. It can be planted in succession at weekly intervals to extend the harvest period. It must be harvested prior to flower stalk formation.

Troublesome insects include aphids (*Aphis gossypii*), which can be carriers of mosaic virus (spinach blight or yellows), and leaf miners (*Pegomya hyoscyami*). Downy mildew (*Peronospora effusa*) is another problem disease.

Suggested cultivars include 'America,' 'American Savoy,' 'Dark Green Bloomsdale,' 'Hybrid No. 7,' 'Long Standing Bloomsdale,' 'Melody Hybrid,' 'Virginia Blight Resistant,' 'Viroflay,' and 'Winter Bloomsdale.'

Fig. 14–39. Spinach 'Long Standing Bloomsdale'. U.S. Department of Agriculture/Ferry Morse Seed Co. photo.

Fig. 14-40. New Zealand spinach. Courtesy of National Garden Bureau, Inc.

SPINACH, NEW ZEALAND

New Zealand spinach is a tender, warm-season crop grown for its edible leaves or greens. It is a substitute for spinach, since it thrives during the warmer weather when spinach does not.

Seed is soaked in water at 120°F (49°C) for 1 or 2 hours to hasten germination. It is sown in place after all danger of frost is past. Soil pH can be from 6.0 to 8.0. Successive harvests on each plant are possible if only partial removal of foliage (tips 3 to 4 inches long) is practiced. Insect and disease pests are minimal.

SQUASH

Squash includes a large group of cucurbits, many of which are vine crops; some are bush forms. All are noted for their edible fruit. Summer squashes include those fruits eaten when immature up to when the rind just starts to harden; the shell should be easily cut with a fingernail. Summer squashes (*Cucurbita pepo*) are bush types. Winter squash (*Cucurbita pepo, C. mixta, C. maxima, C. moschata*) are eaten after the fruit is mature (hard shelled) and is usually a vine crop, although some bush types are available. The division between winter squash and pumpkin (which see) is not absolute. Winter squash has a flesh that is darker orange, sweeter, less fibrous, and higher in dry matter than pumpkins or summer squash.

Squash is a warm-season, frost-sensitive crop grown in most of the United States. It can be seeded directly in place, or 2-week–old transplants can be used if care is taken to avoid root disturbance. Soil pH can be between 5.5 and 7.5.

Insect pests include the squash borer (*Melittia cucurbitae*), squash bug (*Anasa tristis*), squash beetle (*Epilachna borealis*), striped cucumber beetle, white fly (*Trialeurodes vaporariorum*), melon worm (*Diaphania hyalinata*), pickle worm (*Diaphania nitidalis*), and aphid. Powdery mildew and other diseases associated with cucumber (which see) can be troublesome.

Fig. 14-41. Squash. A. 'Early Prolific Straightneck', a summer squash cultivar. Author photo. B. 'Early Butternut', a winter squash hybrid. Courtesy of All-America Selections.

(A)

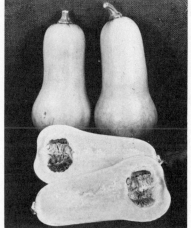

(B)

Cultivars of the summer-squash type include 'Aristocrat Hybrid,' 'Cocozelle Bush' (Italian marrow), 'Golden Courgette,' 'Patty Pan,' 'Prolific Straightneck,' 'Scallopini Hybrid,' 'Summer Crookneck,' 'True Queen,' and 'Zucchini.'

Winter and fall types include 'Banana,' 'Boston Marrow,' 'Bush Table Queen,' 'Butterbush' (bush), 'Buttercup,' 'Butternut,' 'Delicious,' 'Green Striped Cushaw,' 'Hubbard,' 'Royal Acorn,' 'Table Queen,' and 'Waltham Butternut.'

SWEET POTATO

The sweet potato is a warm-season root crop noted for its tuber like storage roots. Since it requires a somewhat long growing season, it is a crop adapted to the South (zones 8 to 10). It can be grown in the milder regions of the North. It should not be confused with the yam (*Dioscorea* sp.), which is mainly confined to the tropics. North Carolina and Louisiana are the primary commercial producers.

Propagation is by rooting of slips forced from the sweet potato in propagation beds, usually of sand. At 75° to 80°F (24° to 26.7°C), rooted plants can be produced in 6 weeks. These are set out after all danger of frost is past. Soil pH should be between 5.2 to 6.7. Later plantings can also be obtained through rooting tip cuttings of growing vines. Temperatures below 55°F (12.8°C) can be detrimental. Sweet potatoes can be harvested any time they reach a usable size.

(A)

(B)

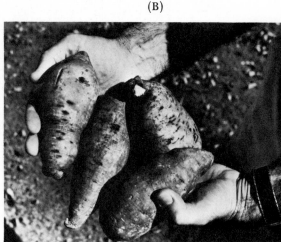

Fig. 14-42. Sweet potato showing foliage (A) and tuberous root (B). U.S. Department of Agriculture photo by Allan Stoner.

Insect pests include the leaf beetle (*Typophorus nigritus*), sweet potato weevil (*Cylas formicarius*), various tortoise beetles, cutworm, and nematode. Troublesome diseases include mosaic, black rot (*Ceratocystis fimbriata*), soft rot (*Rhizopus stolonifer*), stem rot (*Fusarium oxysporium*), and root rot (*Plenodomus destruens*).

Moist or soft-flesh cultivars are 'Centennial,' 'Georgia Red,' 'Gold Rush,' 'Porto Rico,' and 'Velvet.' Dry-flesh cultivars include 'Orlis' and 'Yellow Jersey.' 'Nugget' and 'Nemagold' are intermediate cultivars.

Introduction to Horticultural Production / Part III

Fig. 14-43. Taro. U.S. Department
of Agriculture photo.

TARO

Taro or dasheen is a warm-season crop noted for its edible corms. Its long growing season restricts it to zone 10 in the United States. There are two kinds of taro: wetland and upland. Wetland taro is planted in flooded land and requires the longer growing season of the two; it is grown in Hawaii. Upland taro is grown in conventional soil, requires a shorter growing season, and can be found in zone 10. 'Trinidad' is one good cultivar of upland taro.

Propagation is by corms or corm sections, which can be planted in place or started earlier indoors. Soil pH can be from 6.0 to 8.0. Destruction of the acrid component, calcium oxalate, is accomplished by heating prior to consumption.

TOMATO

Tomatoes are warm-season, frost-intolerant solanaceous crops noted for their edible fruit. They are grown over much of the United States. Fruits vary from small cherry sizes (1 ounce) to large baseball sizes (1 pound or more). Shapes range from plum to pear to round, and colors vary from greenish-white through yellow, orange, pink, and red. California and Florida are the leading states in tomato production.

(A)

(B)

Fig. 14-44. A. 'Pink Panther', a hybrid pink-fruited tomato. Courtesy of George J. Ball, Inc. B. Tomatoes can be grown in tomato "cages," staked, or unsupported. U.S. Department of Agriculture photo.

In the North, tomatoes are set out as 4-to 5-week-old transplants after all danger of frost is past. Seed can be sown directly in place in the South or transplants can be used. Tomatoes also place first in vegetable crops grown commercially in greenhouses. Early, midseason, and late cultivars are planted for an extended harvest. Soil pH can vary from 5.5 to 7.0. Tomatoes of the determinate or self-pruning type are not suited for staking, but can be grown in tomato cages if desired. Indeterminate types have continually growing branches not terminated by fruit set, so they are staked or trellised if space is a problem. Fruit set is poor below 55°F (12.8°C) or above 95°F (35.1°C). Tomatoes may be picked green and full-sized for later ripening if a frost threatens or they are to be shipped a long distance. Otherwise, they are picked in the hard-pink to soft-red stage.

Insect pests include the cutworm, flea beetle, tomato horn worm (*Manduca quinquemaculata*), aphid, tomato fruit worm (also called corn earworm), and white fly. Diseases include bacterial wilt (*Pseudomonas solanacearum*), early blight (*Alternaria solani*), late blight (*Phytophthora infestans*), leaf spot (*Septoria lycopersici*), *Verticillium* wilt, *Fusarium* wilt, anthracnose (*Glomerella phomoides*), and mosaic. Some resistance is possible with some cultivars. Blossom end rot is a physiological disorder caused by acid soil plus an inadequate water supply. Sun scald is also seen.

Cultivars of the small cherry types used in salads include 'Basket Pak,' 'Gardener's Delight,' 'Pixie Hybrid,' 'Red Cherry,' 'Small Fry,' and 'Tiny Tim.' Standard large red tomatoes include 'Ace,' 'Ace 55 VF,' 'Beefsteak,' 'Better Boy,' 'Big Boy,' 'Big Early,' 'Big Girl,' 'Bounty,' 'California 145,' 'Early Girl,' 'Fireball,' 'Floramerica,' 'Heinz 1350,' 'Marglobe,' 'Pearson,' 'Pink Panther,' 'Ramapo Hybrid,' 'Rutgers,' 'Spring Giant,' 'The Juice,' 'Tropic,' 'Valiant,' and 'VF Hybrid,' to name only some.

Paste-type red plum cultivars include 'Roma VF' and 'San Marzano.' Small yellow cultivars include 'Yellow Pear' and 'Yellow Plum.' Large pink cultivars include 'Oxheart' and 'Ponderosa.' Large orange cultivars include 'Jubilee' and 'Sunray.'

TURNIP

The turnip is a hardy, cool-season root crop noted for its edible storage root. Their tops are also used for greens. It is similar to rutabaga, but it is a little less hardy, smaller, has a lighter-colored flesh, little to no neck, requires a shorter growing season, and is less pungent. Turnips are grown as a spring and fall crop in the North, and as a fall, winter, and spring crop in the South. Turnips succeed better in the South than rutabagas.

Fig. 14-45. Turnip. U.S. Department of Agriculture photo.

Turnips are sown directly in place. Soil pH can be 6.0 to 7.0. Turnips are subject to the same insects and diseases as cabbage (which see).

Suggested cultivars include 'Crawford,' 'Early Purple-top Milan,' 'Purple-top White Globe,' 'Shogoin,' 'Seven Top,' 'Six Weeks,' 'Tokyo Cross,' and 'Yellow Globe.'

Fig. 14-46. 'Sweet Favorite', a hybrid watermelon. Courtesy of All-America Selections.

WATERMELON

The watermelon is a warm-season, frost-sensitive vine crop or cucurbit. It is noted for its edible fruit. Since it requires a long growing season, it is usually limited to the South. Florida, Texas, and Georgia are the leading producers. Shorter-maturing cultivars are available for the North. Earlier growth is possible with clear plastic mulch in the North.

Seeds can be started indoors and put out as 2–week–old transplants if root disturbance is minimal. Seeds can also be sown directly in place. All danger of frost must be past. Soil pH can range from 5.0 to 6.0. Insect and disease pests are similar to cucumbers (which see). A ripe melon has a hollow sound when thumped, or low central ridges can be felt.

Early-maturing small cultivars include 'Burpee Hybrid' (seedless), 'Burpee Sugar Bush' (bush type), 'Fordhook Hybrid,' 'New Hampshire Midget,' 'Sugar Baby,' 'Triple Sweet' (seedless), and 'Yellow Baby.' Larger, later cultivars include 'Black Diamond,' 'Charleston Gray,' 'Congo,' 'Crimson Sweet,' 'Dixie Queen,' 'Fairfax,' 'Stone Mountain,' 'Sweet Favorite,' and 'Tom Watson.'

FURTHER READING

Abraham, George, and Katy Abraham, *The Green Thumb Garden Handbook* (3rd ed.). Englewood Cliffs, N.J.: Prentice-Hall, Inc., 1977.

Burrage, Albert C., *Burrage on Vegetables* (rev. ed.), Susan A. Hollander and Timothy Hollander, eds. Boston. Houghton Mifflin Co., 1975.

Crockett, James U., *Crockett's Victory Garden*. Boston: Little, Brown & Co., 1977.

Faust, Joan L., *The New York Times Book of Vegetable Gardening*. New York: Quadrangle/The New York Times Book Co., 1975.

Knott, J. E., *Handbook for Vegetable Growers* (rev. ed.). New York: John Wiley & Sons, Inc., 1962.

Rodale, J. I., ed., *How to Grow Vegetables and Fruits by the Organic Method*. Emmaus, Pa.: Rodale Books, Inc., 1961.

Splittstoesser, W. E., *Vegetable Growing Handbook*. Westport, Conn.: Avi Publishing Co., 1979.

Thompson, Homer C., and William C. Kelly, *Vegetable Crops* (5th ed.). New York: McGraw-Hill Book Co., 1957.

U.S. Department of Agriculture, *Agricultural Statistics*. Washington, D.C.: U.S. Government Printing Office (published annually).

———, "Home Garden Vegetables," in *Gardening for Food and Fun*. Yearbook of Agriculture. Washington, D.C.: U.S. Government Printing Office, 1977.

Ware, G. W., and J. P. McCollum, *Producing Vegetable Crops* (2nd ed.). Danville, Ill.: Interstate Printers and Publishers, Inc., 1975.

Pomology

Pomology is the branch of horticulture that deals with the production, storage, processing, and marketing of fruit. Much of this, except for production, has been discussed in Part Two. This chapter will cover mostly production aspects.

Economics   On the noncommercial level, fruit growing ranks behind ornamentals and vegetables. Part of this results from the space requirements and labor involved with fruit growing. Also, the time available to homeowners or the desire to grow things may be exhausted in meeting the ornamental requirements of the well-landscaped home and the money-saving aspects of a vegetable garden. Those who do practice fruit growing (17 million American households) are rewarded with the beauty of certain fruit plantings and the appealing taste of fresh fruits.

On the commercial level, acreage is under that devoted to vegetable production (2.9 versus 3.4 million acres in 1975). In that same year, 26.4

558

million tons of fruit was utilized from these orchards. Similar figures were observed in 1976. As with vegetables, this places the wholesale value of yearly fruit production in the billion dollar range.

The most important fruits in the United States in terms of commercial horticultural value are as follows, in approximately decreasing order: orange, grape, apple, peach, strawberry, grapefruit, lemon, pear, cherry, avocado, apricot, nectarine, cranberry, tangerine, olive, bushberries, temple orange, tangelo, plum, lime, date, fig, papaya, pomegranate, banana, and persimmon.

In terms of tonnage, citrus fruits lead with 55 percent of the total fruit production during 1975. All citrus fruit production is centralized in Florida, California, Texas, and Arizona. Florida was the largest producer in 1975 of oranges, grapefruits, and tangerines, and the sole producer of limes, tangelos, and temples. California was the largest producer of lemons, second in oranges and tangerines, and third in grapefruits. Texas was second and third in grapefruits and oranges, respectively. Arizona was second in lemons, third in tangerines, and fourth in oranges and grapefruits.

Grapes accounted for 16.3 percent of the total 1975 fruit tonnage. of this, 90 percent was produced in California. Apples accounted for 13.4 percent, and Washington and New York led total production with 31.0 and 12.1 percent, respectively. Peaches accounted for 5.3 percent of the total fruit tonnage; California led with 64.9 percent.

Little change was observed in fruit tonnage for these fruits in 1976.

Fruit Production

Horticulturists usually group fruits on the basis of their climatic requirements. These include the temperate, subtropical, and tropical fruits. Temperate-zone fruits are deciduous and tropical fruits are evergreen; subtropical fruits can be either.

Temperate fruits include the **pome fruits** (apple, pear and quince), **stone fruits** (apricot, cherry, nectarine, peach, and plum), and the small or **berry fruits**: brambles (black raspberry, red raspberry, blackberry, dewberry, boysenberry, youngberry, and loganberry), blueberry, cranberry, currant, gooseberry, grape, and strawberry. The brambles, blueberry, currant, and gooseberry are often categorized as **bushberries**.

Subtropical and **tropical fruits** include avocado, banana, citrus, date, fig, mango, papaya, persimmon, pineapple, and pomegranate. There are many other minor fruits not mentioned, but some of which will be covered later in the chapter.

Basic practices in pomology include the propagation and improvement of the cultivar, site selection, planting and spacing, training and pruning, irrigation, fertilization, weed control, soil management, pollination, thinning, insect and disease control, harvesting, postharvest physiology, and preservation. Most of these practices have been covered in Part Two.

Cultivars should be selected for high productivity and high-quality fruit. Much of the fruit grown exists as a clone and, as such, is vegetatively propagated. Grafting and budding are used extensively because of poor rooting qualities or for dwarf fruit tree production.

Site selection can be important for long-range success. Microclimate and soil conditions of sites are especially important. Spring frosts during or shortly preceding or following bloom are a serious hazard in fruit growing. The role of microclimate in the alleviation of frost hazard could make a site highly desirable. Soil condition is important, since roots of fruit trees must extend 3 feet or more in depth for high productivity. Any soil condition that interferes with root development, such as hardpan, wetness, or high salinity, could reduce yields.

Planting and spacing are determined by two, sometimes conflicting factors: (1) growth, flowering habits, and light requirements, and (2) management practices. The aim is for highest productivity with least labor. Trends are toward dwarf trees and management practices, which are more conducive to mechanized operations. Labor is the largest cost factor in fruit production. Cultivars that require cross-pollination pose special requirements in regard to planting and spacing. Smaller fruits, such as strawberries, can be managed in beds with rows, whereas larger fruits, such as apples, can be managed in orchards.

Many fruit plants need to be pruned for form when young. This type of pruning is called *training*. Generally, trained fruit plants will require maintenance pruning to maintain fruitfulness and ease of other management practices, such as pest control or harvesting. The types of training for fruit trees of standard size (modified leader, central leader, and open center) and their maintenance pruning were covered in Chapter 9. Recommended practices for each fruit will be cited in this chapter.

Dwarf fruit trees are being increasingly trained in a hedgerow system, which is essentially a tree wall. This practice provides maximal trees per acre and facilitates the use of mechanical equipment for pruning, spraying of growth regulators and pesticides, and harvesting of fruit. This is a consequence of the drive to maximize production and minimize labor. Hedgerow trees may be supported by post or three- and four-wire trellis systems, such as used with grapes, Hedgerows may also be left freestanding on stronger-rooted stocks. The need for dwarf trees limits this practice at present to apple and pear.

Soils around fruit plantings can be kept clean through mechanical cultivation and/or chemical weed control. This approach minimizes nonfruit plant competition for nutrients, water, and light, and increases nutrient availability by increasing the rate of breakdown for soil organic matter. Disadvantages include erosion and the loss of organic matter, which causes detrimental changes in soil structure. Alternately, mowed sod may be left. This is not destructive of soil structure, but the sod does compete with the fruit plantings for nutrients and water. Additional fertilizer and water would be needed. However, some organic matter and nutrients are added to the soil by natural decomposition of the sod. Soil structure is improved, and soil moisture fluctuations are less, especially during dry periods. Combinations of the two approaches can be advantageous. Sod may be left in the middle of rows and clean cultivation or mulching practiced directly around the fruit plantings. Winter cover crops may be grown and incorporated into the soil to maintain organic matter and slow erosion. Irrigation, fertilization, and soil pH have been covered in depth in Chapter 7.

Pollination may be enhanced by the introduction of beehives into the fruit plantings during flowering. Insect control must not be detrimental to bees at this time. Hand pollination is not practical because of labor costs. Removal of flowers or young fruit (thinning) is a common practice of fruit growers. Remaining fruit develop larger or more rapidly, and the following year's flowering and fruit set are not affected adversely. This can be done by hand, but mechanical or chemical means are more economical. Mechanical shakers or growth regulating substances can be used, as discussed in Chapter 9.

Topworking is also practiced with fruit trees. This consists of altering the cultivar of an established tree through grafting or budding of a new cultivar on its trunk and/or branches. This is a useful tool to update an orchard to a more recent, improved cultivar, to add a pollinizing cultivar to a portion of an orchard with a poorly pollinizing cultivar, or to alter disease resistance. The latter has been done with pears susceptible to fire blight. The cultivar 'Old Home' is resistant, but not an excellent cultivar for fruit. New orchards can be established by topworking first with 'Old Home' (interstem) followed by more desirable cultivars.

Pest control is probably the most expensive and time consuming aspect of fruit growing. Integrated pest control, the judicious use of biological and chemical controls, is being increasingly utilized. Costs and environmental pressures can be reduced with this approach (see Chapter 10).

Harvesting, postharvest physiology, and preservation of fruits have been covered in Chapter 12. An encyclopedic listing of fruits follows.

Careers in Pomology

Production nurseries propagating fruit trees need breeders and propagators knowledgeable in pomology. Fruit production has managerial and field positions analagous to those described for vegetable production in Chapter 14. Similarly, the sales agencies and positions therein, brokers, buyers, and promotional agencies cited in Chapter 14 have similar opportunities for people experienced in pomology. The various positions described under education and research in Chapter 13 also require people with knowledge of pomology.

(A)

Fig. 15-1. A. Apple, showing foliage and fruit. U.S. Department of Agriculture photo. B. An apple orchard which has sod in the middle of rows and mulch around the trees. This combined approach maintains soil structure, but decreases competition for nutrients/water near the trees. U.S. Department of Agriculture/Soil Conservation photo by J. E. McKittrick.

(B)

APPLE

Apple cultivars (Fig. 15-1) are probably derived from *Malus pumila,* or possibly from hybrids of *M. pumila* and *M. sylvestris.* Cultivars number in the hundreds; the most important is 'Delicious.' Other cultivars of importance are 'Baldwin,' 'Ben Davis,' 'Cortland,' 'Early Harvest,' 'Empire,' 'Golden Delicious,' 'Gravenstein,' 'Grimes Golden,' 'Harlson,' 'Idared,' 'Jonathan,' 'Macoun,' 'McIntosh,' 'Newtown,' 'Red Baron,' 'Rhode Island Greening,' 'Rome,' 'Rome Beauty,' 'Spartan,' 'Stayman,' 'Twenty Ounce,' 'Yellow Transparent,' and 'York.' Many other cultivars are known. Those interested in growing apples should consult local agricultural authorities for the cultivars best suited to their locality.

Much of the United States is suitable for the temperate apple, except for Florida and southern extremities of Texas and California. A cold dormancy period of 1200 to 1500 hours of 45°F (7.3°C) or lower is needed. Winters with temperatures below −20°F (−30°C) are not suitable for good apple production. Commercial apple production is centered in the Northeast, central Atlantic, Ohio basin, North Central and Western states, and parts of the Southwest. Sites should have good, deep, well-drained agricultural soil. Apples can be grown in orchards or by the homeowner. The latter must realize that some effort and expense is required.

Propagation of apples is generally by budding or whip grafting of cultivars onto rootstocks. The latter may be used to control the size of the tree. Full-sized trees, which attain a height of 16 feet or more, are produced with rootstocks grown from cultivars such as 'Delicious' or 'McIntosh.' Other rootstocks are used for size control; these are usually produced by mound layering of suckers produced by plants with the desired size-controlling rootstock. Fully dwarfed trees, about 6 feet tall, are produced with 'East Malling 9' rootstock and semidwarf trees of 8 to 10 feet from 'EM 26,' 'EM 7,' and 'Malling-Merton 106.' There are also several other rootstocks. All have advantages and disadvantages, such as disease resistance or susceptibility, in addition to size-controlling characteristics. A natural mutation, the spur type, can also be used in size control. It is about 12 feet tall, even when grafted on standard rootstocks, or smaller on size-controlling rootstocks. A diagram of respective sizes of rootstocks is shown in Figure 15-2.

Interstem grafting, or the grafting of an interstem onto a rootstock followed by grafting or budding of the desired cultivar onto the interstem, is practiced with apples. Advantages such as earlier ripening, better root growth of a dwarf apple, longer life, or higher productivity can be realized with this type of grafting. 'EM 9,' 'Robusta 5,' and 'Hibernal' are commonly used as interstems.

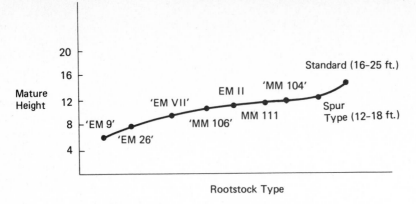

Fig. 15–2. Respective sizes of apple trees produced by various types of rootstocks. The spur type can vary from 12–18 feet and the standard 16–25 feet depending on pruning and growing conditions.

Planting is done in early spring before the soil is dry. In mild climates, autumn planting may be preferable. Standard-sized apples can be planted 35 to 45 feet apart. Spur varieties on standard rootstocks can be planted about 2 to 3 feet closer. Spacing of dwarf trees with standard or spur-type cultivars is determined by the type of rootstock and desired planting density. For example, the minimal spacing with 'EM 9' rootstock is 6 feet between trees in the row and 14 feet between the rows. This gives about 518 trees per acre. For 'EM 7' or 'MM 106,' the spacing can be 8 by 16, 10 by 18, and 12 by 20 feet for planting densities of 340, 242, and 182 trees per acre, respectively. If grafted trees are planted, the bud union must never be planted below the soil surface. Bearing age varies. Dwarf trees on 'EM 9' can begin bearing the second year after planting a 1- or 2-year-old nursery-grown tree. Semidwarf trees may require 4 or more years, and standard sizes up to 10 years.

Pruning should be kept to a minimum the first 3 to 8 years. The main purpose is to train the tree such that it has the desired placement of scaffold branches and proper leader development. Subsequent maintenance pruning is to keep the tree balanced in shape and proportions of old and young wood, such that maximal fruit yield is possible (see Chapter 9). Most pruning is during late winter up to the time active growth starts in spring. Some summer pruning may be needed with young or dwarf trees to improve shape and remove unwanted growth or with mature trees to remove watersprouts, weak wood, and unproductive wood.

Standard-sized trees are usually trained to the modified-leader system. Dwarf trees can be trained that way, also, but the hedgerow system is becoming increasingly popular (Fig. 15–3). 'EM 9' requires staking or trellis support as a hedgerow, whereas better-rooted dwarf trees may be grown in a hedgerow with or without a trellis. The best in terms of yield appears to be a spur cultivar grafted onto a semidwarf rootstock and grown in a hedgerow.

Topworking may be utilized to change cultivars in older apple orchards. Grafts used for this purpose include cleft grafting, bark grafting, and modified kerf grafting. Young trees may be topworked with the tongue or whip graft.

Apple cultivars are self-incompatible, but most cultivars can be pollinated by almost any other cultivar through bee activity. Several varieties should be grown to ensure pollination. It is essential to have overlapping flowering periods for the cultivars, which can differ in their time of blooming. Distances between two cultivars should be less than 100 feet. Thinning will be necessary when overcropping occurs and to prevent biennial bearing. Chemical thinners are available for apples, such as Sevin, naphthylacetamide, naphthylene acetic acid, and 2-methyl-4,6-dinitrophenol.

Up to 400 different insect and disease problems occur with apples. Commercial growers use several sprays of various pesticides, and the home owner may need four to eight treatments, if unblemished high-quality fruit is desired. Pests and treatments vary from area to area, so a local agricultural authority should be consulted. A few of the insect pests are mites (European red and several other), scale, aphids, tarnished plant bug, fruitworms, leafrollers, curculio, codling moth (*Carpocapsa pomonella*), and apple maggot (*Rhagoletis pomonella*). Scientific names are unlisted here, because there are several species for each. Scab (*Venturia inaequalis*) is one of the main fungal diseases.

Fig. 15–3. Dwarf 'Golden Delicious' apple trees grown in a hedgerow (Van Roechoudt System) using trellis support. Courtesy of Stark Bro's Nurseries and Orchards Co.

Average yields of apples for standard orchards are about 6 bushels per tree. Higher yields of 15 to 20 bushels per tree are possible for standard-sized cultivars with optimal conditions. Yields up to 45 to 60 bushels per tree are possible with high-density plantings of spur semidwarf trees grown in hedgerows. Apples should be stored under controlled atmosphere. Home storage is best at 33°F (0.5°C) and high humidity.

APRICOT

There are at least three varieties of the apricot, a temperate fruit (Fig. 15-4). *Prunus armeniaca* var. *armeniaca* is the commercial apricot source. Variety *sibirica* and variety *mandshurica* have greater cold tolerance, but less desirable fruits. Cultivars of commercial importance include 'Blenheim,' 'Moorpark,' 'Royal,' and 'Tilton.' Newer cultivars are 'Alfred,' 'Earliril,' 'Farmingdale,' 'Goldcot,' 'Vecot,' and 'Wilson Delicious.' Less desirable but having maximal winter hardiness are 'Moongold' and 'Sungold.'

Apricots only require a cold dormancy period of 700 to 1000 hours below 45°F (7.3°C). Therefore, apricots bloom early and are highly susceptible to late spring frosts, which restricts their culture to areas with minimal problems of spring frosts. About 95 percent of the commercial crop is grown in California. Cultivars of smaller size and less sweetness than those of commercial use can be grown by the homeowner where peaches succeed, especially if a favorable microclimate exists.

Fig. 15-4. Apricot, showing foliage and fruit. U.S. Department of Agriculture photo.

Apricots are propagated by budding onto apricot seedling rootstocks. Peach and plum rootstocks are sometimes used to overcome problems, such as caused by soil conditions. Mature height is about 30 feet. Dwarf types are not available.

Apricots have a minimal planting distance of 20 feet. They may be planted as 1- or 2-year-old trees in early spring in the East or late autumn in California. The modified-leader system of training is best, although an open center tree as used with peaches is satisfactory. Since fruit is borne on spurs that last about 3 years, pruning is aimed at renewing spur growth. Most apricot cultivars are self-fruitful. Bees are the chief pollinators. Exceptions are 'Earliril,' 'Moongold,' and 'Sungold,' which need another cultivar for cross-pollination. Thinning of flower buds to reduce overbearing is practiced by hand (pruning) or mechanical shakers. Use of chemical thinning sprays is minimal. Fruit thinning is not practiced, since the size increase of the remaining fruit is minimal. The curculio (*Conotrachelus nenuphar*) is a serious pest, and bacterial leaf spot (*Xanthomonas prunii*), brown rot (*Monilinia fructicola*), mushroom root rot (*Armillaria mellea*), and bacterial canker (*Pseudomonas syringae*) can be troublesome diseases.

Fresh apricots do not last long or ship well. Much of the crop is dried, canned, or juiced. Annual yields of 250 pounds per tree are possible.

AVOCADO

The avocado, *Persea americana,* is a tropical fruit (Fig. 15-5). Cultivars are divided into three groups that differ in fruit characteristics and climatic requirements. The West Indian group is strictly tropical and includes early ripening cultivars such as 'Fuchs,' 'Pollock,' 'Simmonds,' 'Trapp,' and 'Waldin.' The Guatemalan group is somewhat hardier; common cultivars are 'Hass' and 'Nabal.' The Mexican group is the hardiest and includes cultivars such as 'Duke,' 'Mexicola,' and 'Topa Topa.' Hybrids also exist between the groups. Important hybrid cultivars are 'Booth,' 'Choquette,' 'Fuerte,' 'Hall,' 'Hickson,' 'Lula,' and 'Simpson.'

The West Indian group can only be grown in the warmest parts of Florida, since injury occurs below 28°F (-2.2°C). The Guatemalan group can withstand temperatures to 25°F (-3.9°C) and can be found in much of zone 10. The Mexican group can stand temperatures as low as 20°F (-6.7°C) and can be grown in southern California (zone 9). Avocados are grown commercially in orchards and by the homeowner.

Propagation is by shield budding and side grafting on seedling avocado rootstocks. This is done in the spring and fall. Shield budding is practiced in California with Mexican group rootstocks and in Florida with West Indian or Guatemalan rootstocks. Mature trees reach a height of 30 feet or more.

Grafted trees are transplanted in the spring or early summer. Spacing is between 20 and 30 feet, depending on desired planting density. Fruit

Fig. 15-5. Avocado, showing foliage and fruit. American Society for Horticultural Science photo.

may be expected in as little as 3 years, although heavy fruiting will not occur for 5 to 8 years.

Growth habits of the avocado are such that pruning required for training and maintenance will be minimal. Several cultivars should be present to ensure cross-pollination. Within a group, pollen production and pistil receptability have minimal overlap. Bees are the main pollinators. Fruit thinning is unnecessary.

Insects that attack the avocado include several thrips, mites, and scales. Disease problems include scab (*Sphaceloma perseae*), avocado root rot (*Phytophthora cinnamomi*), usually associated with wet soils, fruit spot blotch (*Cercospora perseae*), and anthracnose (*Glomerella cingulata*).

Mature trees in orchards can produce from 2 to 6 tons per acre. One to three bushels can be expected from an individual tree. Storage is possible for several weeks, but temperatures should not drop below 42°F (5°C).

BANANA

Common edible bananas are cultivars of *Musa acuminata* (Fig. 15-6). These cultivars include the diploids, 'Blande' and 'Paka,' and the triploids, such as 'Dwarf Cavendish,' 'Grand Nain,' 'Gros Michel,' 'Lacatan,' and 'Valery.' Other cultivars are those derived from the hybrid, *M. acuminata* X *M. balbisiana* (designated as *M. X paradisiaca*). These may be diploid ('Ney Poovan'), triploid, ('French Plantain,' 'King,' 'Nadan,' and 'Silk'), or tetraploid ('Tiparoot').

Bananas are tropical fruits. Few, except for 'Dwarf Cavendish,' are grown in the continental United States. 'Dwarf Cavendish' can be grown outdoors in zone 10 or in the greenhouse elsewhere. In the tropics, bananas are grown commercially on plantations and as backyard plants. Windy sites are to be avoided.

Propagation is by rhizome cuttings with at least two buds (eyes) or by suckers that arise at the base of the pseudostem. Height varies from about 5 feet for 'Dwarf Cavendish' to 30 feet for the standard commercial cultivars.

'Dwarf Cavendish' can be set 8 to 10 feet apart. Other cultivars require 14 to 20 feet. In plantations the pseudostem is cut off after fruiting, since it dies after fruiting once. It is possible in the tropics for the next-oldest pseudostem to produce fruit in 10 months. Rooted suckers can produce fruit in 12 months to 3 years, depending on climate.

Pruning is mainly to remove spent pseudostems. Pollination is not required with the majority of the commercial cultivars, which set their fruit parthenocarpically. Diploid cultivars require pollination. Thinning of fruit is not practiced.

Insect and disease problems are minimal in the United States. Several insect and disease problems are encountered in the tropics. Disease resistance varies among the cultivars.

Bananas are harvested immature and ripened at temperatures above 53°F (11.8°C). Refrigerated storage of ripened bananas causes brown discoloration and off-flavor.

Fig. 15–6. Bananas being harvested. Perforated polyethylene bag covering the bunch was placed when fruit was young to prevent fungal/insect damage, scarring of bananas, and to speed ripening. American Society for Horticultural Science photo.

Fig. 15-7. Blackberries. A. Trailing cultivar. B. Erect cultivar. U.S. Department of Agriculture photos.

BLACKBERRY

Blackberries are deciduous bramble fruits derived from various species of *Rubus,* although the exact lineage between some cultivars and species is not clear (Fig. 15-7). Unlike raspberries, the fruit and receptacle do not separate, so both are eaten. Blackberry cultivars can be broken down into five groups. Erect and nearly erect types exist in the eastern United States; 'Bailey,' 'Darrow,' 'Early Harvest,' and 'Eldorado' are typical cultivars. Trailing types without red cane hairs also exist in the same area and are typified by 'Lucretia.' Trailing types with red cane hairs are found in the southeast ('Floragran,' 'Mayes' and 'Oklawaha,'). Trailing types also exist on the Pacific Coast. Popular cultivars of this last group include 'Boysen' (boysenberry), 'Logan' (loganberry), and 'Young' (youngberry). Semitrailing evergreen types also are found on the Pacific Coast and include cultivars such as 'Evergreen' and 'Thornless Evergreen.' Trailing form blackberries are often referred to as dewberries. For those who object to thorns, the following are thornless cultivars: 'Black Satin,' 'Smoothstem,' 'Thornfree,' and 'Thornfree Black Honey.'

Blackberry culture is limited by cold winters in the North and the Plain States and by heat and drought in the Southwest. Hardiness varies among the cultivars. 'Evergreen,' 'Thornless Evergreen,' 'Boysen,' 'Logan,' and 'Young' are, for example, not hardy in the East. Sites should have sufficient moisture, as blackberries are more prone to wilting damage than other fruits because of higher rates of transpiration. Blackberries are grown both commercially and by homeowners.

Blackberries can be propagated from suckers that arise from the roots, root cuttings, and rooting of cane tips. The latter is preferred for the trailing blackberries. If left unpruned, heights can vary from 3 to 10 feet, depending on cultivars.

One-year-old plants are placed out in the spring as early as possible. Erect types are set about 2 to 3 feet apart in rows 8 to 10 feet apart for mechanical harvest. Hills at 3 to 4 feet apart can be used for home gardens. Trailing and semitrailing types are set 5 to 12 feet apart in rows separated by 7 to 10 feet, depending upon the vigor of the cultivar. First-year canes, primocanes, do not bear fruit; second year, or floricanes, do. Trailing types are conveniently trained to two or more wire trellises about 6 feet high (Fig. 15–8). Erect blackberries require summer topping of the tips of new shoots when they are 18 to 24 inches high. About 3 to 4 inches is removed, and tips should be pinched out afterward. This produces a stout plant and good lateral development. If this is not done, erect types should be topped at 5 feet. Laterals can be cut back in early spring or after blooming to about 18 inches (Fig. 15–9); this practice ensures that an excessive part of the crop is not removed. Floricanes are removed after fruiting.

Fig. 15–8. Trellis systems for blackberries. A. Erect cultivar supported by one-wire trellis. B. Trailing cultivar trained to two-wire trellis. U.S. Department of Agriculture photo.

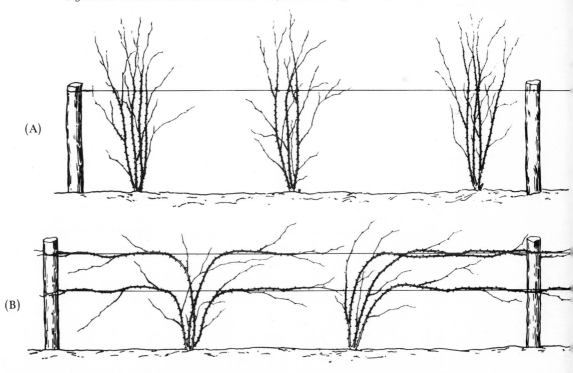

(A)

(B)

Some spring thinning of canes for all groups is usually desirable such that there is one cane every 8 to 12 inches. Canes are removed from trailers after harvest, and new canes are trellis-trained in early spring before growth starts. Some cultivars are self-fertile and others are not. Whether all self-fertile cultivars can pollinate themselves is not completely clear. It would appear that the growing of more than one cultivar and encouraging bees would be beneficial. Fruit thinning is unnecessary.

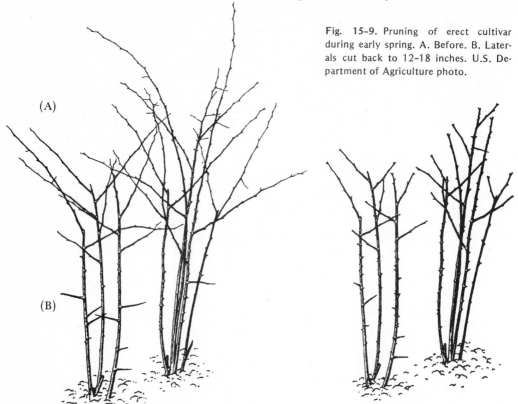

Fig. 15-9. Pruning of erect cultivar during early spring. A. Before. B. Laterals cut back to 12-18 inches. U.S. Department of Agriculture photo.

(A)

(B)

Disease problems include anthracnose (*Elsinoë veneta*), spur blight (*Didymella applanata*), crown gall (*Agrobacterium tumefaciens*), orange rust (*Gymnoconia peckiana* and *Kunkelia nitens*), Verticillium wilt, double blossom (*Cercosporella rubi*), and several viruses (blackberry mosaic, blackberry dwarfing, and so on). Troublesome insects include spider mite, tree cricket (*Oecanthus* sp.) cane borer (*Oberea maculata* and *Agrilus ruficollis*), crown borer (*Bembecia marginata*), consperse stink bug (*Euschistus conspersus*), cane maggot (*Pegomya rubirora*), rose chafer, blackberry sawfly (*Pamphilius dentatus*), and blackberry psyllid (*Trioza tripunctata*).

Yields with average to good care are 2300 to 5000 quarts per acre, or up to 15 pounds per plant. Blackberries are picked when ripe. Precooling and refrigeration are recommended.

Blueberries can be grouped into lowbush and highbush species (Fig. 15-10). Most cultivars are derived from two highbush species, *Vaccinium corymbosum* and *V. ashei*. The former is more important commercially, and some cultivars include 'Angela,' 'Berkeley,' 'Bluecrop,' 'Blueray,' 'Bluetta,' 'Collins,' 'Coville,' 'Darrow,' 'Dixi,' 'Earliblue,' 'Herbert,' 'Jersey,' 'Lateblue,' 'Pemberton,' 'Pioneer,' 'Rubel,' 'Weymouth,' and 'Wolcott.' A few cultivars of *V. ashei* are 'Callaway,' 'Garden Blue,' 'Tifblue,' and 'Woodward.' Lowbush species, although not cultivated, can be harvested from native stands of *V. angustifolium* and *V. myrtilloides*.

Lowbush species are found mostly in New England and maritime Canada. Highbush blueberries grow best in moist acidic soil at pH 4.0 to 5.2, have a cold requirement similar to Elberta peach, cannot withstand temperatures below –20°F (–30°C), and require a growing season of at least 160 days. They can be found from Florida to Maine and west to Michigan. *Vaccinium ashei* (rabbit-eye blueberry) is important in the Southeast; it is more tolerant of soil type, heat, and drought, but quality is not as good as *V. corymbosum*. *Vaccinium ovatum* (evergreen blueberry) and *V. membranaceum* (mountain blueberry) are blueberries gathered in the wild in the Pacific Northwest. Blueberries are grown commercially and by the homeowner.

Blueberries are propagated by rhizome division and hardwood or softwood cuttings. Hardwood cuttings are utilized extensively for nursery production of highbush cultivars, whereas rhizome division is used for lowbush types. Lowbush species vary in height from 1 to 2 feet; cultivars of *V. corymbosum,* 6 to 12 feet; *V. ashei* cultivars, 4 to 8 feet; *V. membranaceum,* 4.5 feet; and *V. ovatum,* 10 feet or more.

Two-year-old highbush plants are usually set out in early spring, although early to late fall is possible in some areas. Plants are placed 4 to 5 feet apart in rows spaced 8 to 10 feet. Plants start to bear well in their third year.

Fig. 15–10. Blueberry highbush cultivar showing foliage and fruit. U.S. Department of Agriculture photo.

Blueberries are borne on wood of the previous season's growth. Pruning is necessary to maintain production of large-sized fruit. Pruning is possible from the time of leaf drop to shortly after blossoms appear. Bushy growth at the base of plants is first removed after the third growing season. Thin bushy wood and old stems that have lost vigor are removed in subsequent years. Lower, drooping branches are removed. Many cultivars overbear, so some thinning of fruit buds by pruning is desirable. Exact pruning requirements vary among cultivars, so local sources should be consulted. Because of possible self-sterility and pollen scarcity, best fruit production is obtained with cross-pollination. The best practice is to interplant with compatible cultivars and to place bee sources at flowering time.

Insects that attack blueberries include the blueberry maggot (*Rhagoletis mendax*), cranberry fruit worm (*Acrobasis vaccinii*), cranberry weevil (*Anthonomus musculus*), cherry fruit worm (*Grapholitha packardi*), blueberry bud mite (*Acalitus vaccinii*), stem gall wasp (*Hemadas nubilipennis*), stem borer (*Oberea myops*), blueberry thrips (*Frankliniella vaccinii*), scale (several spp.) and leaf hopper (*Scaphytopius magdalensis*). Troublesome diseases include stunt (a mycoplasm spread by leafhopper), mummy berry (*Monilinia vaccinii-corymbosi*), red ring-spot virus, stem canker (*Physalospora corticis*), botrytis tip blight (*Botrytis cinerea*), and phomopsis twig blight (*Phomopsis vaccinii*). Disease resistance varies with the cultivar. Birds can be a serious pest.

A mature plant can yield 6 to 8 pints or, under optimal conditions, 20 pints. With good management yields up to 6000 pints per acre are possible. Blueberries should be picked ripe. Blueberries are not as perishable as raspberries.

CALAMONDIN ORANGE (SEE CITRUS)

CHERRY

Most cultivated cherries (Fig. 15-11) are derived from two species. Sweet cherries are from *Prunus avium* and sour cherries from *P. cerasus*. Duke cherries, hybrids between the preceding two, are *P.* X *effusus*. Lesser cherries include the dwarf or western sand cherry, *P. besseyi*, and the Nanking cherry, *P. tomentosa*. The former can be used as a dwarfing stock for plum, prune, and peach. Sweet cherry cultivars include 'Bing,' 'Compact Lambert,' 'Emperor Francis,' 'Gold Sweet,' 'Hedelfinger,' 'Hudson,' 'Lyons,' 'Napoleon,' 'Rainier,' 'Republican,' 'Royal Ann,' 'Schmidt,' 'Stella,' 'Ulster,' 'Van,' 'Vista,' 'Viva,' and 'Windsor.' Sour cultivars include 'Early Richmond,' 'English Morello,' 'Meteor,' 'Montmorency,' and 'Northstar.' Duke cultivars include 'Late Duke,' 'May Duke,' and 'Royal Duke.'

Growth regions vary for these cultivars. Sweet cherries are the most tender, being comparable to the peach. Sour cherries are more hardy. Dormancy requirements are 1100 to 1300 hours below 45°F (7.3°C) and 1200 hours, respectively, for the sweet and sour cultivars. Flowers appear early in the spring and are susceptible to injury from late frosts. The sweet cherry is particularly adapted to the Hudson Valley of New York, the Great Lakes shores, and the Pacific Coast. The sour cherry can be grown in these regions, plus farther north and south. Cherries are grown commercially and to a limited degree by homeowners.

Propagation of sweet and sour cultivars is by budding on seedling rootstocks of *P. avium* (Mazzard cherry) and *P. mahaleb* (Mahaleb cherry). Mazzard is the preferred stock, but where problems of extreme temperatures and drought exist, Mahaleb is better. 'Morello' stock is used where soil is wet or a semidwarfing effect is desired. Sweet and sour cherries have maximal heights of 60 and 30 feet, respectively.

Vigorous 1- or 2-year-old nursery trees can be planted in the fall in milder climates and in early spring planting for colder climates. Sweet cherries can be planted from 20 to 36 feet apart in rows set 20 to 36 feet apart. Sour and Duke cherries require 20 by 20 feet. Bearing may start after 5 to 6 years, with full bearing starting at 8 to 9 years for sour cherries, and starting at 8 to 9 years and becoming maximal at 14 years for sweet cherries.

Pruning to train sweet or Duke cherries is minimal; a spreading habit with several large scaffold branches and height kept to 20 feet are desirable. Sour cherries are trained to either open-center or modified-leader systems; the latter reduces the risk of weak, rotting crotches. Subsequent pruning of both is to encourage spur development. Fruit thinning of sour cherries is unnecessary, and fruit thinning of sweet cherries by chemical means is not dependable.

Sweet cherries are self-unfruitful. All cultivars are not cross-compatible, so sweet cultivar mixtures must be selected with care. Sour cherries are mostly self-fruitful. Duke cherries require cross-pollination. Bees are utilized for all to ensure maximal fruit production.

Fig. 15–11. Sweet cherry showing foliage and fruit. U.S. Department of Agriculture photo.

Troublesome diseases include leaf spot (*Coccomyces hiemalis*), brown rot (*Monilinia fructicola*), scab (*Fusicladium cerasi*), powdery mildew (*Podosphaera oxyacanthae*), virus yellows, ring spot virus, and black knot (*Dibotryon morbosum*). There are many more viruses than listed here that attack cherry; stock should be purchased as certified virus-free. Insect pests are black cherry aphid (*Myzus cerasi*), plum curculio (*Conotrachelus nenuphar*), leaf roller (*Archips argyrospilus*), cherry maggot (*Rhagoletis cingulata*), mites, sawfly (*Hoplocampa cookei*) moth (several spp.), scales (several spp.), slugs, and borers (several spp.). In addition, birds are a serious problem, and fruit splitting can occur if heavy rains coincide with ripening.

Cherries can benefit from precooling. Refrigerated shipping is desirable for long-distance shipping. Storage at 31° to 32°F (0°C) and 85 to 90 percent relative humidity for 10 to 14 days is possible. Unripe cherries can be held in controlled atmosphere conditions (see Chapter 12) for 25 days and ripened later. Sour cherry production can reach 10,000 pounds per acre or 22,000 pounds with optimal management. Sweet cherries produce 8000 pounds per acre. Yields per tree for both are 2 to 3 bushels.

CITRON (SEE CITRUS)

CITRUS

There are a number of evergreen citrus fruits grown commercially and by the homeowner in the subtropic areas (Fig. 15–12). Cultivars of the lime, *Citrus aurantifolia*, include 'Bearss' and 'Tahiti.' Seedling trees of the species are named for their point of origin, that is, Key, Mexican, or West Indian lime. The sour orange, *C. aurantium*, is not widely cultivated in the United States, but makes an excellent marmalade. Cultivars of the lemon, *C. limon*, include 'Eureka,' 'Lisbon,' and 'Villafranca,' which are used commercially. Lemon cultivars of value in the home garden or greenhouse include 'Meyer' and 'Ponderosa.' Otaheite orange, much used as a tub plant, is a form of *C.* X *limonia* (*C. limon* X *C. reticulata*). The citron, *C. medica*, is grown for its rind, which is candied. The tangor, *C.* X *nobilis* (*C. reticulata* X *C. sinensis*) is known for its cultivars 'King' and especially 'Temple.' The grapefruit, *C.* X *paradisi* (*C. maxima* X *C. sinensis*), includes cultivars 'Duncan,' 'Foster,' 'Marsh,' 'McCarty,' 'Ruby,' and 'Thompson.' The tangerine is *C. reticulata* and is known for its cultivars 'China,' 'Clementine,' 'Cleopatra,' 'Dancy,' and especially 'Pankan.' The cultivars of the sweet orange, *C. sinensis*, are divided into four horticultural groups: Mediterranean, Spanish, blood, and navel oranges. The sweet orange is the most important commercial citrus fruit, and the main cultivars are 'Hamlin,' 'Pineapple,' 'Valencia,' and 'Washington.' Others are 'Homasassa,' 'Jaffa,' 'Lue Gim Gong,' 'Parson Brown,' and 'Ruby.' Cultivars of the

tangelo, *C.* × *tangelo* (*C.* × *paradisi* × *C. reticulata*), are 'Minneola,' 'Orlando,' 'Sampson,' 'Seminole,' and 'Thornton.' The calamondin orange, grown as a tub plant is × *Citrofortunella mitis* (a hybrid genus).

Because of their subtropical requirements, citrus fruits are generally restricted to the southern portion of zone 9 and all of zone 10. This includes Southern California, Florida, Rio Grande Valley in Texas, southwest Arizona, and southern parts of the Gulf states. Frost tolerance varies from almost none to limited as follows in order of approximately increasing hardiness: lime, lemon, grapefruit, sweet orange, tangerine, and calamondin orange.

Citrus cultivars are propagated by budding on to seedling stocks of the same or other species. Seedling stocks are 1 to 2 years old at budding time. Seedling stocks commonly used include sour orange (*C. aurantium*), sweet orange (*C. sinensis*), citrange (× *Citroncirus webberi* cv. 'Troyer,' a hybrid genus from *Citrus sinensis* × *Poncirus trifoliata*), trifoliate orange (*Poncirus trifoliata*), rough lemon (*C. limon* cv. 'Rough'), rangpur lime (*C.* × *limonia*), and Cleopatra tangerine (*C. reticulata* cv. 'Cleopatra'). Sweet orange and rough lemon are suitable for light, sandy soils in warmer areas, sour orange for good-quality soils, and trifoliate orange for heavier, moist soils in cooler areas. Height for mature citrus trees varies. For example, the otaheite orange will reach 3 feet, the lemon 25 feet, the citron 16 feet, the sweet orange 40 feet, and the grapefruit 50 feet.

Citrus planting is usually done from March to May in California and in spring through early summer in Florida. Planting distances vary according to the type of citrus. Each of the following has a planting distance range, which depends on cultivar vigor and types of pruning: lime, 15 to 22 feet; lemon, 18 to 30 feet; grapefruit, 24 to 30 feet; tangerine, 20 to 22 feet; sweet orange, 20 to 30 feet; and tangelo, 24 to 30 feet. Bearing ages vary from 5 to 10 years for good bearing with budded sweet orange cultivars and 4 to 7 years for lemon, lime, and grapefruit budded cultivars.

Nursery trees are pruned back at planting time to 18 to 24 inches. Pruning for the next year or so is mainly to remove sprouts arising below the bud union. Subsequent pruning is to select the four or five main branches that will form the main framework. Maintenance pruning is minimal, since the growth habit of citrus produces a shaped, symmetrical head. Pruning may be done to reduce size, but fruitfulness is reduced. Thinning of fruit is not necessary, unless the crop is too heavy for the tree, since the size of the remaining fruit is not greatly affected.

Insects causing the most damage include California red, purple, and black scale (*Aonidiella aurantii, Lepidosaphes beckii, Saissetia oleae,* respectively), whitefly (*Dialeurodes citri* and *D. citrifolii*), and rust mite (*Panonychus citri* and *Phyllocoptruta oleivora*). There are numerous others of lesser importance; local authorities should be consulted for these. Troublesome diseases include scab (*Elsinoë fawcetti*), twig blight (*Diplodia natalensis*), anthracnose (*Glomerella cingulata*), brown rot, and melanose (*Diaporthe citri*). Many others exist, and details are available from local authorities.

Sweet oranges can be harvested over a reasonable period, as the ripening process is not rapidly followed by deterioration. They can even be left on the tree when ripe. Fruit can be held for several weeks to months with refrigerated storage. Lemons and limes can be picked over a long period and have good storage lives. Grapefruits should not be harvested too long after ripening, and storage life is shorter than for the other citrus fruits. Tangerine harvesting should be prompt, and storage life is short as for the grapefruit.

CRANBERRY

Cranberry cultivars (Fig. 15–13) are derived from *Vaccinium macrocarpon.* Cultivars include 'Early Black,' 'Howes,' 'McFarlin,' 'Searless Jumbo,' and 'Stevens.' Recent cultivars are 'Beckwith,' 'Bergman,' 'Franklin,' 'Pilgrim,' and 'Wilcox.'

Requirements for cranberry culture are such that their cultivation is largely restricted to commercial operations in a limited number of areas. An acid bog or swamp (pH 3.2 to 4.5) with nearby water for irrigation and prevention of frost is needed. Flooding is needed in winter (except on the West Coast) to prevent winter dessication, because the cranberry is shallow rooted. Since bogs or swamps are low, late spring frosts are a threat, which used to be reduced by flooding, but now mostly by sprinkler systems. Flooding is also used in insect control. Leading cranberry areas are found in Massachusetts, New Jersey, Wisconsin, Washington, and Oregon.

Cranberries are propagated by cuttings. The mature plant is under 1 foot in height. Cranberries are planted in spring, and a 12-inch spacing for cuttings is typical. Three years are required before the first harvest.

Fig. 15-13. Cranberry, showing foliage and fruit. U.S. Department of Agriculture photo.

Pruning and fruit thinning are not practiced. Excessive runners can be trimmed after harvest. Bees are placed during flowering to assure adequate pollination.

Diseases associated with cranberries include leaf spot (*Glomerella cingulata*), early rot (*Guignardia vaccinii*) and false blossom virus. Troublesome insects include cranberry fruit worm (*Acrobasis vaccinii*), yellow-headed fireworm (*Acleris minuta*), blunt-nosed leafhopper (*Scleroracus vaccinii,* carrier of false blossom), rootworm, (*Rhabdopterus picipes*) gypsy moth, and girdler (*Chrysoteuchia topiaria*).

An acre of cranberries can produce 100 to 200 barrels. Each barrel weighs about 100 pounds. Cranberries can be held at 35°F (1.8°C).

CURRANT AND GOOSEBERRY

Red and white currants are *Ribes sativum.* Red currant cultivars, the most desirable currant, are 'Cherry,' 'Fay,' 'Perfection,' 'Red Cross,' 'Red Lake,' and 'Wilder.' 'Red Lake' is one of the best. The best white currant is 'White Imperial.' Black currants (*R. nigrum*), alternate hosts to the white-pine blister rust, are legally banned in many areas. Black currant cultivars are rarely found in the United States. The gooseberry is *R. hirtellum;* cultivars are 'Downing,' 'Oregon Champion,' 'Pixwell,' and 'Poorman.' Authorities should be consulted before growing currants or gooseberries (Fig. 15-14), as they may be banned in your area.

(A)

(B)

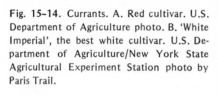

Fig. 15-14. Currants. A. Red cultivar. U.S. Department of Agriculture photo. B. 'White Imperial', the best white cultivar. U.S. Department of Agriculture/New York State Agricultural Experiment Station photo by Paris Trail.

Ribes sp. are very hardy, but do poorly in warm or dry areas. Their culture is best in the northern or higher altitudes in the United States. The gooseberry is less heat sensitive than the currant. Both are grown commercially and by the homeowner, but their acreage is very limited.

Currants are usually propagated from stem cuttings and gooseberries by mound layering. Mature heights are 5 to 6 and 3 feet, respectively. Spring planting is practiced in the colder climates and early spring or fall planting in others. Fall planting is desirable because of early blooming. They may be planted in hills set 5 to 6 feet apart in both directions or 4 to 5 feet apart in rows set at 6 to 10 feet. Nurseries usually sell 1- to 2-year-old plants.

Pruning is used to develop a bush form. Currants produce the best fruit on spurs of 2- and 3-year-old canes, and gooseberries on 1- and 2-year-old canes. Older wood, not being productive, is removed. Thinning of fruit is not practiced. Ribes sp. are self-fruitful, but maximal productivity is dependent upon pollination by bees.

Insect pests include San Jose scale, currant spanworm (Itame ribearia), aphid (Cryptomyzus ribis), and currant fruit fly (Epochra canadensis). Diseases include powdery mildew (several spp.), anthracnose (Pseudopeziza ribis), cane blight (Botryosphaeria ribis), and white-pine blister rust (Cronartium ribicola).

Yields of red currants are around 3 tons per acre and of gooseberries, 300 bushels per acre, or about 3 quarts per bush for each. Currants can only be stored for a short time under cool, dry conditions. Gooseberries last longer. Both should be utilized reasonably soon after harvesting.

Introduction to Horticultural Production / Part III

DATE

The subtropical fruit, the date (Fig. 15-15), is derived from the palm *Phoenix dactylifera*. Cultivars are grouped according to fruit texture: soft, semidry, and dry. Soft cultivars grown in the United States include 'Halawy,' 'Khadrawy,' and 'Saidy.' 'Dayeri' and 'Deglet Noor' are semidry cultivars. Dry cultivars are not popular here. Semidry dates are most popular commercially because they handle better after harvest than do soft cultivars. The latter have higher quality and are found in local markets and backyards.

Dates require high temperatures during growing and ripening, but minimal rain or humidity during ripening. They can stand light frosts. This climate is found in the low desert valleys of Southern California, Arizona, and Texas. Other areas of the United States are not suitable for date culture.

Dates can be propagated by seeds, but offshoots are utilized for cultivars. Offshoots are removed when 3 to 6 years old in the spring. Tops are headed back when planted. Mature height can vary from 50 to 100 feet. Offshoots are planted about 30 feet apart. The first crop comes in 5 to 6 years and full bearing at 10 to 15 years.

Fig. 15-15. Date clusters on tree. U.S. Department of Agriculture photo.

Pruning is restricted to late-summer removal of older fronds that show browning at the tips. Fruit set is greatly improved by hand pollination. Dates are dioecious, but males are kept minimal (1 to 100) in numbers since they yield no fruit. Strands of staminate flowers of 3- to 6-inch length are cut from a freshly opened inflorescence. Two or three pieces are inverted and tied into each pistillate cluster. Pollen may be alternately placed on a cotton ball (2 to 3 inches in diameter) and the ball placed in the pistillate cluster. Pistillate clusters are usually thinned about one third to reduce fruit load, and bunches in excess of 8 to 12 per 80 to 100 leaves are removed.

The main troublesome insects are date mite (*Oligonychus pratensis*), scale, and mealybugs. Black heart rot and false smut are some of the diseases (*Ceratocystis paradoxa* and *Graphiola phoenicis,* respectively).

Each tree can yield 100 to 200 pounds of fruit. Dates can be picked when ripe, but commercial practices are to pick short of ripe and to induce ripening under controlled conditions. Fruits can be held at 30° to 32°F (-1° to 0°C) for several months.

ELDERBERRY

Elderberries are fruits of the common American elder, a deciduous shrub. The species is *Sambucus canadensis,* and a few cultivars are known: 'Adams No. 2,' 'Johns,' 'Nova,' and 'York.' The species *S. caerulea,* blueberry elder, also bears edible fruit, but no recognized cultivars exist.

The American elder grows throughout the eastern half of the United States. Moist soil is best. Clumps spread through stolons. Both backyard and commercial plantings exist, and native stands are also used for fruit production. Its status is that of a minor fruit.

Propagation is by seeds, but preferably by division or cuttings for recognized cultivars. Plants may be planted in spring or fall. Spacing should be at least 12 feet because of their spreading nature. Mature height is 12 feet. Fruit may be produced in 3 years. Pruning is minimal except to remove spent or dead wood. Cross-pollination is needed, so two different cultivars should be planted. No fruit thinning is needed. Insect and disease problems are minimal. Berries are picked ripe and consumed or preserved quickly by canning or drying, or made into wine.

FEIJOA

The feijoa or pineapple guava is derived from *Feijoa sellowiana.* Cultivars include 'Andre,' 'Besson,' 'Choiceana,' 'Coolidgei,' 'Hehre,' and 'Superba.'

Its subtropical requirements limit the feijoa to the southern parts of zones 9 and 10. It is grown in Southern California and Florida and to a lesser extent in the southern parts of the Gulf states. It can withstand light frosts. Commercial and backyard culture is limited, so it is a minor fruit.

Propagation is by young wood tip cuttings, layering, and whip grafting. Seeds are also used, but the former methods are used for cultivars. Spring or fall planting is possible. Plants should be 15 to 18 feet apart. Bearing starts in 3 to 4 years. Minimal pruning is needed, and fruit thinning is unnecessary. Cross-pollination is desirable, so two or more cultivars should be planted. Nine-year-old plants produce 10,000 pounds of fruit per acre. Insect and disease problems are minimal. Keeping qualities are poor. Mature height is 18 feet.

FIG

Fig is a deciduous shrub of the cooler subtropics. Cultivars are derived from *Ficus carica* (Fig. 15-16). They include 'Adriatic,' 'Brown Turkey,' 'Brunswick,' 'Calimyrna,' 'Celeste,' 'Dottato,' 'Kadota,' 'Magnolia,' 'Mission,' and 'Ronde Noire.' These vary in their pollination requirements, and their qualities determine whether they are consumed fresh, dried, canned, or used in jellies.

The frost tolerance of the fig is higher than most subtropical fruits. Many cultivars can be grown as far north as zone 8 and a hardier few in zone 5 with proper winter protection or movement indoors during the winter. Figs are grown both commercially, mostly in California, and by the homeowner. The cold dormancy period needed is 200 hours below 45°F (7.3°C), and the growing season should be at least 190 days. Unless the temperatures in the winter are cold enough to produce dormancy, temperatures of wood kill can be considerably higher than the normal 5°F (-15°C).

Fig. 15-16. Fig, showing foliage and fruit. U.S. Department of Agriculture photo.

Propagation is by dormant hardwood cuttings about 4 to 5 inches long. These are taken in the winter or early spring. Both conventional and air layering are also possible. Rooted cuttings are planted in early spring 18 to 25 feet apart. Mature height is 12 to 30 feet. Bearing can start in 2 to 4 years from rooted cuttings, but commercial bearing starts at 7 years.

Pruning is minimal to extensive, depending upon the cultivar. For example, 'Adriatic' is pruned to remove lower spreading branches and to thin out interior branches when they become too thick. Some top pruning is done annually or biennially to induce new vigorous wood. 'Mission' requires only an occasional thinning of branches. 'Kadota' is trained low and flat to facilitate harvesting. 'Calimyrna' requires cutting back of long upright branches, occasional thinning, and heading back of the top.

Pollination requirements vary. Caprifigs are pollinated by the female fig wasp (*Blastophaga psenes*). When she hatches and emerges from the caprifig, where she has overwintered, she picks up pollen and, in looking for sites to lay eggs, pollinates nearby flowers. Caprifigs are of poor eating quality, but are grown to ensure a supply of wasps. The caprifigs are removed prior to wasp emergence and placed in trees of cultivars requiring pollination (caprification), but which are unsuitable hosts for wasp eggs. The main cultivar requiring this treatment is 'Calimyrna.' The process is also used to some extent with 'Kadota.' 'Kadota' and the other previously listed cultivars can produce fruit by parthenocarpic means. Fruit thinning is unnecessary.

Insects causing problems include the root knot nematode, several fig scales and several fig mites. Troublesome diseases include rust (*Cerotelium fici*), leaf spot (several spp.), anthracnose (*Colletotrichum gloeosporioides*), dieback (several spp.), souring, black smut, and internal rot. The latter three diseases from several fungi are often caused by the fig wasp.

'Calimyrna' yields 1.25 to 1.50 tons of dry figs per acre, 'Mission' and 'Adriatic' 2.0 to 2.5 tons per acre, and 'Kadota' up to 5 to 7 tons of fresh figs per acre. Figs for fresh consumption are picked ripe and hold only for 10 days at 31° to 32°F (0°C) and 85 to 90 percent relative humidity.

GOOSEBERRY (SEE CURRANT)

GRAPES

Grapes (Fig. 15–17), a deciduous vine crop, belong to the genus *Vitis*. Four cultural classes are grown in the United States: American, muscadine, Old World, and French hybrids. American cultivars are derived from northern and northeastern United States species. *Vitis labrusca* is the most important parent species, but others such as *V. rupestris, V. aestivalis,* and *V. riparia* are involved in the parentage. The latter, the frost grape, is responsible for improved cold hardiness and shorter times to maturity for some

Fig. 15-17. Grape 'Steuben' showing foliage and fruit. Courtesy of Stark Bro.'s Nurseries and Orchards Co.

American cultivars. Cultivars of the muscadine grape are derived mostly from the native species *V. rotundifolia.* The Old World cultivars are mostly from the European grape *V. vinifera.* French hybrids are crosses between *V. vinifera* and native American species.

Cultivars of the American type used as table and wine grapes include 'Catawba,' 'Concord,' 'Delaware,' 'Erie,' 'Fredonia,' and 'Steuben.' These are red-purple grapes, and 'Concord' is the leading cultivar. White cultivars include the leading 'Interlaken,' 'Niagara,' and 'Ontario.' Old World grape cultivars include those grown as table grapes: 'Cardinal,' 'Emperor,' 'Flame Tokay,' 'Perlette,' 'Ribier,' and 'Thompson Seedless.' Those grown for red wine include 'Cabernet Sauvignon,' 'Carignane,' 'Grenache,' 'Pinot Noir,' and 'Zinfandel.' Whites include 'Chardonnay,' 'Muscat of Alexandria,' 'Sauvignon Blanc,' and 'White Riesling.' Older muscadine cultivars include 'Hunt,' 'Scuppernong,' and 'Thomas,' and more recent ones are 'Burgaw,' 'Carlos,' 'Chief,' 'Fry,' 'Magnolia,' and 'Pride,' just to name a few. French hybrids, mainly wine grapes, include 'Aurore,' 'Baco Noir,' 'Chancellor,' and 'Verdelet.' These lists are not complete, but cover most cultivars of importance.

The cultural requirements of the four groups define their areas of importance within the United States. American grapes mature in about 165 days with adequate summer heat. Largest acreage is centered around the Great Lakes. They can be grown east of the Rocky Mountains and north of the cotton belt. Muscadine grapes are suited to the hot South; temperatures should not go below 0°F (-17.9°C). Old World grapes are less hardy and need at least 175 days to maturity. They are grown mainly in California, and in Washington and Oregon to a lesser extent. French hybrids of

V. vinifera and American species are suitable for areas where American cultivars exist. Slopes and sites near bodies of water are important in grape culture because of reduction in late spring or early fall frosts. Grapes, of course, are widely grown commercially and by the homeowner.

All cultivars, except muscadines, are mostly propagated by hardwood cuttings, which consist of 12-inch cane prunings with three buds. They are removed in the winter, callused through storage in moist sand or sawdust, and rooted in early spring. They are planted after the first or second season. Muscadine grapes are propagated by layering or softwood cuttings; those techniques can also be used with the other cultivars. Grafting is also used to allow growth of cultivars susceptible to root louse and nematodes in areas having these pests. Cultivars of the European grape are highly susceptible.

One-year-old cuttings or grafted vines are planted in early spring. Planting distance depends upon cultivar and soil. Vigorous American cultivars are planted 7 to 9 feet apart in rows set 9 feet apart in good soil. Muscadine grapes are set 15 to 20 feet apart in rows spread 9 to 10 feet apart. Old World grapes are planted 8 feet apart in rows set 12 feet apart. Fruit production is discouraged for the first 2 years by removal of flowers or young fruit to encourage good vine development. Bearing is not allowed until the third year after the planting.

Training requirements of grapes are determined by the training system selected. Fruit is produced on a shoot of the present season that arose from wood (cane) of the previous season. Shoots arising from older wood are unproductive, so older wood is pruned out. Pruning is usually done in early spring after the hard freezes, but before buds swell. Trellises, usually wire and post, are used to support the vines. Height is about 6 feet, although arbors may be higher. The Kniffin system (Fig. 15-18) with four canes and a single trunk, with canes at 3 feet and 6 feet, and the umbrella system with two upper arms are popular. Another approach is balanced pruning in which the number of buds left is proportionate to the vigor of the vine. There are several other systems. Specialized texts or local authorities should be consulted for extensive pruning details. Fruit thinning practices vary from none to flower cluster removal, berry cluster removal after berries are set, and partial berry removal in the cluster. Correct pruning will eliminate overcropping or, if necessary, removal of some flower clusters.

Pollination requirements vary, as some cultivars are self-sterile and others self-fertile. Early cultivars of muscadine-type grapes were extensively self-sterile, but later ones are not. Information should be sought from local authorities for the specific cultivar.

Insects that attack grapes include grape phylloxera (*Phylloxera vitifoliae*), grape leaf hopper (*Erythroneura* sp.), grape berry moth (*Paralobesia viteana*), red-banded leaf roller (*Desmia funeralis*), grapevine flea beetle (*Altica chalybea*), grape root worm (*Fidia viticida*), nematodes, rose chafer, climbing cutworm, cane girdler (*Ampeloglypter ater*), grape scale,

Fig. 15-18. Grape pruning. A. Kniffin system in early spring prior to pruning. B. After pruning. U.S. Department of Agriculture photo.

(B)

bud mite (*Eriophyes vitis*), and Japanese beetle. Diseases include black rot (*Guignardia bidwellii*), powdery mildew (*Uncinula necator*), downy mildew (*Plasmopara viticola*), dead arm (*Cryptospora viticola*), leaf spot (*Mycosphaerella personata*), Pierce's disease (Rickettsia organism), and anthracnose (*Elsinoë ampelina*).

Yields can vary from 7 to 60 pounds per vine, depending on cultivar training systems and cultural conditions. Yields per acre vary from 3 to 11 tons. American cultivars can be stored at 32°F (0°C) and 80 to 85 percent relative humidity for 3 to 8 weeks. Old World grapes can be stored at 31° to 32°F at 87 to 92 percent relative humidity for 1 to 5 months. Controlled atmosphere conditions will extend storage life.

GRAPEFRUIT (SEE CITRUS)

GUAVA

The guava (Fig. 15-19) is a tropical tree fruit of the genus *Psidium*. The most commonly cultivated species are the common guava, *P. guajava*, and the purple strawberry guava, *P. littorale* var. *longipes*. *Psidium guajava* is more tender and found only in subtropical Florida and Hawaii in the United States; *P. littorale* var *longipes* is more hardy and is seen in southern California and Florida. Only a few cultivars exist at present: 'Red Indian,' 'Ruby,' and 'Supreme.' Commercial and backyard culture of the guava is limited, making it a minor fruit.

Propagation is possible by seed, but not without variation. Propagation of cultivars is by air layering, root sprouts produced by the mother plant, and shield or patch grafting. Late spring and fall planting are practiced in Florida and California, respectively. The common guava reaches a height of 30 feet and is planted at 15 to 25 feet intervals. The strawberry guava reaches 10 to 25 feet in height and is set at 15 to 20 feet. Good bearing can be expected at 8 to 10 years.

Fig. 15-19. Guava, showing foliage and fruit. U.S. Department of Agriculture photo.

Pruning is minimal. Best fruit is borne on 1-to 3-year-old wood. Thinning and heading back of the top can be done at 2- to 3-year intervals. Fruit thinning is not necessary. Self-pollination is possible, but best fruit set is probably with cross-pollination by bees.

Insects and disease are minimal. Yields are 200 to 700 pounds of fruit per tree. Fruit can only be stored briefly under refrigeration.

JUJUBE

The jujube or Chinese date is *Zizyphus jujuba.* Good cultivars are 'Li,' 'So,' 'Tanku Vu,' and 'Yu.' It can be grown in zone 9, especially in the hotter, drier, alkaline parts such as the desert valleys in California. It has minimal importance as a commercial or backyard fruit.

Cultivars are propagated by grafting or root cuttings. These are set about 25 to 35 feet apart in the spring and can reach a height of 40 feet. Bearing can start as young as 4 years. Pruning is mainly to build a good framework and to remove older, unproductive wood. Fruit is borne on growing shoots of the current year. It appears that maximal fruit production occurs with bees and cross-pollination between two cultivars. Insect and disease problems are minimal.

JUNEBERRY

The juneberry, also called shadbush, serviceberry, and sugarplum, are species of *Amelanchier.* Species from which cultivars are derived are *A. alnifolia* and *A. stolonifera.* Mature heights are 25 and 4 feet, respectively. Cultivars of *A. alnifolia* are 'Altaglow,' 'Forestburg,' 'Indian,' 'Pembina,' 'Shannon,' and 'Smoky.' 'Success' is a cultivar of *A. stolonifera.*

Juneberries are very winter hardy through zone 5. Commercial production is essentially nonexistent. Some fruits are gathered from the wild and from backyards. It is classed as a minor fruit. Propagation of the species is by seed. Cultivars can be propagated by softwood cuttings, division, suckers, and grafting. Planting is in the spring. Pruning is not necessary other than to remove dead wood and to maintain shape. Cross-pollination by insects is desirable. Insect pests are leaf miner (*Nepticula amelanchierella*), scales, and mites. The fireblight disease (*Erwinia amylovora*) and birds are also troublesome.

KUMQUAT

Kumquats are subtropical fruits related to citrus. The nagami kumquat, *Fortunella margarita,* and the marumi kumquat, *F. japonica,* are the main two cultivated. The meiwa kumquat is thought to be a hybrid, possibly between these two species.

Kumquats are grown in the same areas as oranges, and their culture is similiar (see oranges under citrus). Commercial production is almost nil; backyard culture exists. Propagation is by grafting on *Citrus limon* cv. 'Rough' and *Poncirus trifoliata* stocks. Spacing is 15 feet. Mature height is about 10 feet. They are grown more as ornamental than as fruit trees. Fruits are used in jellies and marmalades.

LEMON (SEE CITRUS)

LIME (SEE CITRUS)

MANGO

The mango (Fig. 15-20), *Mangifera indica,* is a tropical tree fruit. Cultivars found in commercial and backyard production in the United States include 'Haden, 'Irwin,' 'Julie,' 'Keitt,' 'Kent,' 'Palmer,' and 'Zill.' Because of its tropical nature, the mango is only grown in the southern half of zone 10 in Florida and in a very small section of California.

Fig. 15-20. Mango, showing foliage and fruit. U.S. Department of Agriculture photo.

Cultivars are propagated by chip budding, inarching, or side-veneer grafting. Grafting is done on seedlings 4 to 6 weeks old during middle to late summer. These are set about 40 feet apart. Grafted trees can commence bearing in 3 to 5 years. Mature height may reach 90 or more feet.

Mangos are rarely pruned. The presence of bees appears to aid fruit set. Insect pests include several mites, red-banded thrip (*Selenothrips rubrocinctus*), scale (several spp.), and fruit fly (*Toxotrypana curvicauda*). The disease anthracnose (*Glomerella cingulata*) can be troublesome. Selective harvesting is needed as the fruit softens and falls at maturity. Keeping qualities are poor.

Fig. 15–21. Mulberry, showing foliage and unripe fruit. From Harold C. Bold, *The Plant Kingdom*, 4th edition, © 1977, p. 259. Reprinted by permission of Prentice-Hall, Inc., Englewood Cliffs, New Jersey.

MULBERRY

Mulberries (Fig. 15–21) are borne on deciduous trees of the genus *Morus*. Cultivars of the mulberry specifically planted for fruit in the North are derived from *M. alba*. These are 'Illinois Everbearing,' 'New American,' 'Thornburn,' 'Trowbridge,' and 'Wellington' (possibly same as 'New American'). Cultivars of the more tender *M. nigra* are grown in the South. 'Downing' is one of them. Hardiness zones for the two species are 5 and 3, respectively.

Species are grown from seed. Cultivars may be propagated by hardwood or softwood cuttings and by budding. These are planted 25 to 30 feet apart in the spring. Commercial production is nil; most production is in backyards. Pruning is only needed to remove deadwood and to thin out branches. Cultivars of *M. alba* may reach 80 feet and *M. nigra* 30 feet. The dioecious condition may be encountered with some cultivars. Insect pests are the mulberry white fly and San Jose scale. Bacterial blight can be troublesome. Fruits have poor keeping qualities.

NECTARINE (SEE PEACH)

OLIVE

The olive (Fig. 15-22) is an evergreen tree of the species *Olea europaea*. The variety *europaea* is the typical cultivated common olive. Cultivars include 'Ascolano,' 'Barouni,' 'Manzanillo,' 'Mission,' and 'Sevillano.' Commercial and backyard production of olives occurs in the United States.

Climatic requirements of olives are such that their production is limited to the hot dry areas of the Southwest, especially the hot interior valleys of California. Winter chilling for 12 to 15 weeks with night and day temperatures of 35°F and 60°F (1.8°C and 15.6°C) is needed to initiate flowering. Temperatures below 12°F (-11.1°C) are fatal. Blooming is late enough to miss spring frosts. A long, hot growing season is needed for good fruit development. Bearing is poor in the Southeast, as the late flowering and requirement of a long growing season expose the fruits to fall frost damage there.

Propagation is by rooting leafy cuttings. Grafting or budding is also practiced; the whip graft or side graft is favored. Trees are set 35 feet apart in early spring. Bearing may start 5 years after planting 2- or 3-

Fig. 15-22. Olive, showing foliage and fruit. U.S. Department of Agriculture photo.

year-old nursery stock. Full bearing can start in 12 to 20 years. Mature height is 25 feet or more.

Pruning is mainly to select three to five scaffold branches. Subsequent pruning is minimal, as it reduces yield. Most cultivars are self-fruitful, but cross-pollination probably increases yields. Bees appear to be effective, but wind is the chief vector for pollination. Excess fruit set can occur, resulting in smaller fruit and possibly in alternate bearing. A chemical fruit-thinning agent can be used.

The only serious insect is black scale (*Saissetia oleae*). A gall-causing bacteria, *Pseudomonas savastanoi,* and the fungi, *Verticillium* and *Cyclo-conium oleaginum,* which cause wilt and defoliation, respectively, can be troublesome.

Yields vary from 1 to 5 tons per acre. Olives are usually processed by pickling or for oil.

ORANGE, SWEET AND SOUR (SEE CITRUS)

OTAHEITE ORANGE (SEE CITRUS)

Fig. 15-23. Papaya, showing foliage and fruit. U.S. Department of Agriculture photo.

PAPAYA

The papaya (Fig. 15-23) is a herbaceous tropical fruit. The species is *Carica papaya*. The only cultivar appears to be 'Solo.' Since it can only withstand a few degrees of frost, it is only cultivated in the warmer parts of zone 10 in Florida and California. Commercial and backyard production is limited, making it a minor fruit.

Propagation is by seeds started indoors and placed out in early spring. Plants are set about 10 to 12 feet apart. Bearing may start with a 1-year-old plant, but they only bear for 3 to 4 years. Mature height is 25 feet.

No pruning or fruit thinning is practiced. Some set fruit by parthenocarpic means, others are dioecious requiring a male and female plant, and others are self-fruitful. Best fruits appear to arise with dioecious plants pollinated by bees. Insects and disease problems are minimal. Each plant can bear 12 to 30 fruits.

PASSIONFRUIT

Passionfruit or granadilla are species of *Passiflora*. Edible species include purple granadilla (*P. edulis*), sweet granadilla (*P. ligularis*), yellow granadilla (*P. laurifolia*), giant granadilla (*P. quadrangularis*), sweet calabash (*P. maliformis*), and curuba (*P. mollissima*). These are woody tropical vines, and their culture is limited to the southern parts of zone 10 in California and Florida. Commercial production does not exist in the continental United States, but backyard plantings do.

Propagation is by seeds and cuttings. Plants are set out in early spring from fall-planted seeds. They are placed 6 to 10 feet apart in rows 10 feet apart. They are usually trellised. Plants can start bearing at 1 year and remain productive for 6 years. They are usually pruned back severely after harvesting the fruit. Flowers are self-sterile and some plants self-incompatible. Compatible clones or cultivars must be selected; bees are effective pollinators. Nematodes can be very troublesome.

Yields are from 15,000 to 40,000 pounds per acre, or 40 pounds at most per vine.

PEACH AND NECTARINE

The peach (Fig. 15-24), *Prunus persica,* is a temperate-zone fruit. The nectarine is a smooth-skinned variety of the peach, *P. persica* var. *nucipersica.* Both are grown extensively in commercial and backyard production.

Peaches are a little less hardy than apples, so their range is a bit south of the apple. Zone 5 is the limit for peaches. The chilling requirement for peaches is usually 600 to 1000 hours below 45°F (7.3°C), but values as low as 50 hours and as high as 1200 hours are known for cultivars adapted to the southern and northern limits, respectively, of the peach growing range. Winter temperatures should not fall below –15°F (–26.2°C), late spring frosts can damage the fruit buds, and clear hot summers are best for fruit development. Best conditions are found in the South, MidAtlantic States, West Coast, and those areas near large bodies of water in the northern states, such as the Great Lakes area.

Cultivars are extremely numerous and adapted for regions, time of ripening, and various fruit characteristics such as clingstone, freestone,

(A)

(B)

Fig. 15-24. A. Peach showing foliage and fruit. B. Nectarine. U.S. Department of Agriculture photos.

yellow-fleshed, and white-fleshed. Only a few can be mentioned here. More specialized works and local sources should be consulted for others. In the extreme north of zone 5, cultivars with the hardiest fruit buds are recommended: 'Polly,' 'Prairie Dawn,' 'Reliance,' and 'Veteran.' At the other end in Florida, cultivars with minimal chilling requirements do well: 'Angel,' 'Flordanon,' 'Florida Gem,' 'Honey,' 'Imperial,' 'Jewell,' 'Lutti-chau,' 'Suber,' and 'Waldo.' In New England, New York, and Michigan, suggested cultivars are 'Canadian Harmony,' 'Candor,' 'Cresthaven,' 'Garnet Beauty,' 'Jefferson,' 'Jerseyland,' 'Redhaven,' 'Redskin,' and 'Triogem.' Those grown in the Southeast are 'Biscoe,' 'Blake,' 'Candor,' 'Cardinal,' 'Coronet,' 'Dixired,' 'Early Coronet,' 'June Gold,' 'Maygold,' 'Ranger,' 'Redglobe,' 'Redskin,' and 'Springold.' Midwest cultivars are 'Canadian Harmony,' 'Cresthaven,' 'Garnet Beauty,' 'Glohaven,' 'Harbinger,' 'Loring,' and 'Rio Oso Gem.' In the Pacific Northwest the following are grown: 'Cardinal,' 'Delp Early Hale,' 'Dixired,' 'Earlihale,' 'Early Red-haven,' 'Gold Medal,' 'J. H. Hale,' 'Redglobe,' 'Redhaven,' 'Redskin,' and 'Rio Oso Gem.' Californian cultivars include 'Altair,' 'Bonanza,' 'Cardinal,' 'Desertgold,' 'Fay Elberta,' 'Regina,' 'Redglobe,' 'Redtop,' 'Rio Oso Gem,' 'Saturn,' 'Summerset,' and 'Suncrest.' A few nectarine cultivars are 'Early Flame,' 'Lexington,' 'Mericrest,' 'Pioneer,' 'Redchief,' and 'Silver Lode.'

Peach cultivars are propagated by being budded on seedling stocks. The preferred rootstocks in the North and South are, respectively, 'Siberian C' and 'Nemaguard.' 'Halford' and 'Lovell' are common rootstocks produced from cannery pits. Stocks from sand cherry, *Prunus besseyi,* Nanking cherry, *P. tomentosa,* and plum, *P.* 'St. Julien A,' are used to produce dwarf peaches. Peach cuttings can also be rooted with moderate success. Pits are sown in late summer or in spring after stratification. Buds are inserted late on rootstocks after one season of growth; buds remain dormant then until spring. After one season's bud growth, trees are set 18 to 24 feet apart in the spring in the North and in fall or spring in the South. Bearing can start as early as 3 years, and maximal bearing is from 8 to 12 years. Lifespans are short for peaches, varying from 8 to 20 years. Standard trees reach a height of 24 feet.

Peach trees are usually trained to the open center form and, to a lesser extent, the central-leader system. Maintenance pruning is to keep height within control for harvesting, to keep the center open (for open-center form), and to maintain productivity. Peaches are borne on wood produced the previous season. Most cultivars are self-fruitful, except for 'Alamar,' 'Candoka,' 'Chinese Cling,' 'Hal-berta,' 'J. H. Hale,' 'June Elberta,' and 'Mikado.' Interplanting of another cultivar is needed for these. Bees are the main pollinators. Fruit thinning is usually necessary, unless a late spring frost has reduced the crop. Larger fruits are produced by thinning either by hand, mechanical, or chemical means.

Troublesome insects are the peach borer (*Sanninoidea exitiosa*), oriental peach moth (*Grapholitha molesta*), plum curculio (*Conotrachelus nenuphar*), scale (several spp.), aphid (several species), and tarnished plant bug (*Lygus lineolaris*). Diseases include peach leaf curl (*Taphrina deformans*), brown rot (*Monilinia fructicola*), bacterial leaf spot (*Mycosphaerella* spp.), scab (*Cladosporium carpophilum*), yellows (virus), little peach (virus), phony peach (*Rickettsia* organism), and peach X disease (*Mycoplasma* organism). Numerous viruses attack peach; stock should be purchased virus-free.

Yields are 3 to 4 bushels per tree and 20 to 25 tons per acre in better orchards. Peaches can be held at 32°F (0°C) and 85 percent relative humidity for a month. Controlled-atmosphere storage looks promising for peaches.

PEAR

Pears (Fig. 15-25) can be divided into three groups: European pear (*Pyrus communis*), Asian pear (*P. pyrifolia*), and Eurasian pear (a hybrid between the previous two, *P.* X *lecontei*). The best-known cultivars of the latter are 'Kieffer' and 'Leconte.' Cultivars of the European pear are 'Anjou,' 'Bartlett,' 'Bosc,' 'Clapp,' 'Comice,' 'El Dorado,' 'Flemish Beauty,' 'Seckel,' and 'Winter Nelis.' Commercial pear production is mostly 'Bartlett,' followed by 'Bosc' and 'Anjou.' Cultivars showing resistance to fire blight are 'Kieffer,' 'Magness,' 'Maxine,' and 'Moonglow.' Cultivars of the oriental pear are seldom seen in the United States. Pears are grown in home orchards and produce acceptable fruit with less attention than apples.

Pears have a more limited range than apples. Their northern range, zone 5, is limited by cold temperatures below –20°F (–25°C), and the humidity in the South is limiting. The bacterial disease fire blight, spread by bees and favored by rains, especially limits pear culture. Most of the commercial production centers around the Pacific Coast states, Washington, Oregon, and California. Production occurs in the Great Lakes areas, but is limited. Home plantings occur in these and other areas, but cultivars should be chosen carefully on the basis of local acceptability. About 900 to 1000 hours below 45°F (7.3°C) are required to break dormancy.

Pears are widely propagated by budding of cultivars on seedlings of the European pear, especially 'Bartlett.' *Pyrus calleryana* is used in milder areas. Dwarf pears may be produced by budding onto a Malling quince 'A' (*Cydonia oblonga* 'Angers') root; an interstem, 'Old Home,' is usually used to overcome incompatibility problems and to increase resistance to fire blight. The quince cultivar 'Provence' is also used for dwarfing purposes. Mature standard European pears can reach 45 feet in height.

One- or two-year-old nursery stock is planted in the early spring in the North, and in fall or spring in the South. Pears may be planted 8 to 14 feet apart in rows set 12 to 24 feet apart. Dwarf trees may be set 4 to 8 feet apart in rows set 10 feet apart. Pears may commence bearing after 3 years from planting of nursery stock. Pears may live up to 100 years.

Training is primarily the modified-leader system. Three or four main scaffold limbs are left, except in areas where fire blight is prevalent; six limbs are left in the latter areas in case the blight necessitates some removal. Fruits are borne on spurs that are productive for 7 to 8 years. Maintenance pruning is mainly to remove unproductive spurs and to encourage formation of new ones.

Most pear cultivars are all or partly self-sterile, so two or more cultivars should be interplanted. Bees are recommended for maximal fruit productivity. Unfortunately, they are vectors for the spread of fire blight disease where it is present. Parthenocarpic fruits also occur. Fruit thinning is minimal; it is practiced by hand or chemical and hand treatments when larger fruit is desired. Chemical sprays of hormones may be used to reduce premature droppage of fruit.

Fig. 15-25. Pears, U.S. Department of Agriculture releases. A. 'Magness'. B. 'Dawn'. C. 'Moonglow'. U.S. Department of Agriculture photo.

Troublesome insects include the codling moth (*Carpocapsa pomonella*), plum curculio, pear and tarnished plant bug, pear midge (*Contarinia pyrivora*), pear thrip (*Taeniothrips inconsequens*), blister mite (*Eriophyes pyri*), and pear psylla (*Psylla pyricola*, vector of a probable mycoplasma disease, pear decline). Rodents can cause extensive girdling damage. Diseases of importance are leaf blight (*Fabraea maculata*), fire blight (*Erwinia amylovora*), and pear decline. Yields of 150 to 240 bushels per acre (4 to 5 bushels per tree) are possible, or about an average of 8 tons per acre. Pears should be precooled. Storage is at 30° to 31°F (-1° to -0.6°C). With precooling and these temperatures, storage for 2 months or longer, depending on cultivar, is possible. Controlled atmosphere storage is better, but used very little.

PERSIMMON

Persimmons (Fig. 15-26) grown for fruit include *Diospyros virginiana* and *D. kaki,* the American and Oriental (or Japanese) persimmon, respectively. Only the Oriental persimmon is grown commercially; both are grown in backyard plantings. American persimmon cultivars are 'Early Golden,' 'Miller,' and 'Ruby.' 'Fuyu,' 'Gailey,' 'Hachiya,' and 'Tanenashi' are Oriental persimmon cultivars of major importance.

Fig. 15-26. American persimmon showing foliage and fruit. U.S. Department of Agriculture photo.

The American persimmon is hardy to zone 5 and the Oriental persimmon to zone 6. Best growing conditions for the latter are found in the cotton belt (zone 8), and commercial production is centered in California, Texas, Florida, southern Georgia, and southern parts of the Gulf States. The American persimmon is found primarily in the eastern and southeastern states to Texas.

Propagation of cultivars is by cleft grafting, whip grafting, or shield budding on 1- to 2-year-old seedling stocks. The seedling stock in the East and South is *D. virginiana.* In California, both *D. kaki* and *D. lotus*

Introduction to Horticultural Production / Part III

(date plum) are used. Transplanting is difficult because of the presence of a taproot. Grafted trees are planted in the late fall in warmer areas and in early spring in others. Spacing is 15 to 20 feet between trees. Bearing may commence 2 to 3 years after grafting. Trees tend to be short-lived. Mature height is 45 feet for the Oriental type and more for the American species.

Training is to the modified-leader system. Heading back annually is needed to maintain a manageable size. Fruit is produced on the previous season's wood. Flower types vary considerably; perfect, staminate, and pistillate flowers may even be produced on one tree. Parthenocarpic fruit set also occurs. 'Gailey' is frequently planted as a pollen source. Bees are the primary pollinators. Fruit thinning is not practiced. Insect and disease problems are minimal.

Many cultivars have an astringent taste until fully ripe. 'Fuyu' is an exception. Yields of 50 pounds per tree are seen, although with optimal culture higher yields are possible. Fruits are usually harvested unripe and allowed to ripen afterward. Storage at 30°F (-1°C) and 85 to 90 percent relative humidity is possible for 2 months.

PINEAPPLE

The pineapple (Fig. 15-27), *Ananas comosus,* is a terrestrial bromeliad primarily grown in the tropics. Culture is limited in the continental United States to southern Florida and California. Commercial production is mostly in Hawaii. The main commercial cultivar is 'Smooth Cayenne.' Others are 'Abbaka,' 'Porto Rico,' and 'Red Spanish.'

Fig. 15-27. Pineapple, showing foliage and fruit.
American Society for Horticultural Science photo.

Propagation is mainly by suckers developed along the stem or slips at the peduncle just below the fruit. The crown (leafy part on top of fruit) can also be rooted, but is much slower to produce fruit. Pineapple slips or suckers are placed 8 to 22 inches apart in beds, usually after they have been removed from harvested fruit and dried for 1 to 4 weeks. Mature height is 3 feet. Fall-planted slips can set fruit the second summer following. Fruit formation is usually induced by naphthaleneacetic acid for purposes of uniform harvesting. Ethylene gas from an apple in a plastic bag around the plant can be used to cause flowering and fruit formation in individual plants. Ethylene in water with an absorbent is also sprayed on pineapple fields. Parthenocarpic fruit formation is possible.

Nematodes, thrips, scale and mealybugs are troublesome. The latter two carry viruses that can infect the plant. Wilt is a troublesome disease. Pineapples are best when picked ripe, but are usually picked about a week before full maturity by commercial growers.

PLUM

Plums (Fig. 15-28), species of *Prunus,* can be divided into four groups. The common or European plum is *P. domestica.* This group is the source of prune plums used to prepare prunes by drying. Cultivars in this group include 'Agen' (largest commercial prune source), 'Arctic,' 'Bradshaw,' 'French' (best for jam), 'Grand Duke,' 'Italian Prune,' 'Reine Claude,' 'Stanley' (good home garden type), 'Washington,' and 'Yellow Egg.' Japanese or Oriental plums derive from *P. salicina;* cultivars are 'Beauty,' 'Formosa,' 'Santa Rosa,' and 'Shiro.' Native American plums are derived from a number of species: wild plum, *P. americana* ('De Soto,' 'Forest Garden,' 'Hawkeye,' and 'Wolf'); sand plum, *P. angustifolia* ('Caddo Chief,' 'Strawberry,' and 'Yellow Transparent'); hortulan plum, *P. hortulana* ('Golden Beauty,' 'Miner,' and 'Wayland'); beach plum, *P. maritima* ('Autumn,' 'Eastham,' 'Hancock,' 'Premier,' 'Raritan,' and 'Stearns'); wild goose plum, *P. munsoniana* ('Newman,' 'Robinson,' and 'Wild Goose'); Canadian plum, *P. nigra* ('Cheney,' 'Hasca,' and 'Oxford'); and the Pacific plum, *P. subcordata* ('Sisson'). The last group is of hybrids between the American and Japanese species; these are represented by cultivars such as 'Ember,' 'Kahinta,' 'La Crescent,' 'Monitor,' 'Superior,' and 'Underwood.' The cultivars mentioned are only a partial listing; these and others can be found both in commercial and backyard production. The cultivars used commercially come mainly from *P. domestica* and *P. salicina.*

Because of the wide number of plum species, it is possible to grow plums in just about any part of the United States. Commercial production is centered in California, followed by Oregon, Washington, Michigan, Idaho, Texas, and New York. The European plum is hardy through zone 5, the Japanese plum through zone 8, the American plum as far as zone 2 (*P. nigra*), and Japanese-American hybrids to zones 6 to 7. Local authori-

Fig. 15–28. Plums. A. 'Starking Delicious', a Japanese-type plum. B. 'Blufre', a prune plum. Courtesy of Stark Bro.'s Nurseries and Orchards Co.

ties should be consulted for cultivars best suited to your area. Chilling required to break dormance is 800 to 1200 hours below 45°F (7.3°C) for the European plum and 700 to 1000 hours for the Japanese plum.

Propagation is primarily by budding and cuttings to a much lesser extent. Various rootstocks are used. The cherry plum or Myrobalan rootstock (*P. cerasifera*) is widely used in the East and West for European and Japanese plums. Peach stocks are sometimes used for Japanese plums to be grown in light soils and/or in the South. American plum rootstocks are used when increased hardiness is desired.

One-year-old nursery stock is usually planted in early spring. Trees are set 18 feet apart in rows spaced 22 feet apart. Spacing of 8 by 12 feet is possible with cordon-trained trees. Japanese plum may bear fruit in 3 to 5 years, European plums in 5 to 7 years, and American plums in about 8 to 10 years.

Plums are trained in the modified-leader and open-center systems; choice should be based on growth habits. Spreading types are better suited to the open-center system and upright types to the modified-leader system. Fruit is borne on 1–year old wood and the more vigorous spurs on older wood. Pruning is generally light, with the Japanese plum requiring somewhat more than the European plum. Pruning requirements can vary depending on cultivar. European and Japanese plums can grow to 25 to 30 feet in height. American plums vary from 10 to 30 feet.

European plum cultivars vary from self-fruitful to self-unfruitful. Japanese plums are mostly self-unfruitful. American and hybrid plums vary. Pollination requirements of cultivars must be carefully determined, and cultivars chosen for interplanting must be compatible. Bees are recommended for orchards. Fruit thinning is practiced extensively to increase fruit size and to prevent breakage of the tree. Hand, mechanical, and chemical thinning is practiced.

Troublesome insects include the plum curculio (*Conotrachelus nenuphar*), peach borer (*Sanninoidea exitiosa*), and shot hole borer (*Scolytus rugulosus*). Diseases include black knot (*Dibotryon morbosum*), brown rot (*Monilinia fructicola*), and bacterial leaf spot (*Xanthomonas pruni*).

Yields in the East are 3 to 5 tons per acre or 1 to 1.5 bushels per tree. California yields are 6 to 10 tons per acre, with figures as high as 16 tons reported. Plums are perishable and can be held at 30° to 32°F (-1° to 0°C) for 2 to 4 weeks. Precooling is desirable.

POMEGRANATE

The pomegranate (Fig. 15-29), *Punica granatum,* is a deciduous shrub. Cultivars include 'Nana' (dwarf), 'Papershell,' 'Spanish Ruby,' and 'Wonderful.' Its culture is limited in the United States to the hot desert valleys of California and the Southwest. Good crops require a semiarid condition where heat is present for ripening and water can be supplied to the roots. Commercial and backyard production is limited.

Dormant woody cuttings are used for propagation. Rooted cuttings are placed 12 to 15 feet apart in early spring. Pruning is minimal, except to remove suckers and thinning of dense growth. Fruit is borne on spurs found on 2-year or older wood. Bearing may commence in 5 years. Insects and disease are minimal. Mature height is 16 feet. Production is about 5 tons per acre. Fruit picked prior to maturity can be stored for months in a cool, dry place.

Fig. 15-29. Pomegranate, showing foliage and fruit. U.S. Department of Agriculture photo.

QUINCE

Fruits of the species *Cydonia oblonga* are those of the common quince. *Cydonia sinensis* is also used for fruit. The Japanese or Oriental quince, *Chaenomeles speciosa,* is also used occasionally for its fruit. Cultivars of the common quince include 'Champion,' 'Fuller,' 'Meech,' 'Orange,' 'Pineapple,' and 'Smyrna.'

Quinces are comparable to peach in wood hardiness. *Cydonia oblonga* is hardy through zone 5, but *C. sinensis* is hardy only through zone 6. Since it is susceptible to fire blight, this can be a limiting factor. Production is mainly in backyards; commercial production is very limited to small areas in California, Ohio, Pennsylvania, and New York.

Propagation is by rooting hardwood cuttings or by budding of cultivars onto a quince rootstock, usually 'Angers.' Two- or three-year-old nursery stock is set in early spring about 16 feet apart. Bearing may start 2 to 3 years after planting the nursery stock and reach a maximum in 10 years. Mature height is about 20 feet.

Pruning is minimal after training to the open-center system. Fruits are borne on new growth of the current year. Flowers are self-fruitful. Bees are the main pollinators. Some fruit thinning may be needed in some years if the tree overbears. Insect pests include the Oriental fruit moth, quince curculio, aphid, and lace bug. Diseases include the cedar-quince stem rust, black spot, and fire blight.

Yields are about 1 bushel per tree. Quinces can be stored at 30° to 32°F (-10°C) for about 2 months.

RASPBERRY

Raspberries, a bramble fruit, can be divided into three groups. The red raspberry is *Rubus idaeus;* the most common variety in the United States is variety *strigosus.* The black raspberry (Fig. 15–30), is *R. occidentalis.* Hybrids between the two preceding groups are *R. X neglectus,* the purple raspberry. Cultivars of the red raspberry include 'Amber,' 'Augustred' (everbearing), 'Canby,' 'Cherokee,' 'Citadel,' 'Dormanred,' 'Fairview,' 'Fallgold' (everbearing), 'Fallred' (everbearing), 'Haida,' 'Heritage' (everbearing; see Fig. 15–31), 'Hilton,' 'Latham,' 'Matsqui,' 'Meeker,' 'Milton,' 'Newburgh,' 'Pocahontas,' 'Puyallup,' 'Reveille,' 'Scepter,' 'Sentinel,' 'September' (everbearing), 'Southland,' 'Summer,' 'Sunrise,' 'Taylor,' and 'Williamette.' Cultivars of the black raspberry include 'Allegany,' 'Allen,' 'Black Hawk,' 'Bristol,' 'Dundee,' 'Huron,' 'Jewel,' 'Morrison,' 'Munger,' and 'New Logan.' Cultivars of the purple raspberry include 'Amethyst,' 'Clyde,' 'Purple Autumn,' and 'Sodus.'

The red raspberry is hardy through zone 3, and the black and purple raspberries are hardy through zone 4. Heat and drought are limiting factors for raspberry culture. As such they are cultivated mostly north of the

Fig. 15–30. (left) 'Jewel', a black raspberry. U.S. Department of Agriculture/New York State Agricultural Experiment Station photo by Paris Trail.

Fig. 15–31. (above) Heritage, a red raspberry. U.S. Department of Agriculture/New York State Agricultural Experiment Station photo by Paris Trail.

Mason-Dixon line. Raspberries are produced both in commercial and back-yard plantings; the leading commercial areas are Michigan, Oregon, Washington, and New York. Certain cultivars are recommended for various regions; local authorities should be consulted.

Tip layering is used for the propagation of black and purple cultivars. This is done in middle to late summer, and roots will have formed by autumn. Red cultivars are increased by removing the suckers that are produced by the crown and roots. Purchased plants should be certified as virus free.

One-year-old plants are set out in the spring. Red raspberries are set 2 to 3 feet apart in rows at 7 to 9 feet intervals. Black raspberries are set 3 to 4 feet apart in rows 8 to 10 feet apart, and purple raspberries have the same spacing within the row, but rows are set 10 feet apart. Suckers arising more than 1 foot on either side of the row should not be allowed to grow. Simple two-wire trellises are often placed to contain and support the drooping plants. Canes are biennial; fruit is produced only once on the second-year cane, except for everbearing types, which set fruit twice in one season (June and August). These are suggested for areas where winter cold kills (if growing season is long enough) or injures summer-fruiting canes (two years old), since the fall crop is borne on current season canes.

Black and purple raspberries require summer topping; red ones generally do not. When shoots have reached 18 to 24 inches for black cultivars and 30 to 36 inches for purple cultivars, the top 3 to 4 inches is removed. Thereafter, tips should be removed at weekly intervals. All should have fruiting canes removed after harvesting berries (everbearing raspberries later than others) and some thinning out of the weaker canes produced that season (Fig. 15-32). Thinning may be delayed until early spring. In early spring before buds swell, red raspberries should be cut back to 5 to 5.5 feet. Laterals of summer-topped black and purple cultivars should be cut back to 8 and 10 to 18 inches, respectively. Fruit set is improved substantially by bees. Fruit thinning is not practiced. Reds may be productive for 20 years, while purples and blacks last 5 to 10 years.

Mosaic virus is troublesome, as is leaf curl, streak, and tomato ringspot. Aphids and nematodes are involved in the spread of these viruses. Spur blight (*Didymella applanata*), cane blight (*Leptosphaeria coniothyrium*), powdery mildew (*Sphaerotheca macularis*), fruit rot (*Botrytis cinerea*), *Verticillium* wilt and anthracnose (*Gloeosporium allantosporum*) are other diseases that attack raspberries. Insect pests include the consperse stink bug, crown borers (*Bambecia marginata*), cane borers (*Oberea maculata*), raspberry fruit worm (*Byturus rubi* and *B. bakeri*), Japanese beetle, several mites, tree cricket (*Oecanthus* sp.), raspberry saw fly (*Monophadnoides geniculatus*), and rose chafer.

Yields of purple, black, and red raspberries are, respectively, 4000 to 5000, 3000 to 4000, and 2000 to 2500 quarts per acre. Higher yields are possible with optimal conditions. Yields per bush for all vary from 2 to 4 quarts. Raspberries are highly perishable, and precooling is suggested.

(A)

(B)

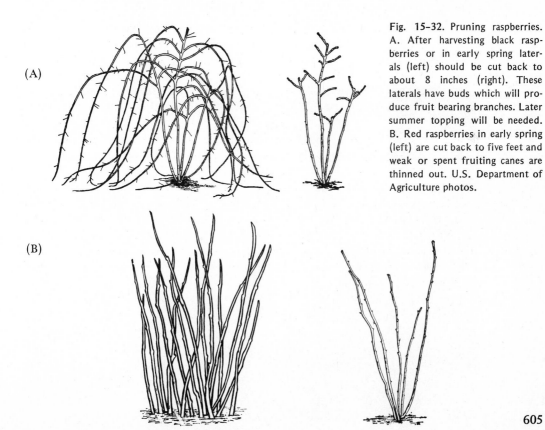

Fig. 15-32. Pruning raspberries. A. After harvesting black raspberries or in early spring laterals (left) should be cut back to about 8 inches (right). These laterals have buds which will produce fruit bearing branches. Later summer topping will be needed. B. Red raspberries in early spring (left) are cut back to five feet and weak or spent fruiting canes are thinned out. U.S. Department of Agriculture photos.

The cultivated strawberry is a hybrid between *Fragaria chiloensis* and *F. virginiana,* and is designated as *F.* ✕ *ananassa* (Fig. 15–33). There are many cultivars of this perennial herb. They vary considerably in horticultural characteristics; local authorities should be consulted for cultivars best suited to your area in the United States. The most important and some of lesser importance at this time are listed: 'Albritton,' 'Aliso,' 'Apollo,' 'Arapahoe' (everbearing), 'Atlas,' 'Badgerbelle,' 'Blakemore,' 'Catskill,' 'Dabreak,' 'Darrow,' 'Delite,' 'Earlibelle,' 'Earlidawn,' 'Earlimore,' 'Fletcher,' 'Florida Ninety,' 'Fresno,' 'Gala,' 'Garnet,' 'Gem' (everbearing), 'Geneva' (everbearing), 'Guardian,' 'Headliner,' 'Heidi,' 'Holiday,' 'Hood,' 'Marlate,' 'Midway,' 'Nisqually' (everbearing), 'Northwest,' 'Ogallala' (everbearing), 'Olympus,' 'Ozark Beauty' (everbearing), 'Pocahontas,' 'Rainier,' 'Raritan,' 'Redchief,' 'Redglow,' 'Robinson,' 'Salinas,' 'Scott,' 'Sequoia,' 'Shuksan,' 'Sparkle,' 'Stoplight,' 'Streamliner' (everbearing), 'Sunrise,' 'Superfection Brilliant' (everbearing) 'Surecrop,' 'Tioga,' 'Titan,' 'Totem,' 'Trumpeter,' 'Tufts,' 'Veestar,' and 'Viking.' A few everbearing cultivars are included; these produce a second crop in the fall and are seen mostly in northern areas (primarily backyard plantings at present).

Commercial production is concentrated in Florida, Louisiana, Arkansas, Missouri, Tennessee, Illinois, New Jersey, California, Washington, and Oregon. Propagation is by new plants produced on runners. New plants (certified virus free) are put out in early spring in the North, and usually in the autumn in the South and California. Late frost pockets should be avoided to avoid frost damage to the flowers in the spring.

Fig. 15–33. Strawberry, showing foliage and fruit. U.S. Department of Agriculture photo.

Fig. 15-34. Matted row system for strawberries. U.S. Department of Agriculture photo.

Spacing depends on the training system (Fig. 15-34) used. In the hill system, where no runners are allowed to develop, plants are set 10 to 12 inches apart in twin rows spaced 12 to 14 inches, with 3.5 to 4 feet between each set of twin rows. This is used in commercial plantings in California, Louisiana, and Florida. The matted row system is used mainly in the other commercial areas. In this system plants are set 15 to 30 inches apart in rows, separated by 3.5 to 5 feet. Runners are allowed to root randomly, but are not allowed to grow beyond a 15- to 24-inch strip. Other approaches are spaced row for cultivars with moderate to weak ability to send out runners, where runners are deliberately placed 4 to 12 inches apart. Specialized growing containers, such as the strawberry pyramid or barrel, are used also. Runner plants should never be allowed to be closer than 3 to 4 inches; excess growth should be removed. Fruit set should be avoided the first summer after the plants are set; this is done by removal of flower stems. Fruit production may last 2 to 6 years. Continuous fruit production is ensured by putting in new plantings every year or two. A winter mulch is suggested in northern areas where the soil freezes.

Most cultivars are self-fruitful. Information suggests that bees could be used to increase volume of fruit and numbers of perfect fruit.

Troublesome insects include beetles (several spp.), leaf roller (several spp.), strawberry weevil (*Anthonomus signatus*), cyclamen mite (*Steneotarsonemus pallidus*), spider mite, spittle bug (*Philaenus spumarius*), white fly (*Trialeurodes* sp.), aphid (several spp.), nematode, strawberry crown borer (*Tyloderma fragariae*), and rootworm (*Paria fragariae*). Virus diseases (transmitted by the aphid) such as crinkle or yellows can be serious. Other diseases include red stele (*Phytophthora fragariae*), leaf spot (several spp.), leaf blight (*Dendrophoma obscurans*), and *Verticillium* wilt.

Yields are about 4.5 tons per acre, or about one quart per plant.

FURTHER READING

Childers, Norman F., *Fruit Nutrition—Temperate to Tropical*. New Brunswick, N.J.: Horticultural Publications (Rutgers University), 1966.

———, *Modern Fruit Science* (7th ed.). New Brunswick, N.J.: Horticultural Publications (Rutgers University), 1976.

Darrow, George M., *The Strawberry—History, Breeding, and Physiology*. New York: Holt, Rinehart & Winston, 1966.

Eck, Paul, and N. F. Childers, *Blueberry Culture*. New Brunswick, N.J.: Horticultural Publications (Rutgers University), 1966.

Janick, Jules, and James N. Moore, *Advances in Fruit Breeding*. West Lafayette, Ind.: Purdue University Press, 1975.

Rodale, J. I., and staff, *How to Grow Vegetables and Fruits by the Organic Method*. Emmaus, Pa.: Rodale Books Inc., 1971.

Scheer, Arnold H., and E. M. Juergenson. *Approved Practices in Fruit and Vine Production* (2nd ed.). Danville, Ill.: Interstate Printers & Publishers, Inc., 1976.

Shoemaker, J. S., *Small Fruit Culture* (4th ed.). Westport, Conn.: Avi Publishing Co., 1975.

Simmons, Alan E., *Growing Unusual Fruit*. New York: Walker & Co., 1972.

Teskey, B. J. E., and J. S. Shoemaker, *Tree Fruit Production* (3rd ed.). Westport, Conn.: Avi Publishing Co., 1978.

Tukey, H. B., *Dwarfed Fruit Trees* (reissued). Ithaca, N.Y.: Cornell University Press, 1978.

U.S. Department of Agriculture, *Fruits and Nuts*. In *Gardening For Food and Fun*. Yearbook of Agriculture. Washington, D.C.: U.S. Government Printing Office, 1977.

Winkler, A. J., and others, *General Viticulture*. Berkeley, Calif.: University of California Press, 1974.

Other Horticultural Crops

Nut trees have much to offer the horticulturist. Nuts are high in food value. The tree itself provides shade and is ornamental in the landscape. However, nut trees are not fully utilized by horticulturists as a food crop or ornamental plant. Plant breeders have not bred nut cultivars to their fullest potential. In time it is hoped we will see the nut tree increase in horticultural prominence.

Commercial nut production is limited to a few states: California, Oregon, Hawaii, Washington, Georgia, Texas, Alabama, and New Mexico. In the United States, commercial shelled-nut production totaled 426 million pounds in 1975. Production overall remained similar in 1976. This consisted predominately of almonds, walnuts, pecans, macadamias, and filberts in order of decreasing commercial value in 1976. Other nuts are grown commercially, but to a much less extent. Commercial value for these major nuts was $333 million in 1975 and $396 million in 1976. Backyard horticulturists grow a wider range of nuts than the commercial growers. Nuts are discussed individually in the following text.

ALMOND

Almonds are deciduous trees grown for their nuts. They belong to *Prunus dulcis; P. dulcis* var. *amara* is noted for oil of bitter almond, and the variety *dulcis* is grown for its edible nut. The flowering ornamental almonds are *P. glandulosa, P. japonica,* and *P. triloba.* Edible cultivars include 'Ballico,' 'Davey,' 'Eureka,' 'IXL,' 'Jordanolo,' 'Kapareil,' 'Merced,' 'Mission,' 'Ne Plus Ultra,' 'Nonpareil' (most widely grown commercial cultivar), 'Peerless,' 'Texas,' and 'Thompson.' Cultivars vary in shell hardness. Soft-shelled cultivars are more important in commercial production.

Hardiness is through zone 7. Actually, hardiness is greater, but flowering is so early that late spring frosts limit the range in terms of nut production. Commercial production is mainly in California, but backyard production is seen in many states. Rainy weather in the spring and summer increases disease problems with blossoms and fruits. Midwinter rains also enhance disease problems.

Propagation is by budding of cultivars. Almond seedlings (variety *amara*) or 'Mission' are used as the rootstock. Peach rootstocks produce short-lived plants. Mature height is about 25 feet.

One-year-old nursery stock is planted in the late fall to early winter. Trees are placed 24 to 30 feet apart. Some nuts may be expected by the third or fourth year, and maximal production will be reached in the eighth year.

Almonds are usually trained to the modified-leader system. Pruning is light and only practiced every second or third year. Fruit thinning is not practiced. Almond flowers are self-incompatible, so two or more compatible cultivars are necessary, and bees are recommended. Two to three thousand pounds of unshelled nuts per acre can be expected, or about 25 to 40 pounds per tree.

Troublesome insects include several mites, shothole borer (*Scolytus rugulosus*), peach twig borer (*Anarsia lineatella*), consperse stink bug, and the navel orange worm (*Paramyelois transitella*). Viral diseases include ringspot, almond mosaic, and almond calico. Other diseases are brown rot (*Monilinia laxa*), shot-hole (*Coryneum carpophilum*), root rot (*Armillaria mellea*), leaf scab (*Cladosporium carpophilum*), crown rot (*Phytophthora* sp.), and canker (*Ceratocystis fimbriata*).

BEECHNUT

Beeches have been grown for their edible nuts, but the practice is extremely limited. They are utilized mostly for ornamental aspects. See the American and European beeches in Chapter 13 (Table 13-8) under *Fagus grandifolia* and *F. sylvatica,* respectively.

BRAZIL NUT

The Brazil nut (*Bertholletia excelsa*) is a tropical evergreen tree found primarily in the Amazon rain forest. Nuts are harvested from natural stands. It is seen occasionally under glass in the United States. Its culture will not be covered here.

CASHEW

The cashew is produced on a tropical evergreen tree (*Anacardium occidentale*). It can be grown in the warmer parts of zone 10, but is seldom seen there. Cashews are produced commercially in India. Their culture will not be covered here.

Fig. 16-1. Cashew, showing fleshy receptacle (accessory part of fruit) and the true fruit, which is roasted before shelling for use as the cashew nut. Roasting destroys the oil which is a dermal irritant. U.S. Department of Agriculture photo.

CHESTNUT

The chestnut is found in the genus *Castanea*. There are several species: American chestnut (*C. dentata*), European chestnut (*C. sativa*), Japanese chestnut (*C. crenata*), and Chinese chestnut (*C. mollissima*). There are other species, but they are not important in terms of nut production. In this country the American and European chestnut have essentially been eliminated by the chestnut blight (*Endothia parasitica*). The Japanese and Chinese chestnuts are somewhat resistant; their cultivars are mostly grown. The recent discovery of a nonvirulent strain of the chestnut blight, which causes disease remission when inoculated into diseased trees offers hope that we may yet see the great American chestnut again. Cultivars and hybrids of the Chinese chestnut include 'Abundance,' 'Clapper,' 'Crane,' 'Eaton,' 'Hemming,' 'Kuling,' 'Meiling,' 'Nanking,' 'Orrin,' and 'Sleeping Giant.'

The Chinese chestnut is hardy through most of zone 5, where the temperature does not go below -15°F (-26.2°C). The Japanese chestnut is hardy through zone 6. Cultivars of the Chinese chestnut are grown for nuts, since combined nut quality and blight resistance are better than in the Japanese chestnut. Commercial and backyard production of chestnuts is limited compared to the more popular nuts. Most commercial orchards are found in Maryland and Georgia. Flower buds can be damaged by late spring frosts.

Cultivars of Chinese chestnuts are usually propagated by the splice or whip graft. Stocks are usually seedlings of the Chinese chestnut. Rooting of cuttings is possible, but difficult. These cultivars may reach 50 feet in height.

Chestnuts are planted in the spring in the North and in the fall or spring in the South. Transplanting is done with care to avoid injury to the taproot system. Spacing can be from 25 to 50 feet, but pruning to keep the tree small will be required at the lesser spacing. Grafted trees may bear nuts as early as the second year after being planted. Two or more cultivars should be planted as maximal production is dependent on cross-pollination.

The most serious insects are the Asiatic oak weevil (*Cyrtepistomus castaneus*) and the chestnut weevil (*Curculio caryatrypes* and *C. sayi*). The chestnut blight is a disease already discussed.

Yields can be from 40 to 73 pounds of nuts per tree. Nuts are stored in ventilated cans or polyethylene bags at 32° to 36°F (0° to 2.3°C).

COCONUT

The coconut is the fruit of the tropical palm, *Cocos nucifera*. Few cultivars exist in true form because of cross-pollination and the fact that coconuts are only propagated from seeds. The only maintained cultivars are

Introduction to Horticultural Production / Part III

the popular 'Golden Dwarf Malay' and the less commonly grown 'Green Dwarf Malay' and the 'Orange Dwarf Malay.'

Coconuts are grown in the tropics and to some extent in southern Florida, Southern California, and Hawaii. Most are grown for ornamental use and the nuts are of secondary value. Commercial operations are essentially nonexistent in the continental United States.

Propagation is from seeds allowed to mature on the tree. The whole fruit is buried about two thirds in a seedbed. Germination requires 4 to 5 months. One- or two-year-old trees are planted about 25 feet apart. They can begin bearing when 6 years old and reach maximal bearing in 12 to 14 years.

Fig. 16–3. Coconut, dwarf showing foliage and fruit. U.S. Department of Agriculture photo.

Trees are not pruned. Both self- and cross-pollination occur. Bees are the main pollinators. Troublesome diseases include bud rot (*Phytophthora palmivora*), lethal yellowing (*mycoplasma* organism), and leaf scorch (*Ceratocystis paradoxa*). Scales, mealybugs, and leaf skeletonizer (*Homaledra sabalella*), are insect pests. A good tree may yield 75 coconuts per year, but 20 to 30 is average.

FILBERT (HAZELNUT)

Filberts are obtained from species of *Corylus.* The European filbert, *C. avellana,* and the giant filbert, *C. maxima,* plus a number of cultivars derived from them, are the main source of nuts. The American hazelnut, *C. americana,* and the beaked hazelnut, *C. cornuta,* are used mainly in breeding programs with the other species to provide increased hardiness and resistance to Eastern filbert blight caused by the fungus *Cryptosporella anomala.* The main commercial cultivar is 'Barcelona,' and 'Daviana' is the pollenizer. Other cultivars include 'Bixby,' 'Buchanan,' 'Cosford,' 'DuChilly,' 'Italian Red,' 'Medium Long,' 'Potomac,' 'Red Lambert,' 'Reed,' 'Rush,' and 'Winkler.'

Chilling requirements of filberts are similar to apples. They are about as hardy as peaches, but their pistils are frequently killed by spring frosts because the pistillate flowers open early. This limits commercial production to areas near large bodies of water, mainly Oregon and Washington. Filberts are grown by homeowners in other areas, but cultivars should be chosen carefully by the backyard grower because of blight and/or hardiness problems.

Species of filberts can be raised easily from seeds. Cultivars are propagated mainly by tip layering, since budding is not satisfactory. Two-year-old plants are placed in the fall or spring about 15 feet apart. In the Northwest, filberts are trained to a single trunk with scaffold branches. Basal suckers are removed. In the Northeast, filberts are allowed to sucker and are grown as shrubs. Rejuvenation pruning is minimal, but does increase yields. Cultivars vary in height from 6 to 30 feet. Trees are monoecious and unfruitful, so a pollinizing cultivar is interplanted among the main cultivar. Filberts may start bearing in 3 years and reach maximal production between 15 and 25 years of age. Commercial yields are from 2000 to 3000 pounds per acre. In the Northeast one tree will produce about 5 to 10 pounds of nuts per year.

Fig. 16-4. Filbert (European), showing fruits still attached to husks (lower left), unattached fruits, and shelled fruits exposing the seed which is the filbert nut. U.S. Department of Agriculture photo.

Troublesome insects include filbert aphid (*Myzocallis coryli*), filbert-worm (*Melissopus latiferreanus*), and filbert bud mite (*Phytoptus avellanae*). Diseases are mainly a fungal filbert blight (*Apioporthe anomala*) in the East and a bacterial filbert blight (*Xanthomonas corylina*) in the Northwest.

HICKORY NUT

Hickory nuts are the fruits of several species of the genus *Carya*. These include the bitternut, *C. cordiformis;* bitter pecan, *C. aquatica;* mockernut *C. tomentosa;* the pecan, *C. illinoinensis* (see pacan); the pignut, *C. glabra;* shagbark hickory, *C. ovata;* and the shellbark hickory, *C. laciniosa.* Desirable nuts, in decreasing order, are the pecan, shagbark hickory, and shellbark hickory. The others vary from poor to inedible. The pecan will be considered under a separate heading in this chapter. Cultivars are numerous, so only a limited number will be mentioned here. Shagbark hickory cultivars of merit include 'Davis,' 'Fox,' 'Glover,' 'Grainger,' 'Hales,' 'Harold,' 'Kentucky,' 'Kirtland,' 'Porter,' 'Wilcox,' and 'Wilson.' For the shellbark hickory, they are 'Keystone,' 'Nieman,' 'Ross,' 'Stephens,' and 'Weiper.' A number of hybrid cultivars exist, such as between the pecan and shagbark or shellbark (a hican); pecan-shagbark cultivars are 'Burton,' 'Henke,' 'Pixley,' and 'Wapello'; Pecan-shellbark cultivars are 'Baress,' 'Bixby,' 'Burlington,' 'Clarksville,' 'Green Bay,' and 'Jay Underwood.' Hickory nuts are not produced commercially, but are grown by noncommercial horticulturists.

The shagbark and shellbark hickories are hardy through zones 5 and 6, respectively. Cultivars should be chosen on the basis of region, since southern cultivars are not apt to mature their nuts in the shorter and less hot growing season of the North.

Species can be propagated from seeds, and cultivars can be propagated by budding or grafting. Seedlings are slow, and great care is required in the latter case. Transplanting must be done with care because of the taproot system. The best approach is to plant the nut in the permanent location in the fall or, if stratified, in the spring. Then the grafting can be done in place. Trees should be placed 60 to 75 feet apart. Because of various degrees of self-unfruitfulness, it is best to plant two or more cultivars. Pruning is only needed to establish lower limb height and to remove dead wood. Seedlings may take 10 to 15 years to bear nuts. Yields are about 50 to 75 pounds per tree. Mature height may be 100 to 120 feet.

Troublesome insects include the hickory bark beetle (*Scolytus quadrispinosus*), several species of aphid, hickory shuckworm (*Laspeyresia caryana*), tussok moth (*Halisidota caryae*), case bearer (*Coleophora laticornella*), walnut caterpillar (*Datana integerrima*), webworm (*Hyphantria cunea*), and the painted hickory borer (*Megacyllene caryae*). Diseases include anthracnose (*Gnomonia caryae*) and canker (several spp.).

Fig. 16–5. Macadamia, showing foliage and fruit. The seed within the fruit is the actual macadamia nut. American Society for Horticultural Science photo.

MACADAMIA NUT

Two evergreen species are grown for the production of the macadamia nut: *Macadamia integrifolia* and *M. tetraphylla*. Cultivars of the first include 'Arcia,' 'Faulkner,' 'Ikaika,' 'Kakea,' 'Keaau,' 'Keauhou,' 'Kohala,' 'Nuuanu,' 'Pahau,' 'Parkey,' and 'Wailua;' cultivars of the second are 'Burdick,' 'Hall,' and 'Santa Ana.' 'Beaumont' is a hybrid cultivar between the two species. These are grown both commercially and in backyards.

Commercial production is confined mainly in Hawaii, but culture is favorable in areas producing citrus and avocado. The macadamia nut in backyard production is seen mainly in zone 10, primarily in Southern California and somewhat in southern Florida.

Rootstocks are propagated from seeds. Cultivars are mostly grafted by the splice method, and budding is by the patch technique. Both *M. integrifolia* and *M. tetraphylla* are reciprocally graft compatible. Trees are placed at 25- to 35-foot intervals, usually in early spring. Trees are trained to the central-leader system. Bearing may start in 5 to 7 years. Mature height is about 50 to 60 feet. Planting of two cultivars for cross-pollination appears to be beneficial. Yields of nuts per acre range from 1.5 to 3.5 tons per acre or about 150 pounds per tree. Insect and disease problems appear minimal.

PEANUT

Peanuts are seeds of an annual, herbaceous legume, *Arachis hypogaea*. Cultivars are 'Improved Spanish,' 'Valencia,' and 'Virginia Runner.' Fruits need a long season of heat to mature. Best areas are zone 6 and south.

Peanuts are grown both commercially and by the homeowner in those areas. Commercial production was 1.5 million acres in 1976 and the value was $750 million.

Seeds (preferably removed from the pod) are planted when the danger of frost is past. Soil pH is best between 5.8 and 6.2. Rows are spaced 30 inches apart. Seeds are spaced 6 to 8 inches apart in the row and covered with 2 to 3 inches of soil. Peanuts grow 12 to 18 inches tall. Flowering will commence 4 to 6 weeks after planting. After pollination the ovary elongates and pushes into the soil. The peanut develops belowground. A growing season of 4 months is required.

Troublesome insects include cornstalk borer (*Elasmopalpus lignosellus*), leaf hoppers, rednecked peanutworm (*Stegasta bosqueella*), fall armyworms (*Spodoptera frugiperda*), velvet-bean caterpillars (*Anticarsia gemmatalis*), and white-fringed beetle (*Graphognathus* spp.). Serious diseases are leafspot (*Mycosphaerella arachidicola*) and southern stem blight (*Sclerotium rolfsii*).

Fig. 16-6. Peanut, showing foliage and underground fruit, the peanut. U.S. Department of Agriculture photo.

PECAN

Of all the hickory nuts, the pecan, *Carya illinoinensis,* is the one most valued by commercial and backyard nut growers. Cultivars are numerous. Some of those grown in the North include 'Busseron,' 'Butterick,' 'Colby,' 'Fritz,' 'Giles,' 'Green River,' 'Indiana,' 'Major,' 'Niblack,' and 'Witte.' Some grown in the South include 'Barton,' 'Caddo,' 'Candy,' 'Curtis,' 'Delight,' 'Desirable,' 'Elliott,' 'Ideal,' 'Kennedy,' 'Mahan,' 'Moore,' 'Onliwon,' 'Schley,' 'Stuart,' 'Success,' 'Texhan,' 'Van Demand,' 'Western Schley,' and 'Wichita.'

Fig. 16–7. Pecans. U.S. Department of Agriculture photo.

Pecans require a frost-free growing period of 150 to 210 days, depending on the cultivar. This determines the grouping of cultivars into northern and southern categories. Cultivars recommended for your region should be chosen. Hardiness is limited to zone 6. Flowering is late enough to be uninjured by late spring frosts. Commercial production is limited to the South as far west as New Mexico.

Propagation is by patch budding or whip grafting of cultivars onto pecan rootstocks. Rootstocks are grown from seeds of 'Riverside,' 'Burkett,' and 'Western' in the West, from 'Curtis,' 'Stuart,' 'Success,' and 'Mahan' in the Southeast, and from 'Giles' in the North. Three- or four-year-old trees are transplanted best in late winter or early spring as the buds begin to swell. Care must be exercised because of the taproot. Trees are placed 50 to 70 feet apart, although spacings of 30 to 40 feet can be used if thinning of trees is practiced when crowding occurs. Bearing can be reasonable by 7 to 10 years. Mature height may be as much as 150 feet.

Pruning to produce a tree with a single trunk with lowest branches at 5 to 7 feet is standard practice. Maintenance pruning is minimal. Trees are monoecious and wind pollinated. Trees are self-fertile, but cultivars are often mixed, since cross-pollination may produce more or higher-quality nuts. A tree may produce 50 to 100 pounds of nuts per year.

Troublesome insects include several mites, several aphids, shuckworm (*Laspeyresia caryana*), nut casebearer (*Acrobasis juglandis* and *A. caryae*), scales, weevil (*Curculio caryae*), and curculio (*Conotrachelus affinis*). Scab (*Cladosporium effusum*) is a serious disease.

PISTACHIO

Pistachios (*Pistacia vera*) are grown both commercially and by backyard horticulturists. They are dioecious. Pollinating cultivars are 'Peters' and 'Chico.' Female cultivars include 'Bronte,' 'Kerman,' 'Red Aleppo,' and 'Trabonella.'

Hardiness is through zone 9. Long, hot summers with low humidity are favorable. Some chilling is required to break dormancy. Commercial and amateur production is limited in the United States and primarily restricted to Southern California.

Propagation of cultivars is by T-budding on seedling rootstocks. *Pistacia atlantica* and *P. terebinthus* are mainly used as rootstocks because of nematode and soil fungi resistance. Container-grown trees have a higher transplanting survival rate than bare-rooted trees.

Container-grown trees are set out in any season at 30-foot intervals. Smaller distances can be used, but interplants must be removed when crowding occurs. Trees may begin bearing in 4 to 5 years. One male cultivar is planted per 10 to 12 female cultivars. Pollination is by wind. Training is minimal and is used to produce a high heading, rather than bushy, spreading tree. The modified-leader system is becoming increasingly important. Yields vary from 15 to 50 pounds of nuts per tree per year. However, production varies because of alternate bearing. Mature height is 15 to 30 feet.

Disease problems include leaf spot (*Phyllosticta lentisci*), root rot (*Phymatotrichum omnivorum*), oak root fungus (*Armillaria mellea*) and *Verticillium* wilt. Scales, aphids, and mites are minor insect pests.

WALNUT

Walnuts are species of *Juglans*. The Persian walnut (sometimes called English walnut) is the only species (*J. regia*) of interest to both commercial and amateur horticulturists. It has both soft- and hard-shell cultivars; the former are favored. The following species are mainly of interest to backyard nut growers: black walnut (*J. nigra*), butternut (*J. cinerea*), and Japanese walnut (*J. ailanthifolia*). Cultivars of the Persian walnut grown on the West Coast include 'Concord,' 'Eureka,' 'Farquette,' 'Gustine,' 'Hartley,' 'Lompac,' 'Mayette,' 'Payne,' 'Pioneer,' 'Placentia,' 'Serr,' 'Spurgeon,' 'Tehama,' and 'Vina.' Those grown in the East are 'Broadview,' 'Colby,' 'Greenhaven,' 'Gratiot,' 'Hansen,' 'Jacobs,' 'Lake,' 'McDermid,' 'McKinster,' 'Metcalfe,' 'Schafer,' and 'Somers.' Some cultivars of the black walnut are 'Elmer Meyers,' 'Huber,' 'Mintle,' 'Michigan,' 'Ohio,' 'Patterson,' 'Snyder,' 'Sparrow,' and 'Thomas.' Butternut cultivars include 'Ayers,' 'Craxeasy,' 'Johnson,' 'Kinneyglen,' 'Love,' 'Thill,' and 'Van Sykcle.' Japanese walnut cultivars include 'Bates,' 'Cardinell,' 'Caruthers,' 'Evers,' 'English,' and 'Wright.' Hybrid cultivars between the butternut and Japanese walnut are 'Corsan,' 'Crietz,' 'Dunoka,' 'Fioka,' and 'Helmick.' These lists are by no means complete. Local authorities should be checked for cultivars best suited to your area.

Persian walnuts are hardy into zone 6 if the proper cultivars are planted (sometimes hardier cultivars are termed Carpathian walnuts). Commercial production is primarily limited to California and Oregon.

Temperatures above 100°F with low humidity can be harmful; so can late spring frosts. Winter chilling is required to break dormancy. A growing season of at least 150 days is required for most cultivars, with some requiring as much as 260 days. Japanese and black walnuts are hardy through zone 5. The butternut is the hardiest, as it is hardy into zone 4.

Persian walnuts in the West are usually propagated on seedling rootstocks by the whip graft or patch bud. The northern Californian black walnut (*J. hindsii*) is usually the rootstock. 'Paradox,' a hybrid between the Persian and black walnut, is also used. Hardy Persian or Carpathian walnuts are propagated onto black walnut (*J. nigra*) rootstocks by patch or chip budding and by bark grafting. Japanese, black, and butternut walnuts can be propagated by bud grafts. The Japanese walnut cultivars can be grafted on black or butternut seedling rootstocks, butternut cultivars on black walnut, and black walnut cultivars on black walnut seedling rootstocks.

Fig. 16-8. Walnuts. U.S. Department of Agriculture photo.

Cultivars of the Persian, black, Japanese, and butternut walnuts are set out as 1-year-old grafts and placed 50 to 70 feet apart. Closer distances can be used if specialized pruning systems are employed. Persian walnuts are planted in late winter in the West. In other areas, early spring or late winter are suitable times for any cultivars. Bearing may start in the fifth or sixth year. Mature heights are 70 feet for the Persian walnut, 150 feet for the black walnut, 60 feet for the Japanese walnut, and 90 feet for the butternut.

Persian walnuts are trained to either an open center with a high head or to the central- or modified-leader system. Subsequent pruning consists of thinning top growth to admit light and to stimulate the production of fruiting wood in the central portion. Cultivars of the other species are trained to a single trunk with the lower branches removed at 5 to 7 feet.

Well-managed orchards of Persian walnuts may yield 1 to 2 tons of unshelled walnuts per acre per year. Homeowners might expect 20 to 35 pounds of nuts per tree per year for the various walnuts discussed here. Trees are monoecious and self-fruitful. However, mixed cultivars are often planted as increased yields can result from cross-pollination. The vector is wind.

Persian walnuts are troubled by the following insects: codling moth (*Carpocapsa pomonella*), walnut aphid (*Chromaphis juglandicola*), several mites, and walnut husk fly (*Rhagoletis completa*). Diseases include canker (*Diplodia juglandis*), walnut blight (*Xanthomonas juglandis*), heart rot (*Fomes* sp.), and leaf spot (several spp.). Black line, once thought to be delayed grafting incompatibility, is now known to be viral induced. For the black walnut, insect pests include black walnut caterpillar (*Datana integerrima*), fall webworm (*Hyphantria cunea*), aphid, and lacebug (*Corythucha juglandis*); anthracnose leaf spot (*Gnomonia leptostyla*) can also be a problem. Butternuts are troubled by canker (*Diplodia juglandis*) and by the butternut curculio (*Conotrachelus juglandis*). All walnuts are susceptible to varying degrees to the witches'-broom or bunch disease, which is suspected to be an insect-transmitted virus.

Culinary Herbs

Herbs are treated here in the horticultural, rather than botanical sense. In the former sense they are horticultural crops utilized as secondary ingredients in food preparation for a source of flavor and seasoning and medicinally in the preparation of pharmaceuticals. Culinary herb crops may be annuals, biennials, or perennials. Most are aromatic or sweet smelling and are found in the mint (Labiatae), parsley (Umbelliferae), or daisy (Compositae) family.

Herbs are usually grown in a small, informal section of the garden, or sometimes in an elaborate formal garden devoted entirely to herbs. Some are decorative and used in borders or in rock gardens. Some are suitable for indoor or greenhouse culture. A discussion of the various herbs follows. Care must be exercised with the more unusual herbs to avoid the danger of poisonous look-alikes or the use of the wrong plant part, which can be poisonous.

ANGELICA (ANGELICA ARCHANGELICA)

Angelica is a biennial or short-lived perennial herb hardy through zone 5. Mature height is 6 feet. Propagation is by seed or division. Seeds are sown late summer to early autumn and covered with ½ inch of soil. Seed viability is limited. Plants should be spaced 30 inches apart. Soil should be moist, but well drained. Exposure should be sunny to partial shade. Life of the plant can be prolonged by topping to prevent flowering. Leaves are used in cooking fish, stems are candied or used to flavor liqueurs, petioles are candied, and roots are used to flavor liqueurs. Flavor is juniperlike.

Fig. 16–9. Anise. U.S. Department of Agriculture photo.

ANISE (PIMPINELLA ANISUM)

Anise is an annual that reaches a height of 2 feet. Seeds are sown in place (18 to 24 inches apart) after the danger of frost is past. Transplanting is difficult. A well-drained soil with full sun is best. Seeds are used in breads, pastries, and many foods. Flowers are used to flavor liqueurs. Leaves are also used to flavor foods. Flavor is licorice-like.

BORAGE (BORAGO OFFICINALIS)

Borage is an annual herb that reaches a height of 2 feet. Propagation is by seeds sown in the spring, cuttings, and division. A well-drained soil and full sun are best. Plants should be spaced 24 inches apart. The tender, upper leaves have a cucumberlike flavor and are used in salads, iced drinks, and pickles.

BURNET (POTERIUM SANGUISORBA)

Burnet is a perennial herb reaching a height of 2 feet and is hardy through zone 4. Propagation is by seeds and division. Full sun and a dry to moist, well-drained soil are best. Leaves are used when young in salads, iced drinks, and vinegar. The flavor is cucumberlike.

CARAWAY (CARUM CARVI)

Caraway is usually grown as a biennial through zone 4. Mature height is 2 feet. Propagation is by seeds sown in spring. A sunny location and soil on the dry side are favored. It is grown mostly for its seeds, which are used to flavor rye bread and other baked goods, as well as processed foods.

Fig. 16-10. Caraway. U.S. Department of Agriculture photo.

CATNIP (NEPETA CATARIA)

Catnip is a perennial herb that is hardy through zone 4. Mature height is 3 feet. Propagation is by seed in the fall or division in the spring. An average soil and full sun are favored. Growth is vigorous, and it can easily become a weed. Its main use is as a tonic for cats.

CHERVIL (ANTHRISCUS CEREFOLIUM)

Chervil is an annual reaching a height of 2 feet. Propagation is by seed sown in early spring or late summer. It is treated as a spring or fall crop, as heat is detrimental. Plants are spaced about 12 inches apart. Culture is simple in most soils and full sun to partial shade. Leaves are ready in 6 to 8 weeks and are somewhat like parsley. They are used as a garnish and to flavor soup, fish, and meat.

CHIVES (ALLIUM SCHOENOPRASUM)

Chives are a perennial herb hardy through zone 3. Mature height may be 1 to 2 feet. Propagation is by seeds, bulbs, or division, usually in early spring. Full sun and average soil are best. Leaves are used to flavor salads, omelettes, sour cream, soups, vegetables, and many other foods. Flavor is oniony. Chives can be grown easily indoors.

Fig. 16-11. Chives. U.S. Department of Agriculture photo.

CORIANDER (CORIANDRUM SATIVUM)

Coriander is an annual herb that has a mature height of 2 to 3 feet. Propagation is by seeds in the spring or fall. Plants should be spaced 10 inches apart. An average soil and full sun are best. Fruits and seeds are candied or used as seasoning in poultry dressing, pickles, beverages, baked goods, and processed meats.

COSTMARY (CHRYSANTHEMUM BALSAMITA)

Costmary is a perennial herb hardy through zone 5 and reaching a height of 2 to 3 feet. Propagation is by division and root cuttings. Dry soil and full sun are best, but partial shade is acceptable. Plants should be 3 to 4 feet apart. Leaves have a lemon-mint flavor and are used to flavor tea or iced beverages.

DILL (ANETHUM GRAVEOLENS)

Dill is a biennial treated as an annual; it reaches a height of 3 to 5 feet. Propagation is by seeds in the spring. Spacing is 12 inches. Average soil and full sun are best. Dill transplants poorly. Leaves are used to flavor cottage cheese, potato salad, sauces, soups, vegetables, and fish. Seed heads are used in pickles, breads, and soups.

Fig. 16-12. Dill. U.S. Department of Agriculture photo.

Fig. 16-13. Fennel. U.S. Department of Agriculture photo.

FENNEL (FOENICULUM VULGARE)

Fennel is a perennial usually treated as an annual. It reaches a height of 5 feet. Propagation is by seed in the spring. An average to dry soil and full sun are best. Plants should be 12 inches apart. Two varieties, *dulce* (Florence fennel or finocchio) and *azoricum* are known. The former is grown for its enlarged, edible leaf bases, which are blanched or eaten raw. The latter is grown for its leaves and seeds used in salads and to flavor cooked foods. Flavor is aniselike.

FIELD MINT (MENTHA ARVENSIS)

Field mint is a perennial hardy through zone 5. Mature height is 2 feet. Propagation is by division and cuttings in the spring. It can become weedy as it spreads vigorously. Average to moist soil and full sun are best. Leaves are used to flavor drinks.

HOREHOUND (MARRUBIUM VULGARE)

Horehound is a perennial hardy through zone 4. Mature height is 3 feet. Propagation is by seeds, cuttings, division, and layering. A dry soil with full sun is best. Spacing is about 12 inches. It can become weedy. Leaves and stems are used to flavor candy and coughdrops.

HORSERADISH (ARMORACIA RUSTICANA)

Horseradish is a perennial usually treated as an annual, in that the roots are harvested in the fall. Mature height is 1.5 to 3 feet. Propagation is by root cuttings in the spring. These are planted 3 to 5 inches deep and 10 to 18 inches apart in the row. Rows are spaced 3 feet apart. An average to moist soil and full sun are best. Mature roots are grated to prepare the familiar relish. It can become a weed if roots are allowed to remain more than one year.

HYSSOP (HYSSOPUS OFFICINALIS)

Hyssop is a perennial hardy through zone 3. Mature height is about 18 inches. Propagation is by seeds, cuttings, and division. A well-drained soil and full sun to partial shade are best. Spacing is 12 inches. Fresh shoot tips are used to flavor vegetables and soups.

LEMON BALM (MELISSA OFFICINALIS)

Lemon balm is a perennial hardy through zone 5. Mature height is 2 feet. Propagation is by seeds, division, or cuttings in the spring. An average soil and partial shade are best. The lemon-scented leaves are used in teas, soups, salads, cold drinks, and many cooked foods.

LOVAGE (LEVISTICUM OFFICINALE)

Lovage is a perennial hardy through zone 6. Mature height is 6 feet. Propagation is by seeds in the fall or by division in the spring. Spacing is about 18 inches. A moist soil with full sun to partial shade is best. Seeds are used in confectionery and cordials. Dry or fresh leaves are used to flavor salads, soups, and sauces. It is used as a celery substitute.

PARSLEY (PETROSELINUM CRISPUM)

Parsley is a biennial that can be treated as an annual. Mature height is 10 to 15 inches. Propagation is by seeds soaked in water prior to sowing in early spring. Plants can be grown easily indoors. Spacing is 6 to 8 inches. An average soil with full sun is best. Fresh leaves are used as a garnish and in soups, salads, and various cooked dishes. The typical variety with curled crisped leaves is *crispum.* The variety *neapolitanum* has flat leaves.

PEPPERMINT (MENTHA X PIPERITA)

Peppermint is a perennial hardy through zone 4; it reaches a height of 1 to 2 feet. Propagation is by division or cutting. An average to moist soil and full sun are best. Leaves are used to flavor foods, candy, and beverages. They are the source of peppermint oil used in confectionery.

ROSEMARY (ROSMARINUS OFFICINALIS)

Rosemary is a perennial hardy through zone 6 and reaches a height of 6 feet. Some winter protection may be necessary in zone 6. Propagation is by seeds (slow to germinate) or cuttings. A well-drained soil and full sun are best. Spacing is 3 feet. Dried or fresh leaves are used to flavor vegetables, meats, fish, poultry, soups, sauces, and many other cooked products.

SAGE (SALVIA OFFICINALIS)

Sage is a perennial hardy through zone 4. Mature height is 2 feet. Propagation is by seeds, cuttings, or division. Spacing is 24 inches. Average soil and full sun are best. Fresh leaves are used to flavor cheese, pickles, and sausage. Dried, powdered leaves are used in stuffings and many cooked foods.

SPEARMINT (MENTHA SPICATA)

Spearmint is a perennial hardy through zone 4. Mature height is 2 feet. Propagation is by division or cuttings in the spring. An average to moist soil and full sun are best. Fresh leaves are used to flavor jellies, vinegar, iced drinks, sauces, soups, candy, cooked dishes, fruit, and ice cream. Spearmint oil, used for flavoring, is derived from the leaves.

Fig. 16–14. Peppermint. U.S. Department of Agriculture photo.

Fig. 16–15. Spearmint. U.S. Department of Agriculture photo.

SUMMER SAVORY (SATUREJA HORTENSIS)

Summer savory is an annual that reaches a height of 18 inches. Propagation is by seeds in the spring. Spacing is 12 inches. An average soil with full sun is best. Leaves are used to flavor meat, fish, poultry, vegetables, sausage, sauce, soup, stuffing, and gravy.

SWEET BASIL (OCIMUM BASILICUM)

Sweet basil is an annual that reaches a height of 2 feet. Propagation is by seeds in the spring. Spacing is 12 to 18 inches. Average soil and full sun are best. Leaves are used to flavor meat, fish, eggs, cheese, soups, sauces, sausage, salads, vegetables, and tomato products.

SWEET CICELY (MYRRHIS ODORATA)

Sweet cicely is a perennial herb hardy through zone 5. Mature height is 3 feet. Propagation is by seeds in the fall or division in the spring or fall. Average soil and partial shade are best. Leaves are used to flavor soups and salads. Dried seeds have a licorice taste and are used to flavor foods.

SWEET MARJORAM (ORIGANUM MAJORANA)

Sweet marjoram is a tender perennial (zone 9) usually treated as an annual. Mature height is 2 feet. Propagation is by seeds after all danger of frost is past. Average soil and full sun are best. Leaves are used to flavor salads, soups, stuffing, sausage, sauces, cooked dishes, meat, and fish.

Fig. 16-16. Sweet marjoram. U.S. Department of Agriculture photo.

SWEET WOODRUFF (GALIUM ODORATUM)

Sweet woodruff is a perennial hardy through zone 5. Mature height is 6 to 12 inches. Propagation is by root division in the spring. Moist soil and partial shade are best. Leaves are used to flavor wines, liqueurs, and cold drinks.

TARRAGON (ARTEMISIA DRACUNCULUS)

Tarragon is a perennial that is hardy through zone 6. Mature height is 2 feet. Propagation is by cuttings. Dry soil and full sun are best. Fresh leaves are used in salads, pickles, and to make tarragon vinegar. Dry leaves are added to meats, poultry, soups, eggs, sauces, dressings, and so on.

THYME (THYMUS VULGARIS)

Thyme is a perennial hardy through zone 6. Mature height is 6 to 15 inches. Propagation is by seeds, division, and cuttings in the spring. Dry soil and full sun are best. Fresh and dry leaves are used to flavor meats, fish, poultry, soup, vegetables, sauces, dressings, pickles, and vinegar.

WILD MARJORAM (ORIGANUM VULGARE)

Wild marjoram is a perennial hardy through zone 4. Mature height is 2.5 feet. Propagation is by seeds, cuttings, layering, and root division in the spring. Average soil and full sun are best. Leaves are used in a manner similar to sweet majoram.

WINTER SAVORY (SATUREJA MONTANA)

Winter savory is a perennial hardy to zone 6. Mature height is 15 inches. Propagation is by seeds, division, and cuttings in the spring. A dry soil and full sun are best. Leaves are used to flavor poultry, meat, eggs, sausage, soups, salad, vegetables, sauces, gravy, and stuffing.

FURTHER READING

Hylton, William H., ed., *The Rodale Herb Book.* Emmaus, Pa.: Rodale Press Book Division, 1974.

Jaynes, Richard A., (ed.), *Nut Tree Culture in North America.* Hamden, Conn.: Northern Nut Growers Association, 1979.

Kadans, J., *Modern Encyclopedia of Herbs* (4th ed.). Englewood Cliffs, N.J.: Prentice-Hall, Inc., 1970.

Krochmal, A., and C. Krochmal, *A Guide to Medicinal Plants of the United States.* New York: Quadrangle/The New York Times Book Co., 1973.

Morton, J. F., *Major Medical Plants: Botany, Culture, and Uses.* Springfield, Ill.: Charles C. Thomas, Publisher, 1978.

Muenscher, W. C., and M. A. Rice, *Garden Spice and Wild Pot Herbs.* Ithaca, N.Y.: Cornell University Press, 1955.

Ramalingham, V., and others, eds., *Medicinal Plants* (3 volumes). New York: MSS Modular Publications, Inc., 1974, 1977, 1977.

Sanecki, Kay N., *The Complete Book of Herbs.* New York: Macmillan, Inc., 1974.

Woodroof, J. G., *Tree Nuts* (2nd ed.). Westport, Conn.: Avi Publishing Co. 1977.

Index *

Abelia, 140
Abscisic acid, 16, *117*, 127
Abscission, 116, 117, 122
Acclimation, 128, 130, 136, 199, 270
Acer (*see* Maple)
Acid tolerance, 20
African violet, 133, *182–83*, 185, 334
Agapanthus, 191
Agricultural experiment stations, 176, 210
Air pollution:
   damage, 152, *153*
   in greenhouse, 264
   tolerance, 152
Ajuga, 34
Algae, 170, 176, 327
   chemical control, 358
   natural control, 342
Algicides, 358
Alkaline tolerance, 20
Alkaloids, 82, *83*, 84
Allamanda, 34, 181
Almond, 22, 47, 292
   cultural data, 610
Alstroemeria, 191

Aluminum sulfate, 225
Amaryllis, 35, 190
Amino acids, 74, 103
Ammonium, 78, 79, 149, 162, 215, 217, 261, 262
Andromeda, 24
Angelica, 621
Angiosperms, 9, 10, 97, *98*, 107, *107*
Animal pesticides, 358
Animal pests (*see also,* Insects, Mites, Nematodes):
   chemical control, 358
   natural control, 331, 339, *339*, 340
   symptoms, 322, *322*
Anise, 622, *622*
Annuals, 20, *21*, 24, 95, 96, 131, 172, 179, 425–30, *426–30*
   as bedding plants, 425
   depth of planting, 429
   frost maps:
     fall, 429
     spring, 428
   name and data, *454–57*
   propagation, 425, 426
   scheduling, 425
   uses, 425, *428*

*See also alphabetical listings of ornamentals and vegetables in tables found in Chapters 13 and 14. Numbers in italics indicate figures, tables, and maps.